14th Annual International Astrophysics Conference 2015

AA002686

Linear and Nonlinear Particle Energization throughout the Heliosphere and Beyond

Journal of Physics: Conference Series Volume 642

Tampa, Florida, USA
20 – 24 April 2015

Editor:

Gary P. Zank

ISBN: 978-1-5108-1469-1
ISSN: 1742-6588

Printed from e-media with permission by:

Curran Associates, Inc.
57 Morehouse Lane
Red Hook, NY 12571

Some format issues inherent in the e-media version may also appear in this print version.

Printed by Curran Associates, Inc. (2016)

For permission requests, please contact the Institute of Physics at the address below.

Institute of Physics
Dirac House, Temple Back
Bristol BS1 6BE UK

Phone: 44 1 17 929 7481
Fax: 44 1 17 920 0979

techtracking@iop.org

Additional copies of this publication are available from:

Curran Associates, Inc.
57 Morehouse Lane
Red Hook, NY 12571 USA
Phone: 845-758-0400
Fax: 845-758-2634
Email: curran@proceedings.com
Web: www.proceedings.com

Table of contents

Volume 642

14th Annual International Astrophysics Conference: Linear and Nonlinear Particle Energization throughout the Heliosphere and Beyond
20–24 April 2015, Tampa, Florida, USA

Papers

Preface

G P Zank
Editor, *Proceedings of the 14th Annual International Astrophysics Conference,*
Center for Space Plasma and Aeronomic Research (CSPAR) and Department of Space Physics, University of Alabama in Huntsville, Huntsville AL 35805, USA

E-mail: garyp.zank@gmail.com

The 14th Annual International Astrophysics Conference was held at the Sheraton Tampa Riverwalk Hotel, Tampa, Florida, USA, during the week of 19-24 April 2015. The meeting drew some 75 participants from all over the world, representing a wide range of interests and expertise in the energization of particles from the perspectives of theory, modelling and simulations, and observations. The theme of the meeting was "*Linear and Nonlinear Particle Energization throughout the Heliosphere and Beyond.*" Energetic particles are ubiquitous to plasma environments, whether collisionless such as the supersonic solar wind, the magnetospheres of planets, the exospheres of non-magnetized planets and comets, the heliospheric – local interstellar boundary regions, interstellar space and supernova remnant shocks, and stellar wind boundaries. Energetic particles are found too in more collisional regions such as in the solar corona, dense regions of the interstellar medium, accretion flows around stellar objects, to name a few. Particle acceleration occurs wherever plasma boundaries, magnetic and electric fields, and turbulence are present. The meeting addressed the linear and nonlinear physical processes underlying the variety of particle acceleration mechanisms, the role of particle acceleration in shaping different environments, and acceleration processes common to different regions. Both theory and observations were addressed with a view to encouraging cross-disciplinary fertilization of ideas, concepts, and techniques. The meeting addressed all aspects of particle acceleration in regions ranging from the Sun to the interplanetary medium to magnetospheres, exospheres, and comets, the boundaries of the heliosphere, and beyond to supernova remnant shocks, galactic jets, stellar winds, accretion flows, and more. The format of the meeting included 25-minute presentations punctuated by two 40-minute talks, one by Len Fisk that provided an historical overview of particle acceleration in the heliosphere (see the paper by Fisk, L., *50 Years of Research on Particle Acceleration in the Heliosphere*, in this volume), and another by Len Burlaga, who presented a summary of the exciting new interstellar medium magnetic field observations being returned by Voyager 1 (see the paper by Burlaga, L., *Voyager Observations of the Magnetic Field in the Heliosheath and the LISM*, in this volume).

The meeting was dedicated to two people close to the Annual International Astrophysics Conference series, to the conference participants, and the conference organizers. Both James (Jim) F. McKenzie and Michael (Mike) Intriligator died in the past year. Brief obituaries of both are presented below. Both will be missed greatly in our community.

We thank Adele Corona and ICNS for her continued excellent organization of the AIAC meetings and her help in providing the logistical support for this volume of papers. I should also like to thank Laxman Adhikari for his help with formatting a number of the submitted manuscripts.

Journal of Physics: Conference Series **642** (2015) 011001 doi:10.1088/1742-6596/642/1/011001

Remembering James "Jim" Fairley McKenzie

Gary P. Zank

Jim McKenzie was the consummate theoretical mathematical physicist – like many of us attracted by the mathematical beauty and abstraction of theoretical physics, Jim's greatest pleasure was to take a blank pad of paper, sit under a tree in his garden in Durban with a glass nearby, and an interesting problem to solve. I well remember Jim exhorting his young apprentices that we need to develop "algebraic muscle" – Jim's algebraic muscle was honed by innumerable calculations across the broad field of electromagnetic theory, fluid and magneto-fluid mechanics, linear and nonlinear waves, space physics, and differential equations in general. Jim was possibly unreasonably proud of his theoretical inclinations. As a student, I occupied an apartment next door to Jim and Karen, Jim's wife, for a year (I was house sitting) and shortly after I arrived, Jim stopped over, not for a social call, but because he needed some light bulbs changed in his house. As he explained, he understood and could calculate almost anything related to Maxwell's equations, but the act of unscrewing and replacing one of Maxwell's most important consequence, the light bulb, was more than he could or wanted to manage!

Unlike Jim's contributions to the running of an orderly household, his contributions to science were important, far-reaching and profound. His first papers addressed radiation from fast moving particles – Cherenkov radiation – and as a young student I was very impressed to discover this work referenced in a number of classic texts, including Clemmow & Dougherty's "Electrodynamics of Particles and Plasmas." After Jim's PhD from the Department of Applied Mathematics and Theoretical Physics, Cambridge in 1964, having been the Sir James Caird Scholar, he moved to the US as a National Academy of Sciences Research Fellow. He spent two years at the NASA Ames Research Center in the north of California, before moving south for a year to the University of California at San Diego. While in the US, he worked on one of his most interesting and far-reaching problems, addressing the interaction of waves with shock waves. This work was seminal and laid the foundations for everything done since on this topic in the field of space physics and astrophysics.

Jim's travels took him from the US to Italy and then, as a Professor, to the University of Khartoum in Sudan for four years. After that, Jim found himself a Research Professor at the Danish Space Research Institute for roughly four years. Although he had spent time with Ian Axford, later to become Sir Ian Axford, in La Jolla at UCSD, it was only in the later 1970s that Jim's very productive and influential collaboration with Ian began to flower. Their first important collaboration was a set of influential papers considering the acceleration of minor ions in the solar wind. These heavy ions tend to surf on the Alfven waves of the solar wind, and Jim and Ian provided the early definitive models to describe the process. Almost all work since reflects the basic approach that they introduced.

In the early 1980s Jim moved to the University of Natal, which became his long-term home, along with a later concurrent appointment at the Max-Planck-Institute for Aeronomy (later renamed as the Max-Planck-Institute for Solar System Research). It was at the University of Natal in Durban that I first encountered the "curious Prof. McKenzie" with the strange accent and somewhat weird sense of humor, at least to the South African undergraduate students. I had the great pleasure of taking Jim's first class – a final year class on "waves in fluids." It was quite simply the best class that I had experienced at the time and it set me on my lifetime professional research path. Jim was a wonderful lecturer, at once profound, amusing, insightful, and always mathematically beautiful in his exposition. For an impressionable undergraduate not quite sure what he wanted to do, Jim made an indelible impression. Not surprisingly, after finishing my honors degree in Math and Applied Math, I began working with Jim towards my PhD, and under his astute guidance finished in record time. Jim was still collaborating closely with Ian Axford, now on the problem of cosmic ray acceleration by shocks. This period led to another series of profoundly important papers. It had long been conjectured that cosmic rays originated somehow from supernova remnants. Ian and Jim were instrumental in establishing that cosmic rays were accelerated at shock waves associated with supernova remnants. It is hard to over-emphasize the importance of this work. Diffusive shock acceleration, as it is now known, is truly a universal process, occurring throughout the universe. Virtually every issue of the Astrophysical Journal or similar will contain some reference to or application of the general theory of diffusive shock acceleration. Jim's contributions laid the foundations for understanding the basic plasma physics of the process and will endure for as long as astrophysics is a discipline.

Jim's contributions did not slow down. He wrote numerous papers on waves in the solar wind, introducing some important advances in interpreting nonlinear waves in multi-ion plasmas. A proceedings paper that had an enormous impact was an Axford & McKenzie discussion and investigation of mechanisms to heat the solar corona and thereby drive the supersonic solar wind. This work introduced the "furnace" idea and the notion of ion-cyclotron damping and thus heating of coronal plasma. The ion-cyclotron mechanism is regarded as one of the two most likely processes for heating the solar corona. Solar Probe Plus, to be launched in 2018, will test this idea by measuring the heated solar atmosphere directly. It's a great pity that Jim did not live to see the possible confirmation of one of his great ideas.

It is impossible to capture the full extent of Jim's research interests and activities. Suffice it to say that Jim's contributions in many fields will continue to live on for years to come, inspiring new directions and ideas, and perhaps most importantly of all, inspiring young people to enter the world of mathematical physics through his beautiful presentations of complicated ideas. Whenever Jim visited me at the University of California, Riverside or the University of Alabama in Huntsville, I would always ask him to present a series of lectures to the students. Not only did the students find his classes inspirational, it brought back memories in me of a simpler and happy time long ago when I learned mathematical physics in the ivory tower that the University of Natal in Durban once was. Jim provided a link to that world and an idyllic time that is forever now extinguished. I, and all my colleagues, mourn Jim's passing but will celebrate his legacy as a scientist of great distinction and renown.

Go in peace, Jim.

Journal of Physics: Conference Series **642** (2015) 011001 doi:10.1088/1742-6596/642/1/011001

Michael Intriligator

Devrie Intriligator (lightly edited by Gary P Zank)

Mike (Michael) Intriligator, husband of Devrie Intriligator, an economist of note, died in Los Angeles on June 23, aged 76. He was an internationally renowned expert in econometrics, health care reform, strategy and arms control, peace and security, and transitioning economic and political systems. Mike met a number of us in the very earliest years of his career, and has since very frequently joined his wife Devrie at space physics meeting, including the AIA conferences. Indeed, Mike was a professor in three UCLA departments (economics, political science, and public policy), and became known for his work in the fields of economic theory and mathematical economics. His influential 1971 book, Mathematical Optimization and Economic Theory, brought previously inaccessible mathematical techniques into the mainstream of the field. "More than forty years later it is still one of the very best introductions to dynamic optimization," writes UCLA Professor John G. Riley. "After the book was published, a generation of economists with standard calculus skills were able to both appreciate and apply these techniques." Later in his career, Mike broadened his focus, applying econometric techniques to health care, strategy and arms control, nuclear proliferation, economic reform, warfare, and terrorism. A prolific writer and editor, Mike Intriligator wrote or edited more than 300 published research papers, journal articles, and books. Together with Nobel laureate Kenneth J. Arrow, he edited the 28-volume Handbook of Economics book series. Mike in fact co-authored a book with physicist David Goodstein in a 2013 book, entitled Climate Change and the Energy Problem. We like to think that this interest may have been stimulated in part by Mike's frequent participation in the space physics conferences that he attended. Indeed, Mike could always be found at the back of the meeting room working on one or more problems or even sometimes listening with interest. Because of this, Mike was regarded as an honorary space physicist by many of us. All of us enjoyed wide-ranging conversation and discussions with Mike, who could convey his analyses and ideas in non-technical language. Dynamic and eloquent, he was an internationally sought-after speaker, and served as the "go to" person on several topics for many news sources. Mike was regularly interviewed on TV and radio, and quoted in newspapers. He contributed articles and op-ed pieces to the New York Times, Huffington Post, Truth-Dig, and other news, commentary, and analysis outlets. Michael D. Intriligator was born in Freeport, New York, February 5, 1938, received a Bachelor of Science degree in economics from MIT in 1959, a Master's degree in economics from Yale in 1960, and a Ph.D. in economics from MIT in 1963. Mike joined the UCLA faculty as an assistant professor of economics in 1963, where he remained for the rest of his career, rising to full professor of economics, with joint appointments in political science and policy studies. Mike served in numerous administrative positions throughout his career and won numerous honors. Notable among these was his election as a Foreign

Member of the Russian Academy of Sciences, the only American economist with that distinction. For all of us who knew Mike personally, he will be remembered as a gentle soul. His warm smile and genuine interest in others made him a pleasure to be with. Mike was outgoing and social, with a contagious joie de vivre. He maintained an immense range of friends, from many walks of life, and in many parts of the world. Mike will be missed by all of us.

Peer review statement

All papers published in this volume of *Journal of Physics: Conference Series* have been peer reviewed through processes administered by the proceedings Editors. Reviews were conducted by expert referees to the professional and scientific standards expected of a proceedings journal published by IOP Publishing.

The transport of low-frequency turbulence in the super-Alfvénic solar wind

L Adhikari[1,2], G P Zank[1,2], R Bruno[3], D Telloni[4], P Hunana[2], A Dosch[2], R Marino[5], and Q Hu[1,2]

[1]Department of Space Science, University of Alabama in Huntsville, Huntsville, AL 35899, USA
[2]Center for Space Plasma and Aeronomic Research (CSPAR), University of Alabama in Huntsville, Huntsville,
AL 35899, USA
[3] (INAF-IAPS Istituto di Astrofisica e Planetologia Spaziali, Via del Fosso del Cavaliere 100, I-00133 Roma, Italy
[4] INAF-Astrophysical Observatory of Torino, Via Osservatorio 20, I-10025 Pino Torinese, Italy
[5] National Center for Atmospheric Research, P.O. Box 3000, Boulder, CO 80307, USA

E-mail: `la0004@uah.edu`

Abstract. Understanding the transport of low-frequency turbulence in an expanding magnetized flow is very important in analyzing numerous problems in space physics and astrophysics. Zank et al 2012 developed six general coupled turbulence transport equations, including the Alfvén velocity to describe the transport of low-frequency turbulence for any inhomogeneous flows, including sub-Alfvénic coronal flows, and super-Alfvénic solar wind flows. Here, we solve the 1D steady state six coupled turbulence transport equations of Zank et al. 2012, and the transport equation corresponding to the solar wind temperature in the super-Alfvénic solar wind flows from 0.29 to 100 AU without the Alfvén velocity. We calculate turbulent quantities corresponding to *Voyager* 2 data sets for three cases; i) a positive and negative sign of B_r; ii) the azimuthal angle $\phi = \tan^{-1}(B_t/B_r)$, and iii) a positive and negative sign of B_t, where B_r and B_t are the radial and transverse components of the interplanetary magnetic field, respectively. We compare our theoretical results to the observational results, and find good agreement between them.

1. Introduction

Turbulence is very common in the astrophysical plasma such as the solar corona, the solar wind, the interstellar medium and so on. Turbulence is created in situ by the presence of gradients in density, pressure, velocity, vorticity, temperature, magnetic fields, and current density. For example, gradients in velocity occur at interfaces between high and low speed solar wind flows. Similarly, gradients in density, pressure, velocity and magnetic field are intrinsic properties of a shock. The solar wind may therefore be considered as a huge laboratory for studying turbulence ([1–3]), and has been studied from the beginning of the space age ([1–7]). The realization that the Reynolds number of the solar wind is very large (of the order of $R_e \sim 10^8$ [8]) also supports the notion that the solar wind is highly turbulent.

Numerous problems in space science and astrophysics require a detailed understanding of the transport and dissipation of low-frequency turbulence in an expanding magnetized flow. The development of turbulence transport model equations started with the first observed fluctuations in the interplanetary medium. Paul J. Coleman studied the spectral properties of plasma fluctuations in the solar wind using Mariner 2 spacecraft magnetometer data [5], and found that the spectra of magnetic fluctuations was similar to that of velocity fluctuations, as predicted by Kriachnan 1965 [9]. On the other hand, [10] and [4] studied the Mariner 5 magnetometer and plasma data set using wave analysis, and found that the radial component of the interplanetary magnetic field (IMF) and the solar wind are highly correlated. Based on observational studies, [10] and [4] introduced the idea that solar wind fluctuations may be regarded as a superposition of non-interacting MHD waves, specifically Alfvén waves. The Alfvénic fluctuations do not damp as rapidly as slow and fast modes MHD waves, and hence Alfvén waves dominate fluctuations observed in the heliosphere.

One of the most interesting characteristics of the solar wind is the observed radial solar wind temperature profile which does not follow the adiabatic cooling profile i.e., $r^{-4/3}$, where r is the heliocentric distance ([11, 12, 13, 14, 15, 16]). The dissipation of turbulence generated in situ is thought to be responsible for heating the solar wind in the inner and outer heliosphere.

Observations show that there exist both inward/backward and outward/forward fluctuations in the solar wind. These oppositely directed propagating modes interact non-linearly such that large scales energy eventually accumulates at small scales, which is eventually dissipated into heat energy. The quantity that defines the energy difference between forward and backward propagating modes is known as the cross helicity. The cross helicity gives us an idea about the characteristics of Alfvén wave propagation. Another important quantity is the residual energy, which measures whether the fluctuating energy resides within the magnetic field or velocity field ([21, 22]). The residual energy is defined as the energy difference between the fluctuating kinetic and magnetic energy. The residual energy is found theoretically and observationally to be negative in the inner and outer heliosphere ([21, 22, 23, 24, 25]), indicating the dominance of fluctuating magnetic energy in the MHD fluctuations.

2. Turbulence transport equations

Since the earliest observations of turbulence in the solar wind, there has been considerable interest in understanding how turbulence is transported. Turbulence transport equations describe the evolution of fluctuations throughout the solar wind and the heliosphere. The traditional approach for modeling the transport of fluctuations was based on the Wentzel-Kramers-Brillouin (WKB) theory ([26, 37, 28]), which is a linearized wave description. The WKB model considers non-interacting propagating waves, and therefore neglects possible mixing or coupling between propagating modes. However, forward and backward propagating modes can typically be coupled via a three wave interaction mediated by a zero frequency mode. Therefore, several models were developed to address the limitations of the WKB model ([29–32]).

For studying turbulence, the use of the sum and difference of the velocity and magnetic fluctuations, $\mathbf{u}$ and $\mathbf{b}$, respectively, is more convenient to use rather than considering each separately. This is because the incompressible MHD equations can be expressed compactly in terms of the Elsässer variables. The sum and difference of the fluctuations form the Elsässer variables, $\mathbf{z}^{\pm} = \mathbf{u} \pm \mathbf{b}/\sqrt{\mu_0 \rho}$ ([33]), where the fluctuating magnetic field is normalized to the Alfvén unit and ρ is the solar wind density. The parameters $\mathbf{z}^{\pm}$ describe two oppositely propagating modes, and their directions are defined with respect to the large-scale magnetic field $\mathbf{B}_0$.

WKB theory assumes that interplanetary fluctuations are Alfvén waves, from which it follows that the variance in the fluctuating magnetic field $\langle b^2 \rangle$ decays as r^{-3}, where r is the heliocentric distance. WKB theory successfully explains the radial evolution of magnetic fluctuations within $\sim$ 8 AU. However, beyond 8 AU, WKB theory fails to explain the observed evolution of the variance in magnetic field fluctuations. The reason is that WKB theory is a linear theory for small amplitude Alfvén waves propagating in the inhomogeneous solar wind. Zank et al 1996 [34] developed two coupled turbulence transport equations, which describe the transport of the fluctuating magnetic energy density E_b $(= \langle b^2 \rangle / \mu_0 \rho)$ and the correlation length λ beyond ~ 1 AU. However, [34] cannot be applied within 1 AU because it assumes zero cross helicity and constant Alfvén ratio. Later, [35], [36], and [37] further developed the turbulence transport models using the similar formalism of [34]. These models can be applied within 1 AU, but they do not include the Alfvén velocity and also cannot describe the residual energy. The Alfvén velocity is very important in sub-Alfvénic coronal flows. Similarly, the residual energy is an important quantity for the MHD fluctuations. To address all these limitations, Zank et al 2012 [38] developed six coupled turbulence transport equations that describe the transport of energy in forward and backward propagating modes, the residual energy, the correlation length corresponding to forward and backward propagating modes, and the correlation length for the residual energy. Since [38] includes the Alfvén velocity, we can use the transport equations to study turbulence in sub-Alfvénic coronal flows as well as study turbulence in the super-Alfvénic solar wind. In the latter case, we neglect the Alfvén velocity in the turbulence transport equations.

The 1D steady-state six coupled turbulence transport equations of [38] in terms of $f = \langle {z^+}^2 \rangle$ and $g = \langle {z^-}^2 \rangle$ and $a = 1/2$ and $b = 0$ and neglecting the Alfvén velocity, in a spherical coordinate system are given by

$$U\frac{df}{dr} + U\frac{f}{r} + U\frac{E_D}{r} - \Gamma E_D = -2\frac{fg^{1/2}}{\lambda^+} + 2\left\langle \mathbf{S}^+ \cdot \mathbf{z}^+ \right\rangle; \tag{1}$$

$$U\frac{dg}{dr} + U\frac{g}{r} + U\frac{E_D}{r} - \Gamma E_D = -2\frac{gf^{1/2}}{\lambda^-} + 2\left\langle \mathbf{S}^- \cdot \mathbf{z}^- \right\rangle; \tag{2}$$

$$U\frac{dE_D}{dr} + U\frac{E_D}{r} + \left(\frac{U}{r} - \Gamma\right)\frac{f+g}{2} = -E_D\left[\frac{f^{1/2}}{\lambda^-} + \frac{g^{1/2}}{\lambda^+}\right] + \left\langle \mathbf{S}^- \cdot \mathbf{z}^+ \right\rangle + \left\langle \mathbf{S}^+ \cdot \mathbf{z}^- \right\rangle; \tag{3}$$

$$U\frac{d\lambda^+}{dr} + \frac{E_D}{f}\left[\frac{U}{2r} - \frac{\Gamma}{2}\right](\lambda_D - 2\lambda^+) = 2g^{1/2} - 2\frac{\left\langle \mathbf{z}^+ \cdot \mathbf{S}^+ \right\rangle}{f}\lambda^+; \tag{4}$$

$$U\frac{d\lambda^-}{dr} + \frac{E_D}{g}\left[\frac{U}{2r} - \frac{\Gamma}{2}\right](\lambda_D - 2\lambda^-) = 2f^{1/2} - 2\frac{\left\langle \mathbf{z}^- \cdot \mathbf{S}^- \right\rangle}{g}\lambda^-; \tag{5}$$

$$U\frac{d\lambda_D}{dr} + \left(\frac{U}{r} - \Gamma\right)\frac{f+g}{E_D}\left(\frac{f\lambda^+ + g\lambda^-}{f+g} - \frac{\lambda_D}{2}\right) = \lambda_D\left[\frac{f^{1/2}}{\lambda^-} + \frac{g^{1/2}}{\lambda^+}\right] - \frac{\lambda_D}{E_D}\left[\left\langle \mathbf{z}^- \cdot \mathbf{S}^+ \right\rangle + \left\langle \mathbf{z}^+ \cdot \mathbf{S}^- \right\rangle\right], \tag{6}$$

where f is the energy corresponding to backward propagating modes, g is the energy corresponding to forward propagating modes, $E_D = \langle u^2 \rangle - \langle b^2/\mu_0 \rho \rangle$ is the residual energy, λ^+ is the correlation length corresponding to backward propagating modes, λ^- the correlation length corresponding to forward propagating modes, and λ_D the correlation length corresponding to the residual energy. Terms such as $\langle \mathbf{S}^\pm \cdot \mathbf{z}^\pm \rangle$ refer to sources of turbulence, which we discuss later. The parameter $\Gamma (= n_i n_j \partial U_j / \partial U_i)$ is the shear mixing term, which we neglect by choosing $\Gamma = 0$.

In Equations (1), (2), and (3), the first terms on the right hand side indicate the dissipation for the energy in backward and forward propagating modes and the residual energy. By combining all three dissipation terms, we can write the evolution equation for the solar wind temperature as [18-21, 25]

$$U\frac{dT}{dr}+(\gamma-1)\frac{2UT}{r}=\frac{1}{9}\frac{m_p}{k_B}\alpha\left[\frac{fg^{1/2}}{\lambda^+}+\frac{gf^{1/2}}{\lambda^-}+E_D\left(\frac{f^{1/2}}{\lambda^-}+\frac{g^{1/2}}{\lambda^+}\right)\right], \qquad (7)$$

where $\gamma = 5/3$ is the adiabatic index, m_p is the proton mass, k_B is the Boltzmann constant, and α is the von Kármán-Taylor constant. We choose $\alpha = 1$ ([18]). The first term inside the square bracket in Equation (7) corresponds to the dissipation term for the energy in backward propagating modes, the second term to the dissipation term for energy in forward propagating modes, and the third term is the dissipation term for the residual energy.

The total turbulent energy E_T, the cross helicity E_C, the fluctuating magnetic energy density E_b, and the Alfvén ratio r_A can be expressed in terms of f and g as ([38])

$$E_T=\langle u^2\rangle+\langle b^2/\mu_0\rho\rangle=\frac{f+g}{2}; \qquad (8)$$

$$E_C=2\langle \mathbf{u}\cdot\mathbf{b}/\sqrt{\mu_0\rho}\rangle=\frac{g-f}{2}; \qquad (9)$$

$$E_b=\left\langle\frac{b^2}{\mu_0\rho}\right\rangle=\frac{E_T-E_D}{2}=\frac{f+g-2E_D}{4}; \qquad (10)$$

$$r_A=\frac{\langle u^2\rangle}{\langle b^2/\mu_0\rho\rangle}=\frac{E_T+E_D}{E_T-E_D}=\frac{f+g+2E_D}{f+g-2E_D}; \qquad (11)$$

$$\langle u^2\rangle=\frac{E_T+E_D}{2}=\frac{f+g+2E_D}{2}. \qquad (12)$$

These turbulent quantities have been previously studied by several authors. However, here we apply a more sophisticated set of turbulence transport model equations to calculate these quantities, and then we compare to *Voyager* 2 observations.

3. Sources of turbulence

There are in principle three types of turbulence sources, stream shear ([5]), shock waves ([39]), and pickup ions ([40]). The first two sources are important within $4-5$ AU, and the third source is important beyond the ionization cavity ($\gtrsim 6-10$ AU). The shear source can be written in a form similar to that of [34] and [25],

$$\langle \mathbf{S}^+\cdot\mathbf{z}^+\rangle=C_{\rm sh}(f)\frac{\Delta U_{\rm shear}}{r}f; \qquad (13)$$

$$\langle \mathbf{S}^-\cdot\mathbf{z}^-\rangle=C_{\rm sh}(g)\frac{\Delta U_{\rm shear}}{r}g; \qquad (14)$$

$$\langle \mathbf{S}^-\cdot\mathbf{z}^+\rangle+\langle \mathbf{S}^+\cdot\mathbf{z}^-\rangle=[C_{\rm sh}(f)+C_{\rm sh}(g)]\frac{\Delta U_{\rm shear}}{r}E_D, \qquad (15)$$

where $C_{\rm sh}(f)$ and $C_{\rm sh}(g)$ are the strengths of shear interaction corresponding to backward and forward propagating modes, respectively, and $\Delta U_{\rm shear}$ ($= 350$ kms^{-1}) is the velocity difference between the fast and slow solar wind speed. It is assumed that shear driving generally supplies energy to the forward and backward propagating modes at an equal rate i.e., $C_{sh}(f) = C_{sh}(g)$. Similarly, the stream source of turbulence due to shocks for backward and forward propagating modes, and residual energy respectively is given by

$$\mathbf{S}^+\cdot\mathbf{z}^+ \;=C_{\rm shock}(f)\frac{\Delta U_{\rm shock}}{r}f; \qquad (16)$$

$$\left\langle \mathbf{S}^- \cdot \mathbf{z}^- \right\rangle = C_{\text{shock}}(g)\frac{\Delta U_{\text{shock}}}{r}g; \tag{17}$$

$$\left\langle \mathbf{S}^- \cdot \mathbf{z}^+ \right\rangle + \left\langle \mathbf{S}^+ \cdot \mathbf{z}^- \right\rangle = [C_{\text{shock}}(f) + C_{\text{shock}}(g)]\frac{\Delta U_{\text{shock}}}{r}E_D, \tag{18}$$

where $C_{\text{shock}}(f)$ and $C_{\text{shock}}(g)$ parametrize the generation of energy for the backward and forward propagating modes, respectively, and ΔU_{shock} is the difference between the upstream and downstream speed of the shock. Finally, the pickup ion source of turbulence for backward and forward propagating modes, and the residual energy is

$$\left\langle \mathbf{S}^+ \cdot \mathbf{z}^+ \right\rangle = \frac{1}{2}\frac{f_D(f) n_H^\infty U V_A}{n_{\text{sw}}\tau_{\text{ion}}}\exp\left[-\frac{\lambda\chi}{r\sin\chi}\right]; \tag{19}$$

$$\left\langle \mathbf{S}^- \cdot \mathbf{z}^- \right\rangle = \frac{1}{2}\frac{f_D(g) n_H^\infty U V_A}{n_{\text{sw}}\tau_{\text{ion}}}\exp\left[-\frac{\lambda\chi}{r\sin\chi}\right]; \tag{20}$$

$$\left\langle \mathbf{S}^- \cdot \mathbf{z}^+ \right\rangle + \left\langle \mathbf{S}^+ \cdot \mathbf{z}^- \right\rangle = 0. \tag{21}$$

Here, $0 < f_D(f) < 1$ and $0 < f_D(g) < 1$, and are functions of V_A/U which determines the fraction of pickup ion energy transferred into excited waves. It is assumed that pickup ions provide energy equally to forward and backward propagating modes i.e., $f_D(f) \sim f_D(g)$. The parameter n_H^∞ is the number density of interstellar neutrals, τ_{ion}^0 is the neutral ionization time at 1 AU, λ is the ionization cavity length scale, and χ is the angle between the observation point and the upstream direction.

4. Results

We solve the seven coupled equations (1)–(7) numerically using a Runge-Kutta 4th order method from 0.29 to 100 AU, and compare the theoretical results to *Voyager* 2 observations. The initial conditions for the solar wind parameters at 0.29 AU are, $f = 753$ $(\text{kms}^{-1})^2$, $g = 13,515$ $(\text{kms}^{-1})^2$, $E_D = -57.07$ $(\text{kms}^{-1})^2$, $\lambda^+ = 0.00143$ AU, $\lambda^- = 0.000779$ AU, $\lambda_D = 0.0204$ AU, and $T = 3.5 \times 10^5$ K. For the observations, we use *Voyager* 2 1 hour resolution data sets from 1977 though 2005 to calculate the solar wind parameters from 1 to $\sim$ 75 AU. The process to find the observed values is as follows; we consider R, T, and N components of the magnetic and velocity fields, the solar wind density, and the solar wind temperature. At first, the sequential 10 hours interval data sets are considered, and then the mean field corresponding to the magnetic field, the solar wind speed, the solar wind temperature, and the solar wind density is calculated. After that the fluctuating magnetic field $\mathbf{b}$ and solar wind speed $\mathbf{u}$ are calculated for each 10 hour interval. Elsässer variables are then calculated by using $\mathbf{z}^\pm = \mathbf{u} \pm \mathbf{b}/\sqrt{\mu_0\rho}$, where μ_0 is a permeability of free space, and ρ is the mean solar wind density. Note that Elsässer variables as well as the turbulent quantities are calculated for three cases; i) positive and negative sign of the radial component of the magnetic field B_r; ii) $0° < \phi < 180°$ and $180° < \phi < 360°$, where $\phi = \tan^{-1}(B_t/B_r)$ is an azimuthal angle, and iii) positive and negative sign of the T component of the magnetic field B_t. Here, $0° < \phi < 180°$ identifies one polarity of the magnetic field, and $180° < \phi < 360°$ the other polarity ([41-43]). In our case, we assume that $0° < \phi < 180°$ represents outwardly directed magnetic field, and $180° < \phi < 360°$ inwardly directed magnetic field.

We use two criteria, i) the mean square fluctuations of the velocity and magnetic field should be smaller than the square of mean fields and ii) each interval should contain at least five good data points, in each 10 hour interval data set. The turbulent quantities such as the Elsässer variables, the residual energy, the solar wind density, the solar wind temperature, the variance in the magnetic field fluctuations and the fluctuating kinetic energy, are calculated only if the criteria ii) is satisfied, otherwise we neglect the interval and move then to next interval. We follow similar procedure until the end of each yearly data sets. We use criteria i) to remove irregular fluctuations from each interval. From the observed Elsässer variables, the variances are calculated. These variances correspond to the energy in forward and backward propagating modes. We then use Equations (8)–(12) to find other turbulent quantities. Note that we calculate the R, T, and N components as well as the total turbulent quantities as a function of heliocentric distance. Finally, we smooth the observed values by taking 50 10-hours interval data sets from an intermediate file, where we use criteria i). The remaining observed quantities are averaged to find the smoothed observed values.

To find the correlation length corresponding to forward and backward propagating modes and the residual energy, 20 hour interval data sets are considered. Elsässer variables are then calculated in each 20 hour interval data sets, which are then used to calculate the auto-correlation and cross-correlation function. The auto-correlation function is used to find the correlation length for the forward and backward propagating modes, and the cross-correlation function for the correlation length corresponding to the residual energy ([25,44]). The correlation length corresponds to lag r where the auto- or cross-correlation function becomes 1/e of the maximum value.

Figure 1 shows the comparison between the theoretical and observed correlation length corresponding to different energy modes. The blue curve denotes the correlation length for forward propagating modes, the red curve for backward propagating modes, and the green curve for the residual energy. Similarly, the scatter circles, triangles, and diamonds symbol identify

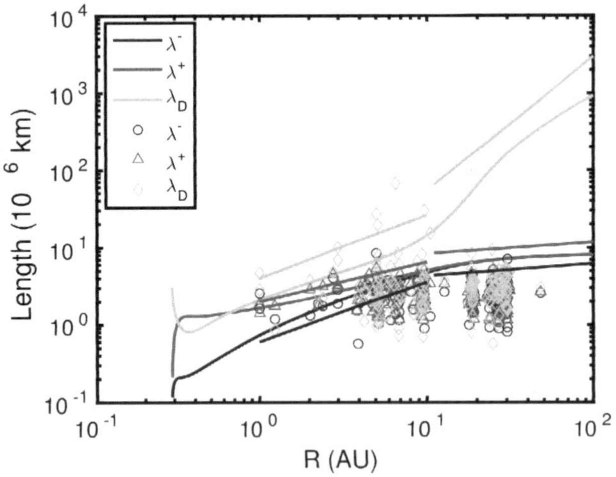

Figure 1. Comparison between the theoretical and observed correlation lengths corresponding to forward and backward propagating modes and the residual energy as functions of heliocentric distance. The solid blue, red and green straight lines guide the theoretical correlation lengths between 1–10 AU and beyond 10 AU (see text for details).

the observed correlation length for forward and backward propagating modes, and the residual energy, respectively. The two correlation lengths (blue and red curves) approach one another beyond ~ 1 AU, and then become approximately equal beyond $\sim 20-30$ AU. On the other hand, the correlation length for the residual energy (green curve) increases beyond 10 AU. The increase in the residual correlation length is due to the assumption that we set the pickup ion source for the residual energy to zero. We believe that pickup ions in the outer heliosphere beyond the ionization cavity generate Alfvénic fluctuations such that there is equipartition between the fluctuating kinetic and magnetic energy. The pickup ion source for the residual energy may need further investigation.

In Figure 1, the solid blue, red, and green straight lines describe the power law fits to different parts of the blue, red, and green curves, respectively, such that between $1 - 10$ AU (non-PUI influenced), and > 10 AU (PUI mediated). The slopes of the blue, red, and green straight lines between $1 - 10$ AU are 0.7772, 0.515, and 0.8166, respectively, and for > 10 AU are 0.16, 0.14, and 1.737, respectively. In the case of blue and red straight lines, the slopes between 1–10 AU are large so that there are significant increments in the correlation lengths, however, there are no significant increments in the correlation lengths beyond 10 AU since the slopes are small. On the other hand, the slope corresponding to the green curve is larger beyond 10 AU than between 1–10 AU, which indicates that the correlation length for the residual energy should increase faster beyond 10 AU than between 1–10 AU.

4.1. Turbulent quantities related to a positive and negative sign of B_r

As mentioned earlier, we calculate the turbulent quantities corresponding to a positive and a negative sign of B_r, and compare them to our theoretical results. The comparisons between the

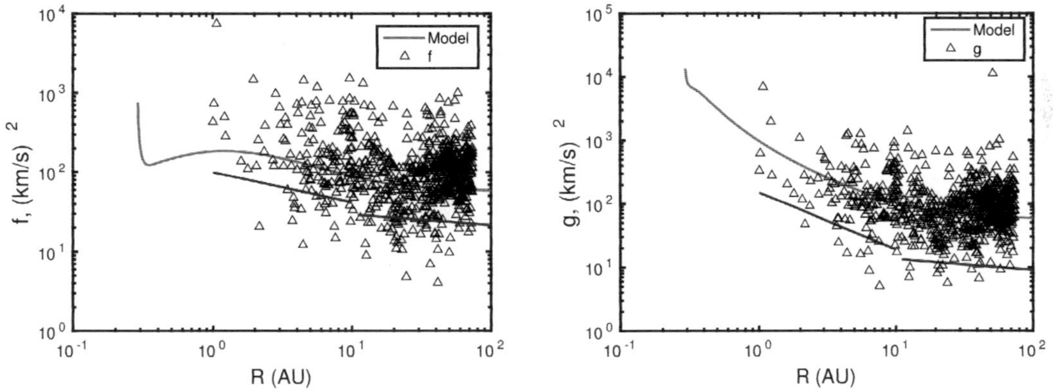

Figure 2. Left: Comparison between the theoretical and observed energy in backward propagating modes as a function of heliocentric distance. Right: Comparison between the theoretical and observed energy in forward propagating modes as a function of heliocentric distance. Solid curves are theoretical results and scatter triangles are total observed values. The solid blue straight lines guide the solid curves between 1-10 AU and beyond 10 AU (see text for details).

theoretical and observed energy in backward and forward propagating modes are shown in the left and right panel of Figure 2, respectively. The left panel of Figure 2 shows that the energy in backward propagating modes decreases sharply at first, increases until $\sim 1 - 2$ AU, and then decreases with increasing heliocentric distance. The comparison between the theoretical and observed f shows that both result follow a similar trend i.e., both decrease gradually as a function of heliocentric distance. The right panel of Figure 2 also shows that both the theoretical and observed g decrease monotonically with increasing heliocentric distance. Furthermore, the left panel of Figure 2 shows an important characteristic in that backward propagating modes are generated within $\sim 1 - 2$ AU. The generation of backward modes is due to the presence of a shear source of turbulence. Furthermore, in the figure, the blue straight lines describe the power law fits to different parts (between 1–10 AU and beyond 10 AU) of the red curves. The slopes of the straight lines between 1–10 AU for f and g are -0.3804 and -0.9012, respectively, and beyond 10 AU are -0.1343 and -0.1725, respectively. The slopes corresponding to different parts of the curves indicate that f and g decrease faster between 1–10 AU than beyond 10 AU.

The left and right panels of Figure 3 illustrate the comparison between the theoretical and observed result for the total turbulent energy E_T and the fluctuating magnetic energy density E_b as a function of heliocentric distance, respectively. To calculate the total turbulent energy and

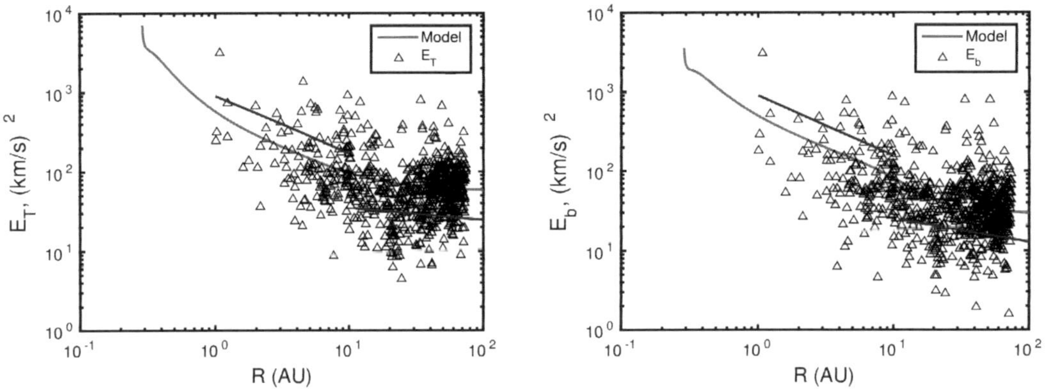

Figure 3. Left: Comparison between the theoretical and observed total turbulent energy as a function of heliocentric distance. Right: Comparison between the theoretical and observed fluctuating magnetic energy density as a function of heliocentric distance. Solid curves are theoretical results and scatter triangles are observational results. The solid blue straight lines guide the solid curves between 1-10 AU and beyond 10 AU (see text for details).

the fluctuating magnetic energy density, we use Equations (8) and (9), respectively. The result shows that the theoretical and observed E_T decrease gradually with increasing heliocentric distance. Similarly, the theoretical and observed E_b decrease monotonically with increasing heliocentric distance. In the left and right plots of Figure 3, the blue straight lines describe the power law fits to different parts of the curves, between 1–10 AU and beyond 10 AU. In the left figure, the slope of the blue line between 1–10 AU is -0.7206 and beyond 10 AU is -0.1538. Similarly, in the right plot, the slopes between 1–10 AU and beyond 10 AU are -0.7565 and -0.2949, respectively. Both E_T and E_b have larger negative slopes between 1–10 AU than beyond 10 AU so that these quantities between 1–10 AU decrease faster in comparison to that of beyond 10 AU.

The left and right panel of Figure 4 illustrate the comparison between the theoretical and observed magnetic field variance $\langle b^2 \rangle$ and fluctuating kinetic energy $\langle u^2 \rangle$ as a function of heliocentric distance, respectively. The left panel of Figure 4 shows good agreement between the theoretical and observed $\langle b^2 \rangle$ from 1 to $\sim$ 75 AU. The right panel of Figure 4 shows that the theoretical $\langle u^2 \rangle$ cannot reproduce the observed $\langle u^2 \rangle$ within $\sim$ 7 AU, however, the theoretical $\langle u^2 \rangle$ is consistent with the observed $\langle u^2 \rangle$ beyond $\sim$ 7 AU. In the left panel of Figure 4, the blue straight lines describe the power law fits to the theoretical results between 1–10 AU and beyond 10 AU. The slopes of the blue lines between 1–10 AU and beyond 10 AU are -2.7565 and -2.2949, respectively. The different slopes indicates that $\langle b^2 \rangle$ decreases with different power law index within 10 AU and beyond 10 AU, such that the decrease in $\langle b^2 \rangle$ is larger within 10 AU than beyond 10 AU.

The left and right panel of Figure 5 illustrate the comparison between the theoretical and observed normalized residual energy and Alfvén ratio as a function of heliocentric distance, respectively. Recall that the residual energy is defined as the difference between the fluctuating kinetic and magnetic energy, and the Alfvén ratio is the ratio between the fluctuating kinetic and magnetic energy. Close to the Sun, the fluctuating kinetic and magnetic energy are

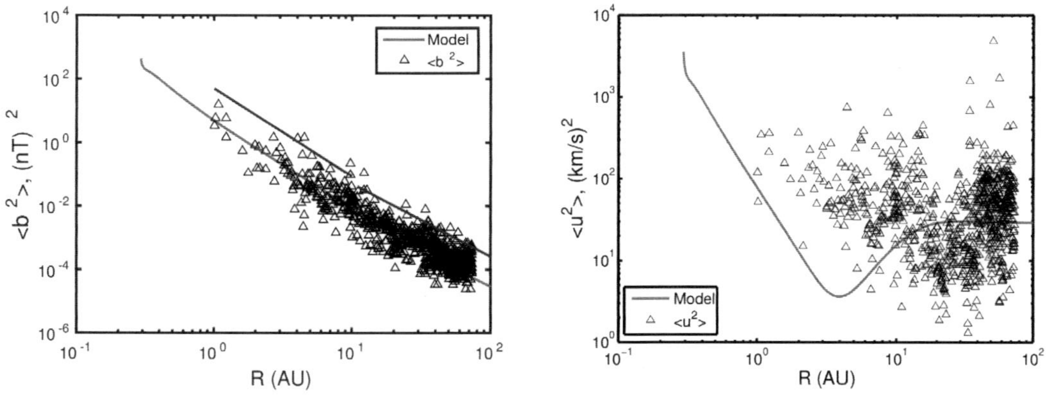

Figure 4. Left: Comparison between the theoretical and observed magnetic field variance as a function of heliocentric distance. Right: Comparison between the theoretical and observed fluctuating kinetic energy as a function of heliocentric distance. Solid curves are theoretical results and scatter triangles are observational results. The solid blue straight lines guide the solid curves between 1-10 AU and beyond 10 AU (see text for details).

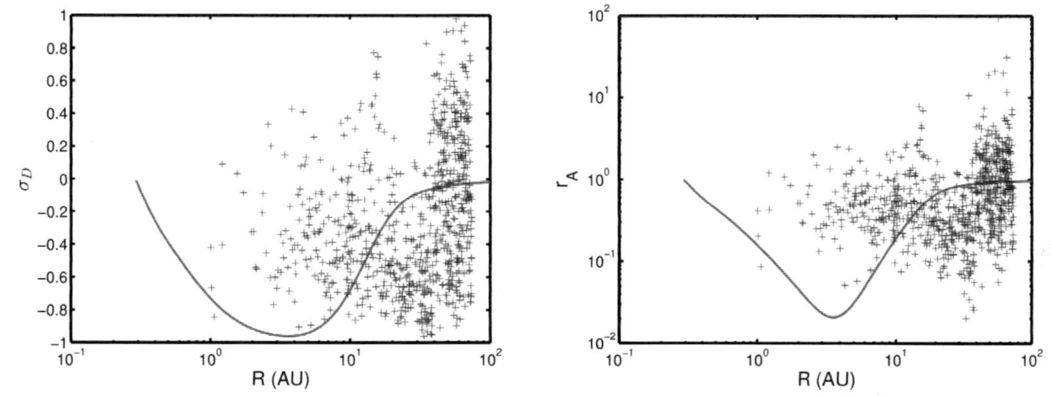

Figure 5. Left: Comparison between the theoretical and observed normalized residual energy as a function of heliocentric distance. Right: Comparison between the theoretical and observed fluctuating Alfvén ratio as a function of heliocentric distance. Solid curves are theoretical results and scatter plus symbols are observational results.

approximately equipartitioned. Therefore, the residual energy is close to 0 at 0.29 AU (left-panel), and the Alfvén ratio close to 1 (right-panel). Both theoretical results for the normalized residual energy and the Alfvén ratio show that the fluctuating magnetic energy in MHD fluctuations increases with increasing heliocentric distance up to $\sim 5-6$ AU, but beyond $\sim 5-6$ AU the fluctuating kinetic energy begins to increase such that the fluctuating kinetic and magnetic energy is approximately equipartitioned beyond $20-30$ AU. The blue plus symbols in the left and right panel of Figure 5 indicate the observed values. There is a large scatter in the observed normalized residual and Alfvén ratio, which may be due to the plasma data.

The left and right panel of Figure 6 illustrate the normalized cross helicity and the solar wind temperature as a function of heliocentric distance, respectively. In the figure, the solid

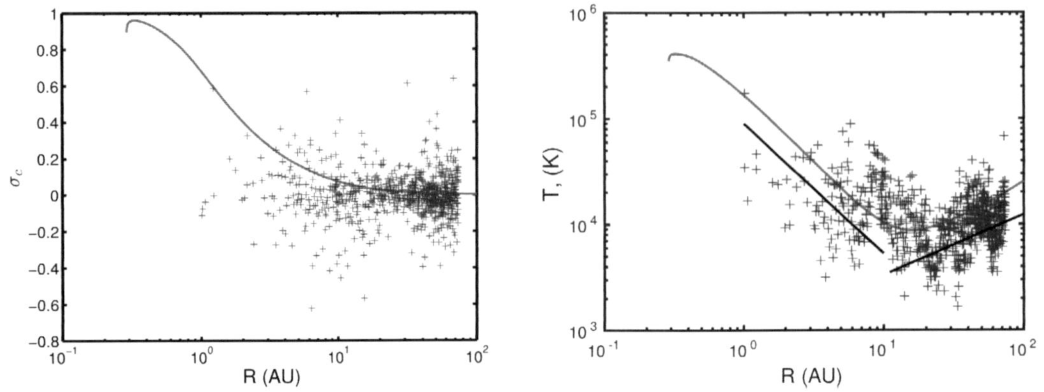

Figure 6. Left: Comparison between the theoretical and observed normalized cross helicity as a function of heliocentric distance. Right: Comparison between the theoretical and observed solar wind temperature as a function of heliocentric distance. Solid curves are theoretical results and scatter plus symbols are observed results. The solid black straight lines guide the solid curves between 1-10 AU and beyond 10 AU (see text for details).

curves are the theoretical results and the scatter plus symbols are the observational results. The left panel shows that the cross helicity decreases gradually with increasing heliocentric distance, and becomes approximately zero beyond $\sim$ 30 AU. The right panel shows that the theoretical and observed solar wind temperature decrease monotonically up to $\sim 10 - 20$ AU, and then increases up to $\sim$ 75 AU. The increase in the temperature beyond $\sim 10 - 20$ AU is due to the presence of pickup ions in the outer heliosphere. In the right panel of Figure 6, the black straight lines describe the power law fits to different parts, between 1–10 AU and beyond 10 AU, of the solar wind temperature (red curve). The slope of the black curve between 1–10 AU is -1.23, and beyond 10 AU is 0.5692, which clearly indicates that the solar wind temperature decreases between 1–10 AU, and increases beyond 10 AU.

4.2. Turbulent quantities related to the azimuthal angle ϕ

The magnetic field in the outer heliosphere is mainly azimuthal, therefore the azimuthal angle ϕ ($=\tan^{-1}(B_t/B_r)$) is used to find the polarity of the magnetic field. The angle ϕ varies between $0°$ and $360°$. Recall that for $0° < \phi < 180°$, the polarity of the magnetic field is outward, and for $180° < \phi < 360°$, the polarity is inward.

The left and right panel of Figure 7 illustrate the energy in backward and forward propagating modes, respectively. The comparison between the theoretical and observed results of f and g shows that both decrease monotonically from $\sim$ 1 AU with increasing heliocentric distance. Moreover, the trends in the observed f and g are similar to the previous observed values obtained by considering the sign of B_r, Figure 2. The left panel of Figure 8 illustrates the total turbulent energy as a function of heliocentric distance, and the right panel the fluctuating magnetic energy density. Figure 8 indicates that the theoretical and observational results of the total turbulent energy and the fluctuating magnetic energy density are in good agreement. These results show that all the energies decay with the expansion of the solar wind.

The left and right panel of Figure 9 describe the variances in the magnetic field and solar wind velocity fluctuations, respectively. The theoretical and observed variance in the fluctuating magnetic field, left panel of Figure 9, show good agreement. On the other hand, the theoretical and observed variance in the fluctuating solar wind velocity, right panel of Figure 9, are consistent

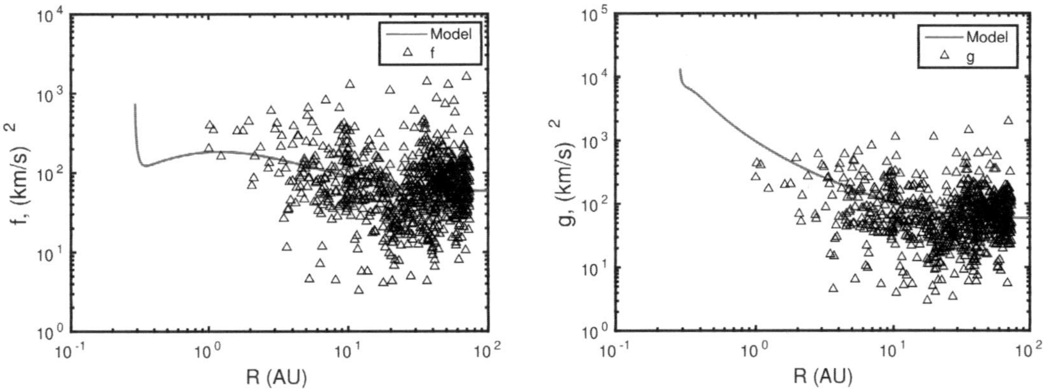

Figure 7. Left: Comparison between the theoretical and observed energy in backward propagating modes as a function of heliocentric distance. Right: Comparison between the theoretical and observed energy in forward propagating modes as a function of heliocentric distance. Solid curves are theoretical results and scatter triangles are observational results.

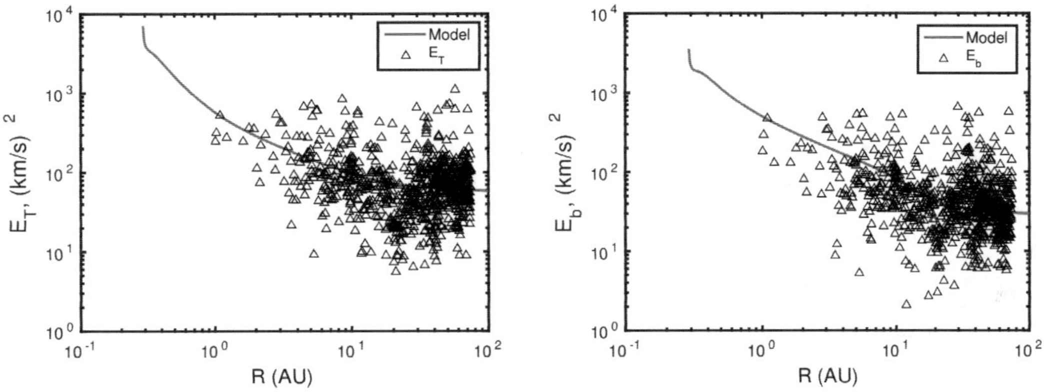

Figure 8. Left: Comparison between the theoretical and observed total turbulent energy as a function of heliocentric distance. Right: Comparison between the theoretical and observed fluctuating magnetic energy density as a function of heliocentric distance. Solid curves are theoretical results and scatter triangles are observational results.

only beyond $\sim 7-8$ AU. The observed values are similar to the previous observed values obtained by considering the sign of B_r, Figure 4.

The left panel of Figure 10 illustrates the normalized residual as a function of heliocentric distance, and the right panel the Alfvén ratio. Similarly, the left and right panel of Figure 11 illustrate normalized cross helicity and the solar wind temperature, respectively. All these observed values are similar to those of the previous observed values as well, Figures 5 and 6.

4.3. Turbulent quantities related to a positive and negative sign of B_t

In this case, we use the sign of B_t to find the polarity of the interplanetary magnetic field. The turbulent quantities are calculated for the positive and negative sign of B_t separately, and then the results are combined.

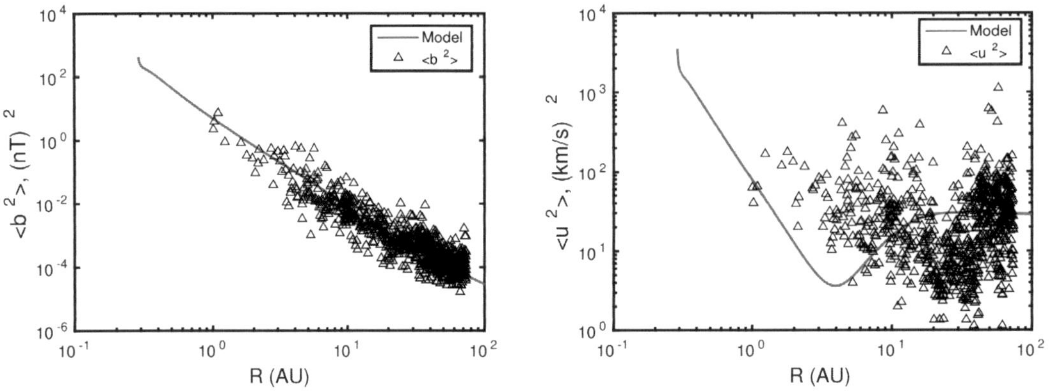

Figure 9. Left: Comparison between the theoretical and observed magnetic field variance as a function of heliocentric distance. Right: Comparison between the theoretical and observed fluctuating kinetic energy as a function of heliocentric distance. Solid curves are theoretical results and scatter triangles are observational results.

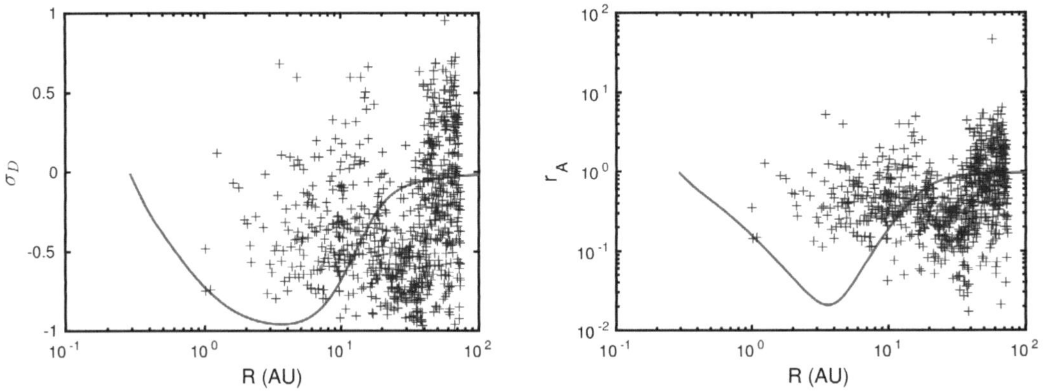

Figure 10. Left: Comparison between the theoretical and observed normalized residual energy as a function of heliocentric distance. Right: Comparison between the theoretical and observed fluctuating Alfvén ratio as a function of heliocentric distance. Solid curves are theoretical results and scatter plus symbols are observational results.

The left and right plots of Figure 12 illustrate the comparison between the theoretical and observed energy in backward and forward propagating modes, respectively. Similarly, the left and right panels of Figure 13 describe the total turbulent energy and the fluctuating magnetic energy density with increasing heliocentric distance, respectively.

The variance in the fluctuating magnetic field and the solar wind velocity as a function of heliocentric distance are shown in the left and right panels of Figure 14, respectively. The normalized residual energy as a function of heliocentric distance is illustrated in Figure 15 (left panel), and the right panel of Figure 15 describes the Alfvén ratio. Furthermore, the normalized cross helicity and the solar wind temperature are shown in the left and right plots of Figure 16. All the comparisons show good agreement between the theoretical and observed results. Moreover, the observed values are also similar to the observed values obtained by considering the azimuthal angle.

Journal of Physics: Conference Series **642** (2015) 012001 doi:10.1088/1742-6596/642/1/012001

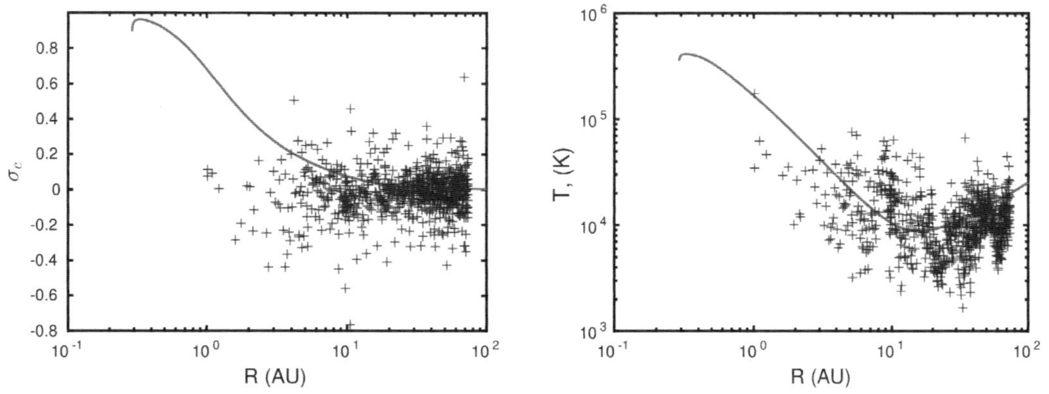

Figure 11. Left: Comparison between the theoretical and observed normalized cross helicity as a function of heliocentric distance. Right: Comparison between the theoretical and observed solar wind temperature as a function of heliocentric distance. Solid curves are theoretical results and scatter plus symbols are observational results.

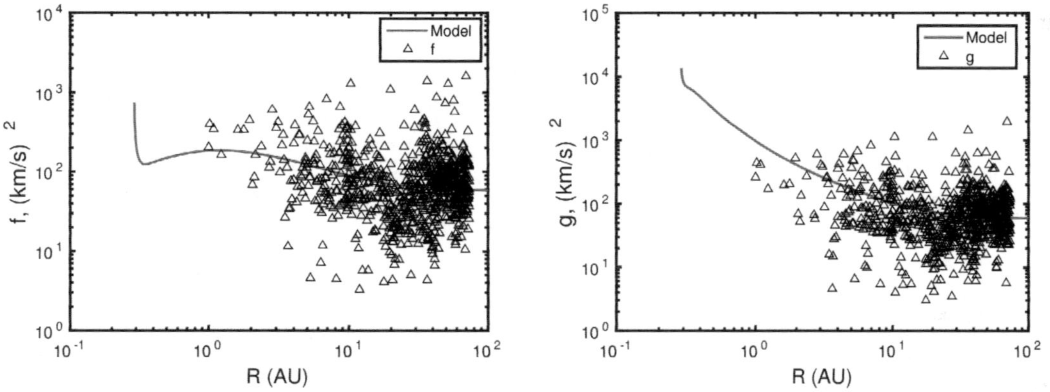

Figure 12. Left: Comparison between the theoretical and observed energy in backward propagating modes as a function of heliocentric distance. Right: Comparison between the theoretical and observed energy in forward propagating modes as a function of heliocentric distance. Solid curves are theoretical results and scatter triangles are observational results.

5. Conclusions

In this paper, we investigated the transport of low-frequency turbulence in the super-Alfvénic solar wind flows theoretically and observationally. We solved the 1D steady state seven coupled turbulence transport equations numerically using a 4th order Runge-Kutta method. We calculated the observed turbulent quantities for three cases; i) a positive and negative sign of B_r, ii) the azimuthal angle ϕ, and iii) a positive and negative sign of B_t. We found that the observed turbulent quantities for the last two cases are the same, and the trends of these observed results are similar to that of the observed results when using the sign of B_r. We found that the theoretical results are consistent with all the observed values. We conclude therefore that the Zank et al 2012 [38] turbulence transport model equations can explain the transport of low-frequency turbulence in the super-Alfvénic solar wind flow.

We considered the evolution of the energy in backward propagating modes from 0.29 to 100

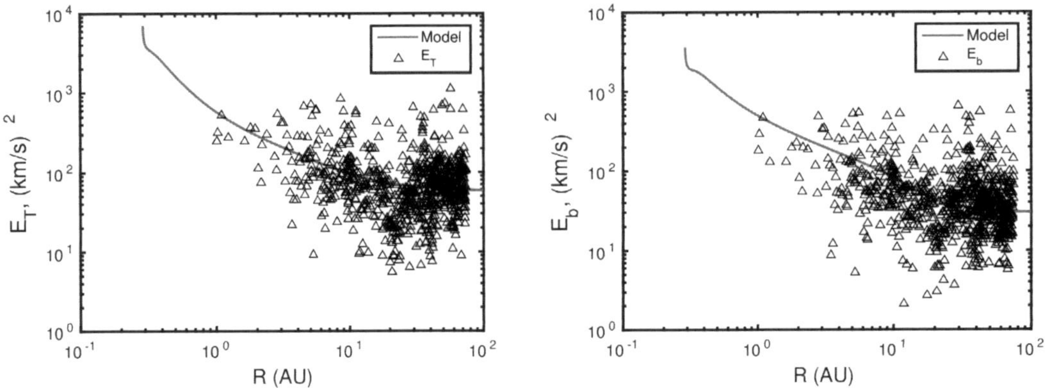

Figure 13. Left: Comparison between the theoretical and observed total turbulent energy as a function of heliocentric distance. Right: Comparison between the theoretical and observed fluctuating magnetic energy density as a function of heliocentric distance. Solid curves are theoretical results and scatter triangles are observational results.

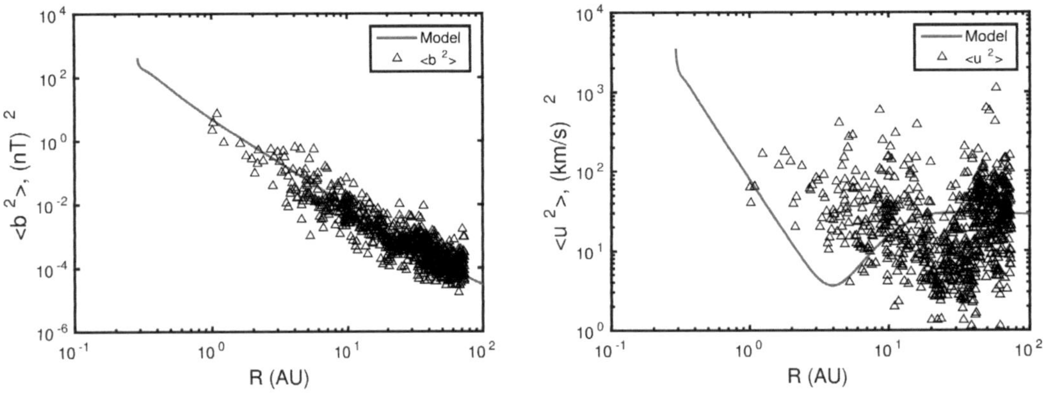

Figure 14. Left: Comparison between the theoretical and observed magnetic field variance as a function of heliocentric distance. Right: Comparison between the theoretical and observed fluctuating kinetic energy as a function of heliocentric distance. Solid curves are theoretical results and scatter triangles are observational results.

AU. Our result shows that backward modes are generated in situ within 1 AU. The creation of backward propagating modes is due to the in situ generation of turbulence by shear driving in the inner heliosphere. We also found that the MHD fluctuations in the solar wind are dominated by the fluctuating magnetic energy between ~ 1 and ~ 10 AU. However, the fluctuating kinetic and magnetic energy is approximately equipartitioned within ~ 1 AU and beyond ~ 10 AU.

We also found that all the energies decay with increasing heliocentric distance at a rate that is slower than would be expected of adiabatic expansion. This is due to the in situ generation of turbulence in the heliosphere. The dissipation of these energies results in plasma heating, which is the reason for the solar wind temperature profile being non-adiabatic. The normalized cross helicity as a function of heliocentric distance is approximately zero beyond ~ 30 AU. It clearly indicates that the energy in forward and backward propagating modes is equipartitioned i.e., $g \sim f$. We find that the correlation lengths corresponding to forward and backward propagating

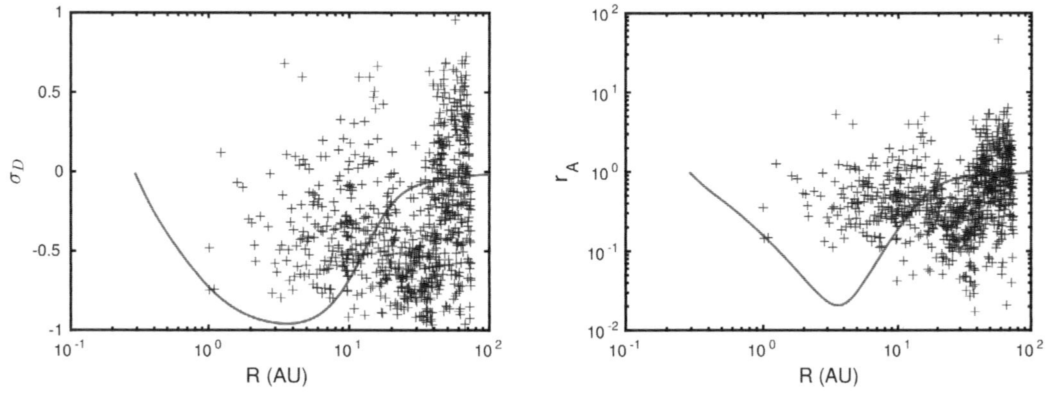

Figure 15. Left: Comparison between the theoretical and observed normalized residual energy as a function of heliocentric distance. Right: Comparison between the theoretical and observed fluctuating Alfvén ratio as a function of heliocentric distance. Solid curves are theoretical results and scatter plus symbols are observational results.

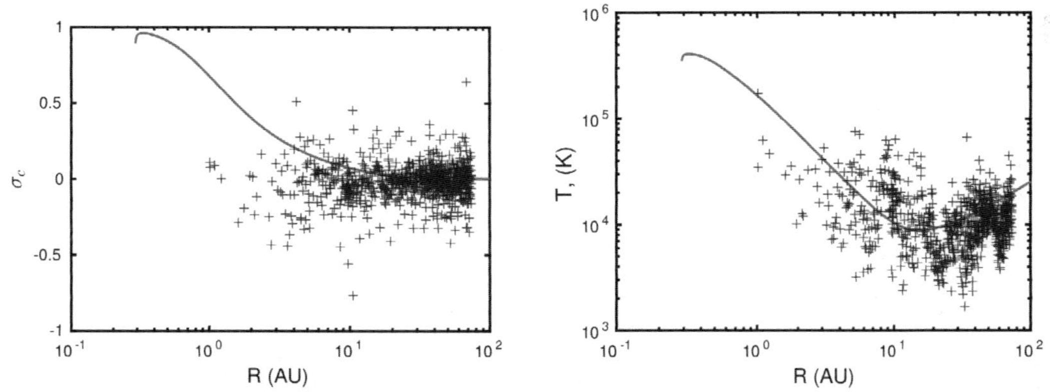

Figure 16. Left: Comparison between the theoretical and observed normalized cross helicity as a function of heliocentric distance. Right: Comparison between the theoretical and observed solar wind temperature as a function of heliocentric distance. Solid curves are theoretical results and scatter plus symbols are observational results.

modes are approximately equal beyond $\sim$ 30 AU i.e., $\lambda^{-} \sim \lambda^{+}$. It suggests that if the energy in forward and backward propagating modes is equal, the correlation length corresponding to these energy modes should also be equal, and vice versa. In this situation, the two energy modes dissipate at the same rate.

Acknowledgments

We acknowledge the partial support of NASA grants NNX09AP74A and NNX14AC08. We also acknowledge the partial support of Alabama EPSCoR GRSP during round 9.

Appendix A. Comparison to the R, T, and N component turbulent quantities related to a positive and negative sign of B_r

In this section, we compare our theoretical results to the R, T, and N component observed values. The top left plot of Figure 17 illustrates the total turbulent energy as a function of heliocentric distance from 0.29 to 100 AU. The top right plot of Figure 17 illustrate the energy

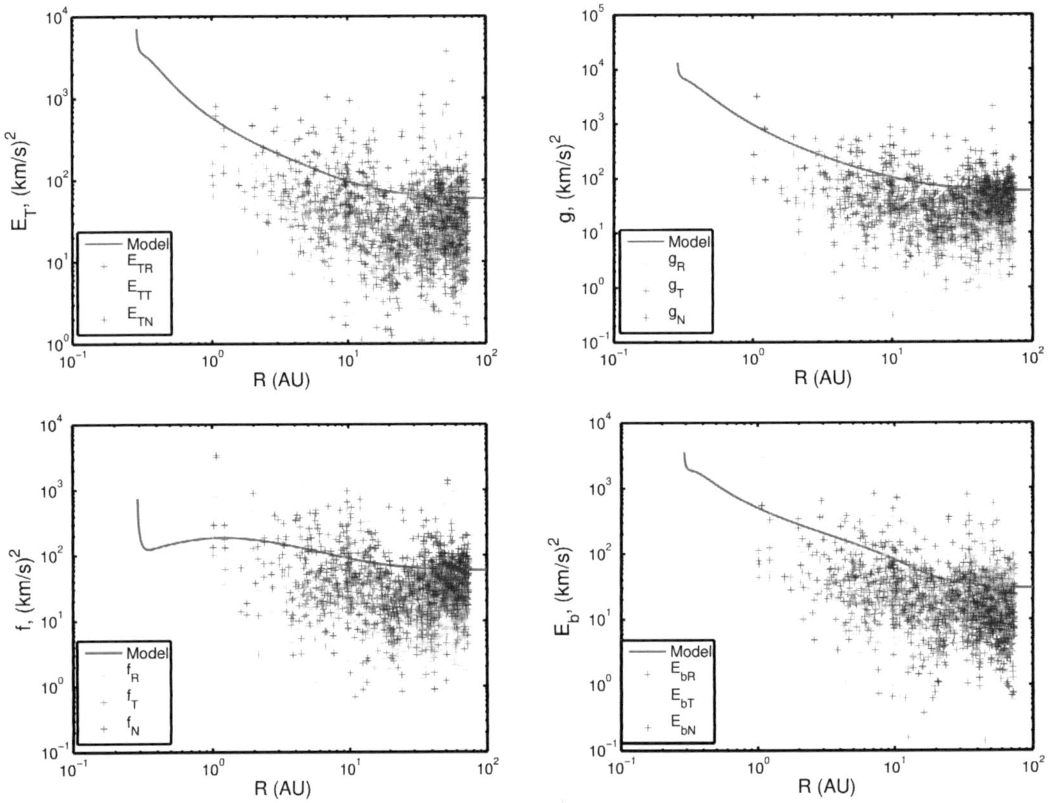

Figure A1. From top left in clockwise order, the plots show respectively the comparison of the theoretical model and the observed total turbulent energy, the energy in forward propagating modes, the fluctuating magnetic energy density, and the energy in backward propagating modes from 0.29 to 100 AU, respectively. The solid curves are theoretical results and the scatter plus symbols are observed values. The suffix R, T, and N indicates the R-component, T-component, and N-component of the observed values.

in forward propagating modes, the bottom left the energy in backward propagating modes, and the bottom right the fluctuating magnetic energy density. In the figure, the red curves denote the theoretical results, and the scatter red, green, and blue plus symbols denote the R, T, and N component observed values. The observed R, T, and N values show that the trend of each component is similar throughout the heliosphere from 1 to $\sim$ 75 AU. These observed values are smaller than the corresponding total observed values, and the trends are similar to the theoretical results as well as to the total observed values.

The left and right panels of Figure 18 illustrate the variance in the fluctuations of the magnetic field and the solar wind speed with increasing heliocentric distance, respectively. Again, in the figure, the red curves denote the theoretical results, and the scatter red, green and blue plus symbols denote the R, T, and N observed values, respectively. The comparison between the

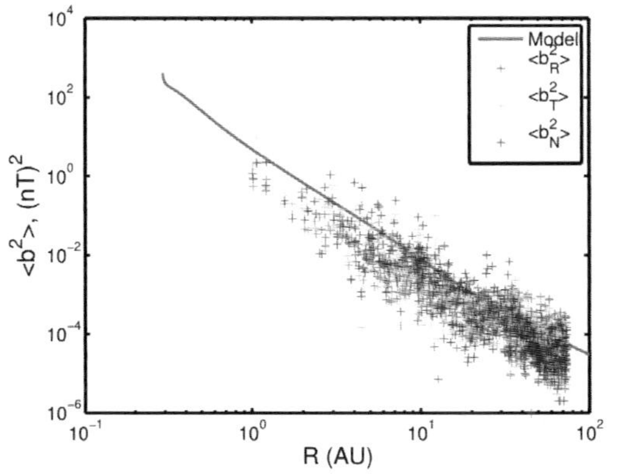

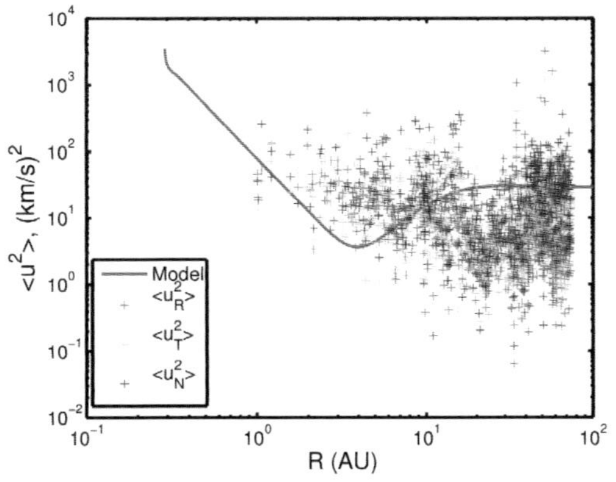

Figure A2. Left: Comparison of the fluctuating magnetic energy between the theoretical model and observations from 0.29 to 100 AU. Right: Comparison of the fluctuating kinetic energy from 0.29 to 100 AU. The scatter plus symbols denotes the observed values derived from *Voyager 2* 1-hr data. The suffix R, T, and N indicates the R-component, T-component, and N-component of the observed magnetic and kinetic energy.

theoretical and observed variance in the fluctuating magnetic field shows that both results are in a good agreement. In the case of the variance in the fluctuating solar wind speed, Figure 18 (right), the observed values within $\sim 5-6$ AU are closer to the theoretical results in comparison to the total observed fluctuating kinetic energy, Figure 14 (right).

Appendix B. Comparison to the R, T, and N component turbulent quantities related to the azimuthal angle

In this case, the R, T, and N component observed values are calculated considering the azimuthal angle ϕ. Note that the turbulent quantities are calculated for $0° < \phi < 180°$ and $180° < \phi < 360°$ separately, and then the results are combined. Figure 19 illustrates the total turbulent energy, the energy in forward propagating modes, the fluctuating magnetic energy density, and the energy in backward modes as a function of heliocentric distance in a clockwise direction, respectively. It is interesting to note that the trend of these observed values is similar to that of the previous observed values, where the sign of B_r was used to calculate the turbulent quantities. The theoretical and observed results also show similar trends from 1 to 75 AU.

The left and right panels of Figure 20 illustrate the variance in the magnetic field and the fluctuating kinetic energy with increasing the heliocentric distance, respectively. Again, these observed results are similar to that of the previous results, left and right panels of Figure 18.

Appendix C. Comparison to the R, T, and N component turbulent quantities related to a positive and negative sign of B_t

We also calculate the R, T, and N component turbulent quantities considering the sign of B_t. The comparison between the theoretical results and the R, T, and N component turbulent quantities with increasing heliocentric distance is shown in Figure 21, where the top left plot describes the total turbulent energy, the top right plot the energy in forward propagating modes, the bottom left plot the energy in backward propagating modes, and the bottom right plot the fluctuating magnetic energy density. It is observed that these turbulent quantities and the previous observed results, Figure 17 are the same. The left and right panel of Figure 22 illustrate

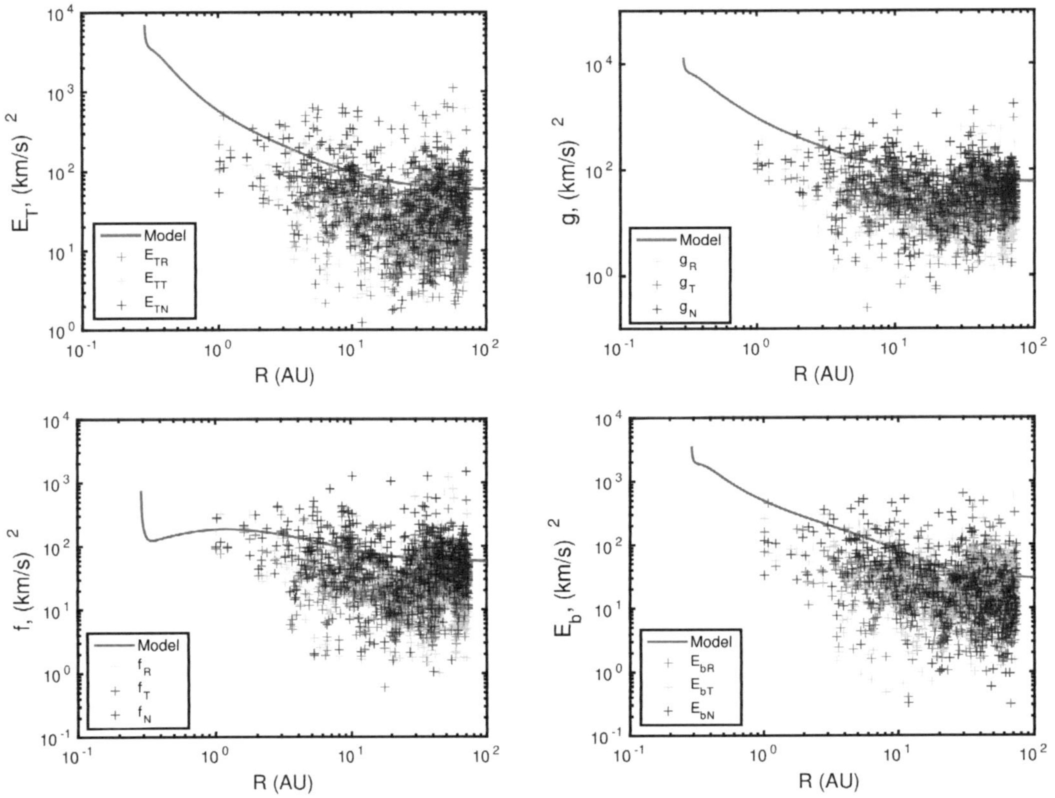

Figure B1. From top left in clockwise order, the plots show respectively the comparison of the theoretical model and the observed total turbulent energy, the energy in forward propagating modes, the fluctuating magnetic energy density, and the energy in backward propagating modes from 0.29 to 100 AU, respectively. The solid curves are theoretical results and the scatter plus symbols are observed values. The suffix R, T, and N indicates the R-component, T-component, and N-component of the observed values.

the variance in the fluctuating magnetic field and the fluctuating kinetic energy as a function of heliocentric distance, respectively.

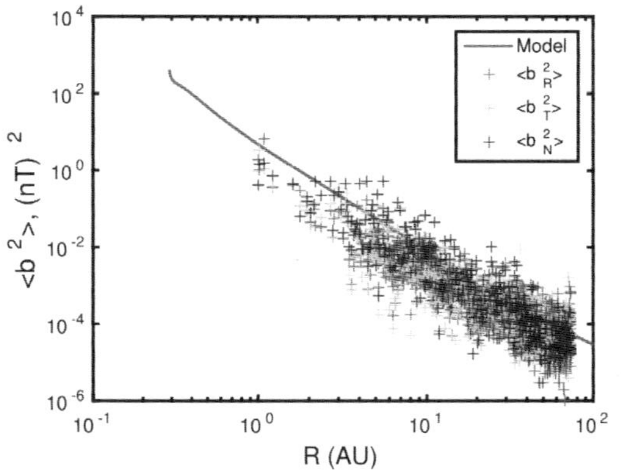

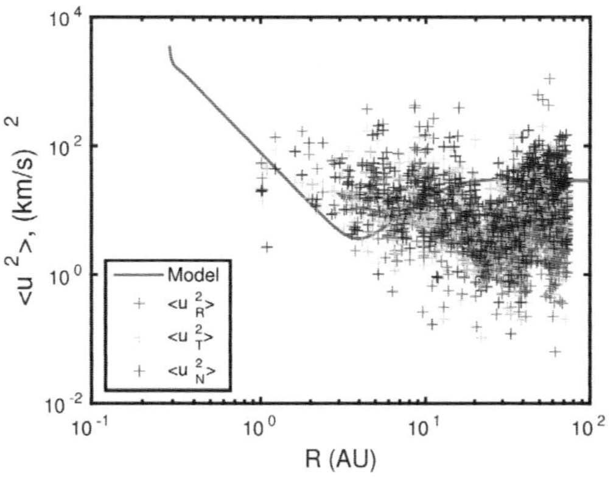

Figure B2. Left: Comparison of the fluctuating magnetic energy between the theoretical model and observations from 0.29 to 100 AU. Right: Comparison of the fluctuating kinetic energy from 0.29 to 100 AU. The scatter plus symbols denotes the observed values derived from *Voyager 2* 1-hr data. The suffix R, T, and N indicates the R-component, T-component, and N-component of the observed magnetic and kinetic energy.

References

[1] Bavassano B, Dobrowolny M, Mariani F, & Ness N F 1982 *J. Geophys. Res.* **87** 3617-3622
[2] Bruno R, & Carbone V 2005 *Living Reviews in Solar Physics* **2** 4
[3] Bruno R, & Carbone V 2013 *Living Reviews in Solar Physics* **10** 2
[4] Belcher J W, & Davis Jr L 1971 *J. Geophys. Res.* **76** 3534
[5] Coleman Jr P J 1968 *ApJ* **153** 371
[6] Goldstein M L, Roberts D A, & Matthaeus W H 1995 *ARA&A* **33** 283-326
[7] Tu C Y, & Marsch E 1995 *Space Sci. Rev.* **73** 1-210.
[8] Zhou Y, Matthaeus W H, & Dmitruk P 2004 *Reviews of Modern Physics* **76** 1015-1035
[9] Kraichnan R H 1965 *Physics of Fluids* **8** 1385-1387
[10] Belcher J W, Davis Jr L, and Smith E J 1969 *J. Geophys. Res.* **74** 2302
[11] Adhikari L, Zank G P, Hu Q, and Dosch A 2014 *ApJ* **793** 52
[12] Breech B, Matthaeus W H, Minnie J, Bieber J W, Oughton S, Smith C W, and Isenberg P A 2008 *Journal of Geophysical Research (Space Physics) 113* 8105
[13] Freeman J W 1988 *Geophys. Res. Lett.* **15** 88-91
[14] Gazis P R 1984 *J. Geophys. Res.* **89** 775-785
[15] Isenberg P A, Smith C W, and Matthaeus W H 2003 *ApJ* **592** 564-573
[16] Isenberg P A 2005 *ApJ* **623** 502-510
[17] Matthaeus W H, Zank G P, Smith C W, and Oughton S 1999b *Physical Review Letters* **82** 3444-3447
[18] Smith C W, Matthaeus W H, Zank G P, Ness N F, Oughton S, and Richardson J D 2001 *J. Geophys. Res.* **106** 825-8272
[19] Smith C W, Isenberg P A, Matthaeus W H, and Richardson J D 2006 *ApJ* **638** 508-517
[20] Smith C W, Vasquez B J, and Hamilton K 2006 *Journal of Geophysical Research (Space Physics)* **111** 9111
[21] Grappin R, Leorat J, and Pouquet A 1983 *A&A* **126** 51-58
[22] Müller W C and Grappin R 2005 *Physical Review Letters* **95** 114502
[23] Chen C H K, Bale S D, Salem C S and Maruca B A 2013 *ApJ* **770** 125
[24] Matthaeus W H, Oughton S, Pontius Jr, D H and Zhou Y 1994 *J. Geophys. Res.* **99** 19267
[25] Adhikari L, Zank G P, Bruno R, Telloni D, Hunana P, Dosch A, Marino R and Hu Q 2015 *The Astrophysical Journal* **805** 63
[26] Hollweg J V 1973 *ApJ* **181** 547-566
[27] Hollweg J V 1973 *J. Geophys. Res.* **78** 3643
[28] Hollweg J V 1974*J. Geophys. Res.* **79** 1539
[29] Marsch E and Mangeney A 1987 *J. Geophys. Res.* **92** 7363-7367
[30] Marsch E and Tu C Y 1989 *Journal of Plasma Physics* **41**479-491
[31] Marsch E and Tu C Y 1990 *J. Geophys. Res.* **95** 8211-8229

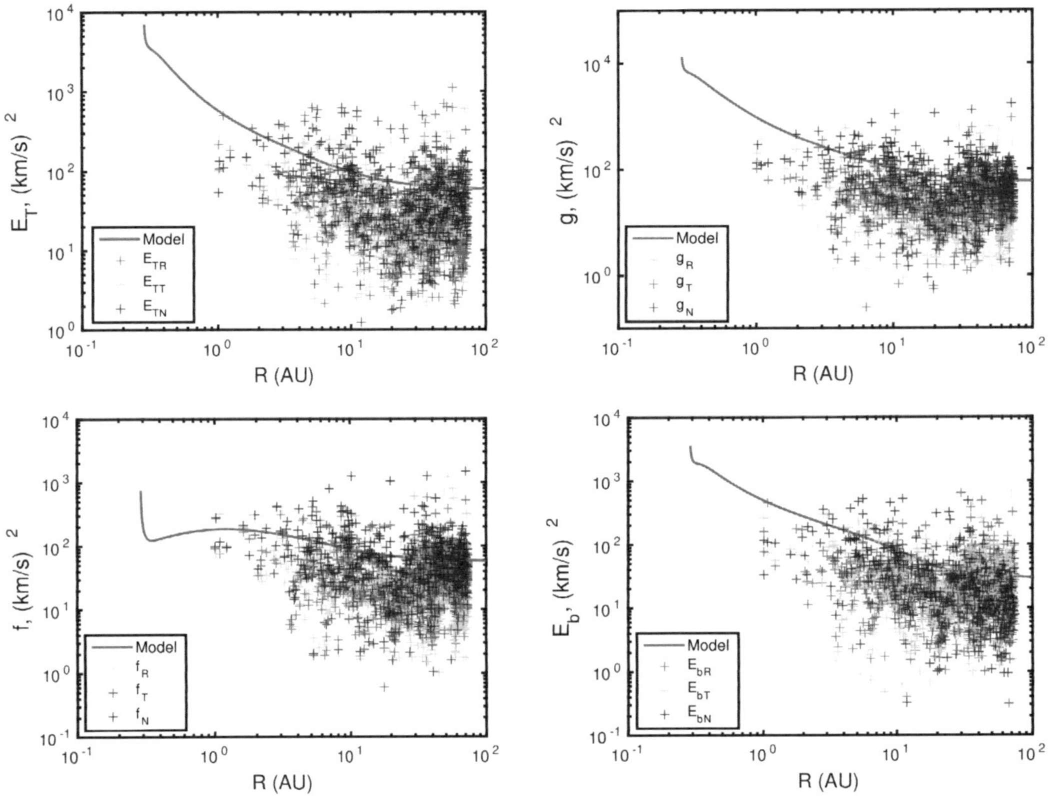

Figure C1. From top left in clockwise order, the plots show respectively the comparison of the theoretical model and the observed total turbulent energy, the energy in forward propagating modes, the fluctuating magnetic energy density, and the energy in backward propagating modes from 0.29 to 100 AU, respectively. The solid curves are theoretical results and the scatter plus symbols are observed values. The suffix R, T, and N indicates the R-component, T-component, and N-component of the observed values.

[32] Marsch E and Tu C Y 1990 *J. Geophys. Res.* **95** 11945-11956
[33] Elsässer W M 1950 *Phys. Rev.* **79** 183-183
[34] Zank G P, Matthaeus W H and Smith C W 1996 *J. Geophys. Res.* **101** 17093-17108
[35] Matthaeus W H, Minnie J, Breech B, Parhi S, Bieber J W and Oughton S 2004 *Geophys. Res. Lett.* **31** 12803
[36] Breech B, Matthaeus W H, Minnie J, Oughton S, Parhi S, Bieber J W, and Bavassano B 2005 *Geophys. Res. Lett.* **32** 6103
[37] Breech B, Matthaeus W H, Minnie J, Bieber J W, Oughton S, Smith C W, and Isenberg P A 2008 *Journal of Geophysical Research (Space Physics)* **113** 8105
[38] Zank G P, Dosch A, Hunana P, Florinski V, Matthaeus W H, and Webb G M 2012 *ApJ* **745** 35
[39] Whang Y C 1991*Space Sci. Rev.* **57** 339-388
[40] Williams L L and Zank G P 1994 *J. Geophys. Res.* **99** 19229
[41] Burlaga L F and Ness N F 1994 *J. Geophys. Res.* **99** 19341
[42] Burlaga L F and Ness N F 1996 *J. Geophys. Res.* **101** 13473-13482
[43] Burlaga L F, Ness N F and McDonald F B 1995 *J. Geophys. Res.* **100** 14763-14772
[44] Matthaeus W H, Dasso S, Weygand J M, Milano L J, Smith C W, and Kivelson M G 2005 *Physical Review Letters* **95** 231101

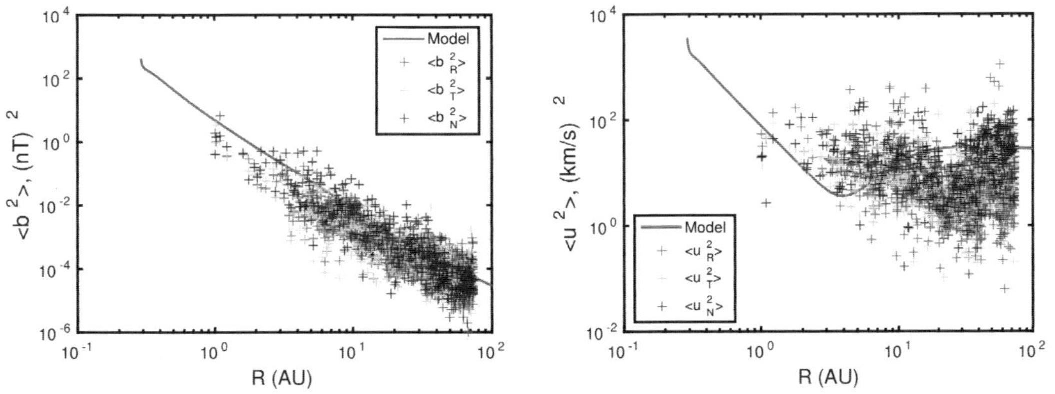

Figure C2. Left: Comparison of the fluctuating magnetic energy between the theoretical model and observations from 0.29 to 100 AU. Right: Comparison of the fluctuating kinetic energy from 0.29 to 100 AU. The scatter plus symbols denotes the observed values derived from *Voyager 2* 1-hr data. The suffix R, T, and N indicates the R-component, T-component, and N-component of the observed magnetic and kinetic energy.

Long-lived energetic particle source regions on the Sun

R Bučík[1,2], D E Innes[1,2], N H Chen[1], G M Mason[3], R Gómez-Herrero[4] and M E Wiedenbeck[5]

[1] Max-Planck-Institut für Sonnensystemforschung, D-37077 Göttingen, Germany
[2] Max Planck/Princeton Center for Plasma Physics, Princeton, NJ 08540, USA
[3] Applied Physics Laboratory, Johns Hopkins University, Laurel, MD 20723, USA
[4] Space Research Group, University of Alcalá, E-28871 Alcalá de Henares, Spain
[5] Jet Propulsion Laboratory, California Institute of Technology, Pasadena, CA 91109, USA

E-mail: bucik@mps.mpg.de

Abstract. Discovered more than 40 years ago, impulsive solar energetic particle (SEP) events are still poorly understood. The enormous abundance enhancement of the rare ^{3}He isotope is the most striking feature of these events, though large enhancements in heavy and ultra-heavy nuclei are also observed. Recurrent ^{3}He-rich SEPs in impulsive events have only been observed for limited time periods, up to a few days which is typically the time that a single stationary spacecraft is magnetically connected to the source active regions on the Sun. With the launch of the two STEREO spacecraft we now have the possibility of longer connection time to solar active regions. We examined the evolution of source regions showing repeated ^{3}He-rich SEP emissions for relatively long time periods. We found that recurrent ^{3}He-rich SEPs in these long-lived sources occur after the emergence of magnetic flux.

1. Introduction

The acceleration of ^{3}He-rich solar energetic particles (SEPs) and their escape into interplanetary space remains an unresolved question in solar physics. Anomalous abundances with several orders of magnitude enhanced ^{3}He ($\sim 10^4$) and ultra-heavies ($\sim 10^2$) suggest a unique acceleration mechanism operating in the active regions on the Sun (see [1], for a review).

Recurrent ^{3}He-rich SEP emissions observed by a single spacecraft during 1-2 days [2, 3, 4, 5] or 2-3 days [6] have suggested a steadier production/release of energetic ions from solar source regions [7]. Multiple spacecraft observations have recently shown that the source regions may produce recurrent SEPs for a much longer time - about a quarter of a solar rotation [8]. It remains unclear what mechanisms in such long-lived sources lead to the repeated particle emission. Is the production more stationary or rather intermittent during such long periods?

To address these questions we examine in this paper a temporal evolution of the three long-lived ^{3}He-rich SEP source active regions (ARs). Two of them were previously reported (AR 11244, 11246) and one is newly identified (AR 11045).

2. Observations

An overview of the source ARs' locations on the solar disk and the spacecraft ecliptic positions during the corresponding ^{3}He-rich SEP events is shown in Figure 1. Left and middle columns

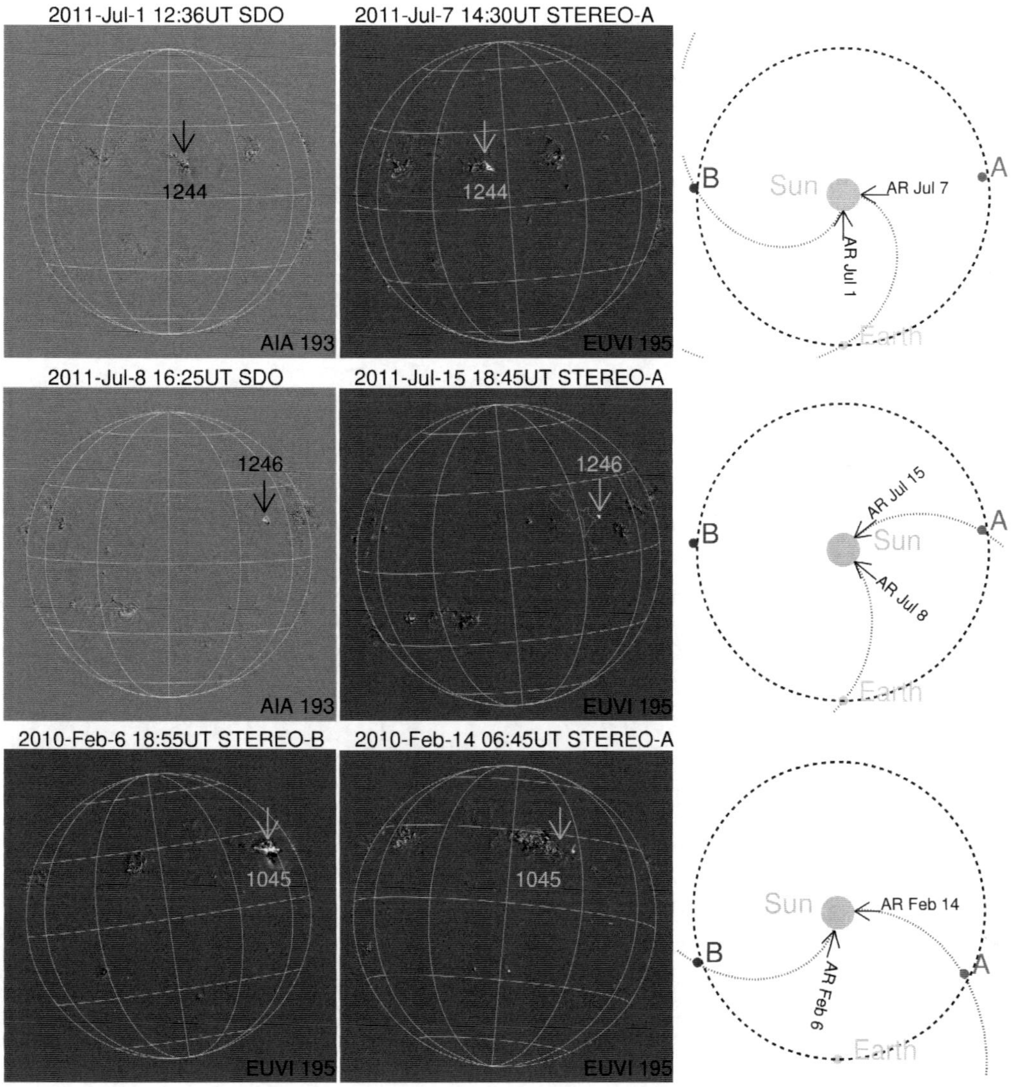

Figure 1. (Left, Middle) EUV 5 minute running difference images of the solar disk around the times of the events associated type III radio bursts. Arrows point to the long-lived ^{3}He-rich SEP source ARs. Red contours indicate approximate coronal hole boundaries as seen in direct EUV images. (Right) Ecliptic positions of STEREO-A and -B spacecraft during the events. Parker spirals are shown to illustrate the spacecraft magnetic connection to the solar sources.

of Figure 1 show AIA [9] and EUVI [10] extreme ultraviolet (EUV) running difference images of the long-lived ^{3}He-rich SEP source ARs 11244, 11246 and 11045 close to the time of events associated type III radio burst onsets. The events from ARs 11244 and 11246 were examined in detail in [8]. AR 11244 was responsible for the 2011 July 1 ^{3}He-rich SEP event observed by STEREO-B and the July 7 event observed at the Earth (L1) by ACE. In both cases AR 11244 was near the west limb, at W97 or W87. Figure 1 (upper row) shows this AR near the central meridian from the Earth (SDO) and STEREO-A views. AR 11246 was responsible for the 2011 July 9 event on ACE and the July 16 event on STEREO-A. This source was located at the

Table 1. ^{3}He-rich SEP events

Start time[a]	Spacecraft	Type III burst	AR	AR location[b]
2011 Jul 1.9	B	Jul 1 12:36	11244	N14W06
2011 Jul 7.9	ACE	Jul 7 14:31	11244	N14W87
2011 Jul 9.0	ACE	Jul 8 16:25	11246	N15W45
2011 Jul 16.4	A	Jul 15 18:46	11246	N15W138
2010 Feb 7.0	B	Feb 6 18:53	11045	N21E15
2010 Feb 14.6	A	Feb 14 06:45	11045	N21W91

[a] at 275 keV nucleon^{-1}; nominal travel time of ions of this energy from Sun to 1 AU is about 0.3 day
[b] from the Earth view

coronal-hole boundary (~W45) and had a quite small size. In contrast, AR 11045 was quite sizeable and was the source of the 2010 February 6 and 7 events on STEREO-B, the February 8 event on ACE [11] and the February 14 event on STEREO-A (see Appendix A). The ^{3}He increase on February 6 contained too much spillover from ^{4}He to be unambiguously identified as a ^{3}He-rich event [11]. However, SIT [12] on STEREO-B shows low energy Fe/O enhanced to ~1 indicating a possible impulsive flare origin of this ^{3}He increase. Table 1 lists ^{3}He-rich SEP events and sources examined in this paper.

Figure 2 shows the evolution of the source ARs in SDO HMI [13] magnetograms (2011 July 1, July 7 and July 9 events) or in STEREO-A EUVI images (2011 July 16, 2010 February 14 events). For each event a sequence of three images is shown. The first image is around the beginning of the region emergence and the third is around the particle injection time from that region. The magnetic field evolution in AR 11045 (the source of the 2010 February 6 event) has been examined in detail in [14].

AR 11244 had started to form by the emergence of a positive polarity magnetic flux in a negative polarity area at the end of 2011 June 29 (see Figure 2a), ~1.5 day before the ion injection in the July 1 ^{3}He-rich SEP event. The potential-field source-surface (PFSS) extrapolations indicate that open field in AR 11244 has negative polarity [8]. Another magnetic flux emergence (of negative polarity) occurred in the same AR 11244 on 2011 July 6 (see Figure 2b), ~1 day before the ^{3}He-rich SEP injection. The event was accompanied by multiple electron events with the first type III burst at 05:10 UT.

The creation of AR 11246 was associated with a negative polarity flux emergence in a coronal hole (of a positive polarity) on 2011 July 7 (see Figure 2c), ~1 day before the ion injection. The emergence covered a small and compact region in contrast to AR 11244. In 2011 July 16 event, the source AR 11246 was out of the Earth view and therefore its evolution could be seen only in STEREO-A EUV images. Figure 2d shows a new brightening at the AR 11246 location on July 14 which may be related to further magnetic flux emergence. This occurred ~1 day before the SEP injection. Note that the event was associated with several type III bursts starting at 12:40 UT on July 15.

AR 11045 emerged on 2010 February 5 with a positive polarity field into a negative polarity area [14], ~1.5 day before the ion injection observed by STEREO-B. The emerging phase of this AR ended on February 8. The PFSS extrapolations along with interplanetary magnetic field (IMF) polarity suggest that ACE was connected to AR 11045 for several days at least until the AR rotated behind the limb on February 14 but no further event on ACE was observed. Li et al. [14] report another magnetic flux increase in AR 11045 on February 12 when the region was close to the west limb. STEREO-A EUV images (see Figure 2e) show a creation of a new flaring

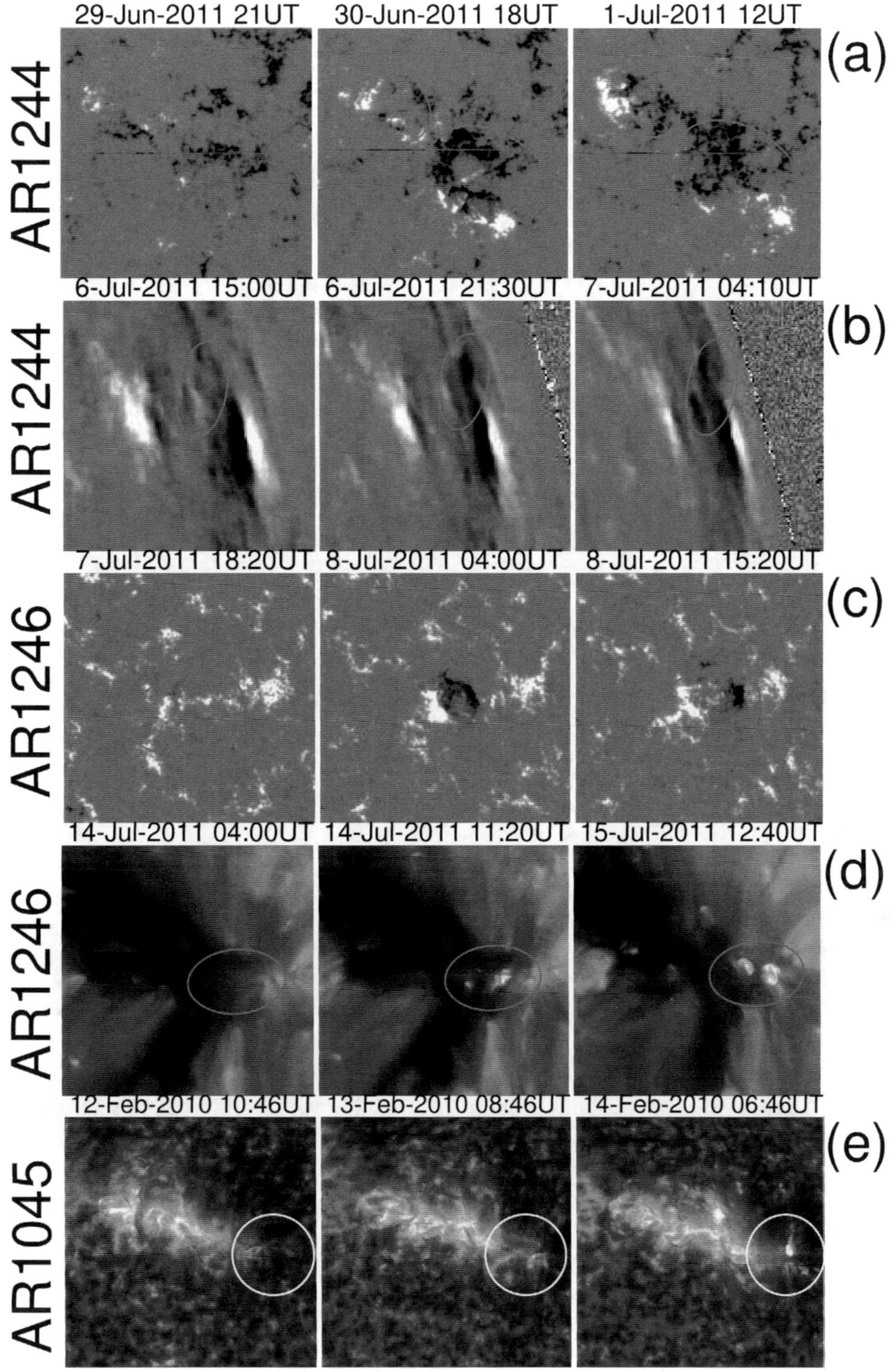

Figure 2. (a, b, c) SDO HMI magnetograms. (d) STEREO-A EUVI 195 Å images. (e) STEREO-A EUVI 304 Å images. The ellipses mark the areas of the magnetic flux emergence (a, b, c) or new flaring activity (d, e).

area in the west edge of AR 11045, perhaps related to the same magnetic flux increase. This emergence occurred ~1.5 day before ^{3}He-rich SEP injection observed by STEREO-A.

3. Summary and discussion

We investigated the temporal evolution of previously reported long-lasting ^{3}He-rich SEP source ARs 11244 and 11246. We also identified a new long-lasting source AR 11045 and examined its evolution. We found that recurrent ^{3}He-rich SEPs in the long-lived sources occur after the emergence of magnetic flux. All these ARs were newly emerging with the first particle emission occurring within ~1-2 days after the emergence. Interestingly, the next ^{3}He-rich SEP event in the same source was associated with additional magnetic flux emergence. The EUV images or PFSS extrapolations indicate that all three sources probably emerged into pre-existing open field.

Our observations are consistent with an earlier example of particle injection in the 2002 December 12 ^{3}He-rich SEP event, which occurred about 1 day after the emergence of magnetic flux in a unipolar area [5]. The authors have suggested that such a process may lead to magnetic reconnection, followed by particle acceleration on open field lines.

The presented long-lasting sources exhibited a wide range of coronal activities. For example, AR 11244 produced several cool, surge-like eruptions, as well as a number of GOES B-class X-ray flares. In contrast, AR 11246 produced only small EUV brightenings. AR 11045 was the most active region with several B-, C- and M-class flares (as well as hard X-rays) and few slow coronal mass ejections. This likely implies that there is no preferred type of ^{3}He-rich SEP source. It appears that any AR can be a ^{3}He-rich SEP source if there is magnetic flux emergence near the open field region.

Acknowledgments

This work was supported by the Max-Planck-Gesellschaft zur Förderung der Wissenschaften. The STEREO SIT is supported by the Bundesministerium für Wirtschaft through the Deutsches Zentrum für Luft- und Raumfahrt (DLR) under grant 50 OC 1301. ACE/ULEIS and STEREO/SIT are supported at APL by NASA grant NNX13AR20G/115828 and NASA through subcontract SA4889-26309 from the University of California Berkeley. R.B. thanks Gary Zank for inviting to the 14th AIAC that led to this paper.

Appendix A.

Figure A1 shows February 14 ^{3}He-rich SEP event observed by STEREO-A. The spacecraft was connected via negative polarity field lines to AR 11045 as indicated by the PFSS extrapolations (Figure A1a) along with IMF polarity. The event shows velocity dispersion (Figure A1c) and occurred during the large gradual event observed by both SOHO and STEREO-A (Figure A1b). Note that high ion counts in SIT spectrograms (Figures A1c, A1d) after the ^{3}He-rich SEP event are due to interplanetary shock passage. Figure A1 indicates that L1 spacecraft were also connected to AR 11045 but ACE did not observe the same ^{3}He injection. Perhaps ACE had a connection to different area in AR 11045.

References

[1] Mason G M 2007 *Space Sci. Rev.* **130** 231
[2] Reames D V and Stone R G 1986 *Astrophys. J.* **308** 902
[3] Mason G M, Mazur J E and Dwyer J R 1999 *Astrophys. J. Lett.* **525** L133
[4] Mason G M, Dwyer J R and Mazur J E 2000 *Astrophys. J. Lett.* **545** L157
[5] Wang Y M, Pick M and Mason G M 2006 *Astrophys. J.* **639** 495
[6] Chen N H, Bučík R, Innes D E and Mason G M 2015 *Astron. Astrophys.* doi: 10.1051/0004-6361/201525618
[7] Pick M, Mason G M, Wang Y M, Tan C and Wang L 2006 *Astrophys. J.* **648** 1247
[8] Bučík R, Innes D E, Mall U, Korth A, Mason G M and Gómez-Herrero R 2014 *Astrophys. J.* **786** 71

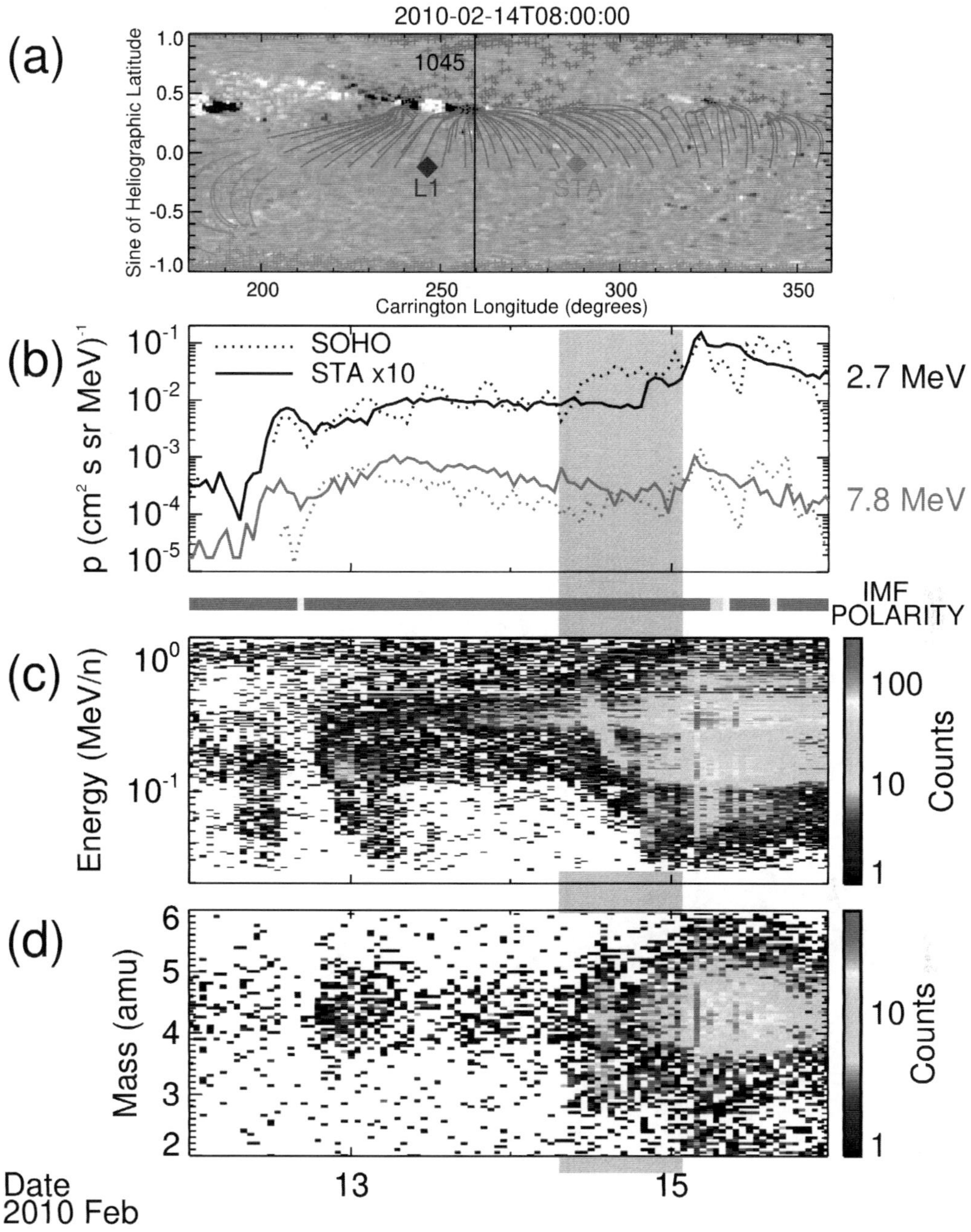

Figure A1. (a) Photospheric magnetic field with PFSS model coronal field lines (red - negative and green - positive polarity) around the injection time in STEREO-A February 14 ^{3}He-rich SEP event. Shown are field lines which intersect source surface at latitudes 0° and ±7°. Diamonds mark L1 and STEREO-A (STA) magnetic foot-points on the source surface. Black vertical line marks west solar limb from the Earth view. (b) SOHO (L1) and STA proton intensities. (c) SIT-A kinetic energy spectrogram of all ions. (d) SIT-A helium mass spectrogram at 0.2-0.5 MeV nucleon^{-1}. Shaded region marks ^{3}He-rich SEP event.

[9] Lemen J R et al 2012 *Sol. Phys.* **275** 17
[10] Howard R A et al 2008 *Space Sci. Rev.* **136** 67
[11] Wiedenbeck M E, Mason G M, Cohen C M S, Nitta N V, Gómez-Herrero R and Haggerty D K 2013 *Astrophys. J.* **762** 54
[12] Mason G M, Korth A, Walpole P H, Desai M I, Von Rosenvinge T T and Shuman S A 2008 *Space Sci. Rev.* **136** 257
[13] Scherrer P H et al 2012 *Sol. Phys.* **275** 207
[14] Li L P, Zhang J, Li T, Yang S H and Zhang Y Z 2012 *Astron. Astrophys.* **539** A7

Journal of Physics: Conference Series **642** (2015) 012003 doi:10.1088/1742-6596/642/1/012003

Voyager observations of the magnetic field in the heliosheath and the local interstellar medium

L Burlaga[1]
NASA Goddard Space Flight Center, Greenbelt, Maryland, Code 673, Greenbelt, Maryland 20771, USA

[1]E-mail: lburlagahsp@verizon.net

Abstract. This paper reviews observations of the magnetic field **B** in the heliosheath by Voyager 1 (V1) and Voyager 2 (V2) and the recent observations of the LISM by V1. The heliosheath is the region between the termination shock (TS) and the heliopause (HP). The TS was identified by both V1 and V2, and the internal structure of the TS was determined by V2. The radial distance of the TS was 94 AU at V1 and 84 AU at V2, suggesting a global asymmetry of the heliosphere. The average direction of **B** in the heliosheath is that of the Parker spiral magnetic field, and non-periodic magnetic sectors were observed by V1 and V2. The heliosheath is disturbed and turbulent, particularly just behind the TS, and it is described by the multifractal spectrum, the q-correlation function and the q-Gaussian distribution. The multifractal spectrum in the heliosheath also varies with the solar cycle. Voyager 1 has been in the LISM since at least August 25, 2012, but the nature of the HP and the time of the HP crossing not been determined. The magnetic field direction is close to, but distinguishable from the Parker spiral magnetic field, and its departure from the Parker spiral magnetic field is slowly increasing with distance from the sun. A large pressure pulse and two shocks or pressure waves have been identified in the LISM, presumably generated inside the heliosphere and transmitted through the heliosheath and heliopause. The interstellar magnetic field lines are draped across the heliopause, and B is approximately 0.5 nT . The fluctuations of B are very small but turbulent on scales 1 day to 468 days, with a Kolmogorov spectrum and an injection scale of approximately 10 pc.

1. Introduction

This paper discusses the magnetic field observations of the heliosheath observed by Voyager 1 (V1) and Voyager 2 (V2) and the recent observations of the interstellar magnetic fields originating in the local interstellar medium (LISM). Earlier reviews of the heliosheath and its interaction with the LISM were published by Seuss [1990], Axford [1990] and Zank [1999]. A recent comprehensive discussion of this topic is given by Zank [2015].

Voyager 1 is currently observing interstellar magnetic fields that are disturbed by the interaction between the heliosphere and the LISM in a region that may extend tens of AU beyond the current position of V1. Although the term LISM is used in many of the papers discussing observations and corresponding theories of the interstellar magnetic fields observed by V1, it is becoming

[1]lburlagahsp@verizon.net.

increasingly clear that one must distinguish between 1) the LISM referred to in astronomical literature on the local Galactic magnetic field [Frisch et al. 2011] and 2) the interaction region between the LISM and the heliosheath on the other. If there is a bow shock in front of the heliosphere this disturbed region could be called the "outer heliosheath". However, McComas et al. [2012] conclude that a bow shock does not exist, and a number of authors refer to the interaction region as the "Very Local Interstellar Medium" (VLISM). In either case, the interstellar magnetic field is draped across the heliosphere and compressed by the interaction, and the LISM is disturbed by shock waves (and possibly other disturbances) produced by the sun and solar wind. This paper uses the term LISM for both the disturbed region near the heliopause and the distant undisturbed region. The magnetic fields and plasmas that are discussed in this brief overview are those observed by V1 and V2 in the heliosheath and by V1 in the LISM.

It is not universally accepted that V1 has entered the LISM. Fisk and Gloeckler [2014] suggests that that the current data are consistent with V1 still being in the heliosheath, based on a model that they developed. In this model, the properties of the outermost region of the heliosheath are distinctly different from the properties described by the observations of the heliosheath discussed in this paper. In the model of Fisk and Gloeckler [2014], the high densities that are inferred from electron plasma oscillations at the largest distances that V1 has explored are considered to be a property of the inner heliosheath, and the strong magnetic fields that are currently observed are solar magnetic fields that result from a diamagnetic effect. Their model predicts that a sector boundary crossing will be observed during 2015 [Gloeckler and Fisk, 2014].

This paper shows that the magnetic field observations made by V1 since mid-2012 are consistent with the known properties of the interstellar medium and current models of the draping of the interstellar magnetic field about the heliosphere as it moves through the LISM.

2. Heliosheath

The supersonic solar wind emanating from the sun terminates at the termination shock (TS), which is the inner boundary of the heliosheath. At least five crossings of the TS were observed by Voyager 2 [Burlaga et al. 2008, Krimigis et al. 2008, and Stone et al. 2008], but only one crossing had a

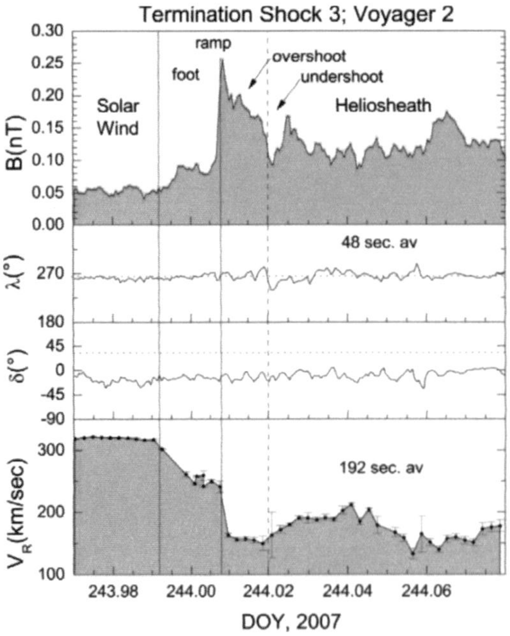

Figure 1. The third crossing of the termination shock by Voyager 2. The magnetic field strength B, the azimuthal angle λ and the elevation angle δ are shown in the top three panels, and the radial component of the solar wind velocity V_R is shown in the lowest panel.

relatively simple structure. The others were complex, because the shock was forming and dissipating on a scale of the order of a few days. Voyager 1 crossed the termination shock [Burlaga et al. 2005, Krimigis et al. 2005, and Stone et al. 2005] did not observe it because V1 moved past the TS during a data gap. Figure 1 shows the termination shock as observed by V2 during 2007. This is a classic supercritical perpendicular shock with a foot, a ramp in which most of the dissipation occurs, an overshoot, and undershoot [Mellott 1985]. The direction of the magnetic field **B** changed very little across the shock.

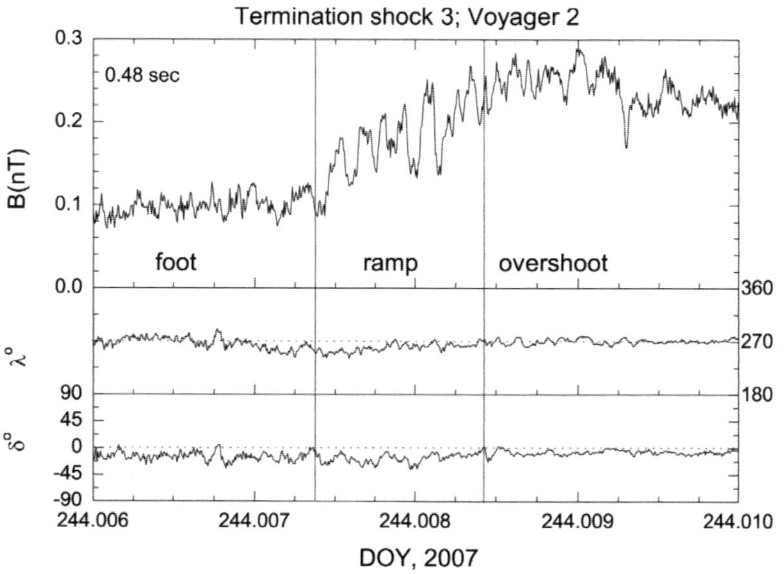

Figure 2. Internal structure of the termination shock 3 observed by Voyager 2 on day 244, 2007. The most significant structure is the ramp, which shows quasi-periodic oscillations.

Figure 2 shows a high-resolution view of the third crossing of the termination shock by V2, showing the foot, ramp, and overshoot. The ramp contains quasi-periodic fluctuations in B with a period of 12.6 seconds, corresponding to a wavelength of approximately 1000 km. The ramp moved past V2 in 90 seconds corresponding to a length scale of approximately 7000 km, which is the ion inertial length c/ω_p, where c is the speed of light and ω_p is the proton gyro-frequency.

An overview of the magnetic field observations in the heliosheath is shown in Figure 3. V1 was in the heliosheath from the time that it crossed the TS [Burlaga et al., 2005 Krimigis et al. 2004, Stone et al, 2015] to the time that it crossed the heliopause (sometime in the middle of 2012), implying that V1 was in the heliosheath for approximately 7.5 years. The magnetic field strength behind the TS was 2-3 times that in the supersonic solar wind prior to the TS crossing. During the first three years after crossing the TS, B fluctuated around a constant value, because the tendency for B to decrease with increasing distance was balanced by the tendency for B to increase with increasing solar activity. As solar activity declined, B decreased and declined until the historically low values were observed at solar minimum, which occurred during 2010-2011 at 1 AU. The magnetic field strength began to increase in 2011 at V1, partly because of the increasing solar activity but also because of the declining solar wind speed observed by Krimigis et al. [2013] between late 2010 and 2012.

Voyager 1 entered the heliosheath in December, 2004. The transition from the heliosheath to the local interstellar medium during 2014 is complex [Burlaga & Ness 2014a]. There does not appear to be a single “discontinuous” change in the magnetic field that can be identified as the heliopause. There is a growing number of theoretical papers that attempt to describe the nature of the transition (e.g., Swisdack et al. [2013], and Strumik 2013, 2014]). Based on the magnetic field and energetic particle observations, it appears that V1 was observing the LISM during 2012 shortly after it crossed a current sheet CS0 that looks like a sector boundary. After crossing this current sheet, the magnetic field differed from the Parker spiral direction by about 20° in the elevation and azimuthal angles as predicted by Pogorelov et al. [2009] as a consequence of the draping of magnetic field lines across the heliopause. The magnetic field strength at this time was larger than observed anywhere in the heliosheath, and the magnetic strength fluctuations were very small.

The cosmic ray intensity increased as V1 moved through the heliosheath toward the source of the cosmic rays (Figure 3d), and then it showed a 2-step structure which is not understood [Webber and McDonald 2013]. The cosmic ray counting rate reached a plateau on August 25, 2012, after which V1 appears to be observing particles and fields in the LISM, as discussed below.

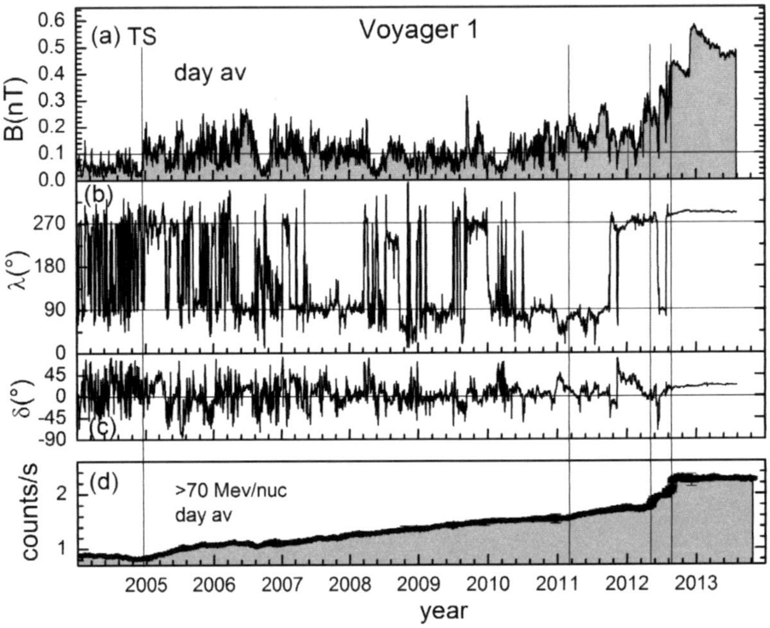

Figure 3. An overview of the magnetic fields and cosmic rays observed by Voyager 1 in the heliosheath. Panel (a) shows the magnetic field strength profile, while (b) and (c) are the azimuthal angle λ and elevation angle δ, respectively. The termination shock, the sunward boundary of the heliosheath, is labelled "T". Panel (d) shows the profile of the >70 MeV/nucleon cosmic rays.

Large-scale features called "GMIRs" are observed in the heliosheath on a scale of one year, particularly near solar maximum [Burlaga, Ness & McDonald 1993]. GMIR's are observed in the supersonic solar wind beyond approximately 15 AU and in the heliosheath. A GMIR in the supersonic solar wind can be preceded by a strong shock, which results from the coalescence of many forward shocks produced by solar activity [Burlaga et al. 2003]. An example of a GMIR produced in the heliosheath observed by V2 is shown in Figure 4 [Burlaga et al. 2011] In this case, the magnetic field strength in the GMIR is twice that in the ambient heliosheath, and it moved past V2 during an interval of 40 days. There is an indication that the GMIR was preceded by a fast-forward shock, but this occurred in a data gao. The strong magnetic fields produced a depression in the cosmic ray intensity, shown in Figure 4b, which is followed by a recovery during the next 200 days. Such a cosmic ray profile is called a long-lasting Forbush decrease. The red curve and Figure 4b is derived from the CR-B relation [Burlaga et al. 2011] that has been observed in the heliosphere that distances greater than or equal to 10 AU.

Burlaga et al. [1993] found that long-lasting Forbush decreases were associated with sustained enhanced solar activity for 1-3 solar rotations which introduce strong fields carried by coronal mass ejection and magnetic clouds as well as shocks associated with faster streams associated with these ejecta Burlaga et al. [2003] showed that such activity could produce organized structures (expanding shells of disturbed magnetic fields and shocks) in the heliosphere at approximately 15 AU characterized by an interval of strong magnetic fields moving past V1 and V2 for 1-3 solar rotations, and persisting out to at least 90 AU.

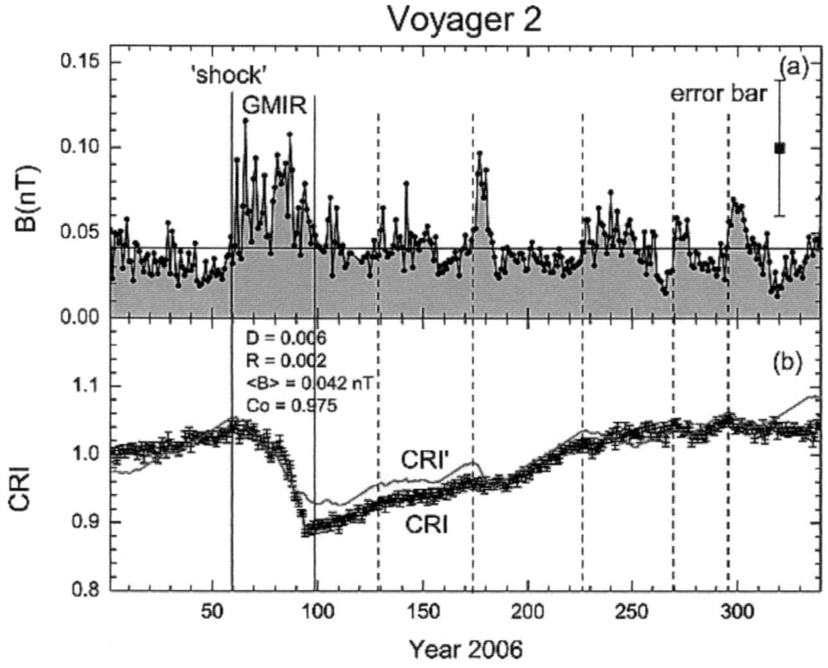

Figure 4. A GMIR observed by Voyager 2 during 2006. The GMIR may have been preceded by a shock wave. The strong magnetic fields in the GMIR produced a depression in the cosmic ray counting rate, after which the cosmic ray intensity recovered in about 200 days, giving a long-lasting Forbush decrease.

The upper left panel in Figure 5 shows typical spiral magnetic field the Parker field which is observed during declining solar activity when the predominant structures are corotating streams near solar maximum. There are times when the sun becomes particularly active for at least a solar rotation

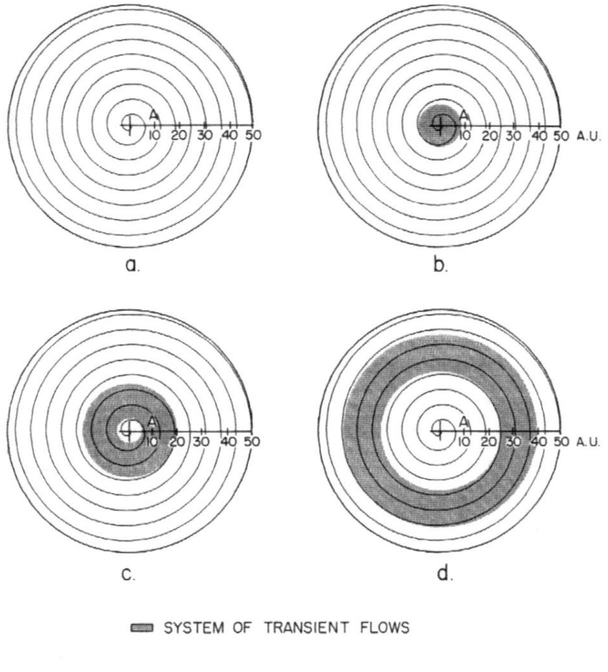

Figure 5. Schematic illustrating the formation and evolution of a Global Merged Interaction Region (GMIR).

or even up to three solar rotations. During this time one observes a number of flares and erupting filaments. The associated coronal mass ejecta and magnetic fields move away from the sun carrying strong magnetic fields with them, as indicated by the shaded circle. This system of transient flows is observed continuously for a few solar rotations and then the activity dies off and a shell containing strong disordered magnetic fields is and carried away from the sun to larger distances. Such shells were discussed many years ago in the context of cosmic ray modulation. The shelves were originally thought to be shells of enhanced turbulence which produce decreases in the cosmic ray intensity on a time scale of a few solar rotations, but B is the primary factor, as shown in Figure 4.

On smaller scales smaller scales (from 1 to approximately 25 days) three types of features, associated with the multifractal structure of the magnetic field in the heliosheath, are observed throughout the heliosheath. These features are illustrated in Figure 6.

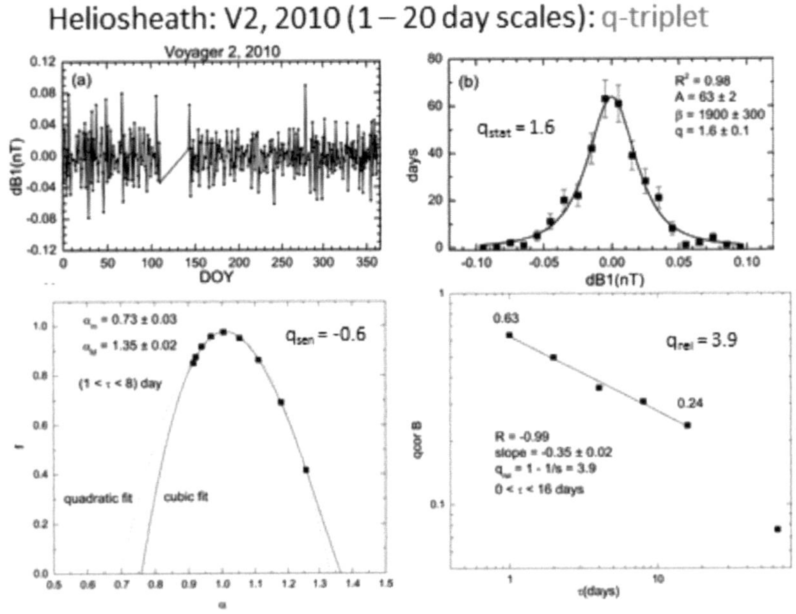

Figure 6. Observations by means of which one can compute the components of the queue-triplet, a concept that was introduced in nonextensive statistical mechanics, and is applicable to open, driven, quasi-equilibrium systems.

The top left panel is a plot of increments of daily averages of B. The distribution of these daily increments of B is shown in the top-right panel. The core of the distribution is Gaussian, but the tail is significantly different from a Gaussian distribution. The curve in this figure is a fit to the data to a q-Gaussian distribution. The q-Gaussian distribution appears in a generalized central limit theorem [Umarov, 2009, 2008]. This distribution is known in non-extensive statistical mechanics as the Tsallis distribution [Tsallis 1988, 2004, 2009], which is derived from an entropy principle, and it appears as one form of the distributions used to fit the observations of nonthermal particle distributions in space physics. The width of the q-distribution is characterized by a number that Tsallis calls q_{stat}. This number would be equal to 1 if the distribution were Gaussian. However, the observed value in the heliosheath is $q_{stat} = 1.6$, indicating these fluctuations are significantly different than Gaussian distributions. The q-Gaussian distributions of increments of B or a general feature of the fluctuations in the heliosheath [e.g. Burlaga et al. 2006d,e, Burlaga et al. 2007, Burlaga and Ness 2009, Burlaga, Ness and Acuña 2009 d, and Burlaga and Ness, 2012a]

A second feature of the fluctuations in the heliosheath on scales from 1 to 25 days is more difficult to explain. It is associated with the multifractal structure of the fluctuations [see Burlaga, 1995]. The lower left panel of Figure 6 shows what is called "the multifractal distribution function". The multifractal distribution function would be a delta function for a Kolmogorov spectrum associated with turbulence. However there are jumps and filaments in the magnetic field strength profile (See Figure 3) which give rise to a multifractal distribution. One way to think of this is to think of a power law distribution as associated with a structure function with an exponent to corresponding to a fractal or a straight line with a certain slope. One can generalize that concept by considering generalized structure functions with exponents different from 2, greater than less than 2. These correspond to increasing the width of the multifractal distribution. The multifractal distribution function gives a number called "q_{stat}", which is related to the difference between the maximum and minimum value of the parameter α [Tsallis 2004]. A multifractal distribution has been observed in the Heliosphere and during by every year that Voyager 1 has been in the heliosheath [e.g., Burlaga et al. [2006], Burlaga and Ness 2012 a, Burlaga & Ness 2013]. These observations were confirmed by Macek [2012, 2013], who also showed that the is the multifractality measure varies with the solar cycle.

A third feature describing the fluctuations in the heliosheath is the correlation function. The lower right panel in Figure 6 shows that the correlation function from V2 during 2010 was a power law, as opposed to the exponential correlation functions that are commonly observed in the laboratory. From the slope of the correlation function of this plot one can derive another parameter "q_{rel}" called the q- relaxation parameter. For the observations by V2 during 2010, q_{rel} = 3.9, indicating long non-Gaussian tails on the distribution.. Similar structures have been seen every year by V1 and V2 in the heliosheath and in the solar wind. The universality of these structures appears to be associated with the non-extensive statistical mechanics that governs the solar wind and heliosheath.

The familiar Boltzmann-Gibbs statistical mechanics deals with equilibrium systems that are usually closed boxes in which the phase space is uniform, there are no correlations, and the distribution functions are Gaussian. Tsallis non-extensive statistical mechanics was designed to describe quasi-stationary or metastable systems that are open, have a hierarchical structure which is described by a multifractal phase space, and contain correlations. Gell-Mann and Tsallis proposed that the $q_{sen} < 1 < q_{stat} \leq q_{rel}$ [Tsallis 2009] This relationship is found in the observations shown in Figure 6, where the numbers are $q_{sen} = -0.6 < 1 < q_{stat} = 1.6 \leq q_{rel} = 3.9$. Essentially the same numbers are obtained in the supersonic solar wind. In fact, this relationship for the q-triplet was first demonstrated using V1 solar wind data [Burlaga & Viñas, 2005, and confirmed by Burlaga and Ness 2013].

3. Transition between the heliosheath and the LISM

We now consider another topic, the transition between the heliosheath and the region of the local interstellar medium that is disturbed by the interaction between the LISM and the heliosphere. During

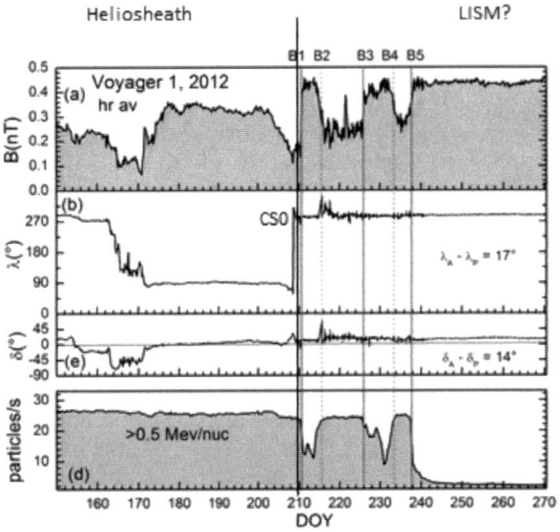

Figure 7. This figure shows the magnetic field strength B, the azimuthal angle λ, the elevation angle δ, and the counting rate of particles with >0.5 MeV/nuc. Figure describes the transition between the heliosheath and t a medium that appears to be the LISM.

2012 there was great excitement in the energetic particle community based on the observations of cosmic rays [Stone et al., 2013] and energetic particles >0.5 MeV/nuc shown in Figure 7 [Krimigis et al.,2013]. Note the dropouts in >0.5 MeV/nuc particles on day 210 and near day 230 and the corresponding recoveries on day 226 and day 238. It was widely thought that these dropouts and recoveries corresponded to crossings of the heliopause, which implies that there were five crossings of the heliopause into and out of the LISM. Later, it was confirmed that there was an increase (decrease) in the magnetic field strength corresponding to each dropout (recovery) in the energetic particles [Burlaga et al., 2013], as shown in Figure 7.

There is a problem with interpreting these dropouts as crossings of the heliopause and entry into the LISM. The solar magnetic field is generated in the sun, and the solar wind magnetic field has a spiral geometry owing to the rotation of the sun and the convection of the magnetic field by the radially following plasma [Parker 1958, 1963]. The interstellar magnetic field on the other is produced by supernovae [Ferriere 1992] and the galactic dynamo [Parker 1971a, b]. One expects the magnetic field direction to change at the heliopause. However this is not what is seen in the observations shown in Figure 7. Neither λ nor δ have any of the discontinuities, except perhaps one that might be that is obscured by noise. The change in the direction in the magnetic field across the discontinuities is < 2°, which is essentially no change that can be detected by the magnetometer. Thus, it is very unlikely that these discontinuities represent crossings of the heliopause.

The magnetic polarity that has been observed by V1 until at least day 271, 2014 was first observed when V1 crossed a current sheet labelled CS0 in Figure 7. The direction of **B** rotated within CS0 during the interval from approximately day 207.2271 - 209.7087, 2012. Burlaga & Ness [2014a] examined the boundary associated with CS0 whose normal **n** is given by the elevation angle δ = 23.9° and the azimuthal angle λ = 177.1°. They analyzed the structure of this rotation using the minimum

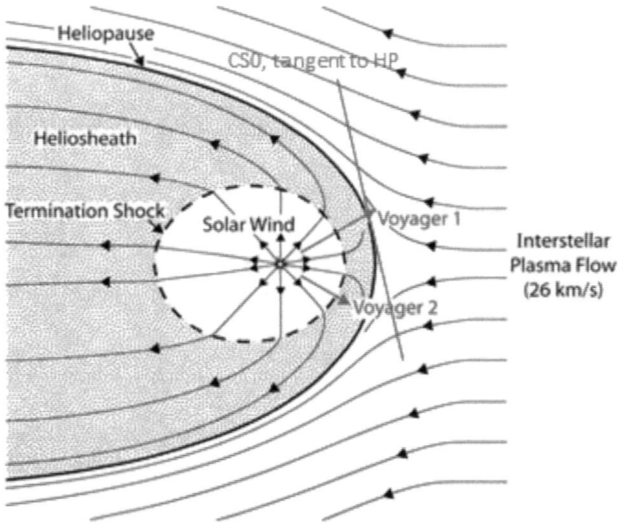

Figure 8. Schematic of the heliopause interacting with interstellar plasma flow. The tangent to the heliopause at the position of V1 is qualitatively consistent with the orientation of CS0

variance method, and found that minimum variance direction was **n** = (-0.91, 0.05, 0.415) in the RTN coordinate system whose origin is at V1. The observed value of the component of **B** in the minimum variance direction is $< B_{min} >$ = 0.095 nT with a standard deviation of 0.057 nT. The current sheet is consistent with a tangential discontinuity within the uncertainties, although we cannot exclude a nonzero component. The minimum variance analysis of the current sheet CS that the minimum variance plane is inclined by an angle i = 100.6° with respect to the solar equatorial plane. The projection of a plane with the normal **n** on a meridional plane passing through the position of V1 at the time that CS0 was observed is shown in Figure 8. This figure has been interpreted as indicating that CS0 is the heliopause. However, two other current sheets (the one prior to CS0 in Figure 7, and another earlier) have been observed with a similar orientation, so that the orientation of the current sheet is not necessarily unique to the heliopause. We regard the nature and location of the heliopause as undetermined and a subject of current research.

There are several theoretical models of the transition between the heliosheath and LISM. Swisdack [2013] and Strumik [2013, 2014] suggested that CS0 is the site of magnetic reconnection which produces magnetic islands, giving rise to a broad and complicated heliopause structure. Indeed, CS0 and the sector boundary before it have a form of a "D-sheet" that Burlaga et al. [1968a,b] suggested might be a site of magnetic reconnection. Smaller events in the solar wind plasma and magnetic field with the same characteristics as a signature as a D-sheets were identified by Gosling [2007] as magnetic reconnection signatures. Opher and Drake [2012, 2013] suggested that magnetic reconnection commonly occurs within the heliosheath. Krimigis et al. [2013] proposed that an interchange instability occurs at the heliopause, allowing interstellar magnetic flux tube to enter the heliosheath. Florinski et al. [2015] suggested that such an instability can account for the dropouts discussed in reference to Figure 7. Liewer [1995] and Zank [1995] suggested that a Rayleigh-Taylor

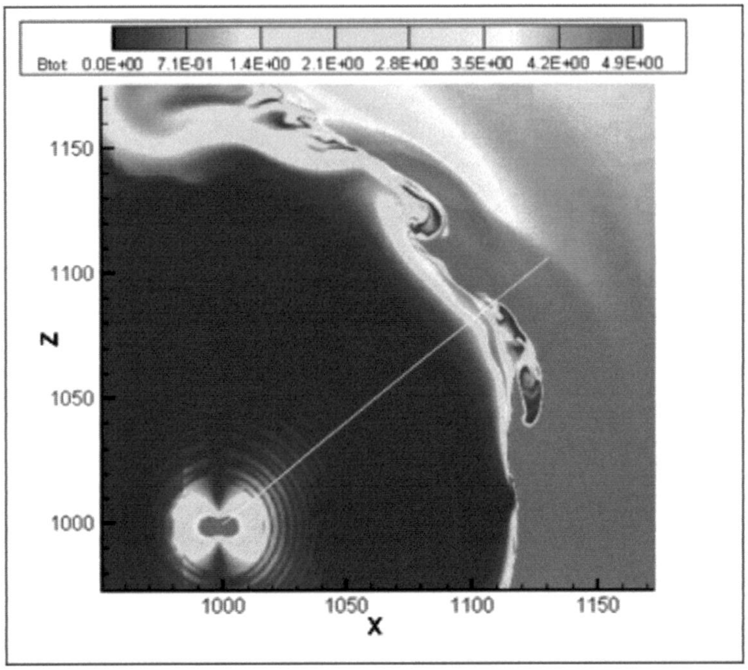

Figure 9. One frame from the model of Borovikov and Pogorelov [2014] showing the result of an instability on the heliopause as a result of the Rayleigh-Taylor instability and possibly also the Kelvin Helmholtz instability. The white line in the figure illustrates a hypothetical Voyager 1 trajectory which passes through three regions of strong magnetic fields which could correspond to the dropouts and >0.5 MeV/nuc particles shown in Figure 7.

instability may occur on the heliosheath as a result of charge exchange, and this process is incorporated in a MHD model presented by Borovikov & Pogorelov [2014]. The results of this model are illustrated in Figure 9, where the dark blue region indicates relatively weak magnetic fields in the heliosheath, and the red/orange regions indicate strong magnetic fields in the LISM in the region near the heliopause. This model can also account for the dropouts discussed in Figure 7.

4. Observations in the LISM.

Strong magnetic fields (>0.4 nT to 0.6 nT) were first observed shortly after the passage of CS0, during the first dropout in >0.5 MeV/nuc particles. They were observed again in association with the second dropout of energetic particles, and they have been observed continually since August 25, 2012 [Burlaga and Ness 2014a; Burlaga and Ness 2014b, Burlaga et al. 2015]. The counting rate of cosmic rays reached a plateau at the expected interstellar value [Stone et al. 2013, Krimigis et al. 2013] as shown in Figure 3. The plasma density reached 0.06 nT, which is characteristic of interstellar values, in October-November, 2012 [Gurnett at al, 2013]. Thus, it appears that V1 has been in the LISM since at least August 25, 2012.

Figure 10 shows observations of the magnetic field observed in the LISM from 2012.65 2 2014.74 [Burlaga and Ness, 2014b]. We shall show that these magnetic field observations throughout this interval, as well as plasma observations observed within the interval, indicate that that V1 is in the LISM, measuring interstellar magnetic fields that are draped across the heliopause and are compressed by the interaction of the heliosphere with the ambient LISM.

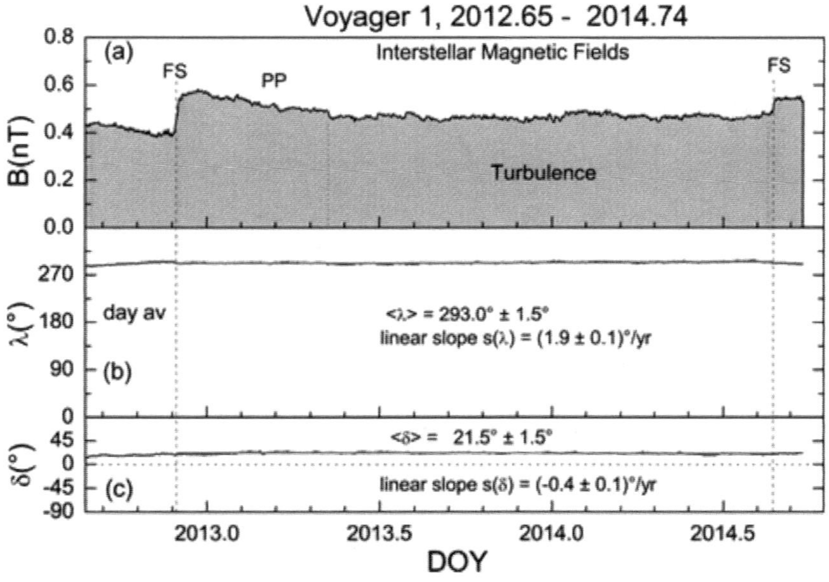

Figure 10. The interstellar magnetic fields after August 25, 2012, following the crossing of a current sheet CS0. (a) shows the magnetic field strength profile, while (b) and (c) should the azimuthal angle λ and elevation angle δ. A region containing both interstellar and heliosheath magnetic fields in flux tubes on interstellar magnetic field lines is labelled "T".

Let us first consider the direction of the magnetic field. The most obvious thing is there are no sector boundaries during this two-year interval, which suggests that V1 was not in the heliosheath. A remarkable feature, which has not been observed in the heliosheath, is the steadiness of the magnetic field direction during the entire interval; there are no visible fluctuations in Figure 10. A third important feature is that the average azimuthal angle λ and the average elevation angle δ are very nearly constant throughout most of the interval, indicated by the red lines in Figure, namely $\langle\lambda\rangle$ = 293.0° ± 1.5° and $\langle\delta\rangle$ = 21.5° ± 1.5°, respectively. These angles are about 21° and 23° from the Parker spiral direction within the uncertainty of ± 1.5°. These numbers were actually predicted as a consequence of magnetic draping by Pogorelov et al. [2009] before these observations were made. These directions are consistent with a interstellar magnetic field that passes through the center of the ribbon observed by IBEX [Grygorczuk et al. 2014]. The ecliptic longitude and latitude (λ, β) of the interstellar magnetic field is (221.9°, 38.4°).

Figure 10 shows that the average magnetic field strength B is 0.5 nTs, and it ranges from 0.4 nT to 0.6 nT. YC Whang's model [2010] of the draping of interstellar magnetic field lines across the heliopause predicts that the interstellar magnetic field strength at the nose of the heliopause should be 1.6 times the strength of the interstellar magnetic field at" infinity". This implies an interstellar magnetic field strength about 0.3 nT or 3 micro-Gauss, which is consistent values for the interstellar magnetic field that are often quoted in the literature.

Compelling evidence that V1 was in the LISM during the interval shown in Figure 11 was published by Gurnett et al. [2015] based on observations from the plasma wave experiment on V1. These observations are shown in Figure 9, which shows three episodes plasma observations as a function of time. The ordinate is the frequency of these fluctuations, which is related to the ambient density of the plasma which is plotted on the ordinate on the right-hand side of Figure 9. From these data, density is found to vary between 0.06 cm^{-3} and 0.11 cm^{-3}. These densities are characteristic of the interstellar medium and are at least 50 times greater than the densities of the heliosheath described above. On the other hand, Fisk & Gloeckler [2014] constructed a model that gives such densities for the distant heliosheath. The model predicts that V1 will cross a sector boundary during 2015 (Gloeckler and Fisk 2014)

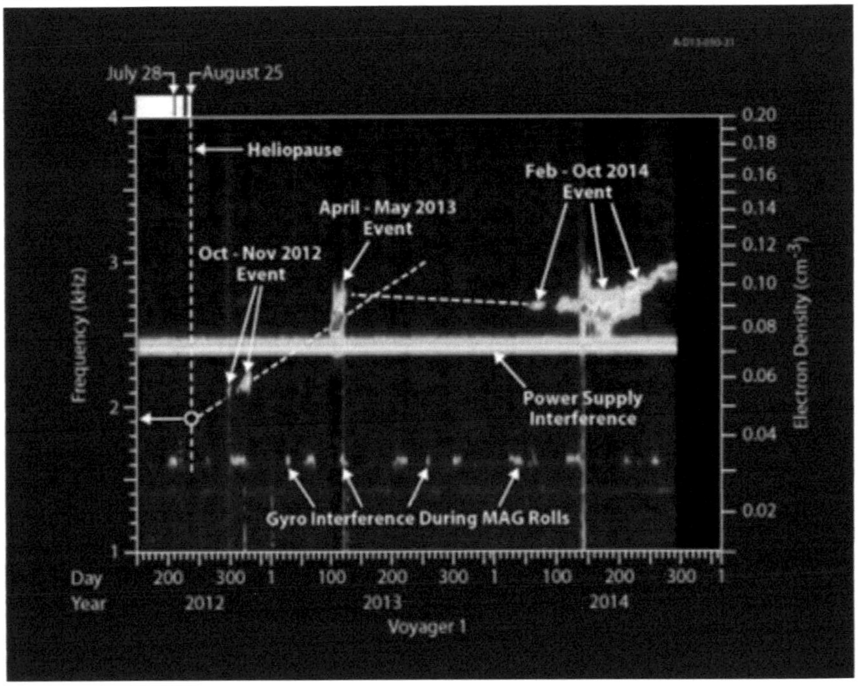

Figure 11. Observations of three bursts of electron plasma oscillations by Voyager 1 between mid-2012 and 2015.0. The densities inferred from these oscillations are shown on the ordinate of the right side of the figure. These densities are characteristic of interstellar densities reported in the astrophysical literature.

Now let us consider the jumps in B labeled "FS" in Figure 10a. Figure 12 from Burlaga et al. [2013a] shows the first jump in B which is associated with electron plasma oscillations observed by Gurnett et al [2013] during 2012 . Gurnett found electron plasma oscillations beginning on day 290 and ending on approximately day 320, 2012, which corresponds to the jump in B. These observations are consistent with the widely accepted view that electron plasma oscillations are accelerated a beam of particles produced by a bump on tail instability [Filbert and Kellogg 1979, Fitzenreiter et al., 1984, Gurnett 1982] generated at the electronic foreshock boundary of a shock wave. There was a small to change in the direction (a few degrees) across this jump, consistent with a present perpendicular shock. By means of the co-polarity theorem Burlaga et al. [2013a] calculated the shock normal

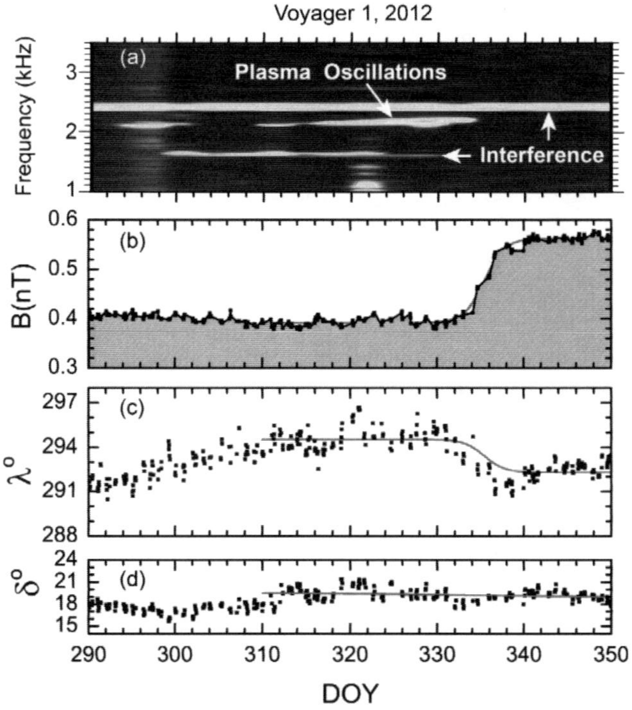

Figure 12. A quasi-perpendicular interstellar shock associated with upstream electron plasma oscillations

which is about 27° from the radial direction. The angle between the upstream magnetic field direction and the shock normal is 85°, indicating that the shock is quasi-perpendicular. The jump in B at the shock is relatively small, 1.4, indicating a relatively weak shock. The upstream density was measured by the plasma oscillations to be 0.06 /cm^{-3} from which one calculates an Alfvén speed V_A = 38 km/s. Assuming an interstellar temperature of 27,000° near the heliopause, one obtains a sound speed of V_S = 17 km/s, so that the medium is magnetically dominated $(V_A/V_S)^2 = 5$). The corresponding magnetoacoustic speed is 40 km/s, indicating that if the jump is associated with a shock, then the shock is propagating at greater than 40 km/s. Such shock speeds have been calculated by Whang and Burlaga [1994] and Washimi [2011]. There is no internal structure of the shock, indicating that it is a laminar shock, in contrast to the highly structured supercritical termination shock discussed in relation to Figures 1 and 2. The thickness of this interstellar shock is very large (> 10^3 c/ω_p or >1000 ion inertial lengths) compared to that of the termination shock, whose thickness was of the order of several ion inertial lengths. The reason for the large difference in thickness is not known, but it may be related to the fact that the interstellar shock in Figure 12 is a resistive, laminar shock in a medium in which the interstellar pickup protons contribute the dominant pressure.

Gurnett et al. [1993] observed radio waves when V1 was in the supersonic solar wind, and they proposed that the radio waves were generated by a shock located beyond the heliopause, in the LISM, by mode conversion of the electron plasma oscillations upstream of the shock. Shortly afterward, Whang & Burlaga [1994] used V2 observations of a strong GMIR shock generated in the supersonic solar wind (similar to that indicated in Figure 4) as input to a calculation based on the method of characteristics. They found that the GMIR shock could propagate through the heliosheath and into the LISM. This process is illustrated in Figure 13, where the trajectory of the shock was computed by the method of characteristics. The panel on the left of Figure 13 shows the termination shock moving slowly inward from 70 AU and the heliopause at 100 AU. We now know that these initial conditions incorrect, but they serve to illustrate the basic process. The GMIR shock enters from the lower left and propagates outward through the heliosheath until it encounters the termination shock, whereupon it decelerates but continues to move outward to the heliopause. The termination shock becomes weaker as a result of the interaction, and it moves slowly outward. When the GMIR shock interacts with the heliopause, it is transmitted into the LISM as a weak slowly moving shock.

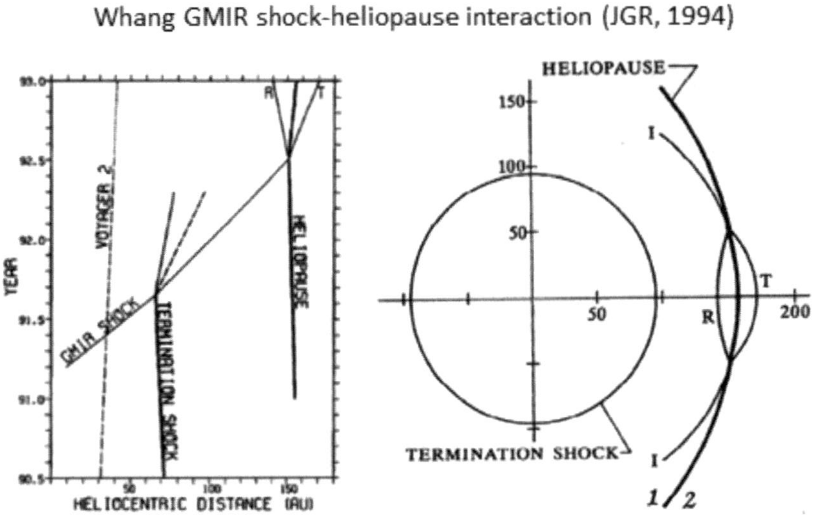

Figure 13. Propagation of a shock from the supersonic solar wind through the termination shock, continuing through the heliosheath, and passing through the heliopause

An inward propagating reverse shock is also generated during this process, and the heliopause is accelerated outward. The result of the interaction with the heliopause is a lenticular region shown on the right panel of Figure 13, which expands as the transmitted shock moves outward and the reflected shock moves inward. The existence of this process was confirmed by an MHD calculation of Washimi et al. [2011].

On smaller scales, Figure 10 shows no evidence of large-scale fluctuations in the LISM like those that are characteristic of the heliosheath. There is no indication of filaments and jumps in B that

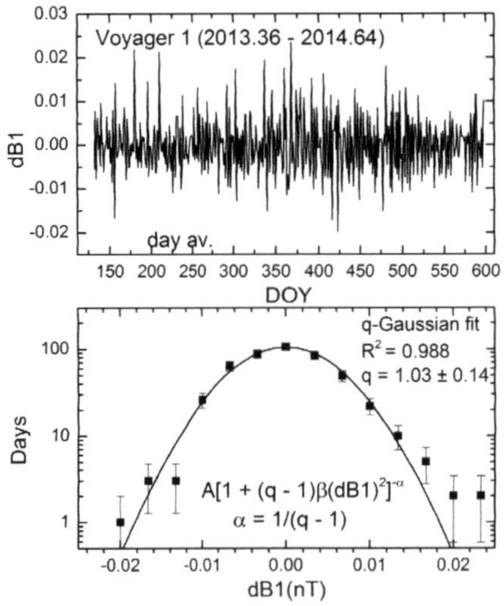

Figure 14. The top panel shows daily increments of the magnetic field strength. The bottom panel shows the distribution function. A q-Gaussian provides a good fit to the observations, which gives q = 1, corresponding to an ordinary Gaussian distribution, which is very different from was observed in the heliosheath.

might be associated with a multifractal structure in the LISM. Macek et al. [2014] found no multifractal structure in B In the LISM after August 25, 2012. The 468 day interval from 213.36 - 214.64 shows a nearly uniform B and consequently a lack of correlations, unlike the heliosheath. The increments of daily averages of B at one day intervals are plotted at the top panel of Figure 14, and the distribution of these increments is shown in the bottom panel of Figure 14. The distribution can be described as a q-Gaussian distribution, but the parameter q_{cor} = 1 (characteristic of a Gaussian distribution) is distinctly different than the value q_{cor} = 1.6 characteristic of the heliosheath.

. The preceding results show that that the large-scale fluctuations observed by V1 after August 25, 2014 are not characteristic of fluctuations in the heliosheath. This is consistent with the view that the latter observations shown in Figure 10 are properties of the LISM.

Finally, we consider the possibility that the very small fluctuations in B observed by V1 from 2013.36-2014.64 might be interstellar turbulence. This question was answered by Burlaga et al. [2015]. After subtracting the trends in B from the observations during the 468 day interval a power spectrum of the IMS magnetic field was computed. The result is shown by the red curve in Figure 15, which is a plot of power spectral density versus wave number k. At the shortest wave numbers, the spectrum levels off to a constant value, which is at the noise level as discussed by Burlaga et al. [2014b]. At the longer wavelengths (smaller wave numbers, corresponding to larger frequencies) the spectrum has the slope of $k^{-5/3}$, consistent with Kolmogorov turbulence.

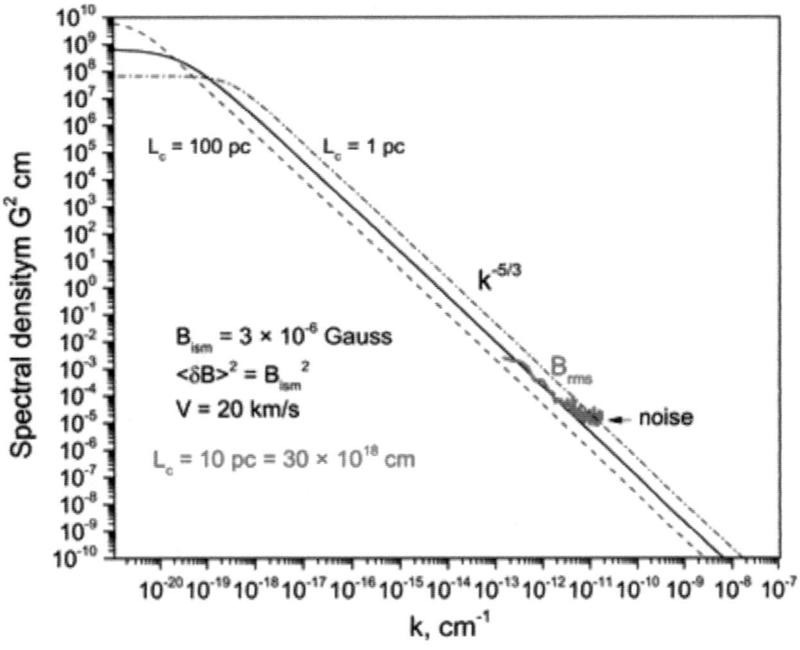

\\

Figure 15. The spectrum of the fluctuations observed by Voyager 1 from 2013.36 to 2014.64 is shown by the red curve. The tail levels out at the largest wave numbers as a result of instrument noise. The three curves are based on a model of Kolmogorov turbulence, with a $k^{-5/3}$ spectrum which levels out at a scale L_C.

Kolmogorov turbulence in the magnetic field of the interstellar medium has been observed with remote-sensing instruments by Minter and Spangler [1996]. Burlaga et al. [2015] considered the relationship between the Voyager 1 observations of the spectrum of magnetic fluctuations and various estimates of the interstellar magnetic turbulence. Three model spectra are plotted in Figure 15 for an outer scale length (stirring scale) of L_c = 100 pc, 10 pc, and 1 pc, based on the assumption that $<\delta B>^2 = B_{ism}^2$. In order to convert from time to wave number, it was assumed that Voyager 1 moved past the fluctuations and 20 km/s. Figure 15 shows that the Voyager 1 spectra are inconsistent with stirring scales of 100 pc and 1 pc, but they are consistent with L_c = 10 pc. In other words, the Voyager 1 spectra measured during 2013 and 2014 are consistent with the interstellar spectra. This supports the view that Voyager 1 is in the LISM.

The preceding discussion assumes that to first order there is only one scale involved in the turbulence, and that the effects of the interaction of the magnetosphere with LISM tend to enhance the turbulence by only a factor of the order of two or three, which would not change our conclusion. Voyager 1 that was observing interstellar turbulence. The discussion also neglects turbulence that might be generated by the fluctuations in the heliosheath acting on the heliopause. However, these fluctuations are known to be very weak [Burlaga et al. 2014 , 2015] and they are expected to be attenuated by the heliopause, so that their contribution to the turbulence might be small. Further studies are needed to confirm the validity of these assumptions.

5. Conclusions

This paper has described and compared the magnetic properties of the heliosheath with those of the LISM in the region close to the heliosheath. It is clear that the properties of these two regions are very distinct in every respect. The simplest explanation of the magnetic field observations after August 25, 2012 is that Voyager 1 has been observing the draped interstellar magnetic field. The Voyager 1 observations of the magnetic field can be explained by generally accepted concepts and numbers for the local interstellar magnetic field. Nevertheless, the model of Fisk and Gloeckler [2014] has not been shown to be invalid up to this time; a definitive observation is predicted to occur during 2015.

Acknowledgments T. McClanahan and S. Kramer provided support in the processing of the data. D. Berdichevsky computed correction tables for the three sensors on each of the two magnetometers. L. F. Burlaga was supported by NASA Contract NNG14PN24P.

References

Axford, W. I. 1990 in *Physics of the Outer Heliosphere*, ed. S. Grzedzielski and D. E. Page (Pergamon, Oxford) p. 7.

Borovikov S N and Pogorelov N V 2014 *ApJ* **783** L16 doi10.1088/2041-8205/783/1/L16

Burlaga L F 1968 *Solar Physics* **4**, 67

Burlaga L F and Ness N F 1968 *Canadian Journal of Physics* **46** S962

Burlaga L F, McDonald F B and Ness N F 1993 *JGR* **98** 1

Burlaga L F 1995 *Interplanetary Magnetohydrodynamics* (Oxford University Press) New York.

Burlaga L F, Wang C, Richardson J, and Ness N 2003 *ApJ* **585** 1158

Burlaga L F and Vinãs A F 2004 *Physica A* Triangle for the entropic index q of non-extensive statistical mechanics observed by Voyager 1 and the distant heliosphere *356* 375

Burlaga L F and Viñas A F 2005 *J. Geophys. Res.* **A07110**

Burlaga L F and Viñas, A F Triangle for the entropic index q of non-extensive statistical mechanics observed by Voyager 1in the distant heliosphere 2005 *Physica A* **356** 375.

Burlaga L F, Ness N F, Acuña M H, Lepping R P, Connerney J EP, Stone E C and McDonald F B 2005 *Science* **309** (5743) 2027

Burlaga L F, Viñas, A, Ness N F and Acuña M H 2006d The Tsallis statistics of the magnetic field in the heliosheath *ApJ* **644** L83

Burlaga L F, Ness N F and Acuña M H 2006e *JGR* **111** A09112

Burlaga L F, Viñas A F, and Wang C 2007 *JGR* **112** A07206 1 – 10

Burlaga L F, Ness N F Acuña M H, Lepping R P Connerney J E P and Richardson J 2008 *Nature* **454** 75

Burlaga L F and Ness N F 2009a *ApJ* **703** 3114
Burlaga L F, Ness, N F and Acuña M H 2009d *ApJL* **691** L82
Burlaga L F Ness, N F Stone E and McDonald F B 2011 *JGR* **116** A12104
Burlaga L F and Ness N F 2012a *ApJ* **744** ,
Burlaga L F and Ness N F 2013 *ApJ* *765* 35
Burlaga L F, Ness N F and Stone, E C 2013 *Sci* **341** 147 doi:10.1126/science.123545
Burlaga L F, Ness N F, Gurnett D A and Kurth W S 2013 *ApJL* **778** L3, doi: 10.1088/2041-8205/778/1/L3
Burlaga L F and Ness N F 2013 *ApJ* **765** 35
Burlaga, L. F., N. F. Ness, D. A. Gurnett, and Kurth, W. S. 2013b, Evidence for a Shock in Interstellar Plasma: Voyager 1, *Astrophys J. Lett.* **778,** L3
Burlaga L F and Ness N F 2014a *ApJ* **784** 146 doi:10.1088/0004-637X/784/2/146
Burlaga L F and Ness N F 2014b Interstellar magnetic field observed by Voyager 1beyond the heliopause, *Astrophys. J. Lett.* **795** L19
Burlaga L F, Florinski V and Ness N F 2015 *Ap. J. Lett* **804** L31
Burlaga L F, Ness, N F, Florinski V, and Heerikhuisen J. 2015b *ApJ* **804** L31
Burrows R H, Zank G P, Webb G M, Burlaga L F, Ness N F Pickup ion dynamics at the heliospheric termination shock observed by Voyager 2 2010 *ApJ*, **715** 1109
Decker R B et al. Voyager 1 in the foreshock, termination shock, and heliosheath 2005 *Science* **309** 2020
Ferriere K 1992 *ApJ* **389** 286
Filbert P C and Kellogg P J 1979 *JGR* **84** 1369
Fisk L A and Gloeckler G 2014 *ApJ* **789** 41
Fitzenreiter R J, Klimas A J, and Scudder J D 1984 *Geophys Res. Lett.* **11** 496 doi:10.1029/GL011i005p00496
Florinski V, Stone, E C, Cummings, A C, le Roux J A 2015 *ApJ* **803**, 47
Frisch, PC, Redfield, S. and Slavin J.D 2011, The Interstellar Medium Surrounding the Sun, *Annual Review of Astronomy and Astrophysics* **49**, 237-279, doi:10.1146/annurev-astro-081710-102613.
Giacalone J and Jokipii J R 1999 *ApJ* **520** 1, 204
Gloeckler G and Fisk L A 2014 *GRL* **41** 5325
Gosling et al. 2007 *GRL* **34** L06102
Grygorczuk J, Czechowski A, and Grzedzielski S 2014 ApJ Lett *789* **L43**
Gurnett D A 1985 in *Collisionless Shocks in the Heliosphere: Reviews of Current Research*, Geophysical Monograph 35, ed. B. T. Tsurutani and R. G. Stone, American Geophysical Union, Washington, DC, pp. 207-224
Gurnett D A, Kurth W S, Allendorf S C and Poynter R L 1993 *Sci* **262** 199
Gurnett D A, Kurth W S, Burlaga L F and Ness N F 2013 *Sci* **2025** doi:10.1126/science.124168115
Gurnett D A., Stone E C, Kurth W S, Cummings A C, Krimigis S M Decker R B, Ness N F., and Burlaga L F *ApJ* submitted, 2015
Isenberg P, Forbes T G, and Möbius E. 2015 *ApJ* (submitted)
Krimigis S M, Roelof E C, Decker, R B, Hill M E 2011 *Nature* **474** 359
Krimigis S, Decker B, Roelof E et al. 2013 *Science* **341** 144
Liewer P C, Karmesin S R and Brackbill J U 1996 *JGR* **101** 17119
Macek W M, Wawrzaszek A, and Carbone V 2012 *JGR* **117** A12101
Macek W M and Wawrzaszek A. 2013 *Nonlinear Processes Geophys* **20** 1061
Macek W M, Wawrzaszek A. and Burlaga L 2014 *ApJL* *793* L1
McComas D, Alexashov D, Bzowski M, Fahr H, Heerikhuisen J, Izmodenov V, Lee M, Möbius E, Pogorelov N, Schwadron N, Zank, GP 2012, *Science* **336** 1291
Mellott M M 1985, Subcritical Collisionless Shocks, Collisionless Shocks in the Heliosphere: Reviews of Current Research, Eds. B. T. Tsurutani and G. Stone, Geophysical Monograph 35, American Geophysical Union, Washington D. C., Page 131

Minter A H and Spangler S R 1996 ApJ 450 194
Opher M and Drake J F 2013 *ApJ* **778** L26 doi:10.1088/2041-8205/778/2/L26
Opher, M, Drake J F, Velli M, Decker R B and Toth G. 2012 *ApJ* *751* 2 80
Parker, E 1958 *ApJ* **128** 664
Parker, N 1963 Interplanetary Dynamical Processes (Interscience Publishers, New York)
Parker, E. 1971a *ApJ* **163** 255
Parker, E. 1971b, *ApJ* **163** 279.
Pogorelov N, Borovikov S, Zank G and Ogino T 2009 *ApJ* **696** 1478
Richardson J, Jasper J, Wang C, Belcher J and Lazarus J 2008 Cool heliosheath plasma and deceleration of the upstream solar wind at the termination shock *Nature* **454** 63
Stone E C et al. Voyager 1 explores the termination shock region and the heliosheath beyond, 2007 *Science* **309** 2017
Stone E, Cummings A, McDonald F, Heikkila B, Lal N, and Webber W 2008, *Nature* **454**, 71
Stone E, Cummings A, McDonald F , et al. 2013, *Science* **341** 150
Stone E , Cummings A, McDonald F , et al. 2013, *Science* **341** 150; Krimigis S, Decker, R, Roelof E et al. 2013 *Sci* **341**, 144; Burlaga L , Ness N and Stone, E 2013 *Science*, **341**, 147, doi:10.1126/science.123545
Pogorelov N V, Borovikov S N, Zank G P and Ogino T 2009 *ApJ* **696** 1478
Strumik M, Czechowski, Grzedzielski S, Macek W M and Ratkiewicz R 2013 *ApJL* **773** L23, doi:10.1088/2041-8205/773/2/L23
Suess, S. 1990 *Rev. Geophys* **8** 97
Swisdak M, Drake, J F and Opher M 2013 *ApJ. Lett* **774** L8
Tsallis C 1988 *JSP* **52** 479
Tsallis C 2004 *Physica A* **340**, 1
Tsallis C 2009, Introduction to Nonextensive Statistical Mechanics (New York: Springer)
Umarov S, Steinberg S and Tsallis, C 2007, in AIP Conf. Proc. 965, Complexity, Metastability, and Nonextensivity: An International Conference, ed. S. Abe et al. (Melville, NY: AIP), 34
Umarov S, Tsallis C and Steinberg S 2008 *Milan J. Math* **76** 307
Washimi H, Zank G, Hu Q, et al. 2011 *MNRAS* **416** 1475
Webber W and McDonald F 2013 *GeoRL* **40** 166520
Whang Y and Burlaga, L 1994 *JGR* ***99*** 21457
Whang Y 2010 *ApJ* **710** 936
Zank 1999 *SSRv* **89** 413
Zank G and Müller H R 2003 *JGR* **108** A6 1240
Zank G 2015 in press

Journal of Physics: Conference Series **642** (2015) 012004 doi:10.1088/1742-6596/642/1/012004

Testing a theory for type II radio bursts from the Sun to near 0.5 AU

Iver H. Cairns and J.M. Schmidt

School of Physics, University of Sydney, Sydney, NSW 2006, Australia

E-mail: i.cairns@physics.usyd.edu.au and jschmidt@physics.usyd.edu.au

Abstract. Type II solar radio bursts have resisted detailed explanation for over 60 years despite being the archetype for collective radio emission associated with shocks. Type II bursts are important because they involve fundamental physics and because most large space weather events at Earth are associated with large, fast, coronal mass ejections (CMEs) and so with type II bursts. Here we present strong evidence for the accurate and quantitative simulation of a type II burst from the deep corona (near 3 solar radii) to near 0.5 AU. The event was observed by the widely separated STEREO A and B spacecraft between 29 November and 1 December 2013. To do so we combine data-driven three-dimensional magnetohydrodynamic simulations (the BATS-R-US code) for the CME and plasma background with an analytic quantitative kinetic model for electron reflection at the shock, transfer of electron energy into Langmuir waves and radio emission, and propagation of radiation to an arbitrary observer. The intensities and frequencies of the radio emissions vary by factors $\approx 10^4$ and $\approx 10^2$, respectively. The theory predicts the intensities, frequencies, and timings of the multiple islands of type II emission very well, with the theory typically in error by less than a factor of 10, 20%, and less than an hour, respectively, for both STEREO A and B. This agreement is strong evidence for the type II theory itself and for accurate prediction by BATS-R-US, when carefully initialised with available data, of the background plasma and magnetic field configurations and of the CME's properties and motion.

1. Introduction

Type II solar and interplanetary radio bursts are bands of emission, often occurring in pairs that differ in frequency by a factor close to 2.0, that drift from frequencies of order 1-200 MHz to 30 kHz and below as their sources move over several days from the corona through the solar wind to Earth and beyond [1, 3, 2, 4, 5]. They are the second most intense class of solar radio bursts, differing in time scale and exciter from the more intense type III radio bursts which typically last about 1 hour.

Type II bursts are the archetype for coherent radio emission produced upstream of shocks: electrons accelerated at the shock form beams in velocity space, which drive high levels of Langmuir waves near the electron plasma frequency f_p, which then couple nonlinearly to produce radio emission near f_p and $2f_p$ (so-called fundamental and harmonic "plasma emission", respectively). The shocks are produced upstream of fast coronal mass ejections (CMEs). Type II bursts thus address fundamental science, since shocks and the electron beam - Langmuir wave systems are two of the most fundamental and important classes of phenomena in plasma physics, space physics, and astrophysics. In addition, the plasma emission mechanism is one of only 4 known coherent radio emission mechanisms.

Despite being discovered over 60 years ago a widely accepted quantitative theory for type II bursts, their fine structures, and their associations with other solar and interplanetary phenomena does not exist [2, 6, 5]. However, we have made great progress recently by combining a kinetic radiation theory [7, 8, 9, 10, 6, 11, 12] with event-specific 3D MHD simulations of CME evolution using the BATS-R-US code [13, 14]. These works demonstrate good agreement for type II emission in the high corona at 1 - 14 MHz for several events with no interplanetary extension [15, 16, 17]. The present paper is a first step towards demonstrating that this theory/simulation capability can accurately predict type II emission from the deep corona to 1 AU. It is the first interplanetary type II burst detected in situ (with associated Langmuir waves and energised electrons) by one of the two STEREO spacecraft (STEREO A, specifically), with both STEREO A and B observing the remote type II emission and the CME itself. The event occurred 29 November - 1 December 2013, with S.D. Bale first noticing the event and [18] first publishing the detailed Langmuir wave observations. Figure 1(left) and 1(right) show snapshots of the radio event and the associated CME, respectively, derived from NASA's CDAWeb facility courtesy of N. Gopalswany and colleagues. As discussed more below, the figure shows that the type II is produced only after the CME passes about 10 solar radii (R_S).

The advent of a combined kinetic radio - MHD CME simulation capability is potentially vital for predicting space weather at Earth. The reasons are that over 90 per cent of large space weather events at Earth are driven by large and fast CMEs [19, 20] and that type IIs are preferentially produced by large and fast CMEs [2, 21, 5]. Thus, if we can accurately simulate specific type II bursts from the Sun to 1 AU, with radio data-theory comparisons establishing confidence in both the radio theory and the MHD simulation code, then prediction of whether the CME will hit the Earth's magnetosphere and what the B_z profile is will ideally allow quantitative prediction of the CME's space weather effects, if any, with adequate lead-time for mitigation.

This paper proceeds by summarizing the radio theory and initialisation of the MHD simulation (Section 2) and presenting the radio predictions and observations in Section 3, restricted to the range $10-0.1$ MHz. Section 4 contains the discussion and a summary.

2. Radio theory and MHD simulation

The radiation theory [7, 8, 9, 6, 11] involves 4 main physical steps, all of which are described quantitatively with analytic theory. The spatially varying shock location, shock normal, and upstream magnetic field and plasma parameters required to predict the radio emission are taken from the underlying MHD simulation, here the BATS-R-US code [13, 14, 22, 23, 24, 25, 26]. A more detailed description of the physics is elsewhere

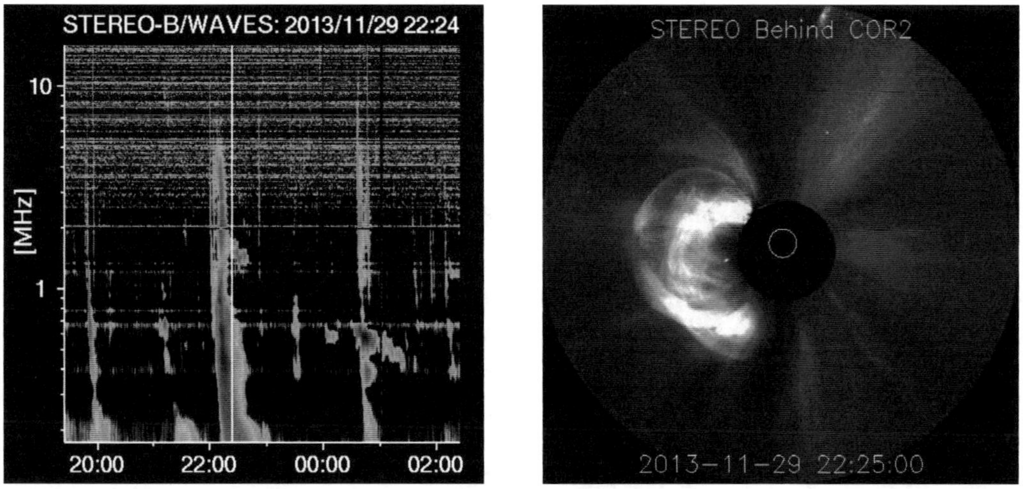

Figure 1. (Left) STEREO B radio observations in the form of a frequency-time spectrogram with amplitude colour-coded for the approximate frequency range 0.18 - 14 MHz from 1945 UT 29 November to 0215 30 November in 2013. The type II is visible as the gently sloping bands of strong emission (red - organge colours) near $1.5 - 1.9$ MHz near 2215 UT and 0.4 -0.7 MHz for 0030-0130 UT. The intense fast drifting emissions over almost the whole frequency range near 2000, 2215, and 0040 UT, for example, are type III bursts. Horizontal lines are interference while the vertical white line near 2225 UT is a marker for the time of the coronagraph image in the righthand panel. (Right) STEREO B COR2 coronagraph image of the CME at time 2225 UT on 29 November 2013. The CME's flux rope is visible as the approximately semi-circular bright feature with a radius near $10R_S$, directed to the left of the Sun in this view, with interior bright white filaments visible at smaller distances. The inner black circular domain corresponds to the coronagraph's occluding disk while the white circle corresponds to the solar photosphere.

[5].

First, some superthermal ambient electrons are reflected back upstream of the shock by the shock's magnetic mirror, resulting in energy gains, a loss cone, and an electron beam. The corresponding acceleration process is shock-drift acceleration, sometimes known in this context as magnetic mirror reflection since the electron gyroradii are typically small compared with the spatial scale of the shock's ramp [27]. The energy gain depends on the angle θ_{bn} between the (local) shock normal and the (local) upstream magnetic field vector $\mathbf{B}$, with energy gains up to a factor ≈ 10 for large $\theta_{bn} \approx 90\,\mathrm{deg}$ and strong shocks with magnetic jumps ≈ 4. Conservation of magnetic moment and energy in the de Hoffman - Teller frame are used to relate the initial and final electron velocities.

Second, a superthermal electron beam develops upstream of the shock because electrons require a minimum speed v_c parallel to $\mathbf{B}$ in order to escape upstream of the shock [28, 7, 29]. In the plasma rest frame this "cutoff" or "escape" speed is

$$v_c = V_{sh} \sec \theta_{bn} \ , \tag{1}$$

where V_{sh} is the shock speed normal to the shock surface. The electron beam speed is $\approx v_c$. Liouville's theorem is used to predict analytically the distribution function of electrons everywhere upstream of the 3D shock, assuming superposition of electrons reflected from the shock's magnetic mirror and ambient electrons with a kappa distribution (in velocity)[11]. Interestingly, the escape speed mechanism corrects the physics compared with the earlier time-of-flight mechanism [28] but yields very similar v_s [7, 29].

Third, if fast enough, dense enough, and cold enough, the electron beam will drive Langmuir waves with frequencies just above f_p via the standard electron beam or "bump-on-tail" instability [28, 2, 7, 29]. As the waves grow, they take energy from the beam electrons, relaxing the distribution function towards a so-called "quasilinear plateau" for which the electron distribution function is locally flat and the resonant Langmuir waves have zero growth or damping. The energy flux into the Langmuir waves can be estimated using quasilinear theory as [8]

$$\frac{dW_L}{dt} = m_e n_b v_b^2 \frac{\Delta v_b}{3l} \tag{2}$$

The righthand side corresponds to an energy density multiplied by a fraction with the unit of inverse time. Here n_b, v_b, and Δv_b are the beam's n umber density, speed, and spread in speed, respectively, at a specified location a distance l fron the shock, all calculable analytically for κ electron distributions [11]. Clearly regions with large n_b, large v_b (corresponding to $\theta_{bn} \approx 90$ deg and large V_{sh} from Eq. 1) and small l are favoured for large energy fluxes into (and so usually large electric fields for) Langmuir waves.

The fourth step is to convert some Langmuir wave energy into radio emission [2, 8, 5]. Sufficiently intense Langmuir waves undergo electrostatic decay $L \longrightarrow L' + S$ of beam-driven Langmuir waves L into backscattered Langmuir waves L' and ion acoustic waves. Stimulated by these S waves, the electromagnetic decay $L \rightarrow T(f_p) + S'$ of beam-driven L waves generates fundamental radio emission $T(f_p)$ near f_p and ion acoustic waves S'. Coalescence $L + L' \longrightarrow T(2f_p$ of the L and L' waves produces harmonic radio emission $T(2f_p)$ near $2f_p$. Known conversion efficiencies for these processes [30, 8, 5] are then combined with the calculated power flux into the Langmuir waves and the simulated electron beam and plasma parameters to analytically calculate the radiated power per unit volume into f_p and $2f_p$ radiation as functions of frequency, time, and location upstream of the shock [8, 11, 12]. Summing the emission from all active foreshock sources and taking into account the inverse square dependence of the flux on the distance between each cell and the observer, straight-line propagation from each cell to t he observer (but with propagation blocked when the path includes locations with higher f_p than the radiation frequency f), and simple propagation time-delays (assuming $f \gg f_p$), leads to the predicted dynamic spectrum $I(f, t)$. The conversion efficiencies for f_p and $2f_p$ radiation depend differently on the beam and plasma parameters [30, 8, 11], so that not all source regions on the shock will produce both observable f_p and $2f_p$ radiation.

Figure 2 shows the relative locations of STEREO A and B, the Sun, and Earth for the type II burst of 29 November - 1 December 2013. STEREO A and B are both on the backside of the Sun. Together, Figures 1 and 2 are consistent with the CME being directed primarily towards STEREO A. (Consistent with this STEREO A observed the CME-driven shock but STEREO B and near-Earth spacecraft did not.) Accordingly, and since STEREO A and B are almost 1 AU apart and the radio flux from a source depends on the inverse distance squared to the observer, we expect that STEREO A and B will measure significantly different dynamic spectra and so provide two independent tests of our type II theory/simulation capability.

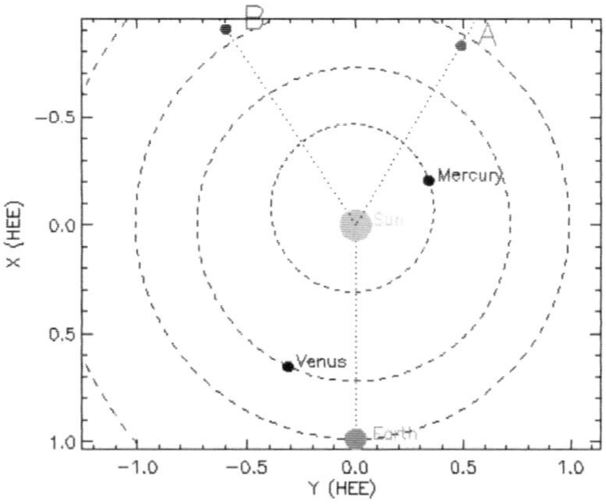

Figure 2. Relative locations of STEREO A and B, the Sun, and Earth on 29 November 2013.

We use the 3D MHD code BATS-R-US code, developed at the University of Michigan [13, 14], to develop a realistic, event-specific, 3D model for the observed CME and to predict the evolution of the CME and its driven shock as they propagate through a realistic, event-specific, 3D plasma and magnetic field model for the background corona and solar wind. BATS-R-US solves the resistive MHD equations with a block-adaptive Riemann solver scheme.

The standard setup routines within BATS-R-US are used very carefully to develop the required models for the background corona and solar wind and for the CME. Magnetic fields in the background corona and solar wind are reconstructed using a potential field solar surface extrapolation from the Wilcox Observatory's photospheric magnetograms for this specific solar rotation [22, 23]. The plasma is simulated using the standard BATS-R-US setup [24, 25, 26]with a refined Wang-Sheeley-Arge model driven by this solar rotation's ACE spacecraft density data, plus spatially-varying polytropic indices for the solar wind acceleration and empirical heating functions for compression regions. A short run with these initial conditions allows an MHD-stable background to develop. The electron and proton plasma densities, temperatures, and flow velocities are assumed to equal the corresponding MHD quantities.

The initial CME is an analytic Titov and Demoulin [31] flux rope that is dimensioned using STEREO coronagraph data and has field directions given by the magnetogram data (not shown). The flux rope is superposed on the MHD background, with a short run yielding an MHD-stable background.

The CME release is via a standard current pulse method, whose characteristics are tailored iteratively using multiple short simulation runs) so that the predicted and observed CME height-time diagrams match below 5 solar radii. After CME release the existence and location of the CME-driven shock is established by looking for the surfances with maximum entropy gradients. Rankine-Hugoniot analyses are then used to determine the shock normals, shock velocity, and plasma and magnetic field parameters upstream and downstream of the shock. Crucially, we emphasise that with the characteristics of the plasma and shock determined in position and time from the MHD simulation, and a value of $\kappa = 2.5$ [32] chosen for the ambient electrons, there are no free parameters in the predictions for the radio emission (or the shock and CME).

3. Radio results for 0.1 - 10 MHz: STEREO A and B event of 29 November to 1 December 2013

Figure 3 compares the theoretical predictions with the observations for both STEREO A and B in the domain $0.1 - 10$ MHz. (The results below 0.1 MHz, including the local radio onset, Langmuir waves, shock crossing, and CME characteristics observed at STEREO A are discussed elesewhere [18, 33]. Figure 3's top and middle panels show the dynamic spectra predicted for STEREO A and B (left and right, respectively) without and with the instrumental background, respectively. The type II is predicted to consist of multiple islands of emission, both fundamental (lower frequency, more intense emissions) and harmonic, with intensities that vary by a factors of over 10^6 without background and of over 10^4. For this event the dominant emission is predicted to be fundamental in this frequency range, with the harmonic being essentially at or below the instrumental background. (Although not shown here, harmonic emission is predicted and observed to dominate radio emission just before the CME-driven shock crosses STEREO A.) It is clear that the dynamic spectra predicted for STEREO A and B are recognisably different, with multiple islands of emission being detectable only for STEREO A and almost all islands more intense for STEREO A. This is expected for a source closer to STEREO A than B, as expected based on the Figure 2 and the CME and its shock crossing STEREO A but not STEREO B.

The bottom panels in Figure 3 superpose the envelopes of the predicted radio signals from the middle panels (with white borders) over the STEREO wave data. All six panels use the same absolute colour bars. It is evident that the theroretical predictions agree very well with the observations, since the white envelopes are very close in frequency and time to the observations. Moreover, the islands agree very closely in colour and so intensity. For instance the first island of fundamental emission is almost precisely predicted in frequency ($\approx 1.3 - 2.0$ MHz), time ($\approx$ 2200 - 2230 UT), and intensity ($\approx 10^{-18}$ W m^{-2} Hz^{-1}) and the second island is displaced upwards by $\approx 10{-}20\%$ but has the correct colour. The predictions and observations are also consistent for the remaining islands, although interference from type III bursts complicates detailed comparisons. A very weak signal near 3 MHz and time 2220 UT may be harmonic emission just above

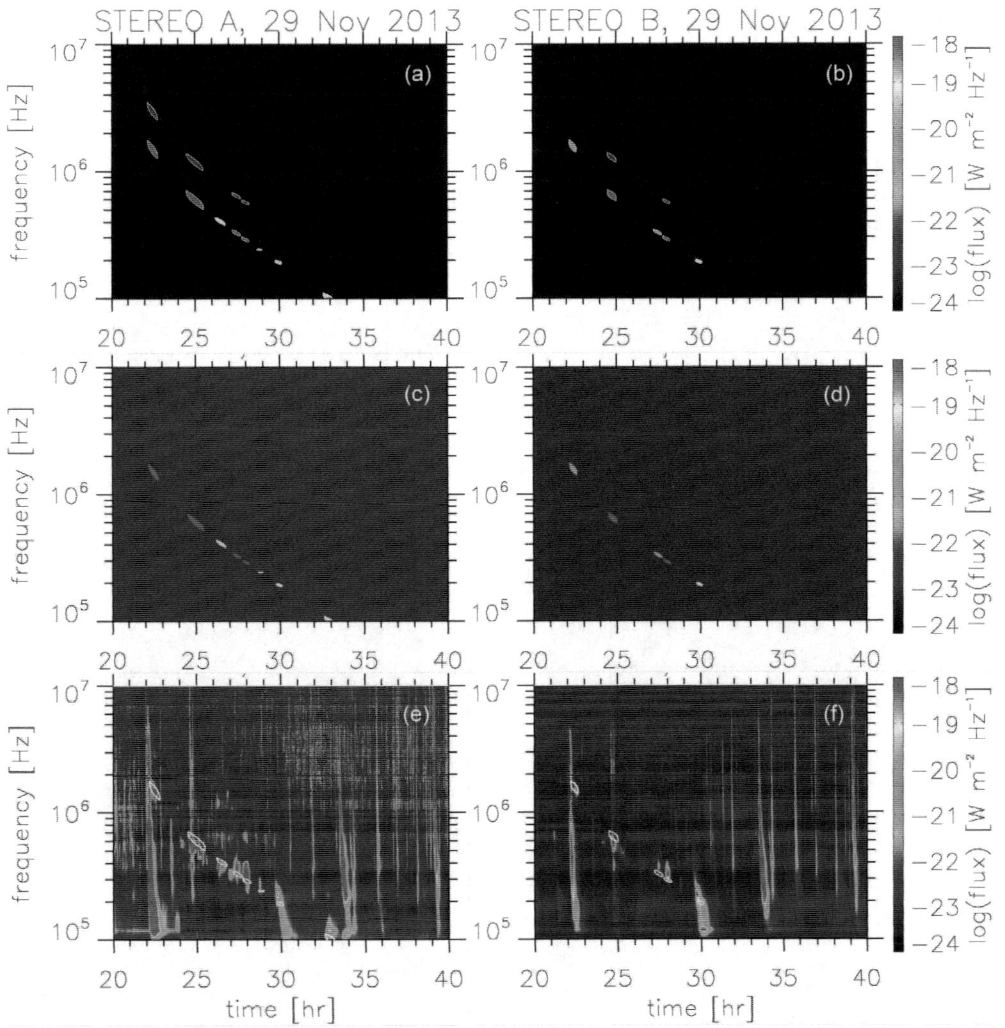

Figure 3. Dynamic spectra predicted and observed for (a), (c), and (e) STEREO A and (b), (d), and (f) STEREO B at coronal frequencies. (a) and (b): pre- dicted F and H emission without background; (c) and (d): predicted F and H emission with background added; (e) and (f): white outlines of predicted emission islands plotted onto the observations.

the instrumental background, rather than just below the background as predicted. It is appropriate to conclude that the theory predicts the observed radiation well, being within a factor of 10 in intensity over range of at least a factor of 10^4, within 20% in frequency, and within better than an hour in time.

An important point to make, in addition, is that significant emission is not predicted where it is not observed. Specifically, the theory does not predict a metric type II burst or emission above about 4 MHz and none is observed.

4. Discussion and Summary

It is important to understand the island structure of the observed radio emissions, the absence of emission above about 4 MHz, and where the shock develops. Figure 4(Left) presents the ratio of the simulated magnetic field strengths just downstream and upstream of the shock as a function of radial distance r (and so of time) along the direction determined by the velocity vector of the CME's centroid, corresponding to the shock's nose location (or maximum distance from the Sun) . This ratio $B2/B1$ must exceed 1 for a shock to exist. Since $B2/B1 > 2$ for $r > 2R_S$, Figure 4(Left) shows that a moderately strong shock forms below $2R_S$ and that the shock only strengthens by about 40% with increasing r and t. With Figure 1 showing that no type II emission is observable until the CME reaches about $10R_S$ and Figure 4(Left)'s showing that the shock evolves low in the corona and remains moderately strong from early times, it is clear that this event's type II emission does not turn on quickly after the shock forms or strengthens suddenly. Accordingly the start of type II emission cannot be interpreted routinely as the time of the shock's formation or as a time when the shock suddenly strengthens.

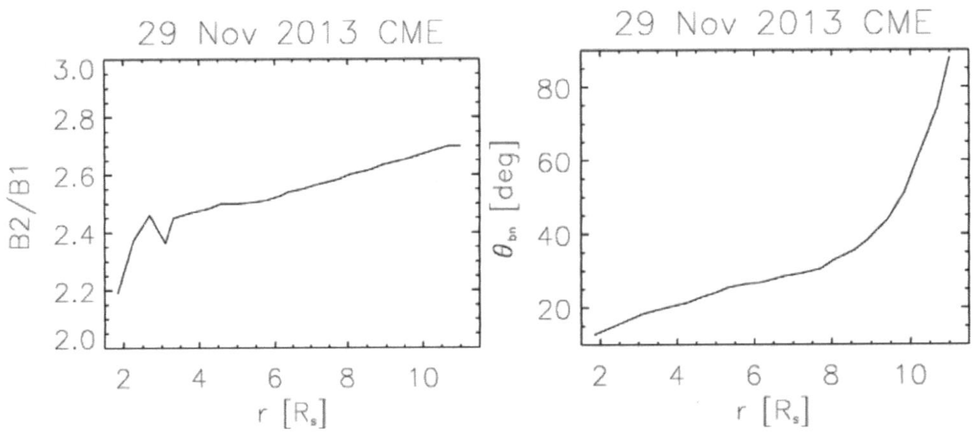

Figure 4. (Left) The ratio $B2/B1$ of the magnetic field downstream and upstream of the nose of the simulated shock (corresponding to the Rankine-Hugoniot prediction for the shock) as a function of heliocentric distance r. (Right) The value of θ_{bn} at the shock's nose as function of r.

Figure 4(Right) shows θ_{bn} as a function of r along the velocity vector of the CME's centroid. Crucially, θ_{bn} remains small compared with 85 deg until $r \geq 11R_S$ and Figure 1 shows that the radio emission starts when the shock nears $11R_S$. Thus, Figures 1, 3, and 4 provide strong evidence that the first island of radio emission starts when the shock's nose reaches a spatial region where the magnetic field orientation is close to perpendicular to the shock normal (almost parallel to the local shock surface). This is where the theory in Section 2 (cf. also [34, 2, 7, 5]) predicts that mirror reflection can produce fast electrons and a large energy flow into the Langmuir waves should occur. Put another way, the radio emission starts when sufficiently large regions of the shock reach regions where $\theta_{bn} \approx 85 - 90\,\text{deg}$. Similar results were found in our earlier work on purely decametric type II bursts [15, 16, 17]. The simulation results are therefore internally self-consistent and consistent with the theory. Furthermore, the very good agreement between the observations and the predictions provides strong evidence that the theory and simulation account very well for the observations and so represent a viable solution for the "type II problem".

It is emphasized that the agreements in frequency, timing, and intensity of the islands of observed and predicted type II emission in Figure 3 are amazingly good. This demonstrates not only that the kinetic radio theory is performing extremely well but also that the predicted time-varying 3D shock location, velocities, and jump conditions and also the predicted 3D magnetic fields (both their orientations and magnitudes) and density structures are predicted very well by our data-driven BATS-R-US simulation.

It is typical in works on solar and interplanetary radio bursts to regard the density at any given location to vary by a factor of 10 up or down with time (e.g., from day to day), corresponding to factors of 3 in f_p and 3 to 6 in the radiation frequency (the latter for $2f_p$ radiation). Similar comments follow for the density as a function of solar latitude and longitude and r. Thus to correctly predict the frequency of the radiation to within $\approx 20\%$ and whether the radiation is produced near f_p or $2f_p$ are major theoretical feats. Similarly, the predicted electron beam, Langmuir wave, and radio emission parameters depend sensitively on θ_{bn} upstream of the 3D shock, as shown in Section 2 and the papers quoted therein, so to accurately predict the observed islands of emission within a factor of 10 in intensity, 20% in frequency, and fractions of an hour in time requires very accurate prediction of the 3D CME and plasma properties by the data-driven BATS-R-US simulation. To be blunt, for these reasons the authors expected to only obtain qualitative agreement between observations and predictions when they started this work. Since they have found similarly good semi-quantitative agreement for their 3 published events [15, 16, 17] and the current event, the agreement found does not appear to be chance but instead to show that carefully initialised, data-driven, BATS-R-US simulation runs appear to be a very fruitful approach to modelling CMEs and related triggering of space weather events. Indeed, although not shown here, the current run predicts very well the time-varying B_z field and the shock's arrival time (within about 1 hour) at STEREO A [33]. Thus, the agreement found can be regarded as strong evidence that BATS-R-US when properly initialised is surprisingly accurate.

The current analyses have multiple limitations. One is that we assumed that the background corona and solar wind remained constant over a solar rotation when setting up the simulation. This ignores time variations during the event-centered solar rotation. Others are the multiple, but standard, assumptions made in the BATS-R-US setup for the corona and solar wind [22, 23, 24, 25, 26]. Another is that a simple Titov and Demoulin [31] model is assumed for the flux rope, with parameters inferred from coronagraph images and by comparing the predicted and observed CME height-time variations. Nevertheless, the very good agreement found between the predicted and observed radio emissions (and the shock arrival time and $B_z(t)$ variations at STEREO A [33]) suggests that these limitations are not severe.

Multiple avenues for further work are apparent. One is to directly predict the time-varying locations of the active radio sources and to compare these with STEREO and other observations. Another is to model one or more metric type II bursts, as well as other type IIs that extend into the interplanetary medium. Another is to refine the theory, for instance along the lines in Ref [5]. Another, perhaps more important, one is to couple the radio code with the output from other 3D MHD codes for the current event and other events.

In summary, this paper presents strong evidence for the accurate and quantitative simulation of a type II burst from the deep corona (near 3 solar radii) to near 0.5 AU. The analysis combines data-driven and event-specific 3D MHD simulations with the BATS-R-US code for the CME, shock, and plasma background with an analytic quantitative kinetic model for electron reflection at the shock, transfer of electron energy into Langmuir waves and radio emission, and propagation of radiation to an arbitrary observer. The intensities, frequencies, and timings of the islands of radio emission are predicted very well by the simulation, with errors typically by less than a factor of 10 (over a range of $\approx 10^4$), 20% (over a range of 100), and less than an hour, respectively, for both STEREO A and B. The radio source turns on when sufficiently large regions of the shock enter plasma domains where $\theta_{bn} \geq 80\,\text{deg}$, as expected theoretically, and not when the shock first develops. In this case the type II emission is predicted and observed to start when the shock nose is near $11R_S$, whereas the shock forms below $2R_S$. Together with previous results for three decametric type IIs, the very good agreement between the predictions and observations provides strong evidence for the type II theory itself and for the accurate prediction by BATS-R-US (when carefully initialised, at least) of the CME's 3D properties and motion and of 3D plasma and magnetic field environment through which the CME-driven shock moves.

Acknowledgments
The authors acknowledge financial support from the Australian Research Council's Discovery grant DP140103933 and thank S.D. Bale and their STEREO/Waves colleagues for informing them of this type II event.

References
[1] Wild, J P and Smerd S F 1972 *Ann. Rev. Astron. Astrophys.* **10** 159
[2] Nelson G J and Melrose D B 1985 *Solar Radiophysics: Studies of Emission From the Sun at Metre Wavelengths* (New York:Cambridge Univ. Press) 333
[3] Cane H V and Stone R G 1984 *Astrophys. J.* **282** 339
[4] Bale S D, Reiner M J and Bougeret J-L 1999 *Geophys. Res. Lett.* **26** 1573
[5] Cairns I H 2011 *The Sun, the Solar Wind, and the Heliosphere*, eds M.P. Miralles and J. Sanchez Almeida (New York: Springer) 267 Sun, the Solar Wind, and the Heliosphere", Eds. M.P. Miralles and J. Sanchez Almeida, Springer, 267 (73 pages), 2011.
[6] Schmidt J M and Gopalswamy N 2008 *J. Geophys. Res.* **113**, A08104
[7] Cairns I H 1986 *Proc. Astron. Soc. of Aust.* **6** 444
[8] Knock S A, Cairns I H., Robinson P A, and Kuncic Z 2001 *J. Geophys. Res.* **106** 25,041
[9] Knock S A, Cairns I H, Robinson P A and Kuncic Z 2003 *J. Geophys. Res.* **108** 1126
[10] Knock S A, Cairns I H, and Robinson P A 2003b *J. Geophys. Res.* **108** 1361
[11] Schmidt J M and Cairns I H 2012a *J. Geophys. Res.* **117** A04106
[12] Schmidt J M and Cairns I H 2012b *J. Geophys. Res.* **117** A11104

[13] Powell K G, Roe P L, Linde T J, Gombosi T I, and De Zeeuwet D L 1999 *J. Computational Phys.* **154** 284
[14] Tôth G B, van der Holst B, Sokolov I V, De Zeeuw, D L, Gombosi T I et al. 2012 *J. Computational Phys.* **231** 870
[15] Schmidt J M, Cairns I H, and Hillan D S 2013 *Astrophys. J. Lett.* **773** L30
[16] Schmidt J M and Cairns I H 2014 *J. Geophys. Res.* **119** 69
[17] Schmidt, J M, Cairns I H, and Lobzin V V 2014 *J. Geophys. Res.* **119** 6042
[18] Graham D B and Cairns I H 2015 *J. Geophys. Res.* **120** 4126
[19] Richardson, I.G. et al., 2006, J.Geophys. Res., 111, A07S09
[20] Lee, J-O, Moon, Y-J, Lee K-S, and Kim R-S 2014 *Sol. Phys.*, **289** 2233 318 10.1007/s11207-013-0466-6.
[21] Gopalswamy N, Kaiser M L, Thompson B J., Burlaga L F et al. 2000 *Geophys. Res. Lett.* **27** 1427
[22] Roussev I I, Gombosi T I, Sokolow I V et al. 2003 *Astrophys. J. Lett.* **595** L57
[23] Roussev I I, Sokolov I V, Forbes T G et al. 2004 *Astrophys. J. Lett.* **605** L73
[24] Cohen O, Sokolov I V, Roussev, I. I., Arge C N, Manchester W B et al. 2007 *Astrophys. J. Lett.* **654** L163
[25] Cohen O, Sokolov I V, Roussev I I, and Gombosi T I 2008 *J. Geophys. Res.* **113** 3104
[26] Downs C, Roussev I I, van der Holst B, Lugaz N, Sokolov I V, and Gombosi T I 2010 *Astrophys. J.***712** 1219
[27] Wu C S 1984 *J. Geophys. Res.* **89** 8857
[28] Filbert, P C and Kellogg, P J 1979 *J. Geophys. Res.* **84**, 1369
[29] Cairns I.H. 1987 *J. Geophys.Res.* **92** 2315
[30] Robinson P A. and Cairns I H 1998 *Sol. Phys.* **181** 363
[31] Titov V S and Démoulin P 1999*Astron. Astrophys.* **351** 707
[32] Maksimovic M, Pierrard V, and Riley P 1997 *Geophys. Res. Lett.* **24** 1151
[33] Schmidt J M and Cairns I H 2015 *Geophys. Res. Lett.*, Quantitative prediction from the Sun to 1 AU of type II solar radio emission, a coronal mass ejection, and a magnetic trigger for space weather, submitted
[34] Holman G D and Pesses M E 1983 *Astrophys. J.* **267** 837

Journal of Physics: Conference Series **642** (2015) 012005 doi:10.1088/1742-6596/642/1/012005

Ion pickup observed at comet 67P with the Rosetta Plasma Consortium (RPC) particle sensors: similarities with previous observations and AMPTE releases, and effects of increasing activity

A.J. Coates[1,2], J.L. Burch[3], R. Goldstein[3], H. Nilsson[4], G. Stenberg Wieser[4], E. Behar[4] and the RPC team

1. Mullard Space Science Laboratory, University College London, Dorking, UK

2. Centre for Planetary Sciences at UCL/Birkbeck, London, UK

3. Southwest Research Institute, San Antonio, Texas USA

3. Swedish Institute of Space Physics, Kiruna, Sweden

a.coates@ucl.ac.uk

Abstract. Rosetta's unique trajectory is allowing exciting measurements of the development of cometary activity between ~3.6 and 1.2 AU for the first time. For a few months following Rosetta's arrival at comet 67P in August 2014, data from the Rosetta Plasma Consortium (RPC) particle instruments (the Ion and Electron Spectrometer (IES) and the Ion Composition Analyser (ICA)), have shown that the low activity cometary environment was initially dominated by the solar wind. This was expected in the early stages of the mission. In addition to the solar wind and related He^+ populations, a low energy pickup ion population is seen intermittently in the early phase of the mission near the comet. The population is very time dependent, but at times reaches higher energy approaching the solar wind energy. During these intervals, ICA data indicate that the composition is mainly water group ions. The rising energy signatures of these ions observed at times indicate that they are in the early phases of the pickup process, initially accelerated by the electric field ('early phase pickup'). Here, we compare these exciting pickup ion measurements with Giotto measurements at the relatively weak (compared to Halley) comet Grigg-Skjellerup, where early phase pickup was seen including non-gyrotropic cometary ions, and with the AMPTE lithium and barium releases. Our results reveal some striking similarities with the AMPTE releases, particularly the rising energy signature related to early pickup, and a momentum balance between the pickup ions and the deflected solar wind. There is also evidence for momentum transfer between the pickup ions and the solar wind, with less velocity change seen in the solar wind alpha particles compared to the protons; this was also observed in an AMPTE lithium release. We discuss the effects of increasing activity observed between 3.6 to 1.8 AU, including the increasing dominance and energisation of pickup ions, increasing ionospheric effects and the decreasing effect of the solar wind.

1. Introduction

As Rosetta accompanies its comet from ~3.6 AU to perihelion (~1.2 AU), it provides an unique opportunity to follow the solar wind-comet interaction from relatively dormant to well-developed (e.g. [4]). Initial pickup ion measurements from Rosetta at comet 67P indicate the importance of the solar wind in the early measurements [1,2], as had been expected before the mission [3,4]. Pickup ions in the early stages of the pickup process, accelerated along the electric field, were seen. Here, we will compare the early Rosetta results with pickup ion measurements from earlier cometary missions, pointing out the likely importance of non-gyrotropic ion distributions [5] and comparing with the AMPTE lithium [6,8] and Barium [7] releases.

The solar wind interaction with a comet has been observed at comets Giacobini-Zinner (GZ, e.g. [9]), Halley (e.g. [10]), Grigg-Skjellerup (GS, e,g.[11]), Borrelly (e.g.12) and now Churyumov-Gerasimenko (CG, 67P, e.g. [1,2]). In addition, before the 'Halley Armada' in the mid-1980s, relevant processes were studied by the US-German-UK AMPTE mission [13,14] that released neutral lithium and barium atoms in the solar wind in the vicinity of the Earth, giving an early chance to study cometary plasma physics and processes [6,15,16,17].

Unlike the earlier missions, the Rosetta orbiter stays with 67P as the comet nears the Sun, and observations were made starting at ~3.6 AU and are planned until perihelion (1.2 AU) which the comet will reach in August 2015, with some post-perihelion measurements until December 2015 in the nominal mission; this has recently been extended to the end of September 2016 when the spacecraft will be at 3.8 AU.

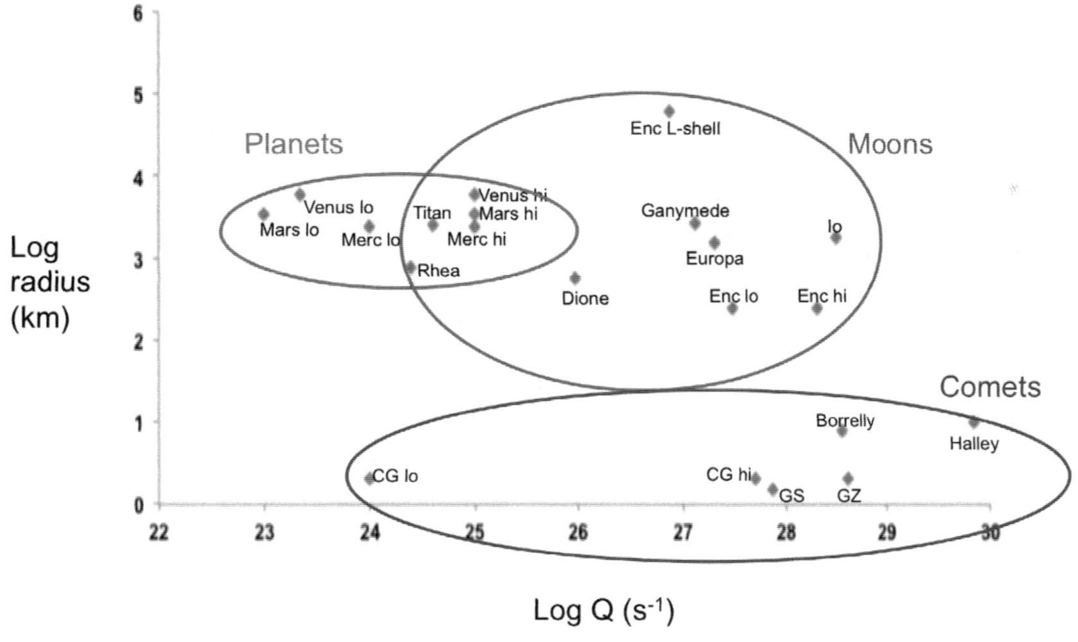

Figure 1 - Outgassing rates, Q for comets, planets and moons (adapted from [20])

As a comet nears the Sun, the nucleus warms and water molecules and other neutrals sublimate producing the neutral coma. The outflow of neutrals also brings dust. The neutral atoms and molecules ionize in sunlight and via charge exchange. On ionization the ions feel the effect of the solar wind electric field given by $\mathbf{E} = -\mathbf{v}_{sw} \times \mathbf{B}$. They are then initially accelerated along the electric field, and then spiral around the magnetic field, to form a cycloidal trajectory in real space. This corresponds to a ring in velocity space, which is unstable and plasma waves are formed. Far from the Sun, the comet-solar wind interaction starts relatively simply as an inert body with relatively few pickup ions, but closer to the Sun boundaries such as the bow shock and contact surface are present in a 'developed' interaction [e.g. 4]. At a developed comet-solar wind interaction such as Halley, ring distributions of pickup ions are seen upstream [e.g. 30] while a bispherical shell is observed upstream of the bow shock [32]. The maximum energy available (in the frame of the neutral particles) is $E_{max,\,shell} = 4m_{amu}E_{sw}$, corresponding to up to ~70keV for a water group ion in the solar wind [e.g. 30]. In the case of the moon, neutral reflection can give an extra 'kick' to the pickup process, giving an energy up to $9m_{amu}E_{sw}$ [18,19].

Comet 67P is of lower activity than any of the other comets visited so far (see Figure 1, from [20]). Consequently, a developed solar wind interaction is not expected except near perihelion, at the closest distances between 67P and the Sun (see also simulations by Koenders [21] and Rubin [22]). The Rosetta Plasma Consortium (RPC) of 5 sensors and electronics is designed to follow this interaction [3,23]. RPC includes two particle instruments, the Ion Composition Analyser (ICA, [24]) and the Ion and Electron Sensor (IES, Burch et al., [25]) that measure suprathermal ions up to 22 keV. In addition, the Rosina instrument is designed to study the neutral particle environment and thermal ions [26].

The first particle data from 67P were published by Nilsson et al. [1] and Goldstein et al. [2] from the ICA and IES instruments respectively.

In this paper, we briefly review the ICA and IES data available so far from the Rosetta mission, and compare these with measurements from the Giotto mission at GS and the AMPTE lithium and barium releases. We find that several results from 67P have some similar features to those found from earlier missions, but are more complete; we also note that the increasing activity is already affecting the data (see Figure 6 and related discussion).

2. Pickup ions previously observed at comets

From the earlier encounters at active ($Q > 5x10^{27}$ s^{-1}) comets, observations showed that:

(a) Pickup ions produce mass loading (a decrease in the solar wind speed) [e.g. 9,10,12].

(b) Bow shocks are eventually formed, that under exceptionally high Q (> $5x10^{28}$ s^{-1}) may become 'mass loading shocks'.

(c) Upstream of the shock, nongyrotropic rings [5], rings [30,31] and bispherical shells [32] are commonly observed.

(d) Pickup ion acceleration (formation of suprathermal tails) takes place. Such acceleration was observed at comets Giacobini-Zinner (e.g. [9,33]), Halley (e.g. [10,34,30,32]) and Grigg-Skjellerup (e.g. [11,35,5,36]) and reviewed [45,46,4].

(e) Pickup ions play a key role in cometary bow shocks [27,28]. During some of the observed shock crossings, particularly the wide, quasi-parallel outbound shock observed by Giotto at comet Halley, additional ion pickup was observed within the shock structure; this is described as a 'mass loading shock' [29].

At 67P, much less developed interactions are expected, with perihelion and immediate post perihelion interactions at a production rate approaching that of comet GS.

3. Early results from 67P[1]

Initial results after achieving orbit showed the early appearance of low fluxes of low energy (~10s of eV) water group pickup ions 100 km from 67P, with the comet at 3.6 AU from the Sun, on 7 August 2014, (Fig 1a of [1]]). This was expected since, with 67P at 3.5 AU, its outgassing rate was $<10^{25}$ s^{-1} (see Figure 1, also [48]). Water group ions were seen moving at ~90° to the solar wind direction. Additional observations from 11 August showed some acceleration up to ~100eV (Fig 1b of [1])..

[1] The first particle data from 67P were published by Nilsson et al. [1] and Goldstein et al. [2] from the ICA and IES sensors respectively.

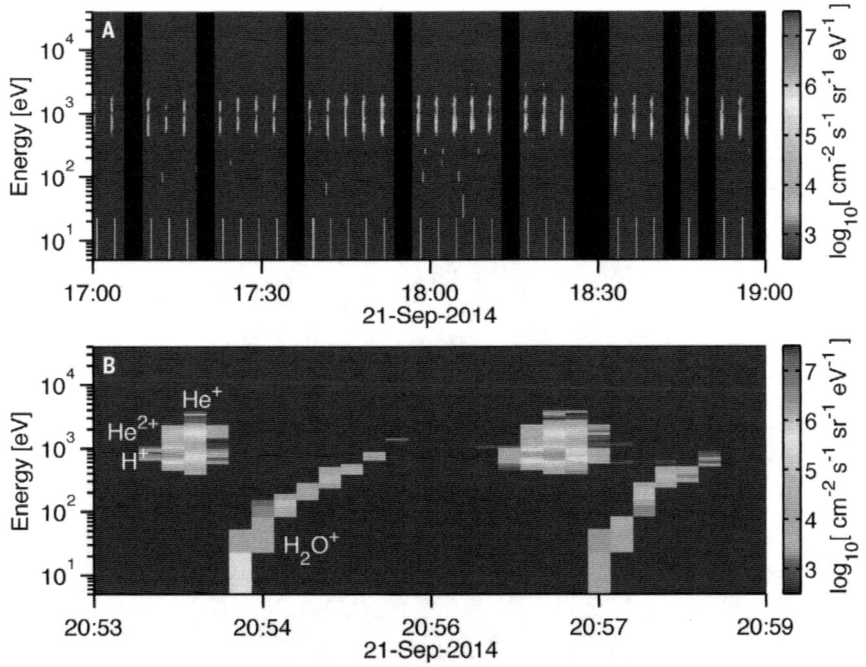

Figure 2 - Rosetta ICA data from 21 September 2014 at ~3.3 AU (Figure 2 from [1]). Panel A shows the solar wind at ~1keV and pickup ions at ~10eV Panel B shows solar wind ions H^+ and He^{++}, and He^+ created from charge exchange between He^{++} and H_2O [1]. The rising energy signature of H_2O^+ in the lower panel is caused by internal instrument stepping of the elevation angle, resulting in sampling of pickup ions from different parts of the comet's coma [1].

On 21 September 2014 Rosetta was at 28km from 67P and ~3.3 AU from the Sun, and the solar wind was seen to be deflected [1]. The deflection was ~20° for H^+, 10° for He^{++}. Rising energy signatures were noted at up to ~800eV (see Figure 2) – these water group ions were accelerated along the electric field in the early phase of pickup. The different energies were interpreted as coming from different polar deflection angles in the instrument sequence – due to the instrument field of view this corresponds to different locations in the comet's atmosphere [1] and different points on the early part of the cycloidal trajectory of pickup ions. The rising energy signatures, which were identified as water group ions in the ICA composition data, are shown in Figure 2.

The IES data on 10 September 2014, when Rosetta was 28km from the comet and 3.4 AU from the Sun, also showed early pickup ions at ~10eV [2]. These were correlated with enhancements in the IES electron fluxes up to ~200 eV, interpreted as electrons from the photoionization of neutrals, and with pressure data from the Rosina COmetary Pressure Sensor (COPS) (Fig 1 of [2]).

At 28km from the comet on 21 September 2014 and at a comet –Sun distance of ~3.3 AU, pickup ions were seen by IES at higher energies up to ~1keV, at 90° to the solar wind direction [2].

On 6 Jan 2015, the data showed a clear deflection of the solar wind protons, and it was shown that these were more deflected than He^{++}) (see Figures 4 and 5 of [2]). On this day the pickup ions were at 45° to solar wind. A month later on 6 Feb. 2015, the pickup ions (PU) and solar wind (SW) directions are shown in the top panel of Figure 3, and an anti-correlation in their deflection was seen(bottom panel). When the pickup ions have moved to right, the solar wind ions moved to the left – indicative of possible effects resulting from momentum conservation.

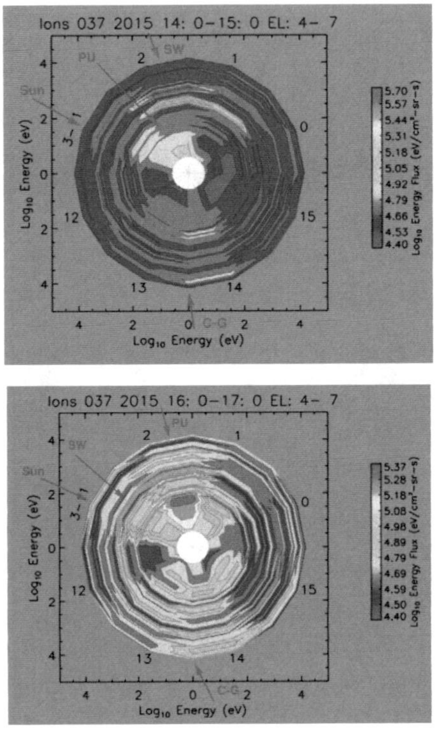

Figure 3 - Rosetta IES sectored spectrograms recorded on 6 February 2015 (14:00-15:00 top, 16:00-17:00 bottom) showing deflection of the solar wind (Figure 6 from [2]). Directions of motion of the solar wind (SW), pickup ions (PU) and the comet (C-G) are indicated. Note that between the plots, while the pickup ions move to the right, the solar wind ions move to the left.

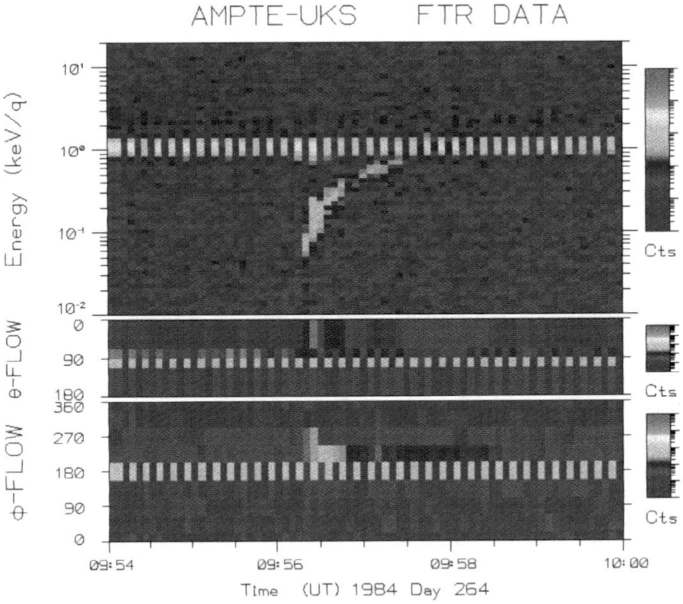

Figure 4 - AMPTE-UKS ion data from 11 September 1984 showing a solar wind lithium release (from [6]). Note the rising energy signature of lithium pickup ions.

4. Results from AMPTE

The Active Magnetospheric Particle Tracer Explorers (AMPTE) mission was a joint US-German-UK mission involving 3 satellites – the Charge Composition Explorer (CCE), Ion Release Module (IRM) and UK subsatellite (UKS). The latter two were in an 18.7 R_E apogee orbit that placed them in the solar wind for some portions of their orbits. On September 11 and December 26 1984, ~10^{25} neutral lithium and ~$5x10^{24}$ barium atoms respectively were released [14].

The lithium release (see Figure 4) as was recorded by UKS and IRM, showing the formation of the early part of a ring distribution as the cloud of neutral atoms receded from the spacecraft, while being ionized by sunlight [6,17]. These observations were used to measure the speed at which the neutral cloud expanded. The spectrogram shown in Figure 4 is very reminiscent of the early Rosetta results.

In addition, it was possible to measure the change in solar wind and pickup ion velocities [6,8], see Figure 5. These were used to calculate the momentum interchange in the interaction between the solar wind and lithium ions [8].

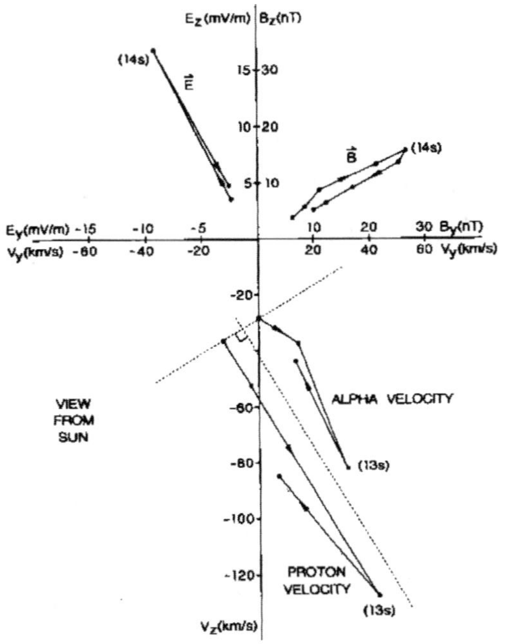

Figure 5 - B and E directions, and observed solar wind deflection (from [8]) 13-14s after the AMPTE lithium release on 11 September 1984. The pickup ions move along E and the solar wind deflection is in the opposite direction showing momentum conservation. The solar wind alpha deflection is less than that of the protons.

The barium release also showed that the momentum of the deflected barium ions was opposite to that of the deflected solar wind [7]. The momentum transfer was also modeled and calculated theoretically [37].

5. Non-gyrotropic ions at Grigg-Skjellerup

The Giotto encounter with comet Grigg-Skjellerup provided the closest comet production rate conditions (~$7.5x10^{27}$ s^{-1} [11])) to the most active phase of 67P (see Figure 1). One of the interesting results was the observation that near to the comet the pickup ion distributions were nongyrotropic. This was interpreted to be the result of the variation of the density gradient of the injected ions over a cycloid, between cusp and cusp [5]. This causes additional instabilities (e.g. [38,39,40]). It is possible that a similar effect was measured at 67P, which may contribute to the observed magnetic fluctuations from related plasma instabilities; these magnetic fluctuations have been called a 'singing comet' [see also 41].

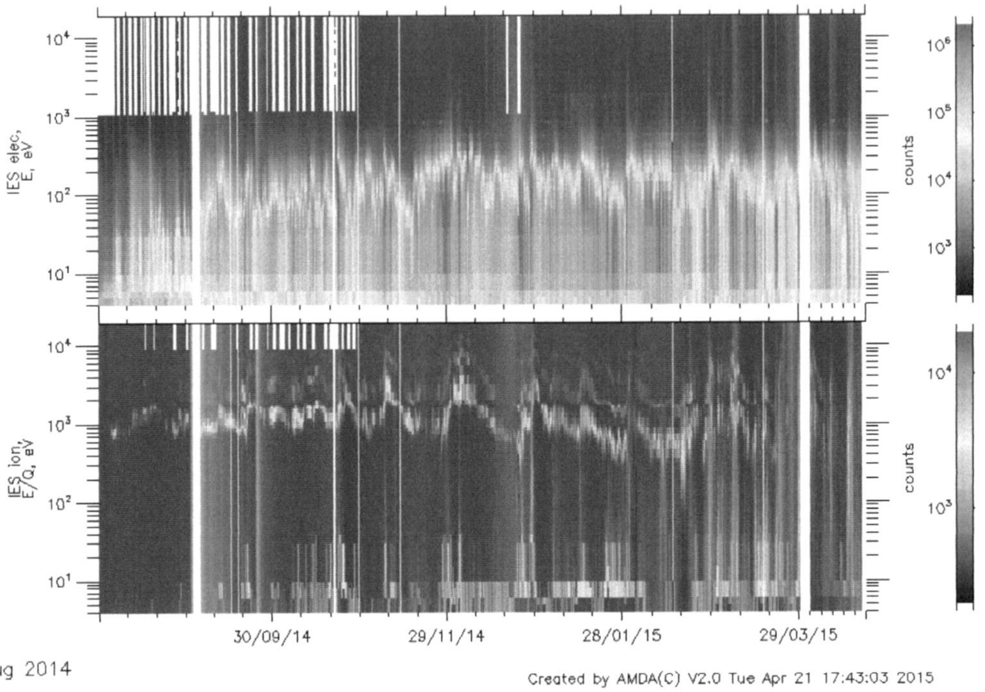

Figure 6 - Rosetta IES data from a few days before orbit insertion at 67P (1 August 2014, comet-sun distance 3.62 AU) to 21 April 2015 (comet-sun distance 1.81 AU). During this time, the Rosetta orbiter has been between ~10-300 km from the nucleus of 67P. The top panel shows electrons, the bottom panel ions. The electron data show a general increase with time indicating higher activity. The ion data show the variable solar wind protons (~1 keV), also the alpha particles at 2x solar wind proton energy/q and He^+ at 4x solar wind proton E/q. The increasing pickup ion population is seen at ~10eV, with variable intensity that increases during the plot. The pickup ion energy also increases towards the right of the plot, indicating additional acceleration (see also [47]).

6. Discussion and new results

There are several similarities between the early Rosetta results and the AMPTE lithium and barium releases discussed above. In particular:

1. Rising energy spectrograms associated with varying neutral densities with distance from the observation point (see Figures 2 and 4)
2. Deflection of the solar wind in an opposite direction from the pickup ion deflection, indicating momentum conservation
3. A more enhanced deflection for solar wind protons than for alpha particles.

These are all features of the early stage of the pickup process.

As the comet nears the Sun, a more developed solar wind interaction is expected. RPC is beginning to see this in the raw data (see Figure 6). Increased numbers and energies of pickup ions, as well as less prominent solar wind, are seen in the plot. In addition, electron fluxes are prominent indicating ionospheric effects. A diamagnetic cavity may be expected near to the nucleus [21,22], and if operations at ~10^3 km away from the nucleus occur, it may be possible to distinguish features such as a Mach cone and formation of a bow shock (e.g. [21,22]).

Regarding bow shock identification, previous cometary missions confirmed the presence of a weak cometary bow shock (e.g. [27,28,42]) using several features. These included calculations of the Mach number (M) [43], conditions where the ion implantation rate exceeds a critical value (e.g. [44]), and thermalisation of the key species. Heavy ions played a key role in the shock structure: mass loading causes the shock to be low M~1-2 [43,27,28]. The shock position was observed close to the predicted point in the flow (a 3d version of 1d critical point [44]). The shock structure was complex, and cometary ions dominated the pressure. Speed dips, similar to a subshock, were observed, with scale lengths varying from H^+ to O^+ gyroradii (for perpendicular shocks), and wider (for parallel shocks). In some cases 'shocklets' were seen. Also observed was a wide, parallel 'mass-loading shock' structure seen in the Halley outbound data [29]. Thermalisation was seen for different species at different times, with an indication that the thermalisation begins in the shock foot (associated with a density increase for heavy ions). Rosetta may enable the first studies of bow shock onset at a weak comet [42].

7. Conclusions

Rosetta is producing unique measurements of the solar wind-comet interaction between ~3.6 AU and 1.2 AU. The early results from Rosetta RPC shows pickup ions at an early phase as they start their gyration following acceleration along the electric field. During this early phase, momentum balance between the new ions (propagating along **E**) and the deflected solar wind is seen [1,2]. We have seen that the AMPTE lithium and barium releases saw the effects of similar physics (e.g.[6,8,7]). The Giotto measurements at GS showed non-gyrotropic ions [5] that may be relevant at 67P. This non-gyrotropy may cause plasma instabilities and additional waves.

Rosetta is also observing the evolution of the comet-solar wind interaction between relatively dormant and developed interactions (see Figure 6, and ICA measurements reported in [47]). The pickup ions are intensifying and becoming more energetic, and the cometary electrons related to the ionosphere have become more prominent.

Observations of a diamagnetic cavity and a cometary bow shock are anticipated – these will be identified by examining the Mach number, the point in the flow, and the thermalisation of all species. It is also important to note that kinetic effects will be important, particularly pickup ion gyration. We look forward to the next few months of unprecedented measurements near the heart of a comet approaching the activity level of GS.

Acknowledgments

AJC acknowledges support from the UCL-MSSL consolidated grant. The work on RPC-IES was supported, in part, by the U.S. National Aeronautics and Space Administration through contract #1345493 with the Jet Propulsion Laboratory, California Institute of Technology. The work on RPC-ICA was supported by the Swedish National Space Board. We thank the teams at Imperial College London and ESA who have been responsible for the operation of RPC.

References

[1] Nilsson H *et al* 2015 Birth of a comet magnetosphere: A spring of water ions *Science* **347** aaa0571

[2] Goldstein R *et al* 2015 The Rosetta Ion and Electron Sensor (IES) measurement of the development of pickup ions from comet 67P/Churyumov-Gerasimenko *Geophys. Res. Lett.* doi:10.1002/2015GL063939 in press

[3] Trotignon J G *et al* 1999 The Rosetta Plasma Consortium: Technical Realization and Scientific Aims *Adv. Space Res.* **24** 1159

[4] Coates A J 2004 Ion pickup at comets *Adv. Space Res.* **33** 1977

[5] Coates A J *et al* 2003 Velocity space diffusion and non-gyrotropy of pickup water group ions at comet Grigg-Skjellerup *J. Geophys. Res.* **98** 20,985

[6] Coates A J, Johnstone A D, Smith M F and Rodgers D J 1986 AMPTE-UKS ion experiment observations of lithium releases in the solar wind *J. Geophys. Res.* **91** 1311

[7] Coates, A J *et al* 1988 Development of the first artificial comet: UKS ion measurements *Adv. Space Res.* **8** 15

[8] Johnstone A D, Coates A J, Smith M F and Rodgers D J 1985 Lithium tracer ion energisation observed at AMPTE-UKS Ion Acceleration in the Magnetosphere and Ionosphere, ed T Chang Geophysical monograph 38, (Washington: American Geophysical Union) pp.186-190

[9] Hynds R J *et al* 1985 Observations of energetic ions from comet Giacobini-Zinner *Science* **232** 361

[10] Johnstone A D *et al* 1986 Ion flow at Halley's comet *Nature* **321** 344

[11] Johnstone A D *et al* 1993 Observations of the solar wind and cometary ions during the encounter between Giotto and comet Grigg-Skjellerup *Astron. Astrophys.* **273** L1

[12] Young D T *et al* 2004 Solar wind interactions with Comet 19P/Borrelly *Icarus* **167** 80

[13] Acuna M H *et al* 1985 AMPTE-Mission Overview *IEEE Trans. Geosci. Remote Sensing* **GE-23** 175

[14] Bryant D A, Krimigis S M and Haerendel G 1985 Outline of the Active Magnetospheric Particle Tracer Explorers (AMPTE) Mission *IEEE Trans. Geosci. Remote Sensing* **GE-23** 177

[15] Valenzuela A *et al* 1986 The AMPTE artificial comet experiments *Nature* **320** 700

[16] Rodgers D J *et al* 1986 UKS plasma measurements near the AMPTE artificial comet *Nature* **320** 712

[17] Möbius E *et al* 1986 Observation of lithium pick-up ions in the 5- to 20-keV energy range following the AMPTE solar wind releases *J. Geophys. Res.* **91** 1325

[18] Saito Y *et al* 2008 Solar wind proton reflection at the lunar surface: Low energy ion measurement by MAP-PACE onboard SELENE (KAGUYA) *Geophys. Res. Lett.* **35** L24205

[19] Coates A J 2012 Ion pickup and acceleration: measurements from planetary missions, Physics of the Heliosphere: a 10-year Retrospective, Ed Jacob Heerikhuisen, Gang Li, Nikolai Pogorelov, Gary Zank Proc.10th Annual Astrophysics Conference AIP proceedings volume 1436, (Melville, NY: AIP) pp.89-102

[20] Coates A J *et al* 2015 Distant ionospheric photoelectron energy peak observations at Venus Planet. Space Sci. doi:j.pss.2015.02.003 in press

[21] Koenders C *et al* 2015 Dynamical features and spatial structures of the plasma interaction region of 67P/Churyumov-Gerasimenko and the solar wind *Planet. Space Sci.* **105** 101

[22] Rubin M *et al* 2014 Plasma environment of a weak comet - Predictions for Comet 67P/Churyumov-Gerasimenko from multifluid-MHD and Hybrid models *Icarus* **242** 38

[23] Carr C *et al* 2007 RPC: The Rosetta Plasma Consortium *Space Sci. Rev.* **128** 629

[24] Nilsson H *et al* 2007 RPC-ICA: The Ion Composition Analyzer of the Rosetta Plasma Consortium *Space Sci. Rev.* **128** 671

[25] Burch J L *et al* 2007 RPC-IES: The Ion and Electron Sensor of the Rosetta Plasma Consortium *Space Sci. Rev.* **128** 697

[26] Balsiger H et al 2007 Rosina Rosetta Orbiter Spectrometer for Ion and Neutral Analysis *Space Sci. Rev.* **128** 745

[27] Coates A J et al 1990 Plasma parameters near the comet Halley bow shock *J. Geophys. Res.* **95** 20,701

[28] Coates A J, Mazelle C and Neubauer F M 1997 Bow shock analysis at comets Halley and Grigg Skjellerup *J. Geophys. Res* **102** 7105

[29] Neubauer F M *et al* 1990 Giotto magnetic field observations at the outbound quasi-parallel bow shock of comet Halley *Ann. Geophys.* **8** 463

[30] Coates, A J *et al* 1989 Velocity space diffusion of pick-up ions from the water group at Comet Halley *J. Geophys. Res.* **94** 9983

[31] Neugebauer M *et al* 1989 The velocity distributions of cometary protons picked up by the solar wind *J. Geophys. Res.* **94** 5227

[32] Coates A J *et al* 1990 Bulk properties and velocity distributions of water group ions at comet

Halley: Giotto measurements *J.Geophys Res* **95** 10249
[33] Gloeckler G *et al* 1986 Cometary pick-up ions observed near Giacobini-Zinner *Geophys. Res. Lett.* **13** 251
[34] McKenna-Lawlor *et al* 1986 Energetic ions in the environment of comet Halley Nature **321** 347
[35] Coates A J *et al* 1983 Pickup water group ions at comet Grigg-Skjellerup *Geophys. Res. Lett.* 20,483
[36] McKenna-Lawlor *et al* 1993 Energetic ions at Comet Grigg-Skjellerup measured from the Giotto spacecraft *Nature* **363** 326
[37] Chapman S C and Dunlop M W 1986 Ordering of momentum transfer along **v** x **B** in the AMPTE solar wind releases *J Geophys Res* **91** 8051
[38] Motschmann U and Glassmeier K-H 1993 Nongyrotropic distribution of pickup ions at comet P/Grigg-Skjellerup: A possible source of wave activity J. Geophys. Res. **98** 20,977
[39] Neubauer F M, Glassmeier K H, Coates A J and Johnstone A D 1993 Low frequency electromagnetic plasma waves at comet P/Grigg-Skjellerup: analysis and interpretation *J. Geophys. Res.* **98** 20,937
[40] Cao J B, Mazelle C, Belmont G and Rème H, 1995, Nongyrotropy of heavy newborn ions at comet Grigg-Skjellerup and corresponding instability *J. Geophys. Res.* **100** 23379
[41] Richter I *et al* 2015 submitted to *Ann. Geophys.*
[42] Coates A J 2009 Bow shocks at comets, Shock Waves in Space and Astrophysical Environments, Ed X. Ao, R. Burrows and G.P.Zank Proc 8th Annual International Astrophysics Conference, AIP proceedings volume 1183, (Melville, NY: AIP) pp 121-130.
[43] Smith E *et al* 1986 Analysis of the Giacobini-Zinner bow wave, Proc 20th ESLAB symposium on the exploration of Halley's comet, ESA SP-250, Vol 3 p461
[44] Schmidt H.U. and Wegmann R 1982 Plasma flow and magnetic fields in comets Comets Ed L.L.Wilkening (University of Arizona press: Tucson) pp. 538-560
[45] Neugebauer M 1990 Spacecraft observations of the interaction of active comets with the solar wind *Rev. Geophys.* **28**, 231
[46] Coates A.J. 1991 Cometary plasma energisation *Ann. Geophys.* **9** 158
[47] Nilsson H at al 2015 Evolution of the ion environment of comet 67P/Churyumov-Gerasimenko: Observations between 3.6 and 2.0 AU *Astron. Astrophys.* In press
[48] Snodgrass C *et al* Beginning of activity in 67P/Churyumov-Gerasimenko and predictions for 2014-2015 *Astron. Astrophys.* **557** A33

Journal of Physics: Conference Series **642** (2015) 012006 doi:10.1088/1742-6596/642/1/012006

Type III-L Solar Radio Bursts and Solar Energetic Particle Events

R T Duffin[1], S M White[2], P S Ray[3], and M L Kaiser[4]

[1] Laney College, 900 Fallon St, Oakland, CA 94607, USA
[2] Space Vehicles Directorate, Air Force Research Laboratory, Hanscom AFB, MA 01731-3010, USA
[3] Space Sciences Division, Naval Research Laboratory, Washington D. C. 20375–5352, USA
[4] NASA Goddard Space Flight Center, Greenbelt, MD 20771, USA

E-mail: robert.t.duffin@gmail.com

Abstract.
A radio–selected sample of fast drift radio bursts with complex structure occurring after the impulsive phase of the associated flare ("Type III-L bursts") is identified by inspection of radio dynamic spectra from 1 to 180 MHz for over 300 large flares in 2001. An operational definition that takes into account previous work on these radio bursts starting from samples of solar energetic particle (SEP) events is applied to the data, and 66 Type III-L bursts are found in the sample. In order to determine whether the presence of these radio bursts can be used to predict the occurrence of SEP events, we also develop a catalog of all SEP proton events in 2001 using data from the ERNE detector on the SOHO satellite. 68 SEP events are found, for 48 of which we can identify a solar source and hence look for associated Type III-L emission. We confirm previous work that found that most (76% in our sample) of the solar sources of SEP events exhibit radio emission of this type. However, the correlation in the opposite direction is not as strong: starting from a radio–selected sample of Type III-L events, around 64% of the bursts that occur at longitudes magnetically well–connected to the Earth, and hence favorable for detection of SEPs, are associated with SEP events. The degree of association increases when the events have durations over 10 minutes at 1 MHz, but in general Type III-L bursts do not perform any better than Type II bursts in our sample as predictors of SEP events. A comparison of Type III-L timing with the arrival of near–relativistic electrons at the ACE spacecraft is not inconsistent with a common source for the accelerated electrons in both phenomena.

1. INTRODUCTION

Solar energetic particle (SEP) events are one of the most dangerous features of space weather. Large events can cause satellite failure due to particle strikes on computer chips, and are a radiation hazard to the health of astronauts. SEP events may be the single most hazardous aspect of manned interplanetary space travel, because it is not feasible to launch the mass of shielding required to protect astronauts from penetrating radiation. For this reason, there continues to be great interest in understanding the causes of SEP events and identifying phenomena that can forecast their occurrence.

SEP events are primarily energetic protons accelerated in the Sun's corona to energies of tens to hundreds of MeV. Since the conditions that lead to acceleration of protons are also likely to be conducive to electron acceleration, and electrons are more efficient than protons at producing electromagnetic radiation that will reach the Earth faster than any proton can, it is natural to look for an electromagnetic tracer of proton acceleration. In particular, electrons are prolific emitters of radio waves, and there has been a long tradition of looking for associations between radio bursts and particle events [e.g., 1; 2].

Recently a class of solar radio bursts now known as "Type III-L" bursts has been reported to show a high degree of association with SEP events [3–6]. These radio bursts are fast–drift bursts (i.e., with frequency–time drift rates that are similar to conventional Type III bursts) that occur in groups *significantly after* the impulsive phase of a flare. A reliable association of such radio bursts with SEP events would have two consequences: (i) study of the properties of the electron acceleration mechanism and site may reveal information about the mechanism and location of proton acceleration; and (ii) the radio bursts may be useful as a predictor of SEP events. Both possibilities have high value for space weather studies.

To date, most studies of this class of radio burst have started with lists of SEP events and looked for associated radio properties. This is appropriate for the first application described above. However, the resulting samples do not represent a test of the predictive power of these bursts: if Type III-L bursts are to be useful as a predictive diagnostic, it is crucial that we be able to identify them from the radio domain alone, and then study the association of a radio–selected sample of Type III-L events with SEP events. The purpose of this paper is to carry out such a study.

An immediate issue is the identification of Type III-L bursts. Type II, isolated Type III and Type IV bursts have very different physical origins that produce such different properties in the frequency–time domain that they can usually be reliably identified in dynamic spectra of radio bursts. However, as their label suggests, Type III-L bursts are related to Type III bursts, and their identification requires a distinctive operational definition that will both avoid confusion with conventional Type III bursts, and provide a process that all observers can apply and agree on. To date, the definitions used in studies of Type III-L bursts have varied from one study to the next in order to improve the degree of correlation with SEP events. In this paper we develop an operational definition based on the previous work, and apply this definition to all large flares in 2001 for which radio dynamic spectra are available (which is all but a small fraction of the events). This produces a large sample of radio–selected Type III-L bursts (acknowledging that not everyone is likely to agree with the definition used to obtain the sample), and we compare this list with the occurrence of SEP events and other phenomena.

In addition to the predictive possibilities of Type III-L bursts, they are important for what they may reveal about the acceleration site. If we can assume that electrons and protons are accelerated in the same region of the corona (but not necessarily by the same mechanism), then the highest frequency at which we observe Type III-L emission indicates the minimum electron density in the acceleration site (since Type III bursts radiate by the plasma emission mechanism). This is important for addressing the ongoing debate over the relative contributions to SEP events of particles accelerated in the low corona, possibly in the main flare site, and particles accelerated by coronal mass ejections (CMEs) at much greater heights. We address this question later in the paper. The next section reviews previous work on Type III-L bursts, and we then go on to justify the definition that we use. The sample of flares is presented, and examples of events are shown. The sample of SEP events in 2001 is described, and we then proceed to summarize the results of correlation studies.

2. TYPE III-L BURSTS

2.1. Previous definitions

The interest in Type III-L bursts began with the observation by [7] of groups of Type III bursts appearing to emerge out of Type II bursts in radio dynamic spectra (frequency–versus–time plots). Type II bursts are slow–drift narrowband features that are nearly horizontal in dynamic spectra below 30 MHz, while Type III bursts are fast–drift features that appear nearly vertical, at least at higher frequencies (above 50 MHz). [3] present an excellent summary of typical radio burst characteristics of solar flares. As in other wavelength domains, the concept of a separation into impulsive and gradual events has been applied to the radio burst properties exhibited by flares. In the radio domain, this separation is often portrayed as the difference between "III/II" events that show a bright Type III burst at the onset of the impulsive phase of the flare followed by a Type II burst several minutes later, and "II/IV" events that do not exhibit an impulsive–phase Type III burst: rather, they produce a Type II event, usually close in time to the soft X–

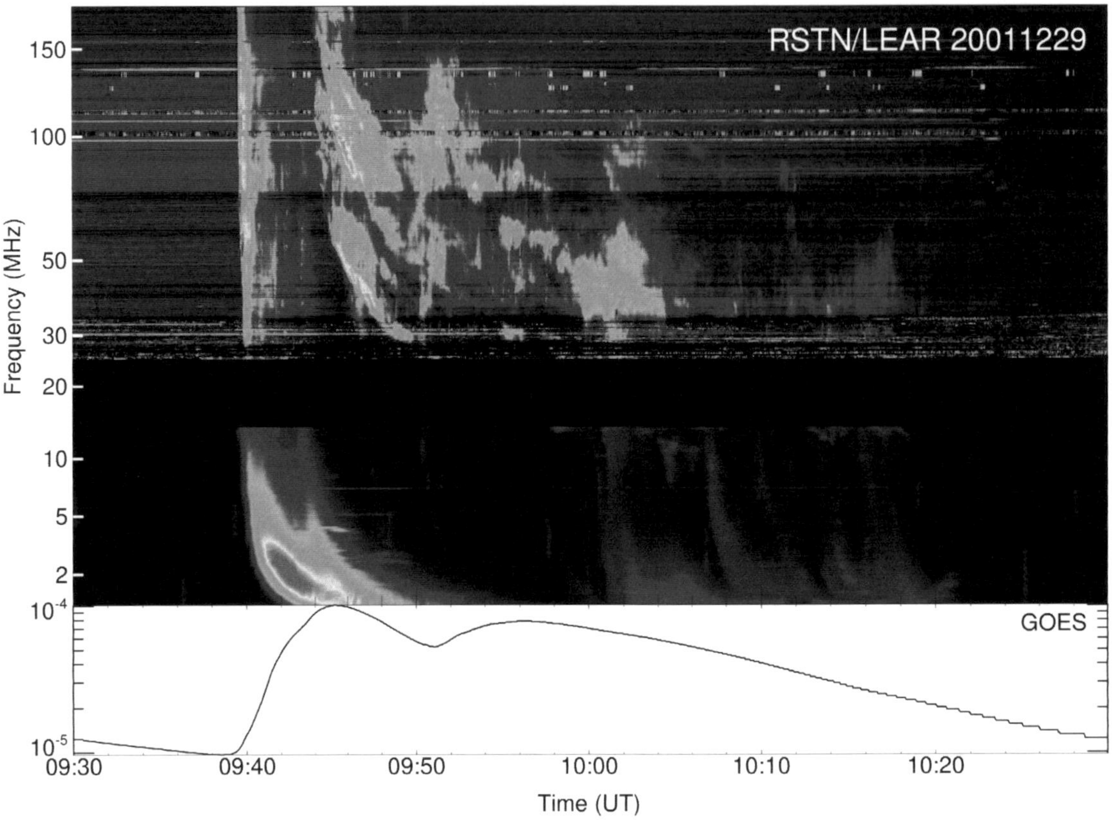

Figure 1. A radio burst from a GOES class M9 flare on 2001 December 29. The displayed image is a radio dynamic spectrum produced by merging data from the USAF/RSTN network in the frequency range 25–180 MHz and data from the WAVES/RAD2 instrument on the WIND satellite in the 1.1-14 MHz range. The label in the top right corner indicates the date and the RSTN observatory supplying the ground–based data (in this case, Learmonth). The frequency axis has a nonlinear scaling (channel widths scale as frequency$^{2/3}$), and there is a gap in the spectrum between 14 and 25 MHz where we have no data. The bottom panel shows the GOES 1–8 Å soft X–ray light curve. In the range above 25 MHz, Type III bursts have a fast frequency drift rate and appear as nearly vertical features, while Type II bursts have a slower drift rate and appear as diagonal features.

ray peak, followed by a Type IV burst (broadband structured continuum emission typically between 10 and 100 MHz that may continue for an hour or more). However, Type IV bursts are much less common than Type III or Type II bursts, and there are different classes of Type IV burst that can cause confusion. Many large flares produce no low–frequency radio bursts at all.

Figure 1 shows an example of an event similar to those studied by [7]. This is a classic "impulsive Type III/II" event (except for the double–peaked structure of the GOES light curve), i.e., it features a fast–drift Type III burst at the onset of the impulsive phase at 09:40 UT, followed by slower–drift Type II emission starting at about 09:44 UT (fundamental at about 65 MHz, second harmonic at 130 MHz, with both bands showing splitting) and drifting downwards in frequency with time. Type III bursts appear in the WAVES spectrum (below 14 MHz) after 10:00 UT and their upper frequency limit appears to coincide with the frequency of Type II emission (unfortunately in the data gap), as if they emerge out of the Type II burst.

The importance of the suggestion by [7] that the later–appearing Type III bursts did in fact originate in

Journal of Physics: Conference Series **642** (2015) 012006 doi:10.1088/1742-6596/642/1/012006

the Type II burst was the implication that the energetic electrons producing the Type III-L emission were accelerated by the shock responsible for the Type II burst (speeds inferred for Type II bursts based on their frequency drift rate and coronal density models are of order 1000 km s^{-1}, similar to the Alfvén speed in the corona). [7] linked the observation of these Type III bursts at frequencies above 20 MHz to the controversial "SA" bursts observed with spacecraft at much lower (kilometric) frequencies [2 MHz, i.e., much further out in the solar wind; 8–13]. The controversy surrounding SA events involved their origin: the low–frequency bursts first identified as SA events showed no evidence for Type III emission at higher frequencies, and [8] argued that they were the low–frequency extension of "herringbone structure" (fast–drift but usually short–lived features known to originate in Type II bursts and therefore likely to be due to shock–accelerated electrons). However, [9] argued that the SA events were more closely associated with microwave bursts that originate in the flare sites, suggesting instead that SA events are produced by electrons accelerated in the flare itself. [13] later argued that some of the SA events occurring later in the flare could be shock–accelerated.

[3], in the paper that coined the term "Type III-L", noted that there were numerous events in which the kilometric emission had properties identical to those studied by [7] but where the highest frequency observed for the fast drift bursts was actually higher than the frequency of the associated Type II at the time, i.e., they had to be accelerated at a location in the corona that was at a higher density and therefore likely to be physically below the height of the Type II shock. [3] concluded that the electrons producing these fast–drift bursts could not be accelerated in the Type II shock but were more likely to be produced in reconnection regions associated with post–flare loop systems below the associated CMEs.

[3] also established a connection between Type III-L events and SEPs: they started with a sample of SEP events and found that nearly all were preceded (presumably by a period corresponding to the time required for the protons to propagate to Earth at nonrelativistic velocities) by Type III-Ls. The characteristics of these radio bursts were

- at kilometric wavelengths (i.e., frequencies below 0.3 MHz) they are long–lasting and very intense fast–drift bursts;
- they originate at around the time that Type II and/or Type IV bursts are visible above 20 MHz, some 5 to 10 minutes after the start of the associated soft X–ray flare;
- they have starting frequencies in the 100–500 MHz range, well above the corresponding Type II frequency;
- and they can occur even in the absence of Type II bursts.

These features again focus attention on the location of the acceleration site of the electrons radiating in Type III-Ls and, by implication, of SEPs if they prove to be associated with such bursts: the high starting frequency of the Type III-L bursts points to an acceleration site low in the corona, well below any shock (CME–associated or not) present at that time.

The importance of these results led to a dedicated study as part of a Living With a Star Coordinated Data Analysis Workshop in 2002, summarized by [4]. The study started with a list of 47 intense SEP events in the period 1997–2001 and investigated associated data. [4] looked for criteria that would produce a high degree of association between Type III-L bursts and SEP events and derived the following conditions:

- a duration at 1 MHz in WIND/WAVES RAD1 data exceeding 10 minutes;
- a peak intensity in 100–200 kHz WIND/WAVES RAD1 data exceeding 20 dB above background;
- Type III appearance in RAD2 more complex than just multiple individual Type III bursts, i.e. presence of bursty features or narrowband features with little frequency drift.

Note an important distinction between this set of criteria and the previous discussions summarized above: unlike [3], who infer a connection between Type III-Ls and low altitudes in the solar atmosphere, these criteria refer to phenomena only visible from space at frequencies well below the ionospheric cutoff, i.e., phenomena at heights well out into the solar wind (the 1 MHz plasma level is typically at $10R_\odot$ or higher in solar atmospheric density models). The set of criteria used by [4] is compatible with acceleration at a shock well away from the flare site, but does not address the possibility of acceleration low in the corona since it ignores emission at higher frequencies. Note that [14] independently studied radio events identified as complex and long–lasting at 1 MHz, and argued that they were likely due to shock acceleration in a secondary process rather than being primary flare particles.

Two more recent studies have used similar criteria to identify Type III-L bursts. [5] focus on the radio properties of 26 SEP events from 1997 to 2002: their events were chosen to have durations longer than 20 minutes at 1 MHz, and to exhibit at least two components at 14 MHz. They investigated the spatial properties of these events using data from the Nançay Radio Heliograph at five frequencies from 164 to 432 MHz. [6] started from a list of SEP events and looked separately at the radio properties of impulsive and gradual events at 1 MHz and lower as seen by the WAVES/RAD1 receiver on WIND, but did not try to "type" the radio emission. They found that for the same time–integrated level of 1 MHz emission (i.e., fluence), gradual events produced far more energetic protons than did impulsive events. On the other hand, [15] studied 3 events associated with WIND/WAVES Type III-L events and argued that since one of these events did not produce SEPs, it casts doubt on the use of Type III-Ls as predictors of SEP events.

The other important feature of most previous studies is that they have started from lists of SEP events, usually intense, and looked for suitable features in radio dynamic spectra ([3] did a radio–selected survey by inspecting dynamic spectra for 89 more-or-less randomly chosen days, but their radio-selected sample consisted of only 6 Type III-L events, of which 5 were associated with SEPs). If Type III-Ls are to be a useful diagnostic for Space Weather purposes, then we must understand the properties of radio–selected samples of these events. That is the goal of this paper.

2.2. Operational definition

Previous studies have employed the L in the Type III-L label to connote the following properties [3; 4]:

- *Late*: the initial Type III–like emission clearly starts after the impulsive phase of the flare. Thus Type III-L events must be distinguished from Type III bursts that commonly occur at the onset of the impulsive phase of a flare.
- *Long*: the emission is long–lasting and accordingly "complex"; and
- *Low* frequency: the emission extends well into the WIND/WAVES range, down to at least 1 MHz.

Using these properties and the previous definitions as a guide, and in conjunction with inspection of the data (in particular, the occurrence of impulsive Type III bursts at the onset of many flares that must be distinguished from Type III-L bursts), we chose the following operational definition for this study:

- The onset of the impulsive phase of each flare is identified as the beginning of the impulsive rise in the GOES 1–8 Å soft X–ray channel. In complex cases, we generally estimate the time of maximum second derivative of the light curve.
- In cases where the soft X–ray light curve is unavailable or confusing, impulsive–phase Type IIIs, if present, can also be used to identify the onset of the impulsive phase.
- After inspection of data, we decided that a suitable (but arbitrary) means to distinguish a Type III-L burst from an impulsive–phase Type III is to require that the Type III-L begin no less than 4 minutes after the onset of the impulsive phase as defined above (the "late" condition).
- To avoid confusion with long–lasting impulsive phase Type III emission, we require that there must be at least a 1 minute gap between any impulsive phase Type IIIs and any candidate Type III-L burst at frequencies above 10 MHz.
- A Type III-L burst must be clearly distinguishable from an individual but late Type III burst: it must have a longer duration and show complexity (e.g., must be clearly not just two individual Type IIIs).

- We require that a candidate Type III-L clearly be associated with the event being investigated, and not possibly a random coincidence in time. In effect this criterion is applied if suitable radio emission occurs very late in the decay phase of the associated flare, or if there is an ongoing storm of Type III bursts of similar intensity that can explain the observed emission independent of any flare or CME.

These criteria clearly still leave room for two observers to disagree on the nature of an event: in particular, the condition that a Type III-L burst show complexity and not be a simple late Type III burst proves to be a difficult distinction in some cases. For this reason, we will include both "definite" Type III-L bursts and "marginal" Type III-L bursts in the final sample. In practice observers may also disagree on the correct onset time indicated by the GOES light curve, particularly for very gradual events, when several events are in progress simultaneously, or when there is a precursor soft X–ray event.

Unlike [4], we do not impose an intensity condition. [4] used such a condition to improve the association of Type III-L emission with large SEP events, but since we wish to define a radio–selected sample we will use the intensity as a secondary rather than as a defining characteristic. The requirement that events be identified by eye, rather than in an automated fashion as in [4], is undesirable for an operational forecast tool but appropriate for this initial survey in view of the need to satisfy (and investigate) the discrimination criteria given above.

3. THE EVENT SAMPLE

3.1. Data Preparation

Since, as far as we know, all major SEP events can be tied back to flares (although frequently behind the limb and then difficult to identify), and because our criteria for Type III-L bursts focus on flares, it is logical to choose our core sample to be based on a set of flares. We use NOAA's list of GOES M- and X–class flares for 2001: we know from previous statistical studies that both radio emission and SEP events are more likely to be associated with large flares.[1] This sample gives us a large number of events without becoming unmanageable. We choose events in 2001 because that year was close to the peak of solar activity in the last cycle, and because radio dynamic spectra are available in digital form from each of the US Air Force Radio Solar Telescope Network (RSTN) sites at Holloman (New Mexico), Palehua (Hawaii), Learmonth (Western Australia) and San Vito (Italy) for essentially the entire year. Digital spectra are not currently as widely available in adjacent solar maximum years. The RSTN dynamic spectra cover the range 25–180 MHz in 802 channels with a time resolution of 3 seconds. The observing sites are far from interference-free, and horizontal lines at fixed frequencies of interfering signals are present in the spectra. We supplement the RSTN data in a few cases with data from the OSRA spectrograph (40–200 MHz at 0.1 s time resolution) at Tremsdorf, Germany, and from the Culgoora spectrograph operated by Australia's Ionospheric Prediction Service. In order to cover low frequencies (below the ionospheric cutoff) we use the WAVES/RAD2 data from the WIND satellite, which cover 1.1–13.8 MHz in 256 channels at 16 second time resolution. Unlike other studies we do not use the more sensitive RAD1 data [4] because RAD1 data have poorer time resolution (1 minute): since "complexity" is believed to be a condition of Type III-L emission, the higher time resolution of RAD2 is essential to identify Type III–like structure within a Type III-L burst.

There were 307 M–class flares and 22 X–class flares in NOAA's flare list for 2001. Of the total of 329 flares, 322 (98%) have "good" WIND/WAVES RAD2 data, 3 have contaminating interference in the RAD2 band that prevent its use, and 4 had no RAD2 data. Of the 322 MX events with good RAD2 data, 298 had coverage from RSTN ground–based radio data, 6 flares had coverage from OSRA ground–based radio data (40-200 MHz), and 5 flares had coverage from Culgoora. Thus we have radio dynamic spectra from 1 to 180 MHz for 92% of all MX flares in 2001. We will also consider the 13 events for which

[1] We are aware, as discussed further below, that a selection based on soft X–ray emission will necessarily exclude some large flares behind the limb that produced SEPs but little soft X–ray emission above the limb, such as the famous 2001 April 18 event described by [16], but this choice does provide a well–defined sample.

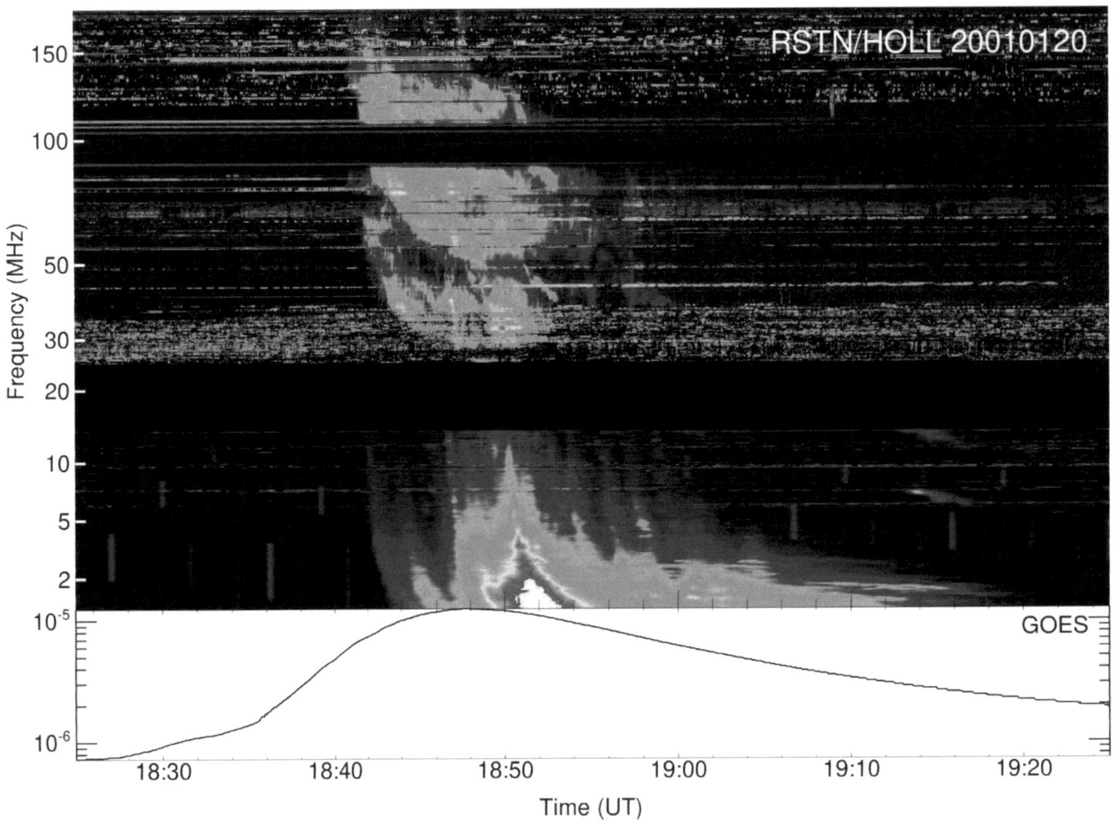

Figure 2. The M1 event of 2001 January 20, with clear Type III-L emission in the WIND/WAVES data below 14 MHz but less obvious in the ground–based data above 25 MHz.

we have RAD2 data but no ground–based data. All 322 RAD2 events had GOES SXR (1–8 Å) data coverage.

For each of these 322 MX events, a one–hour composite plot (of the form shown in Fig. 1) using merged ground-based and satellite–based radio data (upper panel, with a gap between 14 and 25 MHz) and GOES 1–8 Å SXR light curves (lower panel) was produced with each data window beginning roughly 5 minutes before the onset of the impulsive phase of the flare. In the case of very extended events for which a 1–hour window did not encompass all likely flare activity, we extended the plotted window appropriately. The original radio spectra were regridded onto a frequency axis with channel widths scaling as the 2/3 power of frequency, in order to be able to see structure at both high and low frequencies adequately.

3.2. Type III-L Events Associated with MX Flares

Two authors (RTD and SMW) both independently applied the Type III-L defining criteria described above to the 322 event images in order to identify Type III-L bursts. The criteria were applied separately to the upper and lower portions of the radio spectrum, in order to be able to address the question of whether ground–based radio observations alone can be used to identify Type III-L bursts, or whether, as argued by [4], the 1 MHz properties alone suffice. No information other than the event images was used in the identification. Note that this survey does not rely on reports from diverse observers as to the timing of Type III emission or the start and stop times of the flare soft X–ray emission, but rather from inspection of dynamic spectra and direct comparison with GOES light curves for every event in

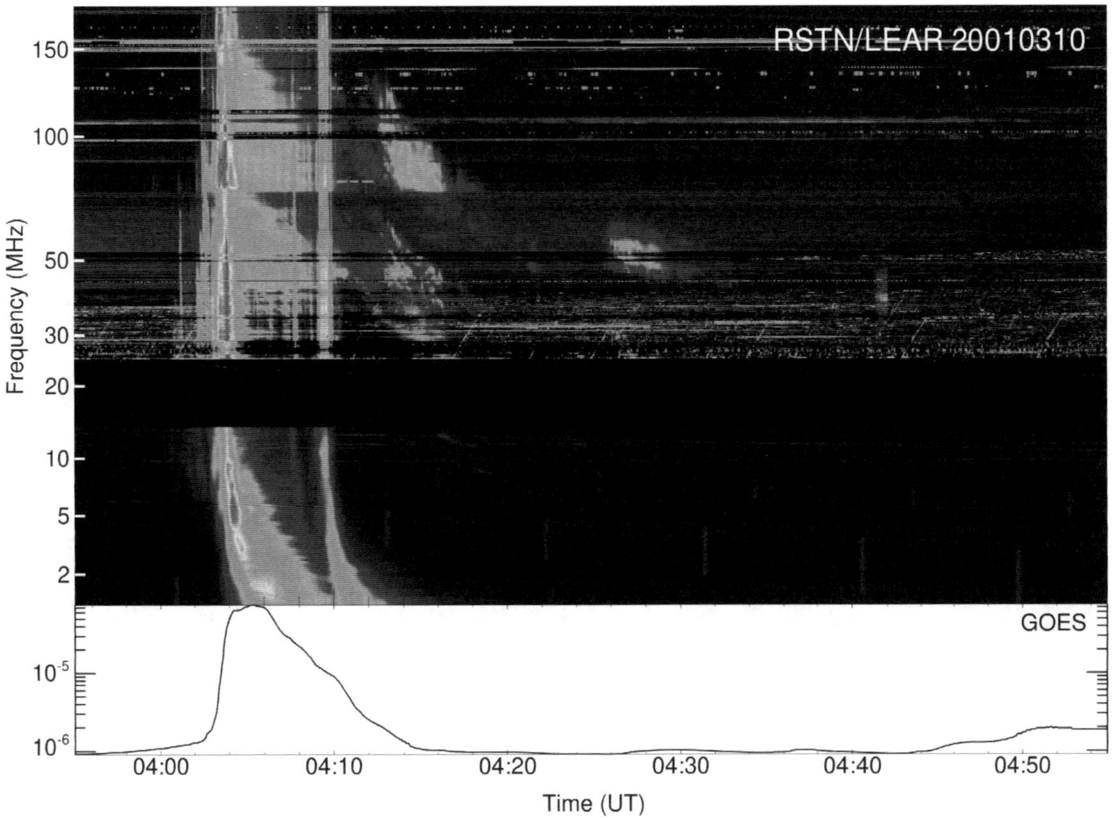

Figure 3. The M7 event of 2001 March 10. In this event fast–drift emission is seen at 04:10 UT, 6 minutes after the impulsive Type III burst at 04:04.

the sample, resulting in a uniform and consistent survey of properties. The two authors achieved 90% agreement on their Type III-L lists. Where there was disagreement or a definitive yes/no answer was not possible, we labeled the event as a marginal Type III-L event. The resulting list of Type III-L bursts from MX–class flares in 2001 is presented in Table 1.

Figures 1–5 present a number of examples of bursts to illustrate the Type III-L identification process, and the issues that typically arise. Figure 2 shows an event with no impulsive Type III burst. We adopt the inflection point in the GOES light curve at 18:35 UT as the start of the impulsive phase. Poorly defined Type II emission starts above 30 MHz at 18:42 UT: it contains narrowband spines of emission with the appropriate drift rate for a Type II, harmonic structure in the RSTN Holloman frequency range, and a faint continuation of both fundamental (6.5 MHz) and harmonic (13 MHz) can be seen in the WAVES data from 19:14 to 19:18 UT. Type III–like emission begins below 14 MHz at about the time that we first see the Type II emission (about 7 minutes after the onset of the impulsive phase), and continues for about 20 minutes with fast–drift features superimposed on a complex background. The Type III-L emission in the ground–based spectrum above 25 MHz is not as clear due to the numerous interference features in the spectrum, but we can see evidence for it from 18:55 to 18:58 UT. We usually found it more difficult to identify Type III-L structure in the ground–based data above 25 MHz because of the combination of more complex solar emissions there than at lower frequencies, and the presence of interference. This event produced SEPs.

Figure 3 shows a classic impulsive III/II event that would not be regarded as Type III-L emission if we imposed a duration limit on the 1 MHz emission as did [4]. The fast–drift burst at 04:10 UT is more

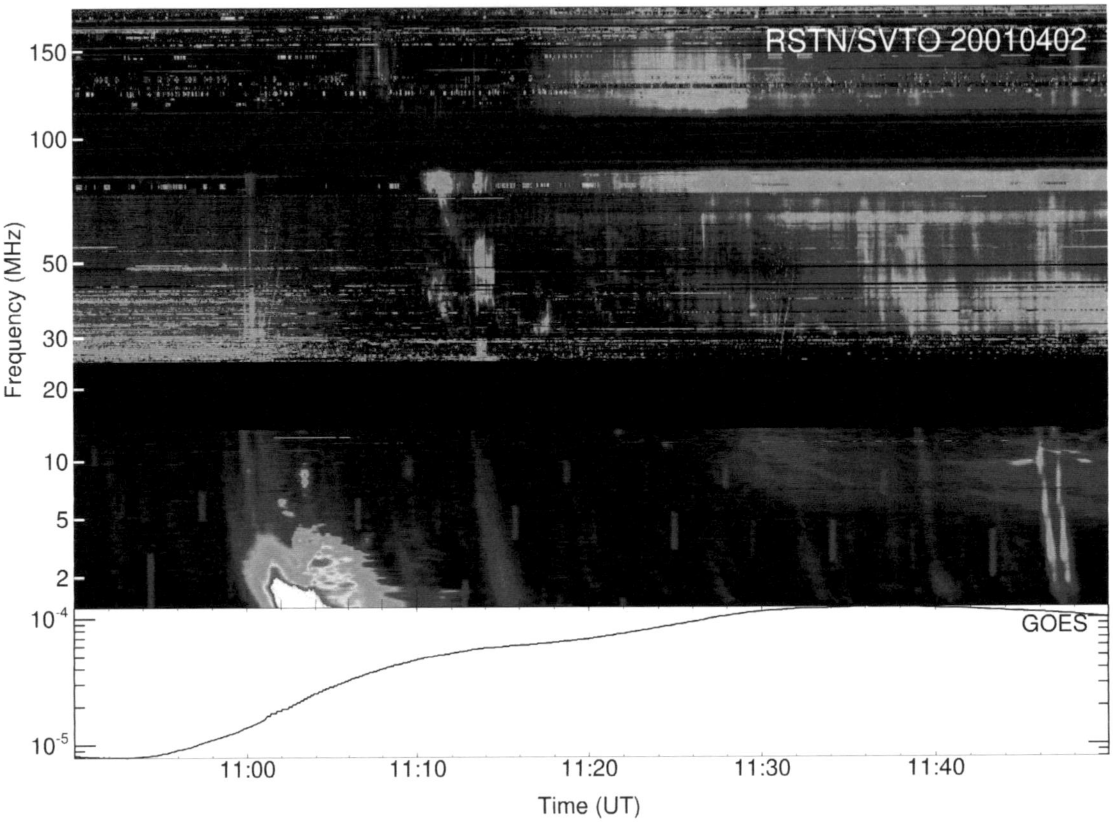

Figure 4. The X1 event of 2001 April 2. This is a very complex event with Type II emission (11:10–11:18), Type IV emission (starting above 100 MHz at 11:18 UT), and fast–drift bursts. Both the emission starting at 10:59 UT and the emission at 11:13 UT are potential Type III-L emission.

than 4 minutes after the impulsive–phase Type III burst at 04:04 UT, and while it could easily be a single Type III burst in the WAVES data, the ground–based data show definite structure, and this satisfies our criterion that the emission clearly be complex. This event produced SEPs.

Figure 4 shows a complex event that we classified as marginal, and one that illustrates the difficulty of classification. This is essentially a "II/IV" event but with only short–lived Type II emission in the ground–based data (starting at 11:10). In the WAVES data the emission drifting into the WAVES frequency range starting at 11:27 UT could be a diffuse Type II emission or an extension of the Type IV emission seen in the San Vito RSTN data. The fast–drift emission starting at 10:59 UT could satisfy our Type III-L criteria if we takes the GOES impulsive onset to be 10:54, but the GOES light curve is complex and there is an inflection point at 10:59 UT that would exclude the early radio emission from consideration. At 11:13 UT more fast–drift emission is seen, clearly consisting of multiple components in the San Vito data, but relatively short–lived in the WAVES data. Note that the duration of the 1 MHz emission here is also less than 10 minutes. This event produced SEPs.

Figure 5 is another complex event featuring many fast–drift bursts. The problem with this event is one of timing: Type II emission is seen from about 22:32 to 22:36, before the impulsive onset implied by the GOES light curve (around 22:45 UT). This suggests that the Type II emission is not associated with the X1 flare but rather with some earlier event, and casts doubt on the appropriate association for other features. However, the CME seen in conjunction with this event is definitely produced by the X1 flare and not any earlier event. We identify the low–frequency emission from 23:10 to 23:30, with a mix

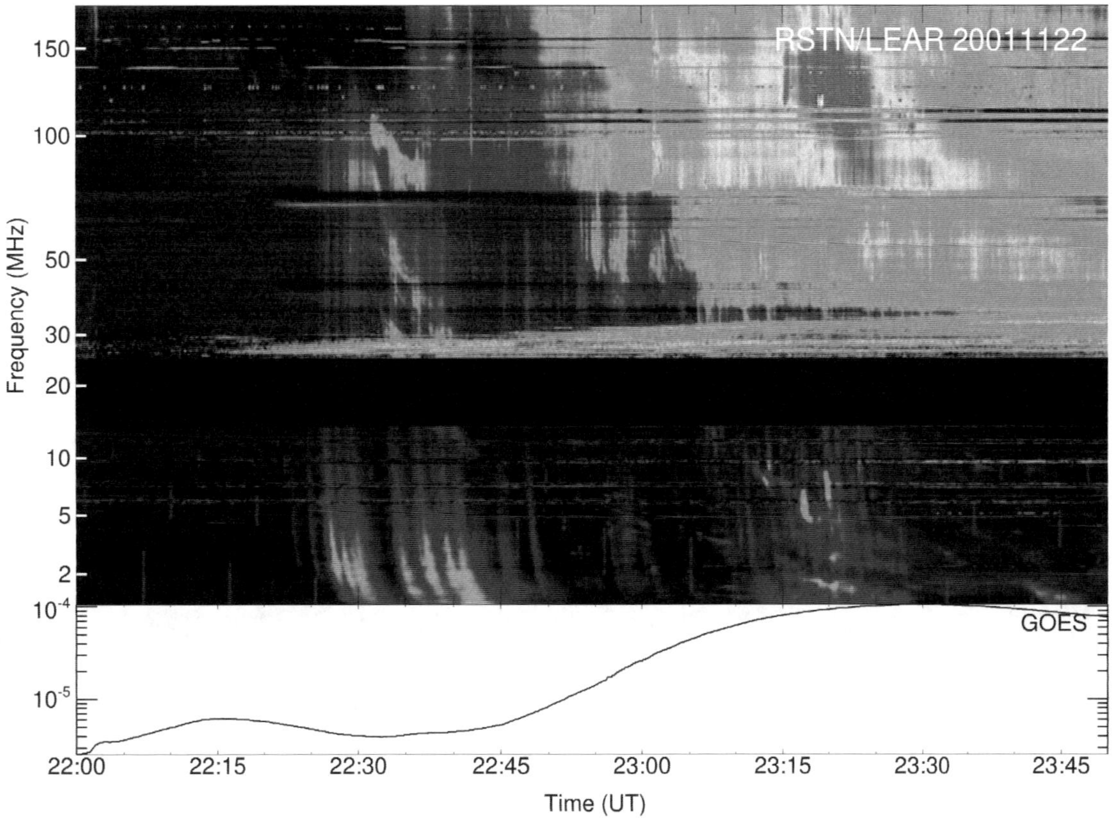

Figure 5. The X1 event of 2001 November 22. This event also has a complex GOES light curve and is odd in that Type II emission appears to precede the onset of the impulsive phase, but the likely Type III-L emission follows the impulsive phase.

of fast–drift and slow–drift features below 14 MHz and coincident with the Type IV emission above 25 MHz, as a candidate Type III-L based on our criteria, but it is listed as marginal because of the possibility that it may not be associated with the X1 flare. This event produced SEPs.

These examples and the notes in the table demonstrate the unsurprising result that coming up with suitable criteria to serve as an operational definition for Type III-L bursts that can be taught, e.g. to a forecaster, is far from easy. This follows from the fact that Type III-Ls have not yet been demonstrated clearly to constitute a phenomenon that is physically distinct in the way that Type II, Type III and Type IV bursts are clearly of different origins: Type III-L bursts appear to contain multiple Type III bursts, and making the distinction between Type III-L and Type III bursts will always be problematic. Nonetheless, two individual authors could agree on about 90% of the classifications. The result is a list of 52 "good" Type III-L events and 18 "marginal" Type III-L events (Table 1). Thus we find that 22% of all MX flares show a form of radio emission that can be classified as Type III-L using our criteria, including 17 of the 22 X flares in 2001 (77%). We include SXR class, timing and location, Type III-L timing, and Type II occurrence for both ground and WAVES data[2], duration at 1 MHz and association with a CME in the Type III-L table. Marginal Type III-L events are indicated by printing the date and time in italics.

[2] Note that the Type II identifications used in Table 1 were carried out by author SMW using the event images and are not the same as the NGDC lists compiled from observatory reports. In particular, the NGDC list is for ground–based observations only, whereas we include those for which we can identify Type II bursts in the WAVES RAD2 data.

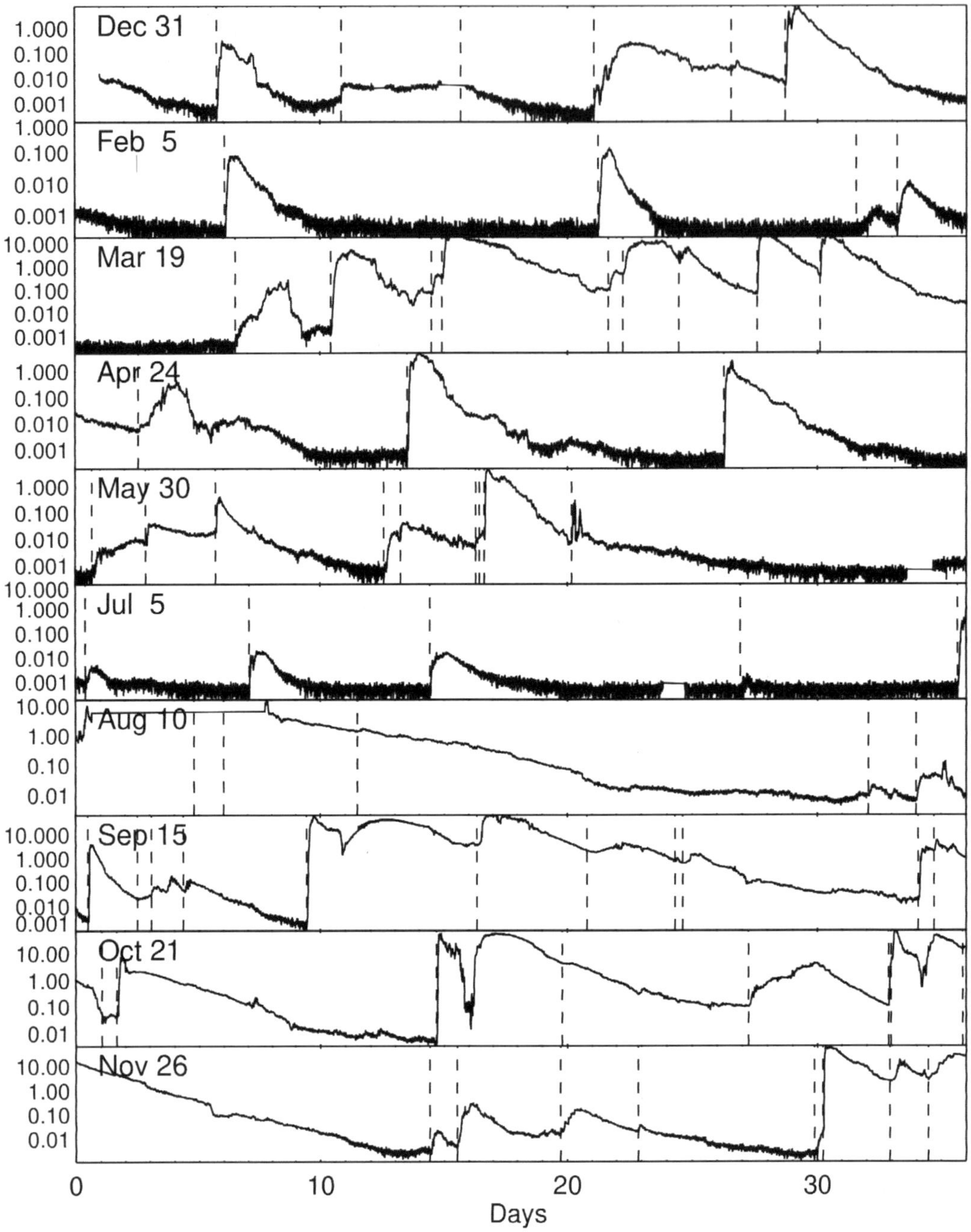

Figure 6. The 13–40 MeV proton count rate for 2001 measured by the ERNE telescope on the SOHO satellite. Each panel covers 36 days starting on the date indicated by the label; the entire year except for the period March 13-19 (no events) is plotted. Dashed vertical lines indicate SEP events identified from the combination of these data and energetic electron data.

Journal of Physics: Conference Series **642** (2015) 012006 doi:10.1088/1742-6596/642/1/012006

3.3. Solar Energetic Particle Events

In order to determine the association between Type III-L bursts and SEP events, we need a list of all SEP events in 2001. Most existing lists focus on large events, whereas we need the most comprehensive list possible: excluding small SEP events from our study can only reduce the degree of association between Type III-L bursts and SEP events. Once an SEP event has been identified, we also need to establish the corresponding solar onset time if we are to compare it with the times of the events in our flare–based Type III-L list. The identification of SEP events and their associated solar phenomena has a long and tortured history, largely due to the fact that we observe protons when, and if and only if, they reach the Earth. Direct propagation from the Sun to the Earth along the Parker spiral magnetic field takes a little over 1 hour for a 10 MeV proton, but this only applies to particles from events that are well connected to the Earth (i.e., those beyond about W30 heliographic longitude, in the range where the magnetic field line in the solar wind that intersects the Earth extrapolates back to the Sun). Western events often show very rapid rises in the proton flux at the Earth about an hour after a large solar event, corresponding to the prompt arrival of the fastest protons. However, solar events in the Sun's eastern hemisphere are poorly magnetically connected to the Earth: protons accelerated there must diffuse across, not propagate along, solar wind magnetic field lines to reach the Earth, and this produces very gradual time profiles whose onset time can be very difficult to determine.

To generate a list of SEP events we primarily used proton flux data in the 13–40 and 40–130 MeV bands measured by the ERNE telescope on the SOHO satellite at the L1 Lagrangian point between the Earth and the Sun (the 13-40 MeV data are shown in Figure 6). We supplement the ERNE data with the less sensitive GOES proton flux data during ERNE data gaps such as 2001 August 9–17. We identify SEP events as clear coherent peaks lasting usually 10 hours or longer in the proton flux histories, excluding fluctuations due e.g. to a shock passage at the Earth. Frequently the rise of a new SEP event will be superimposed on the decay of a prior event: in identifying onset times (of SEPs arriving at the Earth) we attempt to fit and subtract the decay of the previous event. We also used 175–315 keV electron data from the EPAM instrument on the ACE satellite to refine onset times (the electron data also contain events not associated with SEP proton events and thus we do not use them as a primary source of event identifications: several possible events that are clear in the electron data but not evident in the > 13 MeV proton data are excluded from our list). The electron data are useful for defining the onset times: a 200 keV electron typically arrives faster than a 10 MeV proton, with propagation times of order 10-20 minutes [17], so the electrons typically start to rise before the protons, closer in time to the originating event at the Sun. The time resolution in the electron data used here is 5 minutes in most cases, supplemented in suitable cases by 12-second data averaged to 2–minute resolution; times have been corrected for the known clock error in EPAM data.

The resulting list of 68 events is presented in Table 2. We regard 15 of these 68 events as being marginal identifications due either to very low intensity or short durations, and these are identified by printing the date and time in italics. Note that our list of proton SEP events and the ACE/EPAM list of "beam–like electron events" (highly anisotropic electron events beamed along the solar wind magnetic field, see http://sd-www.jhuapl.edu/ACE/EPAM/BeamEventList.pdf, 18 events in 2001) have only 5 events in common (04/26 13:20, 06/15 15:50, 09/24 10:50, 10/09 07:50, 11/04 16:30). All 5 of these events are relatively large proton events, although the 10/09 event occurs during the decay of a larger preceding event and is more pronounced in the electron data than in the proton data. Our list has no events in common with the list of ^{3}He–rich impulsive (small) SEP events studied by [18], and hence is dominated by events that are now typically called "gradual" or proton–rich. Our list is very similar to but larger than that of [19], who searched for events associated with CMEs.

An important issue for identifying SEP events is the fact that during large events the proton flux can remain at extreme levels for many days. We can identify SEP events as small as .002 particles $cm^{-2}s^{-1}sr^{-1}$in the ERNE 13–40 MeV data when the background level is low, while large events may reach many hundreds of particles $cm^{-2}s^{-1}sr^{-1}$ and last for days. Since the flux level can fluctuate due to local changes in propagation conditions or other conditions in the local solar wind, it is not possible to identify small new SEP events against such high backgrounds. For this reason, we will impose blackout periods for the comparison of Type III-L and SEP events: when the ERNE 13–40 MeV level exceeds 3 particles $cm^{-2}s^{-1}sr^{-1}$, we assume that we cannot detect any but very large events during these periods, and exclude Type III-L events occurring during these periods from the comparison. The excluded periods are listed in Table 3. The threshold is applied from onset of SEP events above the threshold until the levels have decreased below the threshold, with gaps to allow inclusion of obvious large SEP events. This exclusion, always occurring during periods of high solar flare activity, removes a number of MX events from consideration, including 4 Type III-L events in our sample. Note that the fact that we find 68 SEP events despite the highly variable detection threshold over the year suggests that many more events would be detected if the background level were uniformly low.

Table 2 includes our best estimate of the onset time at Earth for each event, but these times are not of uniform quality: for large events with a very rapid rise and good energetic–electron count rates, the onset time is accurate to better than 3 minutes[3], but for events with a very gradual rise it is difficult to distinguish the event onset from the background, and the uncertainty may be of order many hours. Since our purpose is to identify the corresponding solar events, we have used the earlier electron onset times wherever they clearly precede the onset of protons: the electron onset times are typically at most 10 minutes earlier than the proton onsets for well–connected western events, but can be an hour earlier than the proton onsets detectable in the data for events with a gradual rise. The second last column of Table 2 gives a qualitative indication of the rate of rise of the event (fast or slow), as an indication of the likelihood that an event is well connected (western events, fast rise) or poorly connected (eastern events, slow rise) to aid in identifying the solar source. Note that in this paper we do not address the distinction between "impulsive" and "gradual" SEP events: these terms have come to reflect differences in SEP composition (gradual events have normal solar wind composition while impulsive events have high charge states of Fe and high ratios of $^3He/^4He$) rather than differences in the rate of change of the count rates [e.g., 20]. The three SEP events found using GOES proton flux data instead of ERNE data are indicated by italics in the comments column: their rates are GOES 10-50 MeV rates and cannot be compared directly with the ERNE rates.

As noted earlier, the onset time of SEPs arriving at Earth may be as little as 10 minutes after they are accelerated at the Sun, or as much as 12 hours, depending on the energy of the first–arriving protons and whether they have propagated 1.2 AU directly along a well–connected field line, or random–walked their way across the solar wind magnetic spiral from the east side of the Sun. This large uncertainty in extrapolating the arrival time at Earth back to a departure time from the Sun leads to major ambiguities in identifying the solar event that produced the SEP. In most of the cases involving large events with a very rapid rise, there is no ambiguity in identification, but events with a gradual rise, even when large, have a large uncertainty in their solar origin because a number of flares and CME eruptions may occur in a several hour window preceding the nominal arrival time of the first energetic protons. (We only need to identify the solar event in order to look for Type III-L bursts, so we can ignore the ongoing debate about the exact particle injection times at the Sun, e.g., see the review by [21].)

To identify the solar event producing each SEP, we looked at both the event lists compiled by NOAA and the CDAW CME catalog[4]. Where there was ambiguity in the association, we generally favored large events and/or bright fast CMEs as the solar origin because they were generally associated with SEPs in events where the origin was clear. It is necessary to use CME lists in order to catch solar events that may have occurred well behind the solar limb and therefore produced little or no detectable soft X–rays, such as the 2001 April 18 event. This does of course bias the associations in favor of events that also produced CMEs; e.g., for the large 2001 May 7 SEP event we regard the 1200 km s^{-1} CME from beyond the west limb as a more likely source than the C4 disk–flare attribution of [22]. Associations of the SEP events

[3] We find that our onset times from the EPAM electron data are consistent with the onsets derived by [17] using higher–energy 0.3-1.2 MeV electrons detected by the COSTEP instrument on SOHO for the 14 events common to both studies.

[4] The CME catalog is generated and maintained at the CDAW Data Center by NASA and The Catholic University of America in cooperation with the Naval Research Laboratory. SOHO is a project of international cooperation between ESA and NASA.

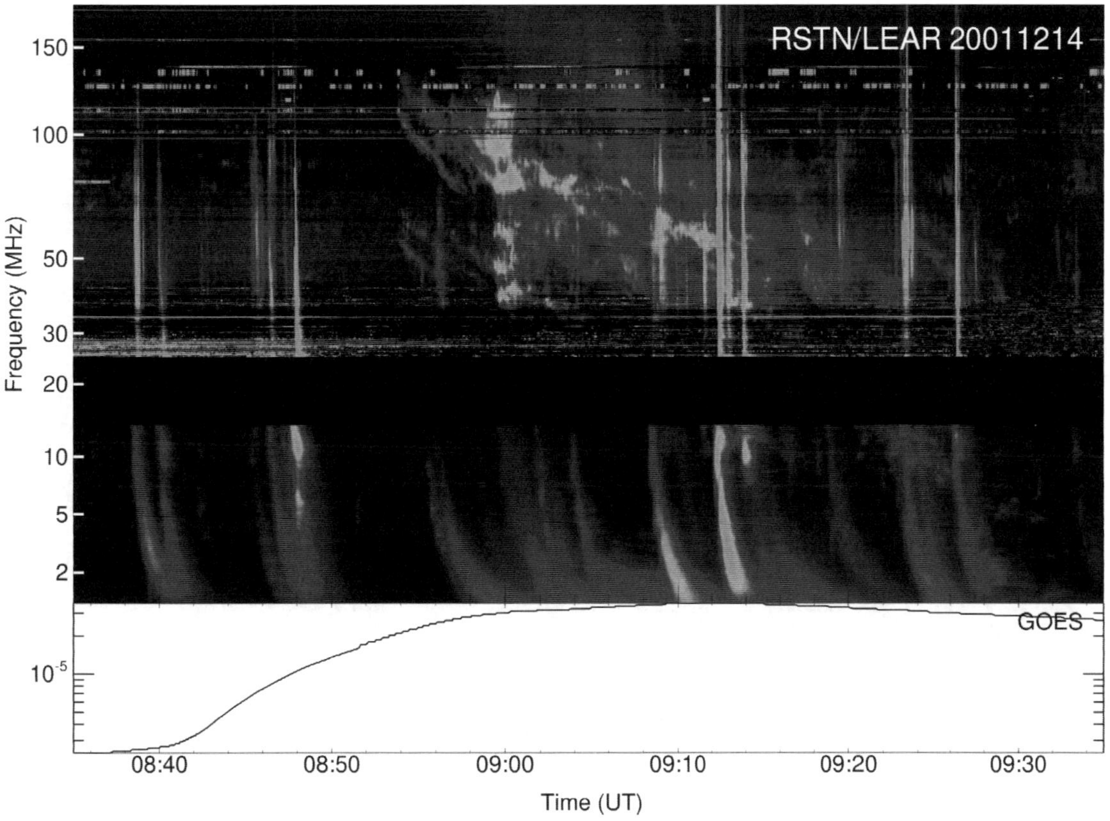

Figure 7. The M4 event of 2001 December 14, a candidate source for the particle event starting on December 15. Although this event shows many Type III-like features that could be Type III-L bursts, we do not classify it as a Type III-L event because numerous bright Type III bursts occur throughout the 06:30 – 09:30 UT period, and so we cannot unambiguously identify any feature with the flare itself.

that we regard as possible but uncertain are classified as marginal associations, denoted by italic type in the association column (column 3) of Table 2. We have produced radio dynamic spectra of each event with an association identical to those used for the Type III-L identification, and use these to also report the presence of Type II emission in the ground and WAVES frequency ranges in the table. The speed of any associated CME (taken from the CDAW catalog) is also given in the table.

Figure 7 shows an example of an event where strict application of our Type III-L criteria rules out a likely–looking Type III-L-SEP association: this flare occurred in the middle of a 3–hour period when numerous bright Type III bursts can be seen in the WIND/WAVES data. Due to the possibility that that III-like bursts occurring in conjunction with the Type II burst could be coincidental, we do not classify this as a Type III-L burst.

Of the 68 SEP proton events listed in Table 2 we find possible solar event associations for 62 events, but of these we regard 14 as marginal associations. The nature and location of the associated event is also listed in Table 2: 35, or just over half, are GOES class M or X flares and thus were included in our core sample, while 17 have no flare association but we can find appropriate CMEs. This implies that a quarter of all SEP events in 2001 originated in flares occurring beyond the solar limb (with most, 14 out of 17, beyond the west limb and thus having a better magnetic connection to Earth): this is not surprising because the detection of eastern–limb events shows that SEPs can diffuse very large distances in azimuth around the Sun. In 10 events the most plausible association was with a smaller GOES class C

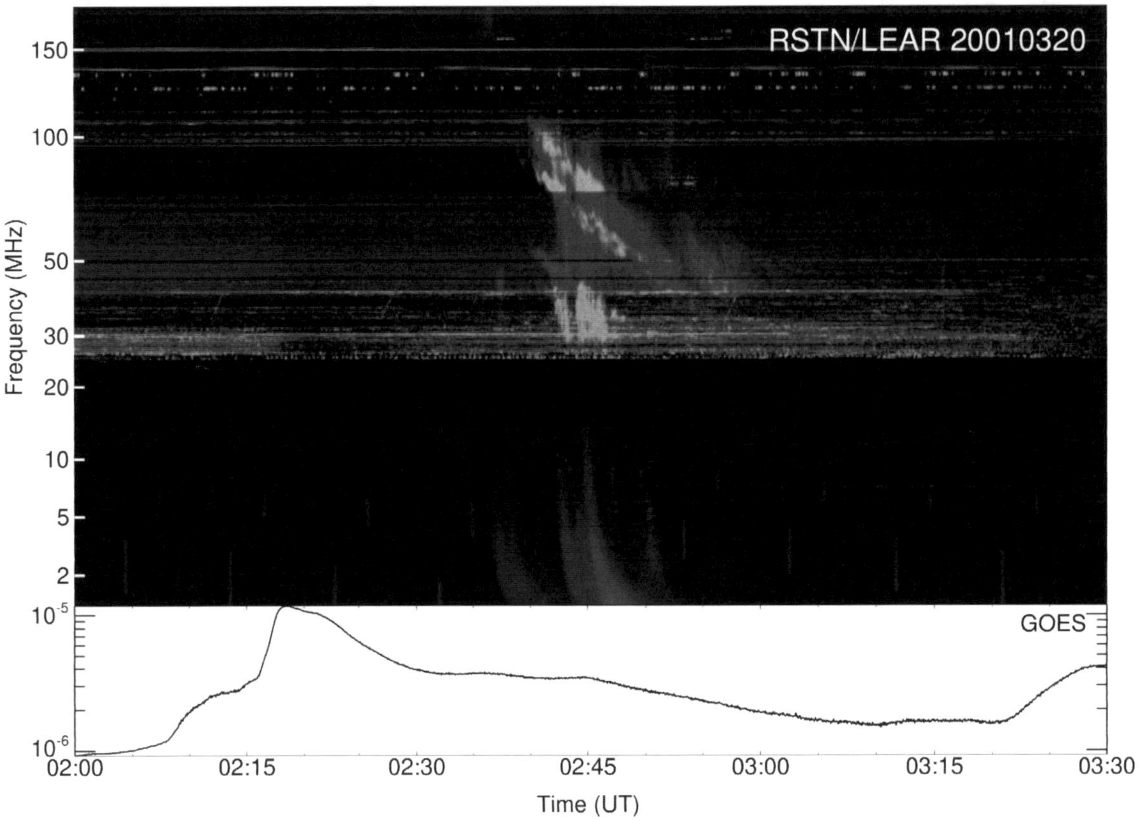

Figure 8. The M1 event of 2001 March 20. This is a long–duration event with a complex GOES light curve. The Type III-L and Type II emission are coincident in time but occur late relative to the soft X–ray peak.

flare, some of which (01/26, 02/11) produced fast CMEs with speeds of order 1000 km s^{-1} or faster. In addition, 4 of the 10 C class flares linked to SEP events occurred at the limb and may have been larger flares that were partially obscured in soft X–rays. We regard 4 of the 10 C–flare associations as marginal: the strong associations all exhibit very bright and/or fast CMEs, consistent with our explicit bias towards such events. 12 of the 22 X–class flares in 2001 are associated with SEP events, including all 5 of the X–class flares that occurred in the W30–W90 longitude range.

4. RESULTS

After the SEP high–background exclusion, we had 50 "good" and 16 "marginal" Type III-L events for 2001. The results of associating these 66 events with the SEP list are given in Table 4. We break the associations down according to whether the Type III-L is "good" or "marginal", and then whether the SEP event is "good" or "marginal", or has a marginal association. Further, since SEP events occurring in the western hemisphere are more likely to be detected at Earth than those to the east, we break the association down by east (heliographic longitudes E30-90), center (E30-W30) and west (W30-W90) locations. We also report the association of Type II events in the MX flare list with SEP events as a comparison.

As expected, the Type III-L list shows no strong longitude preference (radio emission does not care about the Parker spiral), whereas about 60% of the SEP events occur limbwards of W30 (charged–particle propagation *does* care about the Parker spiral). There is very little association between Type III-L bursts (or Type II bursts) with SEP events having marginal associations, so we focus here on associations with "good" SEP events with reliable solar event identifications. As shown in the first line of Table 4, of the 13 "good" Type III-L bursts in the W30–W90 range, 7 (just over 50%) are associated with SEPs, while 3 out of 9 "marginal" Type III-L bursts are associated with SEPs. A further 1 "good" and 3 "marginal" Type III-L bursts in the W30–W90 range are associated with marginal SEP events, so that 8 out of 13 (62%) "good" and 6 out of 9 (67%) "marginal" well–connected Type III-L bursts are associated with SEP events. In the E30–W30 longitude range the degree of association is poorer as expected, although the statistics are not compelling: 8 out of 21 "good" Type III-L bursts are associated with SEP events. These numbers suggest that the presence of a Type III-L burst from a flare in the well–connected region of the solar surface is a good but not a compelling predictor of an SEP event. Figure 8 shows an example of a well–connected event (longitude W54) that produced clear Type III-L emission in conjunction with a Type II burst, but no particle event was detected.

Furthermore, in our sample Type III-L bursts do not appear to be a significantly better predictor of SEP events than Type II bursts: 10 out of 17 (59%) MX flares in the W30–W90 longitude range exhibiting Type II bursts were associated with SEP events. However, these associations for Type II bursts are poorer if we remove the restriction to MX events: we find that 14 out of all 26 reported Type II bursts in the W30–W90 longitude range in 2001 and 13 out of 35 Type II bursts in the E30–W30 range are associated with SEPs. This reflects the fact that smaller flares are quite capable of producing Type II bursts, but are less likely to produce detectable SEP events.

On the other hand, we find that SEP events are quite well associated with Type III-L bursts (Table 5), as the previous studies have reported. In 31 out of 41 (76%) "good" SEP events with a reliable solar event association, the solar event (flare or CME) shows radio emission that fits our Type III-L criteria. For all levels of SEP event with an association, 41 out of 62 show Type III-L emission (66%). However, SEP events are equally well associated with Type II bursts: 32 out of 41 (78%) "good" SEP events with "good" associations and 42 out of all 62 SEP events with any associations (68%) show Type II emission.

The degree of association improves if, as argued by [4], we restrict our attention to Type III-L bursts with durations in excess of 10 minutes at 1 MHz: 11 out of 13 Type III-L events in the W30-W90 range and 9 out of 15 Type III-L events in the E30-W30 range with 1 MHz durations over 10 minutes are associated with SEP events (Table 4). The reverse association is also quite strong: in 33 out of 41 SEP events associated with Type III-L bursts, the 1 MHz radio emission has a duration of 10 minutes or more.

5. ANALYSIS AND DISCUSSION

In the previous sections we have demonstrated that an operational definition of Type III-L bursts, satisfying the previously–inferred (by other authors) properties of *l*ateness, *l*ength and *l*ow–frequency, can be described and applied to suitable data to produce a catalog of such events. There remains considerable subjectivity in the classification, but this is true of other classes of solar radio burst as well.

We compared our list of Type III-L bursts with the list of all SEP events detected in 2001 by the ERNE instrument on the SOHO spacecraft at proton energies above 13 MeV. For well–connected events originating in the western hemisphere, the majority of Type III-L bursts are indeed associated with SEP events, but not all of them, consistent with the arguments by [15]. Furthermore, they do not seem to be better predictors of SEP events from MX–class flares than is the presence of a Type II burst, and SEP events are just as likely to exhibit Type II bursts in their radio spectra as they are Type III-L bursts.

The association between Type III-L bursts and SEP events does improve if we restrict attention to events in which the 1 MHz radio emission lasts 10 minutes or longer. However, the simplest interpretation of this collection of properties appears to be "big flare" syndrome (BFS). [23] describes this condition as follows: "the larger the energy release in a flare, the larger the statistically expected magnitude of any measured flare energy manifestation". An alternative statement of BFS is that "big flares have everything", i.e., correlations between any two properties will be large in big flares because any given property is more likely to be detected in a large flare and correlations will appear even though two properties may not be physically connected. In the case of our study, the hypothesis is that SEP detection is predominantly a function of flare size and favorable magnetic connectivity from the acceleration site to Earth. The fact that Type III-L bursts and Type II bursts show similar degrees of association with SEP events, and vice–versa, is then explained as the result of SEP events generally being large flares that exhibit all phenomena, rather than Type III-L and Type II emission being physically linked to the production of energetic protons. We expect that 1 MHz emission will have a longer duration in large flares, and so the improved correlation with longer Type III-L emission at 1 MHz is also consistent with BFS. The issue that this discussion does not address, however, is what parameter is appropriate to measure the "size" of a flare: soft X–ray class, while useful, is not definitive since it measures the peak SXR flux rather than the fluence, does not represent the energy or speed of any associated mass ejection, and further is not reliable for over–the–limb events.

Nonetheless, the fact that we can find Type III-L emission distinct from the Type III bursts that typically occur at the onset of an impulsive flare (roughly one–quarter of the MX flares in our sample show such impulsive Type III bursts, and about 40% of the Type III-L bursts are preceded by impulsive–phase Type IIIs) is a significant result and potentially highlights a phenomenon that is physically distinct, in the same way that Type IV emission (occurring at frequencies above the ionospheric cutoff, typically after any Type II burst and with longer duration) is clearly the result of a physical mechanism that is distinct from the impulsive Type III bursts and Type II emission. More work is clearly needed to establish such a case. One issue that occurred in classifying Type III-L bursts and is relevant to this question is the difficulty of identifying Type III-L emission in the ground–based dynamic spectra: frequently such emission was confused by both bright flare continuum emissions (Type II, Type IV or other continua) and by interference, e.g., Figs. 2 and 8, where Type III-L emission below 14 MHz clearly occurs at the time of Type II emission above 25 MHz, but it is difficult to discern whether the Type III-L emission is also present above 25 MHz. This makes it difficult for us to address the issue of the acceleration site for the electrons producing the Type III-L emission, but there clearly are a number of cases where the Type III-L emission starts at frequencies above that of concurrent Type II emission, implying an acceleration site relatively low in the corona.

6. FAST–ONSET EVENTS

An issue for the association between Type III-L bursts and SEP events is that if the acceleration of the Type III-L–emitting electrons is coincident with the acceleration of the SEPs, then the SEPs cannot leave the Sun *before* the Type III-L burst occurs. While inspecting the plots for the events in Table 2, we noticed that there are several events in which 175–315 keV electrons arrive at Earth very quickly, close in time to the Type III-L burst onset. Given the extra propagation time required by the electrons to reach Earth, this could pose a problem for the interpretation of a common acceleration site.

We note that there continues to be a debate about the source of the near–relativistic electrons measured by the EPAM instrument. This question is irrelevant to our main purpose of identifying the solar sources of SEPs in order to look for associated Type III-L emission, but is relevant to the issue of the relationship between the electrons emitting Type III-L bursts and damaging protons in SEP events. The debate has arisen largely because of studies [e.g., 24; 25] that found a delay between the launch time of the near–relativistic electrons at the Sun and Type III bursts attributed [26] to 2–20 keV electrons in the corona. Most of these studies concern "beam–like electron events" rather than the large proton–dominated SEP events that make up Table 2, and it is argued that the delays mean that the near–relativistic electrons measured at 1 AU cannot be the high–energy tail of the Type III–emitting electrons and thus require a different source. The delays can be attributed to a CME–driven shock–acceleration source for the near–relativistic electrons [27], to propagation conditions [28], or to an origin in low–coronal phenomena occurring after the impulsive phase [e.g., 29]. An association with the Type III-L bursts occurring after the impulsive phase would fall into the latter category. [30] has reviewed the nature of the solar sources of energetic electron events and has concluded that some appear to arise in the impulsive phase of flares while others (particularly those with relativistic energies) appear to be better associated with acceleration

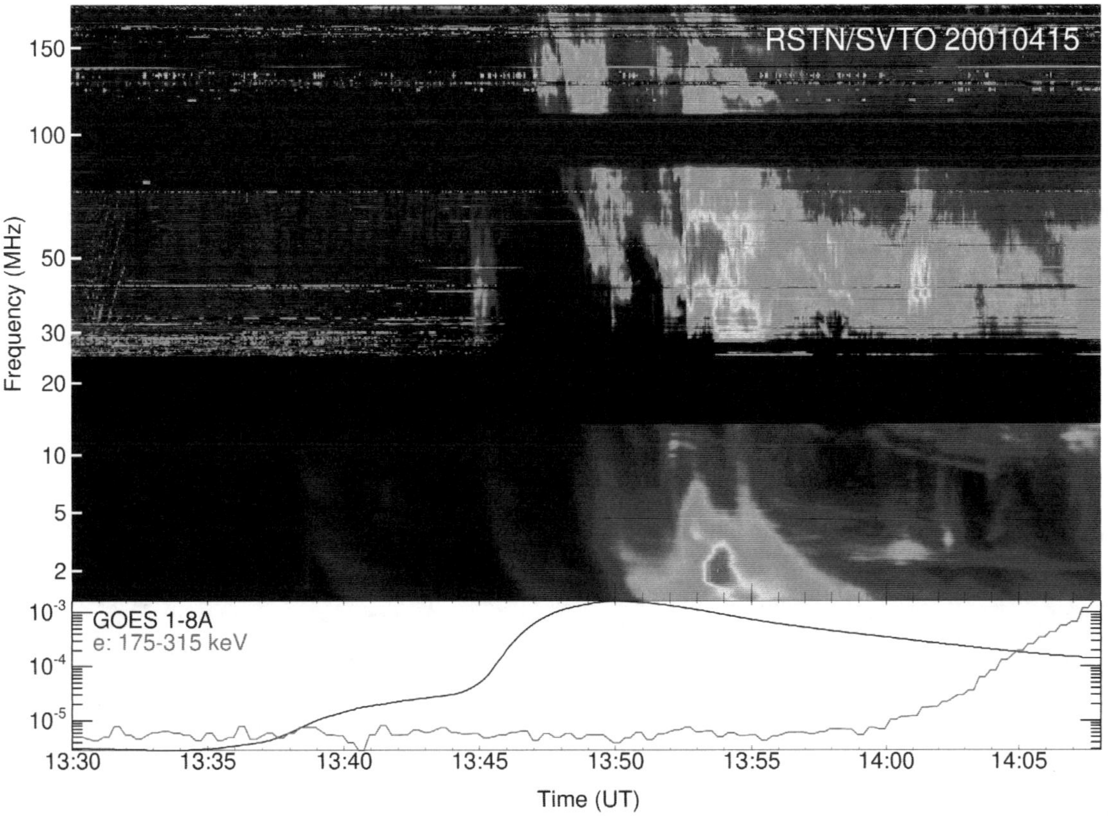

Figure 9. The X14 event of 2001 April 15. The figure has the same format except that the light curve of 175-312 keV electrons (from ACE/EPAM) is shown in the lower panel at 12 second resolution (red curve, logarithmic in the vertical axis, scaled to fit). The ACE/EPAM curve has *not* been shifted in time, i.e., it represents the arrival time of electrons at ACE at 1 AU).

by a CME–driven shock.

Table 6 lists 8 events from our sample where the latest time that the first–arriving electrons could have left the Sun is very close to the onset time of associated Type III-L emission. If a 200 keV (kinetic–energy) electron follows the minimum 1.2 AU–long Parker spiral trajectory from the Sun to the Earth with zero pitch angle and no scattering (fastest possible circumstances), it requires 860 seconds to propagate from the Sun to the Earth, while a 300 keV electron requires 770 seconds. The light travel time of radio waves (1.0 AU) is 500 seconds, so we can assume that electrons in the ACE/EPAM energy range of 175–315 keV leaving the Sun are delayed by a minimum of 6 minutes in arriving at Earth compared with the arrival of electromagnetic radiation departing the Sun at the same time. Table 6 reports the latest possible departure time for the first–arriving electrons leaving the Sun (measured with Earth clocks) by subtracting 6 minutes from the observed time of arrival of the first electrons at 1 AU: in practice we expect that electrons must have been accelerated before the time shown. All 8 events are beyond longitude W50 and hence well connected. In 5 of 8 events the latest SEP launch time is within 12 minutes of the (nominal) SXR onset. In one case, 2001 April 15 (shown in Figure 9), the SEP launch time is within a few minutes of the nominal SXR onset for the X14 flare but this event has a complex soft X–ray light curve with a possible precursor event starting at 13:37 UT, roughly 10 minutes before the latest SEP launch time. In 5 of 8 events the SEP launch time is well (typically 15 minutes) before the first observation of the CME associated with the event in the LASCO/C2 field of view, but in the case of

2001 August 15 event beyond the SW limb a CME is detected significantly before the onset time of the SEPs: radio emission observed at 23:30 UT suggests that the actual event start time was as early as 23:25 UT, well before the first CME observation and more than 40 minutes before the first electrons arrive at ACE.

In none of the events listed in Table 6 does the latest possible SEP electron launch time precede the onset of the Type III-L-burst, so these data do not rule out a common acceleration site for the Type III-L-emitting electrons and the SEP electrons. However, given that our correction for the difference in travel time between electrons and radio photons is very conservative, it is still possible that in some cases (e.g., April 15), SEP electrons may have left the Sun before Type III-L emission is observed. On the other hand, as noted above, it may also be the case that the SEP electrons have an origin quite different from that of the SEP protons that are the main concern for space weather effects.

7. CONCLUSIONS

We have two goals in carrying out this study: (i) to determine whether Type III-L–bursts can be used to help to predict SEP events; and (ii) to determine whether Type III-L–bursts can help us to identify the location of the acceleration site for SEP particles.

We have established a set of criteria that may be used operationally to identify Type III-L bursts and distinguish them from the impulsive–phase Type III bursts. We find that the requirement of a 4–minute delay between the onset of a flare in soft X–rays and subsequent Type III-L emission serves well in most cases to make this distinction. Applying the defining criteria to the set of 322 GOES–class M and X flares in 2001 for which we have radio dynamic spectra, we found that two observers can agree about 90% of the time in their classifications, and a set of 52 "good"' and 18 "marginal" Type III-L bursts resulted. After exclusion of high particle background periods (listed in Table 3), we have 22 Type III-L bursts occurring in the well–connected W30–W90 longitude range, and about 64% of these bursts were associated with SEP events, including 100% of the X flares in this range with Type III-L emission. The association also improves if we require that emission have a long duration at 1 MHz: 11 out of 13 Type III-L bursts in the W30–W90 range are associated with SEP events.

We identified 53 "good" and 15 "marginal" SEP proton events during 2001, and were able to reliably associate solar sources (flares and/or CMEs) with 48 of these events. Of the 41 "good" SEP events with reliable associations, 31 (76%) show radio emission that fits our Type III-L criteria. For comparison, [4] found that 20% of the large SEP events in their sample did not exhibit suitable Type III emission at 14 MHz, although 91% exhibited emission lasting longer than 10 minutes at 1 MHz. Thus a large fraction of SEP events, but not all, do exhibit "late" Type III bursts. This raises the possibility that there is indeed a link between the acceleration mechanism that produces SEPs and the acceleration of the electrons in the beams that produce Type III-L emission. Investigation of the timing of arrival of energetic electrons at the ACE satellite at the L1 Lagrangian point indicates that we cannot rule out the scenario in which they are accelerated at the same time as the Type III-L-emitting electrons.

The more dangerous SEP protons need not be accelerated at the same site and with the same mechanism as the electrons, but we note that in all but a few of the 68 proton events in Table 2, 175-312 keV electrons were detected by ACE/EPAM: electron acceleration almost invariably accompanies ion acceleration. While there are numerous potential sources of accelerated electrons, as discussed earlier, the prevailing view is that the energetic protons in the "gradual" SEP events that dominate Table 2 have no contribution from flare processes in the low corona and are instead accelerated by shocks driven by CMEs [e.g., 20], and indeed many of the solar sources that we identify for the SEP events are associated with CMEs faster than 1000 km s^{-1} (of the 19 CMEs in the western hemisphere in 2001 with speeds faster than 1000 km s^{-1} and widths larger than 100°, at least 14 were associated with detectable SEP protons, and 20 out of 28 halo CMEs faster than 1000 km s^{-1} were associated with detectable SEP protons irrespective of the flare location). If this is the case, it is difficult to understand how there can be any causal association between CMEs and Type III-L bursts originating in the low corona at plasma frequencies above 50 MHz tens of minutes after the impulsive onset of a flare, when any associated fast

CME is generally several solar radii above the surface. The association of 1 MHz emission with CMEs is more plausible in terms of height and timing, but again for those fast–drift bursts originating in the low corona this implies that somehow the fact that the CME shock is accelerating protons enhances the plasma emission of electrons at 1 MHz, typically 5–10 $R_\odot$ above the solar surface, even though the electrons themselves were accelerated far below.

Unfortunately, however, the associations between SEP events and Type III-L bursts are almost identical to those between SEP events and Type II bursts, and it is well established that Type II bursts are a poor predictor of SEP events [e.g., 2; 31] because quite small flares with no detectable SEPs can easily produce Type II bursts. This leads us to speculate that the high degree of association of Type III-L emission with SEP events may be due to big–flare syndrome: the detection of SEPs is largely a function of the total energy released by the flare, and large flares tend to exhibit all forms of solar activity, so that high degrees of correlation will appear even in the absence of a causal connection between two phenomena.

Acknowledgments

RD and SW acknowledge research support from George Mason University and the University of Maryland, respectively. This research was supported by NASA Living–With–a–Star grant NNX06AC18G. Basic research in radio astronomy at the Naval Research Laboratory is supported by the Office of Naval Research. We are grateful to the US Air Force for supplying the RSTN radio data, and to Helen Coffey and NOAA/NGDC for making the data available on the web. Special thanks to: Hilary Cane for extensive discussions and helpful advice; Steve Kahler for valuable comments; Henry Aurass at the Astrophysical Institute Potsdam for supplying OSRA data; and the SOHO/ERNE and ACE/EPAM teams for making their data available for use.

References

[1] Wild J P, Smerd S F and Weiss A A 1963 *Ann. Rev. Astron. Astrophys.* **1** 291

[2] Kahler S W 1982 *Astrophys. J.* **261** 710

[3] Cane H V, Erickson W C and Prestage N P 2002 *J. Geophys. Res.* **107** 1315–1333

[4] MacDowall R J, Lara A, Manoharan P K, Nitta N V, Rosas A M and Bougeret J L 2003 *Geophys. Res. Lett.* **30** 6–1

[5] Pick M and Maia D 2005 *Advances in Space Research* **35** 1876–1881

[6] Cliver E W and Ling A G 2009 *Astrophys. J.* **690** 598–609

[7] Dulk G A, Leblanc Y, Bastian T S and Bougeret J L 2000 *J. Geophys. Res.* **105** 27343–27352

[8] Cane H V, Stone R G, Fainberg J, Stewart R T and Steinberg J L 1981 *Geophys. Res. Lett.* **8** 1285

[9] Kundu M R and Stone R G 1984 *Advances in Space Research* **4** 261–270

[10] Kahler S W, Cliver E W and Cane H V 1986 *Advances in Space Research* **6** 319–322

[11] MacDowall R J, Kundu M R and Stone R G 1987 *Solar Phys.* **111** 397–418

[12] Kahler S W, Cliver E W and Cane H V 1989 *Solar Phys.* **120** 393–405

[13] Kundu M R, MacDowall R J and Stone R G 1990 *Astrophys. Space Sci.* **165** 101–110

[14] Reiner M J, Karlicky M, Jiricka K, Aurass H, Mann G and Kaiser M 2000 *Astrophys. J.* **530** 1049–1060

[15] Gopalswamy N and Mäkelä P 2010 *Astrophys. J. Letters* **721** L62–L66

[16] Hudson H S, Kosugi T, Nitta N V and Shimojo M 2001 *Astrophys. J. Letters* **561** L211–L214

[17] Posner A 2007 *Space Weather* **5** S05001

[18] Nitta N V, Reames D V, DeRosa M L, Liu Y, Yashiro S and Gopalswamy N 2006 *Astrophys. J.* **650** 438–450

[19] Kahler S W 2005 *Astrophys. J.* **628** 1014–1022

[20] Reames D V 1999 *Space Science Reviews* **90** 413–491

[21] Kahler S and Ragot B R 2006 *Astrophys. J.* **646** 634–641

[22] Laurenza M, Cliver E W, Hewitt J, Storini M, Ling A G, Balch C C and Kaiser M L 2009 *Space Weather* **7** S04008

[23] Kahler S W 1982 *J. Geophys. Res.* **87** 3439

[24] Krucker S, Larson D E, Lin R P and Thompson B J 1999 *Astrophys. J.* **519** 864–875

[25] Haggerty D K and Roelof E C 2002 *Astrophys. J.* **579** 841

[26] Lin R P 1985 *Solar Phys.* **100** 537

[27] Simnett G M, Roelof E C and Haggerty D K 2002 *Astrophys. J.* **579** 854

[28] Cane H V 2003 *Astrophys. J.* **598** 1403–1408

[29] Maia D and Pick M 2005 *Advances in Space Research* **35** 1876–1881

[30] Kahler S W 2007 *Space Sci. Revs.* **129** 359–390

[31] Cliver E W, Kahler S W and Reames D V 2004 *Astrophys. J.* **605** 902–910

Table 1. **Type III-L bursts from MX flares in 2001**

SXR Onset[1] (Date UT)	SXR Class	Location	III-L Ground	III-L WAVES	Type II Ground	Type II WAVES	CME km/s	1 MHz dur.[2]	Comments
20010120 / 1835	M1	S07 E40	1846	1846	y	y	840	34	patchy Type II, prototype Type III-L
20010120 / 2108	M8	S07 E46	2114	2115	y	y	1510	18	
20010128 / 1540	M1	S04 W59	1558	1558		y	920	20	faint emission
20010202 / 2351	M2	N09 E72	0004		y	y	640	9	
20010310 / 0403	M7	N27 W42	0409	0409	y	y	820	6	
20010320 / 0216	M1	S05 W54	0242	0242	y		480	16	
20010320 / 1458	M2	S05 W61	1502	1503			410	6	weak, individual bursts?
20010324 / 0135	M1	S14 W42	0141	0141	y		310	7	impulsive-phase IIIs bright, Type III-L weak
20010325 / 0415	M2	N17 E51	0445	0445				8	Type III-L is very late
20010325 / 1103	M3	N21 E59	1115	1115			580	6	
20010326 / 1306	M2	N15 E27	1323	1323			450	6	long impulsive IIIs, weak III-L
20010328 / 0156	M1	N14 E05	0210	0210				7	long impulsive IIIs, weak III-L
20010328 / 2222	M2	N17 W01	2240	2240				16	GOES onset time uncertain
20010329 / 1000	X2	N20 W19	1009	1009	y	y	940	15	confused by ongoing continuum, not clearly fast-drift emission
20010402 / 0526	M1	N13 W63		0533			630	4	nothing above 14 MHz, weak III-L
20010402 / 1006	X1	N17 W60	1016	1016				5	weak III-L, confused by strong SWF[3] and by an ongoing Type III storm
20010402 / 1054	X1	N15 W64	1113	1113	y	y	990	16	GOES onset time uncertain, identify later IIIs as III-L
20010402 / 2134	X17	N14 W82	2152	2147	y	y	2500	19	
20010405 / 0200[4]	M3	W Limb		0207			1860	7	weak III-L, mild SWF[3]
20010406 / 1914	X6	S21 E31		1923		y	1270	16	no RSTN data
20010409 / 1523	M8	S21 W04	1529	1529	y	y	1190	17	
20010410 / 0509	X2	S23 W09	0514	0514	y	y	2410	22	
20010411 / 1259[4]	M2	S22 W27	1315	1316	y	y	1100	16	multiple III-L candidates
20010412 / 1015	X2	S19 W43	1030	1031	y	y	1180	14	multiple III-L candidates
20010415 / 1344	X14	S20 W85	1347	1349	y	y	1200	24	
20010420 / 0504	M1	N18 E63	0520	0520				5	very weak III-L, possibly storm
20010423 / 2020	M4	N14 E23		2028				3	bright III-L over ongoing storm
20010425 / 1342	M3	N18 W09	1348	1348	y		860	6	short II, multiple IIIs
20010512 / 2326	M3	S17 E00	2334	2334	y	y	nd[5]	13	multiple candidate III-Ls
20010520 / 0602	M6	SW Limb	0608	0609	y		550	15	strong SWF[3]
20010604 / 0806	M2	S18 E57	0813	0813			760	6	III-L very similar to impulsive III group
20010605 / 0444	M2	S18 E44	0452	0452			840	5	III-L very similar to impulsive III group
20010613 / 1134	M8	S29 E66		1140			1110	12	no RSTN data
20010615 / 1004	M6	S26 E41		1022	y		1090	7	no RSTN data, multiple III-L candidates
20010623 / 0623	M1	N10 E21	0636	0636				4	could be 2 individual IIIs
20010824 / 2358	M3	S18 E44	0008	0008			480	23	weak but long in WAVES
20010825 / 1624	X5	S17 E34	1631	1631	y		1400	20	GOES onset uncertain
20010827 / 0634	M2	E Limb	0702	0702				2	very late, weak
20010828 / 1556	M1	N15 E65	1606	1606	y		480	12	
20010831 / 1037	M2	N15 E37	1041	1041	y			4	GOES onset uncertain, short
20010831 / 2240	M3	N14 E25	2247	2247	y		480	5	gradual and complex rise phase
20010903 / 1824	M2	E Limb	1829	1829	y	y	1350	11	GOES onset uncertain

Table 1—Continued

SXR Onset[1] (Date UT)	SXR Class	Location	III-L Ground	III-L WAVES	Type II Ground	Type II WAVES	CME km/s	1 MHz dur.[2]	Comments
20010907 / 1528	M1	N19 W65	1541	1541	y		490	4	weak III-L
20010909 / 1511	M3	S32 E22	1516	1516	y			8	extended impulsive IIIs
20010918 / 0005	M1	S32 W78	0034	0034				9	III-L is very late
20010920 / 1815	M1	N09 W11	1838	1838		y	450	9	III-L late, high-frequency cutoff suggests over the limb, not from M1?
20010924 / 1018	X3	S16 E23	1034	1034		y	2400	30	very gradual rise, onset uncertain
20011001 / 0456	M9	W Limb	0520	0511			1410	18	
20011009 / 1048	M1	S28 E08	1056	1056	y	y	970	14	multiple III-L candidates
20011017 / 1119	M1	S17 W22	1144	1144			470	8	very late III-L
20011019 / 0054	X2	N16 W18	0104	0104	y	y	560	12	artefacts in WAVES
20011019 / 1622	X2	N15 W29	1628	1628	y	y	900	30	SWF[3]
20011021 / 0512	M1	W Limb	0522	0522			530	4	weak SWF[3] , individual III?
20011022 / 0034	M1	N17 W57	0045	0045			770	6	very short III-L
20011022 / 1445	M7	S21 E18		1501	y	y	1340	27	
20011022 / 1747	X1	S18 E16	1754	1754			620	25	long-lived narrowband feature at 1 MHz
20011026 / 1430	M2	N07 E18	1434	1434				5	GOES onset uncertain
20011029 / 1110	M4	N12 E25	1114	1114	y		600	5	confused by storm
20011031 / 2136	M1	W Limb	2146	2146				6	much brighter than storm
20011104 / 1605	X1	N06 W18	1618	1618	y	?	1810	36	
20011109 / 1820	M2	S21 W42	1824	1825	y	y	nd[5]	18	
20011117 / 0442	M3	S13 E42	0504	0504	y	?	1380	32	bright at 1 MHz
20011122 / 2020	M4	S25 W67	2024	2024	y	?	1440	26	very similar to 1117 event
20011122 / 2245	X1	S20 W40		2306	y	y	1440	25	radio emission mostly from C6 precursor, appropriate GOES onset uncertain, CME definitely from X1 flare not C6.
20011128 / 1630	M7	N04 E16	1634	1635	y		500	16	horizontal features in WAVES data
20011213 / 1424	X6	N16 E09		1437	y		860	10	
20011226 / 0504	M7	N08 W54	0514	0514	y	y	1450	20	1 MHz features not clearly fast-drift
20011228 / 2002	X3	E Limb	2008	2008	y	y	2220	21	very gradual
20011229 / 0939[4]	M9	S07 W85	1000	1000	y	y	630	20	two peaks in GOES
20011229 / 1154[4]	M1	S26 E87	1204	1204				4	weak III-L

[1]Type III-L events in italics are marginal events.

[2]Duration of continuous emission at 1 MHz in minutes, estimated from figures. *This may be longer than the duration of the Type III-l emission at 1 MHz.*

[3]"Short–wave fadeouts" (SWF) occur when a strong flare overhead produces high ionization in the lower layers of the atmosphere and frequencies above the normal ionospheric cutoff are strongly absorbed.

[4]Event excluded from comparisons due to high SEP background level.

[5]No LASCO CME data at the appropriate time.

Table 2. **Solar Energetic Proton Events in 2001**

SEP Onset[1] (Date UT)	SEP max[2]	SEP assoc.[3]	SXR/CME (Date UT)	Location SXR/CME	Type II Ground	Type II WAVES	CME ($km\ s^{-1}$)	Comment[4] SEP type	Type III-L[5]
20010105 / 1740	3	CME	0105 / 1700	W Limb			830	Slow rise	m
20010110 / 2000	0.08							Fast rise	
20010115 / 1600	0.01	CME	0114 / 0630	NW Limb			945	*Slow rise*	n
20010121 / 0200	3	M8	0120 / 2108	S07 E46	2112	2122	1510	Slow rise	y
20010126 / 1300	0.06	C2	0126 / 1149	S23 W57	1157	1206	930	Slow rise	m
20010128 / 1620	8	M2	0128 / 1540	S04 W59		1558	920	Fast rise	m
20010211 / 0120	0.9	C7	0211 / 0054	N24 W57	0105		1180	Fast rise	y
20010226 / 0610	2	CME	0226 / 0530	W Limb			850	Fast rise	n
20010308 / 1330	0.002	*M6*	0308 / 1116	N30 W18	1126		630	Slow rise	n
20010310 / 0500	0.01	M7	0310 / 0403	N27 W42	0409	0416	820	Slow rise	y
20010325 / 1200	0.3	M3	0325 / 1103	N21 E59			580	Slow rise	y
20010329 / 1015	60	X2	0329 / 1000	N20 W19	1003	1012	940	Fast rise	m
20010402 / 1140	0.4	X1	0402 / 1054	N15 W64	1110	1122	990	Fast rise	m
20010402 / 2155	300	X17	0402 / 2134	N14 W82	2151	2159	2500	Fast rise	y
20010409 / 1600	0.7	M8	0409 / 1523	S21 W04	1527	1542	1190	Fast rise	y
20010410 / 0600	20	X2	0410 / 0509	S23 W09	0513	0522	2410	Slow rise	y
20010412 / 1050	8	X2	0412 / 1015	S19 W43	1020	1035	1180	Fast rise	y
20010415 / 1359	100	X14	0415 / 1344	S20 W85	1347	1357	1200	Fast rise	y
20010418 / 0230	20	CME	0418 / 0213	W Limb	0217	0255	2470	Fast rise	y
20010426 / 1320	0.3	M8	0426 / 1306	N17 W31	1335		840	Fast rise	n
20010507 / 1215	5	CME	0507 / 1206	NW Limb			1220	Fast rise	n
20010520 / 0615	20	M6	0520 / 0602	SW Limb	0605		550	Fast rise	y
20010530 / 1600	0.004	CME	0530 / 0000	E Limb		0012	2090	Slow rise	y
20010601 / 2030	0.05	*CME*	0601 / 1730	W Limb			800	Fast rise	n
20010604 / 1635	0.4	C3	0604 / 1622	N24 W59	1622		460	Fast rise	y
20010611 / 1300	0.4	*CME*	0611 / 0445	SE Limb			1650	Slow rise	n
20010612 / 0545	0.05							Fast rise	
20010615 / 0700	0.01	*C4*	0615 / 0634	S26 E48			No	Fast rise	n
20010615 / 1040	0.02	M6	0615 / 1004	S26 E41			1090	Fast rise	y
20010615 / 1550	5	CME	0615 / 1530	W Limb		1601	1700	Fast rise	y
20010619 / 0340	0.3	CME	0619 / 0354	NW Limb	0335		820	Fast rise	y
20010705 / 0930	0.005							Slow rise	
20010712 / 0130	0.2	CME	0712 / 0006	SW Limb			740	Fast rise	n
20010719 / 1020	0.2	M2	0719 / 0956	S08 W62			1670	Fast rise	n
20010731 / 2200	0.002	*C6*	0730 / 2039	E Limb	2045		no data	Slow rise	n
20010809 / 1610	1	*CME*	0809 / 1030	W Limb			480	Fast rise	n
20010814 / 1200	15						no data	*Slow rise*	
20010816 / 0010	500	CME	0815 / 2354	SW Limb	2331		1600	*Fast rise*	y
20010821 / 1120	2	C3	0821 / 1036	SW Limb			620	Slow rise	n
20010911 / 0130	0.02	M3	0911 / 0048	S28 E09			300	Fast rise	n
20010912 / 2330	0.06	C9	0912 / 2106	SW Limb	2139		670	Fast rise	n
20010915 / 1150	5	M2	0915 / 1106	S21 W49	1129	1207	480	Fast rise	n
20010917 / 1200	0.03	M2	0917 / 0820	S14 E04	0828	0834	1010	Slow rise	n
20010918 / 0145	0.08	M1	0918 / 0005	S32 W78				Fast rise	m
20010919 / 0830	0.1	C7	0919 / 0650	SW Limb	0700		420	Slow rise	m

Table 2—Continued

SEP Onset[1] (Date UT)	SEP max[2]	SEP assoc.[3]	SXR/CME (Date UT)	Location SXR/CME	Type II Ground	Type II WAVES	CME ($km\ s^{-1}$)	Comment[4] SEP type	Type III-L[5]
20010924 / 1050	70	X3	0924 / 1018	S16 E23		1038	2400	Fast rise	y
20011001 / 0800	70	M9	1001 / 0456	W Limb			1405	Slow (!) rise	m
20011005 / 1900	3	*CME*	1005 / 1000	SW Limb			1540	Slow rise	y
20011009 / 0750	1	*C7*	1009 / 0726	W Limb	0737		530	Fast rise	y
20011009 / 1500	2	*M1*	1009 / 1048	S28 E08	1055	1110	970	Slow rise	y
20011019 / 0140	5	X2	1019 / 0054	N16 W18	0101	0117	560	Fast rise	y
20011019 / 1700	6	X2	1019 / 1622	N15 W29	1624	1639	900	Fast rise	y
20011022 / 0100	0.05	M1	1022 / 0034	N17 W57			770	Slow rise	m
20011022 / 1530	8	M7	1022 / 1445	S21 E18	1453	1505	1340	Fast rise	y
20011104 / 1630	60	X1	1104 / 1605	N06 W18	1614	1627	1810	Fast rise	y
20011109 / 1900	3	M2	1109 / 1820	S21 W42	1837	1841	no data	Slow rise	y
20011117 / 0600	4	M3	1117 / 0442	S13 E42	0450		1380	Slow rise	y
20011122 / 2045	8	M4	1122 / 2020	S25 W67	2022		1445	Fast rise	y
20011122 / 2315	70	X1	1122 / 2245	S20 W40	2231	2235	1435	Fast rise	m
20011125 / 2040	11							Slow rise	
20011210 / 1000	0.2	*C9*	1210 / 0933	S25 W19	0940		No	Slow rise	n
20011211 / 1220	0.3	*CME*	1211 / 1000	W Limb			890	Slow rise	n
20011215 / 1730	0.2	*M4*	1214 / 0842	N06 E90	0853		1510	Slow rise	n
20011218 / 2100	0.02	*CME*	1218 / 1830	W Limb			1030	Slow rise	n
20011225 / 2100	0.2	*CME*	1225 / 1130	SE Limb		1123	1770	Slow rise	m
20011226 / 0525	700	M7	1226 / 0504	N08 W54	0455	0517	1450	Fast rise	y
20011228 / 2200	20	X3	1228 / 2002	E Limb	1959	2025	2200	Slow rise	y
20011230 / 1120	30							Slow rise	

[1]Events in italics are marginal SEP events. The onset time is the earliest time that either energetic protons or electrons are detected at Earth distance, with an uncertainty of order 10 minutes for events with a rapid rise but typically much larger uncertainties for events with a gradual rise.

[2]Maximum count rate measured by ERNE in the 13–40 MeV range in $cm^{-2}\ s^{-1}\ sr^{-1}$

[3]Associations in italics are marginal associations. The times reported for associated events are the GOES onset time for flares, as described in the text, or the first reported detection in the case of CMEs.

[4]Comments in italics are events during ERNE data gaps that were determined from GOES proton data.

[5]"y" denotes a "good" Type III-L event, "m" denotes a marginal event, "n" denotes absence of a Type III-L.

Table 3. **SEP exclusion periods**

Interval	Interval
Jan-28 23:47 to Jan-29 12:23	Sep-24 12:50 to Sep-25 21:01
Mar-29 19:57 to Mar-30 00:56	Sep-25 23:25 to Sep 30 20:00
Mar-30 02:05 to Mar-30 13:22	Oct 01 12:00 to Oct-05 11:45
Mar-30 17:37 to Mar-30 17:46	Oct-06 18:35 to Oct-07 23:29
Apr-03 00:10 to Apr-05 22:22	Oct-19 18:14 to Oct-20 13:25
Apr-10 10:00 to Apr-12 06:00	Oct-22 18:38 to Oct-23 00:09
Apr-12 12:00 to Apr-13 01:50	Nov-04 17:46 to Nov-05 16:12
Apr-15 15:02 to Apr-17 02:41	Nov-06 05:35 to Nov-09 12:00
Apr-18 03:31 to Apr-19 12:58	Nov-19 08:50 to Nov-20 06:28
May-07 19:23 to May-08 08:23	Nov-23 00:00 to Nov-24 02:33
Jun-15 17:19 to Jun-15 22:54	Nov-24 08:25 to Nov-25 16:00
Aug-10 08:58 to Aug-10 14:00	Nov-25 23:00 to Nov-28 05:35
Aug-16 01:00 to Aug-18 14:00	Dec-26 06:28 to Dec-28 14:18
Sep-15 13:31 to Sep-15 17:31	Dec-29 02:13 to Dec-30 11:09

Journal of Physics: Conference Series **642** (2015) 012006 doi:10.1088/1742-6596/642/1/012006

Table 4. **Association of Type III-L and Type II[1] bursts with SEP events**

	E90-E30	E30-W30	W30-W90
Association with good SEP events			
Good Type III-L with good SEPs/Total good Type III-L	5/16	6/21	7/13
Marginal Type III-L with good SEPs/Total	0/3	1/4	3/9
Type II with good SEPs/Total Type II	3/16	7/30	10/17
Good SEP (good association) E-W distribution	6	8	27
Association with marginal-association (mA) SEP events			
Good Type III-L with mA SEPs/Total good Type III-L	0/16	1/21	0/13
Marginal Type III-L with mA SEPs/Total	0/3	0/4	0/9
Type II with mA SEPs/Total Type II	1/16	2/30	0/17
mA SEP E-W distribution (good or marginal SEP)	5	3	6
Association with marginal SEP (mSEP) events			
Good Type III-L with mSEPs/Total good Type III-L	0/16	1/21	1/13
Marginal Type III-L with mSEPs/Total	0/3	0/4	3/9
Type II with mSEPs/Total Type II	0/16	2/30	2/17
mSEP E-W distribution (good or marginal assoc.)	2	3	7
Association with all SEP events			
Good Type III-L with all SEPs/Total good Type III-L	5/16	8/21	8/13
Marginal Type III-L with all SEPs/Total	0/3	1/4	6/9
Good Type III-L, 1 MHz duration > 10 min/Total	3/10	8/14	7/9
Marginal Type III-L, 1 MHz duration > 10 min/Total	0/0	1/1	4/4
Type II with all SEPs/Total Type II	4/16	10/30	12/17
SEP E-W distribution	11	13	38

[1]Type II bursts are limited to those observed from the ground and produced by MX flares, excluding the high–particle–background periods of Table 3.

Table 5. **Association of SEP events with radio bursts**

Quality	Good IIIL	Mar. IIIL	No IIIL	IIIL/Total	II/Total[1]
Good SEP/good association	24	7	10	31/41	31/41
Marginal SEP/good association	3	3	1	6/7	4/7
Good SEP/marginal association	2	1	6	3/9	5/9
Marginal SEP/marginal association	1	0	4	1/5	2/5
Totals	30	11	21	41/62	42/62

[1]In this Table Type II bursts include events in which Type IIs occurred either in ground–based data or in WAVES spectra, or both.

Table 6. **Fast–onset SEP events**

Flare/CME (SXR onset)	SEP onset (corrected)[1]	III-L onset	CME - first observation	CME speed (km/s)	Location
20010329 / 10:00	10:09	10:04	10:26	940	N20W19
20010415 / 13:44	13:53	13:50	14:06	1200	S20W85
20010418 / 02:12	02:24	02:20	02:30	2460	Beyond W Limb
20010604 / 16:12	16:29	16:22	16:30	460	N24W59
20010615 / 15:30	15:44	15:35	15:56	1700	W Limb
20010815 / 23:54	00:04	00:02	23:54	1580	SW Limb
20011009 / 07:26	07:44	07:34	08:06	530	W Limb
20011226 / 05:04	05:19	05:14	05:30	1450	N08W54

[1]This is the latest time that the first 200 keV electrons could have left the Sun measured in the Earth observing frame. It results from correcting the actual arrival time at Earth by subtracting 6 minutes to compensate for the minimum possible difference in propagation time between 200 keV electrons following a minimum (1.2 AU) Parker spiral path and direct-path electromagnetic radiation.

Journal of Physics: Conference Series **642** (2015) 012007 doi:10.1088/1742-6596/642/1/012007

Lyman-alpha Radiation Pressure in the Heliosphere: Results from a 3D Monte Carlo Radiative Transfer Simulation

B Fayock[1], G P Zank[1,2], J Heerikhuisen[1,2], C R Gilbert[3] and K Scherer[4]

[1] Center for Space Plasma and Aeronomic Research, University of Alabama in Huntsville, Huntsville, AL 35805, USA

[2] Department of Space Science, University of Alabama in Huntsville, Huntsville, AL 35805, USA

[3] Department of Physics, Georgia Technical Institute, Atlanta, GA, 30332, USA

[4] Institut fur Experimentalphysik II, Ruhr-Universitat Bochum, Universitatsstraβe 150, Bochum, Germany

E-mail: brian.fayock@gmail.com, garyp.zank@gmail.com, jacob.heerikhuisen@uah.edu, cgilbert7@gatech.edu, kls@tp4.rub.de

Abstract. Global models of the heliosphere and its interaction with the local interstellar medium have evolved to fit spacecraft observations since the beginning of their development. It is now understood that neutral hydrogen plays a very important role in the modification of the shape and location of the heliopause. Comparisons of simulations to observations have often led to the need for a more accurate treatment of a particular process. We bring attention to a long-held assumption that radiation pressure on neutral hydrogen atoms due to Lyman-alpha absorption and emission falls off with heliocentric distance at the same rate as solar gravitation. These have generally been combined into a single term. Through our Monte Carlo simulation of Lyman-alpha radiative transfer in the heliosphere, we show that the radiation pressure 1) falls off at a slower rate than r^{-2} and 2) is not isotropic. We find that the radiation pressure in the upwind direction falls off with heliocentric distance with a power of approximately -1.80 between ~6 and 75 AU, while the downwind direction falls of with a much slower rate and more variability over heliocentric distance. We discuss our methods and results as well as the implications below.

1. Introduction

Partially ionized interstellar gas and dust constitute the bulk of the galactic medium through which the sun travels and with which the solar wind interacts, resulting in a complex structure called the heliosphere. The general structure of this interaction is shown in Figure 1. As the plasma and magnetic field of the local interstellar medium (LISM) comes to a pressure balance with that of the solar wind at the heliopause, the neutral hydrogen (H) atoms of the LISM are subject to charge-exchange collisions with plasma and elastic collisions with other H atoms, but do not interact directly with the magnetic field, and thus are able to penetrate into the heliosphere to within several AU of the sun. From the time H atoms come into the picture from the LISM, the charge-exchange collisions with protons create new H atoms with properties of the local plasma distribution. This results in a different population of H atoms for each region, which provides the basis of their separation into components as well as their naming, also shown in Figure 1. Component 0 refers to the neutral H atoms that come directly from the LISM, while neutral components 1, 2, and 3 are created in the outer heliosheath, inner heliosheath, and supersonic solar wind, respectively. We refer the reader to several reviews for a detailed description of the heliosphere and various physical processes [1, 2, 3, 4].

Global models of the heliosphere have been in development for decades. While the interaction of the plasma and magnetic field of both the solar wind and the interstellar medium is responsible for the formation of the heliospheric boundaries, neutral H atoms have been shown to have a significant effect on its overall structure [5, 6]. Therefore, global models must be able to incorporate all interactions relevant to neutral H in order to accurately describe their inclusion. A few of the first models that were able to include neutral H self-consistently were the Monte Carlo code of Ref. [7] and the series of multi-fluid codes by Refs. [8, 9, 10], where coupling to the plasma populations was achieved through source and loss terms.

Since the inception of the self-consistent models, the presence of neutral H has been and

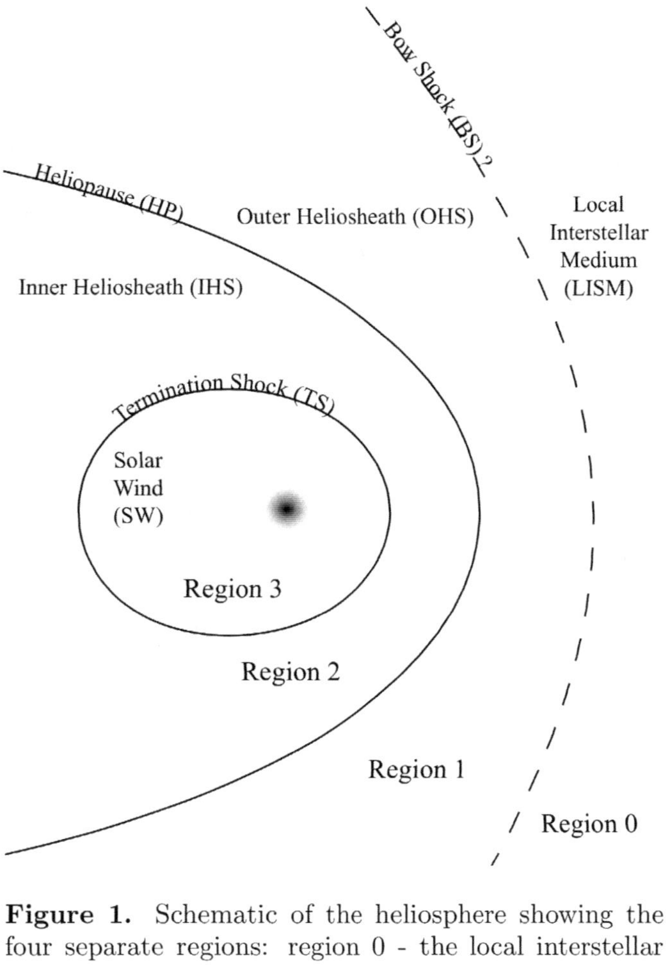

Figure 1. Schematic of the heliosphere showing the four separate regions: region 0 - the local interstellar medium (LISM), region 1 - the outer heliosheath (OHS), region 2 - the inner heliosheath (IHS), and region 3 - the supersonic solar wind (SW).

continues to be studied extensively from both kinetic and multi-fluid perspectives. Some studies with multi-fluid models include, but are certainly not limited to, galactic cosmic ray transport [11], Rayleigh-Taylor and Kelvin-Helmholtz instabilities [12, 13, 14], the coupling of solar wind and interstellar magnetic fields [15, 16, 17], and solar rotation and activity cycle [18, 19]. While the incorporation of time-dependent effects are natural for fluid models, the assumption of being able to treat each population as having a Maxwellian distribution can lead to large deficiencies in the characterization of kinetic energy of neutral H, especially within regions 2 and 3. Differences between the resulting velocity distributions of each region can be seen in Ref. [20], which compares the various structural effects on the heliosphere between kinetic and multi-fluid treatments of 4 neutral H components using the same interstellar parameters.

Further work with kinetic models used κ distributions to describe the velocity distributions of heliosheath protons with the intention of simulating the production and detection of energetic neutral atoms (ENAs) [2, 21, 22, 23] that would be detected by NASA's Interstellar Boundary Explorer [24]. Shortly after launch, the detection of an unexpected and interesting "ribbon" feature within the data required the inclusion of secondary ENA production just outside the heliopause [25], which result from a more complex description of pickup ions in the heliosheath. Due to the influence of the interstellar magnetic field on these secondary ENAs, this led to further investigation of the interstellar parameters with a focus on the strength and orientation of the magnetic field [26, 27] as well as further investigations of pickup ions throughout the heliosphere [2, 28, 29] and a more recent study of solar cycle effects [30].

While there has been substantial progress in the development of global models for both multi-fluid and kinetic treatments of neutral H, there is yet another assumption that requires more attention. Within nearly all works mentioned above, radiation pressure and solar gravitation have been combined into a single term in the treatment of influential forces on neutral H atoms as a ratio, μ. Although variation of this parameter has been used to incorporate the time-varying solar flux, radiation pressure has been assumed to fall off with distance as r^{-2}. We show here through the simulation of Lyman-alpha radiation in the heliosphere using a 3D Monte Carlo radiative transfer code [31] that this assumption is no longer valid. We briefly outline our methods below and discuss the results of radiation pressure as well as their implications for modeling of the heliosphere.

2. Radiative Transfer Code

Lyman-alpha radiation is constantly being absorbed and emitted by every neutral H atom throughout the heliosphere. The probability of absorption depends strongly on the properties of the neutral H distribution. Therefore, observations of the emitted photons can reveal information about the surrounding H distribution and subsequently imply several very important properties about the local plasma distribution. This ultraviolet line has been and continues to be observed by several space-based missions. Our particular interests are the deep-space missions that have been able to measure the trend of Lyman-alpha intensity over an increasing heliocentric distance: *Voyager* [32] and *Pioneer* [33].

In order to study Lyman-alpha radiation in the heliosphere, we developed a 3D Monte Carlo radiative transfer code [31] to simulate the Lyman-alpha backscatter as would be seen by the ultraviolet detectors on board these deep-space missions throughout the heliosphere. This code utilizes a static neutral hydrogen distribution resulting from a 3D MHD-kinetic neutral model of the heliosphere [20, 21]. Within this static code, the LISM is set with the following properties: neutral H density of 0.15 atoms cm^{-3}, 0.05 protons cm^{-3}, 6527 K, and the incoming H flow directed along the x-axis at 26.4 km s^{-1}. The solar wind is set in solar minimum conditions according to the following parameters at 1 AU: a proton density of 7.0 cm^{-3}, velocity of 400 km s^{-1}, and temperature of 100,000 K below a solar equatorial latitude of 35° and 2.6 cm^{-3}, 800 km s^{-1}, and 260,000 K, respectively, above. While the LISM parameters thus far are generally

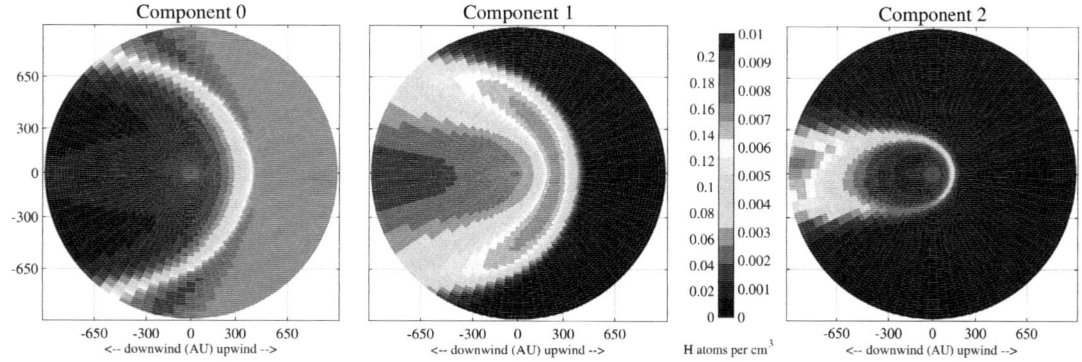

Figure 2. Three components of density for regions 0, 1, and 2 of the neutral hydrogen distribution in the heliosphere. This background data is presented in the Ecliptic plane.

axisymmetric along the x-axis, there is an asymmetry in the magnetic field as its origin is shifted 60° above the Ecliptic plane from the upwind direction. The distribution of density for the first three components of H can be seen in Figure 2. The model also yields the temperature and a velocity vector for each neutral H component.

Our radiative transfer code simulates the absorption and emission of millions of photons within this static neutral hydrogen distribution (without feedback), treating absorption and emission as separate processes with a combination of frame rotations and conservation of momentum. From each interaction we are able to determine the total momentum transferred to the H atoms in the radial direction. The code follows one photon at a time, which is generated only from the sun. Each photon is randomly assigned a direction and a frequency within a range that corresponds to H Doppler shifts within $\pm$ 120 km s^{-1}. This range covers the H velocity distributions for components 0, 1, and 2. The frequencies are also chosen to follow the self-reversed profile developed by Refs. [34, 35]. The same profile is used uniformly over the entire surface so no polar variation is considered here. Each photon is then subject to absorption and subsequent emission until eventual exit from the computational grid.

The absorption is considered as the probability that a single photon will be absorbed by one of the three components of H-atoms within a particular grid cell, each of which are described by a Maxwellian distribution, resulting in the absorption cross section [36]

$$\sigma = \frac{\sqrt{\pi} q_e^2 f_{12}}{m_e c \Delta v_D} exp \frac{(\nu' - \nu_{L_\alpha})^2}{\Delta v_D^2} , \tag{1}$$

where f_{12} is the oscillator strength of the transition (0.4162 for the Lyman-alpha line), q_e is the charge of an electron, $\nu_{L\alpha}$ is the Lyman-alpha frequency, $\Delta v_D = (\nu_{L_\alpha}/c)\sqrt{2k_BT/m_H}$ is the thermal Doppler width, and $\nu' = \nu[1 - (\vec{v}_H \cdot \vec{v}_c)/(c^2)]$ is the Doppler-shifted frequency of the photon in the frame of a H atom having the local bulk velocity of the component in question (each component is calculated separately). The dot product in the calculation of the relative frequency is complicated by the fact that the two vectors are not defined in the same reference frame. The hydrogen vector is defined in Cartesian coordinates in the solar frame, and the velocity of the photon is in its own frame that depends on its position within the solar frame. This issue is solved with frame rotations using quaternions [37] and is described in detail in Ref. [31]. The temperature here is treated homogeneously and is derived from the average kinetic energy associated with random velocities of the hydrogen atoms.

The probability of absorption is then calculated as the complement of the probability of no absorption, which is ultimately described as

$$p_s = 1 - exp[-s(\rho_0\sigma_0 + \rho_1\sigma_1 + \rho_2\sigma_2)] \ , \tag{2}$$

where s is the distance traveled by the photon since its previous interaction or grid-cell boundary crossing and ρ is the density of a particular component. This probability is then compared to a uniformly-generated random variable, t_r between 0 and 1. If $t_r < p_s$, the photon is absorbed. Otherwise, it continues on its path to the boundary of the next grid cell.

Upon absorption, an emission angle is randomly generated from an approximate inverse function [38] of a mixed-dipole phase function [39] in the frame of the H atom. While within the H atom frame, an emission frequency is calculated based on the possibility of either radiative decay or immediate emission, for an absorbed photon with energy greater or less than that of Lyman alpha, respectively. With this information, we are able to calculate the radial contribution of momentum that has been transferred to the hydrogen atom upon absorption as well as the momentum lost to the photon upon emission.

To complete the interaction, we must consider the resulting change in momentum of the H atom before calculating the frequency of the photon in the solar frame. The change in momentum of the H atom is calculated in the photon velocity frame and added to the original velocity, which is then multiplied by the new emission direction as a dot product to find the relative velocity of the hydrogen atom in the new emission frame, thereby allowing for the calculation of the frequency of the emitted photon in the solar frame. The emitted photon is then subject to further absorption and subsequent emission (without limitation) until it exits the computational grid. This process of photon generation, absorption, and emission is repeated for millions of photons while the statistics of each interaction are stored in a 5-dimensional (5D) array, including frequency, direction, and three dimensions of space for the entire computational grid.

3. Direct Results

In order to convince the reader of the reliability of our radiation pressure results, we first compare our simulations of photon backscatter to spacecraft observations. In Figures 3 and 4, we present the direct results from the current work, which has been updated since our first results [40], and compare to *Voyager 1* and *Pioneer 10* data that has been previously reduced by Doyle T. Hall in his thesis in 1992 [41]. In these Figures, we show three values calculated directly from our statistical collection (emission, collection, and total time spent) and one post-processed result.

The statistical values that we present here are rather straightforward. Upon each absorption, the calculation described above is performed and results in an emitted photon. This emitted photon is logged in the appropriate 5D bin, and a new calculation for the probability of absorption must take place. If this photon is then determined to be absorbed within the distance to the next grid cell, the photon is logged for absorption. Otherwise, the photon makes it to the next grid cell, where the calculation is performed again. Whether the photon is absorbed or not, the distance between the current and previous location is added to the running total in the appropriate 5D bin. While this value is logged as distance, the constant value of the speed of a photon allows a direction relationship to the time required to cross that distance. This total value is treated as the probability of a photon being detected. All statistics are then divided by volume to give their value per unit volume. While it is impossible to generate a realistic number of photons within the program, it is the relative value distributed throughout the grid space that we focus on here. As such, each of the statistical values, as well as the spacecraft data, are normalized to 1.0 at approximately 15 AU.

At each point that a photon statistic is logged, its directional velocity is used to determine the appropriate angular bin. In this particular run, there were three possible angular bins, defined by the local polar angle with respect to the radial position vector. An angle of zero is defined to be parallel to the radial vector (anti-solar), while 180° is antiparallel (solar). There are two cones that define the boundaries of these three angular bins such that each is of equal solid angle. There is one anti-solar cone, one cone pointed in the solar direction, and one region for all lateral angles in between. The values shown in Figures 3 and 4 are the results for the angular bin that is directed towards the sun, resulting in backscatter emission, anti-solar collection, total time spent traveling towards the sun, and the post-processed calculation, where the backscatter emission is integrated along the anti-solar line of sight. These values generally follow a similar slope, with the backscatter emission having more variability because it is more dependent on the local parameters of the neutral H distribution.

In terms of our results for radiation pressure, it is more important that the *Voyager 1* results in Figure 4 are acknowledged to match extremely well. Since this data set only goes out to ~40 AU, we can only say with certainty that the results match up to 40 AU, but this is sufficient to establish our point. Our results for radiation pressure can be seen for the upwind and downwind directions in Figures 5 and 6. We plot several values of a constant exponential power for each plot to show how the trends compare to that of solar gravitation. We show linear fits to a single group of points for *Voyager 1* and several groups of points for *Pioneer 10* along with the standard error of the resulting values for each slope. We have not considered uncertainty in either the neutral H data nor the resulting points themselves. The uncertainties in the resulting values are plotted as gray triangles in Figures 5 and 6, which we consider to be very small relative to the line fit and uncertainty calculations and would have no change in our conclusions.

For the upwind direction, the trend of radiation pressure over heliocentric distance between 4.55 and 73.6 AU follows a power of -1.8068±0.0054, which then slowly changes to a steeper slope up to the heliopause, where it transitions rather quickly to a slope that is steeper than -3.00 throughout the hydrogen wall and the remainder of the computational grid at ~1000 AU. The trends in the downwind direction are -1.2126±0.0185 from 2.88 to 18.5 AU and -1.5256±0.0118 from 14.9 to 55.2 AU after which the slope transitions to a power of -2.2287±0.0186 from 88.5 up to nearly 200 AU, where it drops even further over a few tens of AU to -2.6444±0.0052 out to the edge of the computational grid.

4. Discussion

We have presented the results of our 3D Monte Carlo radiative transfer simulation of Lyman-alpha in the heliosphere using a static background of a neutral hydrogen distribution described

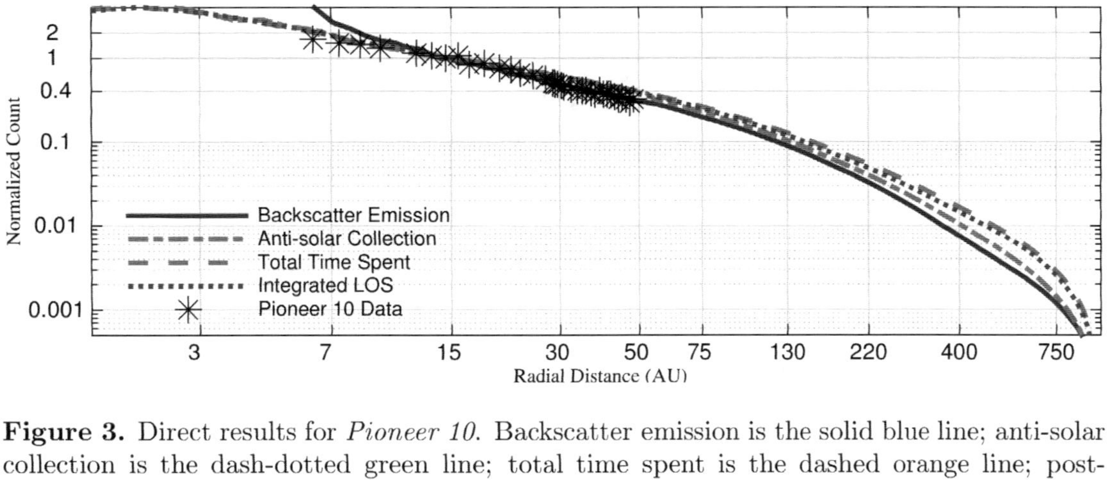

Figure 3. Direct results for *Pioneer 10*. Backscatter emission is the solid blue line; anti-solar collection is the dash-dotted green line; total time spent is the dashed orange line; post-processed integrated line-of-sight is the dotted purple line. The *Pioneer 10* data is shown as asterisks.

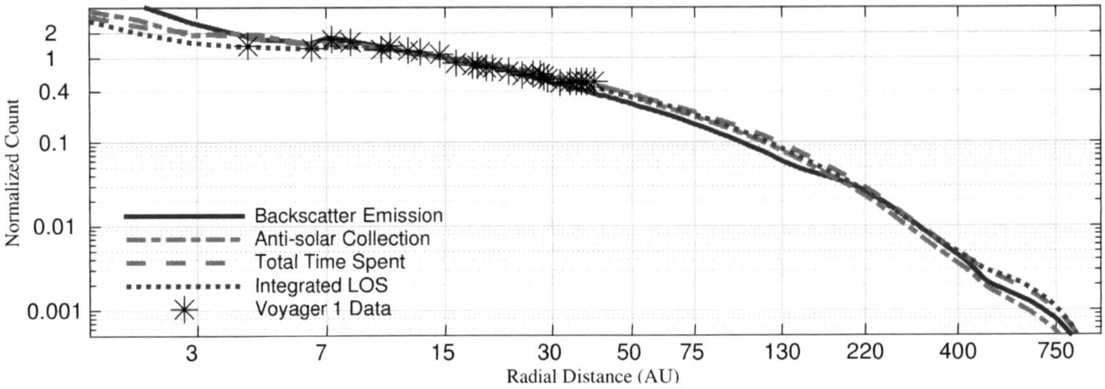

Figure 4. Direct results for *Voyager 2*. Plotted with the same representation as Figure 3.

by three separate components. Our method treats the absorption and subsequent emission of Lyman-alpha photons by H atoms as separate processes. We have shown by comparison that our results are able to reproduce spacecraft data rather well, providing support for the reliability of our radiation pressure results.

We have come to two main conclusions with these results. One conclusion is that the radiation pressure falls off, not as r^{-2}, but at a slower rate within most of region 3 with a variety of slopes elsewhere. This rate is not very much slower in the upwind direction (-1.8068), but is significantly slower in the downwind direction (-1.2126), which brings us to our second conclusion. The radiation pressure is certainly not isotropic. We can see that the results are rather consistent within 75 AU in the upwind direction, but surprisingly much more variable in the downwind direction.

There may be several aspects of the heliosphere that influence the radiation pressure in the manner that we see. We can certainly say that mutliple scattering is incredibly important and provides a much longer lasting source of Lyman alpha radiation in the outer heliosphere than that resuling from an optically thin solution. We also think that the increasing density of neutral H in the upwind direction aids in the decrease in radiation pressure due to an increasing probability of absorption and emission, but the parabolic shape of this increase that culminates

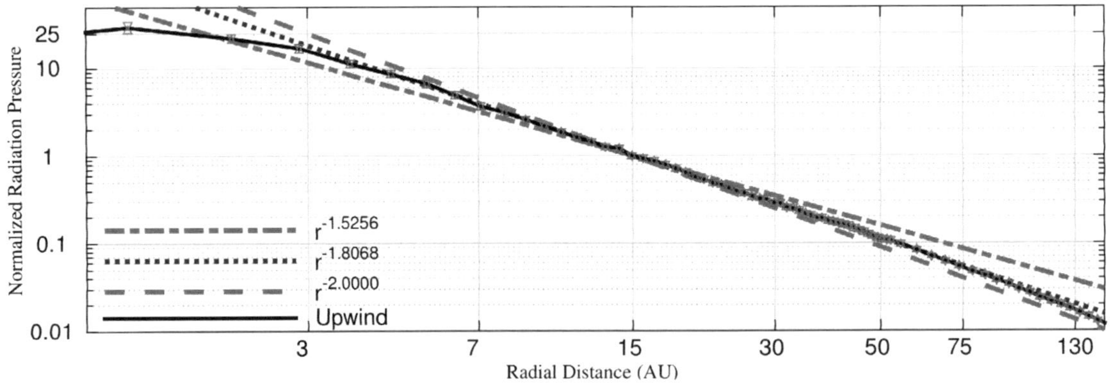

Figure 5. Radiation pressure in the downwind direction. Results show the line fit between 4.55 and 73.6 AU as well as two reference slopes, where -2.00 corresponds to the slope for solar gravitation and -1.5256 is one of the fits for the downwind direction.

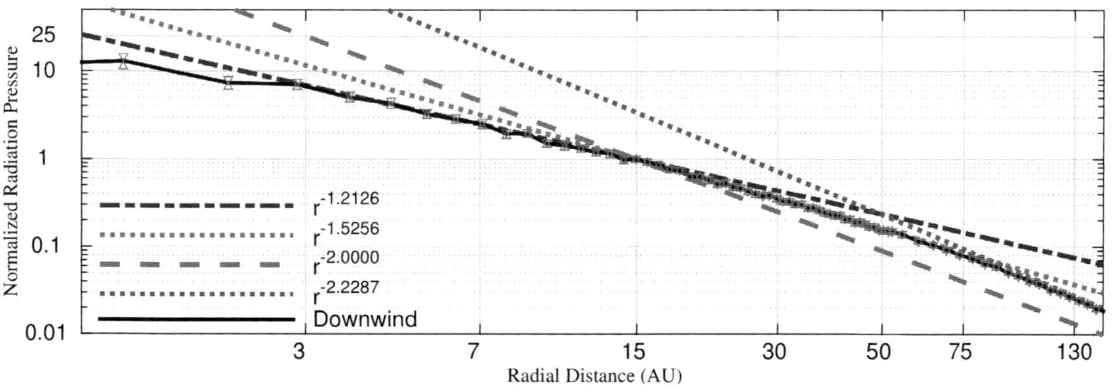

Figure 6. Radiation pressure in the upwind direction. Results show three line fits: one from 2.88 to 18.5 AU, one from 14.9 to 55.2 AU for a little overlap, and another from 88.5 to nearly 200 AU. The reference slope of -2.00 is also shown.

at the heliopause may act as a "focusing beam" that redirects the Lyman alpha toward the downwind direction, providing a much more sustainable source of radiation pressure in the inner heliosphere.

Since the initial heliospheric models describing neutral hydrogen in the 1970's, radiation pressure has been considered to fall off as r^{-2} and combined with solar gravitation as a single parameter, generally in the literature as μ or the ratio of radiation pressure to gravitational force from the sun. Variations of this parameter have been the source of much discussion for heliospheric models, even to this day, along the lines of its variation with solar cycle activity. No matter what value is used for μ, the effect always produces a relation that falls off as r^{-2}, which is not a realistic assumption. We consider the results presented here to be significant, supporting the idea that the heliosphere is not optically thin to Lyman alpha.

There have been a number of radiative transfer studies conducted in the past [38, 41, 42, 43, 44] as well as continuations of those works. However, one or more issues have prevented each from resulting in a match to multiple spacecraft data. One idea that has generally been incorporated is the use of a single function [45] to describe the redistribution of a photon's frequency from absorption to emission by a H atom in the laboratory frame. It is our understanding that this equation was developed for an elegant analytical result, but is not quite necessary for computational calculations. We also treat photons kinetically as opposed to the other option of treating the radiation as a fluid through iterative solutions. In that case, multiple scattering is difficult to handle, especially when the scattering phase function is not isotropic. Another difference between our results and others mention above is the background neutral H distribution. While we have not tested our code on distributions from other groups, our global heliospheric model has been quite successful in comparisons to several types of spacecraft observations. One or more of these differences may be responsible for the agreement of our results with ultraviolet data where other works have been unsuccessful.

5. Conclusions

Heliospheric models have assumed that radiation pressure exerted on neutral H falls off with heliocentric distance at the same rate as that of solar gravitation since their first conception. Through our Monte Carlo simulation of Lyman-alpha radiative transfer in the heliosphere, we are able to statistically calculate the radially-directed momentum that is transferred to neutral H during absorption and emission. We have shown the results for the upwind and downwind directions.

While our calculations have been made within a static heliospheric model under solar minimum conditions, the simple result that the radiation pressure is not isotropic provides enough of an argument to include a more detailed radiation pressure model in the treatment of neutral hydrogen. However, we consider our comparison to spacecraft data to provide the support necessary to conclude that radiation pressure generally falls off slower than r^{-2} except wtihin the heliopause, where it can fall off more rapidly, and cannot be combined into a single parameter with solar gravitation.

Acknowledgments

This work was partially supported by NASA grants NNX08AJ33G, Subaward 37102-2, NNX09AG70G, NNX09AG63G, NNX09AJ79G, NNG05EC85C, NNX09AP74A, NNX10AE46G, NNX09AW45G, NNH09AM47I, and Subcontract A991132BT, and NSF grant ATM-0904007.

References

[1] Zank G P 1999 Interaction of the solar wind with the local interstellar medium: a theoretical perspective *Space Sci. Rev.* **89** 413
[2] Zank G P, Heerikhuisen J, Pogorelov N V, Burrows R and McComas D 2010 Microstructure of the heliospheric termination shock: implications for energetic neutral atom observations *Astrophys. J.* **708** 1092
[3] Zank G P, Heerikhuisen J, Wood B E, Pogorelov N V, Zirnstein E and McComas D 2013 Heliospheric structure: the bow wave and the hydrogen wall *Astrophys. J.* **763** 20
[4] Izmodenov A A, Malama Y G, Ruderman M S, Chalov S V, Alexashov D B, Katushkina O A, and Provornikova E A 2009 Kinetic-Gasdynamic Modeling of the Heliospheric Interface *Space Sci. Rev.* **146** 329
[5] Zank G P, Müller H R and Wood B E 2001 The interaction of the solar wind and stellar winds with the partially ionized interstellar medium *Phys. Plasmas* **8** 2385
[6] Zank G P and Müller H R 2003 The dynamical heliosphere *J. Geophys. Res. (Space Phys.)* **108** 1240
[7] Baranov V B and Malama Y G 1993 Model of the solar wind interaction with the local interstellar medium: numerical solution of self-consistent problem *J. Goephys. Res.* **98** 15157
[8] Pauls H L, Zank G P and Williams L L 1995 Interaction of the solar wind with the local interstellar medium *J. Geophys. Res.* **100** 21595
[9] Zank G P, Pauls H L, Williams L L and Hall D T 1996 Interaction of the solar wind with the local interstellar medium: a multi fluid approach *J. Geophys. Res.* **101** 21639
[10] Pauls H L and Zank G P 1997 Interaction of a nonuniform solar wind with the local interstellar medium: 2. a two-fluid model *J. Geophys. Res.* **102** 19779
[11] Florinski V, Zank G P and Pogorelov N V 2003 Galactic cosmic ray transport in the global heliosphere *J. Geophys. Res. (Space Phys.)* **108** 1228
[12] Florinski V, Zank G P and Pogorelov N V 2005 Heliopause stability in the presence of neutral atoms: Rayleigh-Taylor dispersion analysis and axisymmetric MHD simulations *J. Geophys. Res. (Space Phys.)* **110** 7104
[13] Zank G P 1999 *AIP Conf. Ser.* vol 471, ed S R Habbal *et al* (Melville, NY: AIP), p 783
[14] Avinash K, Zank G P, Dasgupta B and Bhadoria S 2014 Instability of the heliopause driven by charge exchange interactions *Astrophys. J.* **791** 102
[15] Pogorelov N V and Zank G P 2005 Coupling of the interstellar and interplanetary magnetic fields at the heliospheric interface: the effect of neutral hydrogen atoms *Adv. Space Res.* **35** 2055
[16] Pogorelov N V, Zank G P and Ogino T 2006 Three-dimensional features of the outer heliosphere due to coupling between the interstellar and interplanetary magnetic fields. II. the presence of neutral hydrogen atoms *Astrophys. J.* **644** 1299
[17] Pogorelov N V, Heerikhuisen J, Zank G P and Borovikov S N 2009 Influence of the interstellar magnetic field and neutrals on the shape of the outer heliosphere *Space Sci. Rev.* **143** 31
[18] Pogorelov N V, Borovikov S N, Zank G P and Ogino T 2009 Three-dimensional features of the outer heliosphere due to coupling between the interstellar and interplanetary magnetic fields. III. the effects of solar rotation and activity cycle *Astrophys. J.* **696** 1478
[19] Washimi H, Zank G P, Hu Q, Tanaka T, Munakata K and Shinagawa H 2011 Realistic and time-varying outer heliospheric modeling *Mon. Not. Roy. Astron. Soc.* **416** 1475

[20] Heerikhuisen J, Florinski V and Zank G P 2006 Interaction between the solar wind and interstellar gas: a comparison between Monte Carlo and fluid approaches *J. Geophys. Res. (Space Phys.)* **111** 6110
[21] Heerikhuisen J, Pogorelov N V, Florinski V, Zank G P and le Roux J A 2008 The effects of a κ-distribution in the heliosheath on the global heliosphere and ENA flux at 1 AU *Astrophys. J.* **682** 679
[22] Heerikhuisen J and Pogorelov N V 2009 Detecting neutral atoms from beyond the heliopause with Interstellar Boundary Explorer *Astrophys. J.* **695** L58
[23] Zirnstein E J, Heerikhuisen J and Pogorelov N V 2012 *AIP Conf. Ser., Space Weather: The Space Radiation Environment* vol 1500, ed Q Hu *et al* (Melville, NY: AIP), p 262
[24] McComas D J *et al* 2009 IBEX-Interstellar Boundary Explorer *Space Sci. Rev.* **146** 11
[25] Heerikhuisen J, Pogorelov N V, Zank G P, Crew G B, Frisch P C, Funsten H O, Janzen P H, McComas D J, Reisenfeld D B and Schwadron N A 2010 Pick-up ions in the outer heliosheath: a possible mechanism for the Interstellar Boundary Explorer ribbon *Astrophys. J.* **708** L126
[26] Heerikhuisen J and Pogorelov N V 2011 An estimate of the nearby interstellar magnetic field using neutral atoms *Astrophys. J.* **738** 29
[27] Heerikhuisen J, Zirnstein E J, Funsten H O, Pogorelov N V and Zank G P 2014 The effect of new interstellar medium parameters on the heliosphere and energetic neutral atoms from the interstellar boundary *Astrophys. J.* **784** 73
[28] Zirnstein E J, Heerikhuisen J, Zank G P, Pogorelov N V, McComas D J and Desai M I 2014 Charge-exchange coupling between pickup ions across the heliopause and its effect on energetic neutral hydrogen flux *Astrophys. J.* **783** 129
[29] Zank G P, Hunana P, Mostafavi P and Goldstein M L 2014 Pickup ion mediated plasmas. I. basic model and linear waves in the solar wind and local interstellar medium *Astrophys. J.* **797** 87
[30] Zirnstein E J, Heerikhuisen J, Pogorelov N V, McComas D J and Dayeh M A 2015 Simulations of a dynamic solar cycle and its effects on the Interstellar Boundary Explorer ribbon and globally distributed energetic neutral atom flux *Astrophys. J.* **804** 5
[31] Fayock B 2013 *Analysis of solar lyman-alpha scattering in the heliosphere* Ph.D. thesis University of Alabama in Huntsville (*open access* dissexpress.umi.com)
[32] Broadfoot A L, Sandel B R, Shemansky D E, Atreya S K, Donahue T M, Moos H W, Bertaux J L, Blamont J E, Ajello J M and Strobel D F 1977 Ultraviolet spectrometer experiment for the Voyager mission *Space Sci. Rev.* **21** 183
[33] Carlson R W and Judge D L 1974 Pioneer 10 ultraviolet photometer observations at Jupiter encounter *J. Geophys. Res.* **79** 3623
[34] Tarnopolski S 2007 *Expected distribution of interstellar deuterium in the heliosphere* (in Polish) Ph.D. thesis Space Research Centre PAS
[35] Tarnopolski S and Bzowski M 2009 Neutral interstellar hydrogen in the inner heliosphere under the influence of wavelength-dependent solar radiation pressure *Astron. Astrophys.* **493** 207
[36] Rybicki G B and Lightman A P 1979 *Radiative Processes in Astrophysics* (New York: Wiley-Interscience)
[37] Kuipers J B 1999 *Quaternions and Rotation Sequences* (Princeton, NJ: Princeton university press)
[38] Keller H U, Richter K and Thomas G E 1981 Multiple scattering of solar resonance radiation in the nearby interstellar medium. II. *Astron. Astrophys.* **102** 415
[39] Brandt J C and Chamberlain J W 1959 Interplanetary gas. I. hydrogen radiation in the night sky *Astrophys. J.* **130** 670
[40] Fayock B, Zank G P and Heerikhuisen J 2013 Comparison of Pioneer 10, Voyager 1, and Voyager 2 ultraviolet observations with anti-solar lyman-alpha backscatter simulations *Astrophys. J.* **775** L4
[41] Hall D T 1992 *Ultraviolet resonance radiation and the structure of the heliosphere* Ph.D. thesis Arizona University in Tucson
[42] Quémerais E and Bertaux J-L 1993 Radiative transfer in the interplanetary medium at Lyman alpha *Astron. Astrophys.* **277** 283
[43] Scherer H and Fahr H J 1996 H Lyman alpha transport in the heliosphere based on an expansion into scattering hierarchies *Astron. Astrophys.* **309** 957
[44] Gangopadhyay P, Ogawa H S, and Judge D L 1989 Evidence of a nearby solar wind shock as obtained from distant Pioneer 10 ultraviolet glow data *Astrophys. J.* **336** 1012
[45] Hummer D G 1962 Non-coherent scattering: I. The redistribution function with Doppler broadening *Mon. Not. Roy. Astron. Soc.* **125** 21

Journal of Physics: Conference Series **642** (2015) 012008 doi:10.1088/1742-6596/642/1/012008

Transient shocks beyond the heliopause

R. L. Fermo

Department of Space Science and Center for Space Plasma and Aeronomic Research, University of Alabama in Huntsville, Huntsville, AL 35899, USA

N. V. Pogorelov

Department of Space Science and Center for Space Plasma and Aeronomic Research, University of Alabama in Huntsville, Huntsville, AL 35899, USA

L. F. Burlaga

Goddard Space Flight Center, National Aeronautics and Space Agency, Greenbelt, MD 20771, USA

Abstract.

The heliopause is a rich, dynamic surface affected by the time-dependent solar wind. Stream interactions due to coronal mass ejections (CMEs), corotating interaction regions (CIRs), and other transient phenomena are known to merge producing global merged interaction regions (GMIRs). Numerical simulations of the solar wind interaction with the local interstellar medium (LISM) show that GMIRs, as well other time-dependent structures in the solar wind, may produce compression/rarefaction waves and shocks in the LISM behind the heliopause. These shocks may initiate wave activity observed by the Voyager spacecraft. The magnetometer onboard Voyager 1 indeed observed a few structures that may be interpreted as shocks. We present numerical simulations of such shocks in the year of 2000, when both Voyager spacecraft were in the supersonic solar wind region, and in 2012, when Voyager 1 observed traveling shocks. In the former case, Voyager observations themselves provide time-dependent boundary conditions in the solar wind. In the latter case, we use OMNI data at 1 AU to analyze the plasma and magnetic field behavior after Voyager 1 crossed the heliospheric boundary. Numerical results are compared with spacecraft observations.

The Voyager spacecraft were launched in 1977 and are just now reaching the outer edges of the heliosphere and beyond. The termination shock, where the solar wind becomes subsonic, was passed by Voyager 1 in 2004 at 94 AU and by Voyager 2 in 2007 at 84 AU from the sun [1, 2, 3, 4, 5]. Between 28 July and 25 August 2012, Voyager 1 perceived a two-step increase in the galactic cosmic ray (GCR) flux, as well as intermittent drops and ultimately an exponential decay in the anomalous cosmic ray (ACR) flux [6, 7, 8]. Simultaneous jumps in the magnetic field strength were not accompanied by a magnetic field rotation, contrary to expectations [6], but the continuous presence of GCRs from that point to the present day suggests that the 2012 crossing likely was the heliopause. Regardless, it is clear now that the heliopause is not a static boundary. A pair of shocks were observed well into the local interstellar medium (LISM) [9, 10]. Solar transients may play a role in the dynamics of the heliopause.

Two examples of such solar transients are coronal mass ejections (CMEs) and corotating interaction regions (CIRs). A coronal mass ejection is a large-scale release of plasma and magnetic field from the solar atmosphere [11, 12, 13]. Usually modeled as a helical flux rope, these structures first form in the corona but are then ejected out into interplanetary space, where they are subsequently called interplanetary coronal mass ejections (ICMEs) [14, 15, 16, 17]. In contrast, corotating interaction regions form out in the solar wind, as CIRs are created when fast solar wind pushes against regions of slow solar wind. Both CMEs and CIRs can generate shocks and act as sources of particle acceleration [18, 19, 20].

As ICMEs and CIRs propagate outwards from the sun, multiple such solar wind transients may collide with one another and form global merged interaction regions (GMIRs) [21, 22, 23, 24, 25, 26]. These are large, global-scale structures often characterized by enhanced magnetic fields and density fluctuations [22, 27, 28, 29]. However, their ultimate destiny as they approach the outer heliosphere is not yet well understood.

Numerical simulations of the propagation of time-dependent perturbations of the solar wind were first performed using a gas dynamics treatment (without magnetic field, for example) [30, 31, 32, 33, 34, 35], although later considerations included the effects of magnetohydrodynamics (MHD) [36, 37, 38, 39, 40, 41, 42, 43, 44]. The effect of GMIR-like perturbations on the magnetic field distributions in the heliosphere and beyond it in the LISM were considered in Pogorelov and Zank (2005) [45] and Pogorelov et al. (2012) [46]. One-dimensional MHD models suggest that these GMIRs and their associated shocks will have observable effects even at distances as far as Voyagers 1 and 2 [47].

In this paper, we will expand upon such one-dimensional models by simulating solar wind transients in fully three-dimensional MHD models. This will have the advantage of including heliospheric structures not present in one-dimensional models, such as the termination shock and heliopause. In Pogorelov and Zank (2005) [45], it was demonstrated in a three-dimensional MHD-neutral model that GMIRs are able to reach the heliosheath. Fig. 1, taken from Pogorelov and Zank (2005) [45], shows their propagation beyond the termination shock. However, the simulations were not executed for sufficiently long periods of time to describe their propagation through the heliopause boundary into the local interstellar medium. In this paper, we present simulations in which GMIRs are in fact able to penetrate into the LISM.

For our simulations, we use the fully three-dimensional code MS-FLUKSS [48]. The partially ionized plasma is modeled as four separate fluids: the hydrogen plasma, neutral interstellar hydrogen, neutral hydrogen from the heliosheath, and neutral hydrogen from the inner heliosphere (created from within the termination shock). The governing equations for the plasma fluid are the standard ideal magnetohydrodynamics (MHD) equations

$$\frac{\partial \rho}{\partial t} + \nabla \cdot \rho \mathbf{v} = 0 \qquad (1)$$

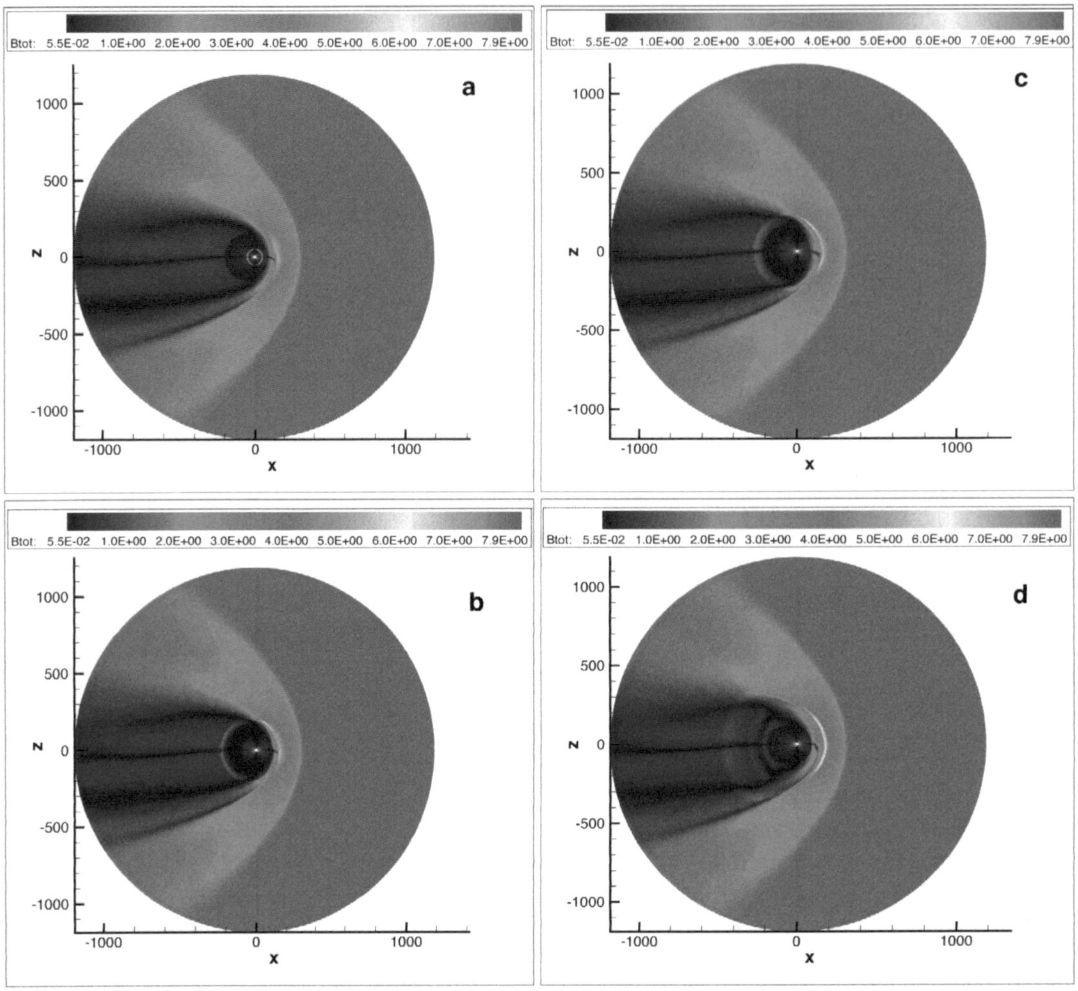

Figure 1. Variations in the magnetic field strength (µG) in the meridional plane (axes in AU) as a GMIR propagates towards the heliopause. Courtesy of Pogorelov and Zank (2005), Fig. 2 [45].

$$\frac{\partial \rho \mathbf{v}}{\partial t} + \nabla \cdot \left(\rho \mathbf{v}\mathbf{v} - \frac{\mathbf{B}\mathbf{B}}{4\pi} \right) + \nabla p = \mathbf{H}^{m}_{p-H} \tag{2}$$

$$\frac{\partial e}{\partial t} + \nabla \cdot \left((e+p)\mathbf{v} - \frac{\mathbf{B}(\mathbf{v}\cdot\mathbf{B})}{4\pi} \right) = H^{e}_{p-H} \tag{3}$$

$$\frac{\partial \mathbf{B}}{\partial t} + \nabla \cdot (\mathbf{v}\mathbf{B} - \mathbf{B}\mathbf{v}) = 0 \tag{4}$$

which are the conservation laws for mass, momentum, energy, and magnetic flux, respectively. Likewise, the Euler gas dynamic equations

$$\frac{\partial \rho_i}{\partial t} + \nabla \cdot \rho_i \mathbf{v}_i = H^d_{H-p,i} \tag{5}$$

$$\frac{\partial \mathbf{v}_i}{\partial t} + \nabla \cdot \rho_i \mathbf{v}_i \mathbf{v}_i + \nabla p_i = \mathbf{H}^m_{H-p,i} \tag{6}$$

$$\frac{\partial e_i}{\partial t} + \nabla \cdot (e_i + p_i)\mathbf{v}_i = H^e_{H-p,i} \tag{7}$$

govern the evolution of the neutral fluids. Here, ρ is the density, $\mathbf{v}$ the flow speed, p the pressure, e the total energy density, $\mathbf{B}$ the magnetic field, and t the time. The subscripts i correspond to each individual neutral population. The source terms H_{p-H} and $H_{H-p,i}$ regulate the rate of charge exchange between protons and the various neutral hydrogen populations.

The inner boundary conditions are obtained from OMNIweb, which cross compares in situ data from several spacecraft in geocentric or L1 orbits, such as ISEE, Wind, and ACE. The resulting 1-hour averaged plasma and magnetic field data are taken as the inner boundary at 1 AU. The magnetic field strength from OMNI is used in conjunction with the solar tilt angle data from the Wilcox Solar Observatory (WSO) to fit to a Parker spiral solution for the solar wind at 1 AU. The outer boundary conditions at r = 1000 AU are taken from the parameters of the LISM consistent with McComas et al. (2015) [49]: $n = 0.08$ cm^{-3}, $n_1 = 0.175$ cm^{-3}, $v = -26.4$ km/s, $B = 3.0$ µG, $T = 8000$ K.

Before performing the time-dependent simulation, we set up a steady-state solution from the solar wind conditions in early 2004 and run the code for approximately 3000 solar rotations. This allows the system to self-consistently generate the basic heliospheric structures, most importantly the termination shock and the heliopause. This steady-state solution is used as the initial condition from early 2004.

In our time-dependent run, solar transients such as CMEs and CIRs during our simulation period will be seen as density and magnetic field enhancements at 1 AU, as reflected by the OMNI observational data. These fluctuations will propagate from 1 AU outwards towards the outer heliosphere. Fig. 2 shows the density in the meridional plane at several times. Fig. 2(a) shows the density at the beginning of the simulation. The ripples within the termination shock of Fig. 2(b) are solar transients such as CMEs or CIRs which were observed in the OMNI data and are propagated out from 1 AU. The color scale has been adjusted in Fig. 2(c) to accentuate the ripples in the heliosheath, where the GMIRs slow down and begin to pile up. The color scale has been adjusted once again in Fig. 2(d) to make more visible the fluctuations in the LISM. The attached multimedia file *density.mp4* in the online version of this publication shows more clearly the outward propagation of these structures through and beyond the termination shock and heliopause.

Two straight lines in Fig. 2 roughly show the directions of the trajectories of the two Voyager spacecraft projected onto the meridional plane. The ripples in density and magnetic field should therefore be seen in cuts along the Voyager 1 and 2 trajectories. Figs. 3(a) and (b) show the density profiles along the Voyager 1 and 2 trajectories, respectively, on 29 October 2012. Likewise, Fig. 3(c) and (d) show the respective profiles

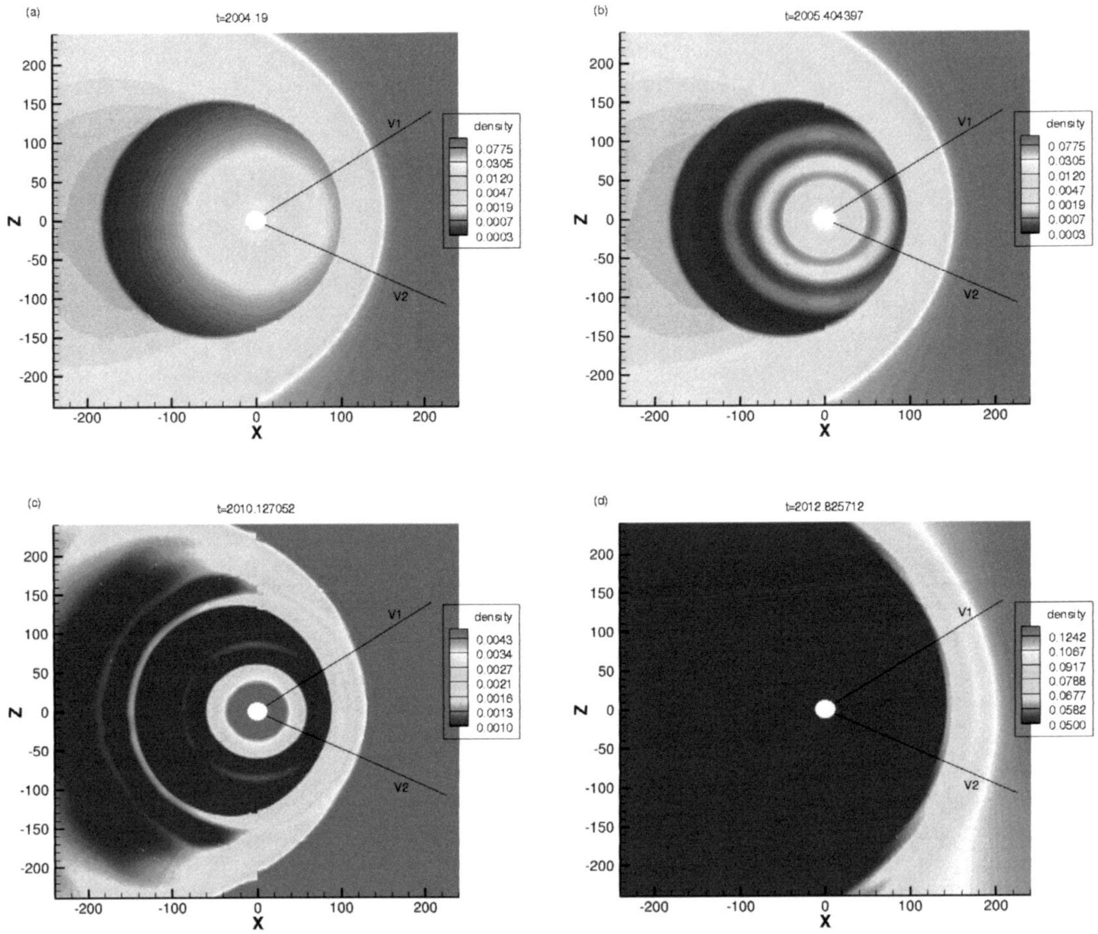

Figure 2. The plasma density (cm^{-3}) in the meridional plane (axes in AU) at various times, including (a) the steady-state solution from which the time-dependent simulation begins on 10 March 2004. Note that the other three plots show the plasma density at different color scales, in order to accentuate the GMIRs (b) within the termination shock on 28 May 2005, (c) in the heliosheath on 16 February 2010, and (d) in the LISM on 29 October 2012.

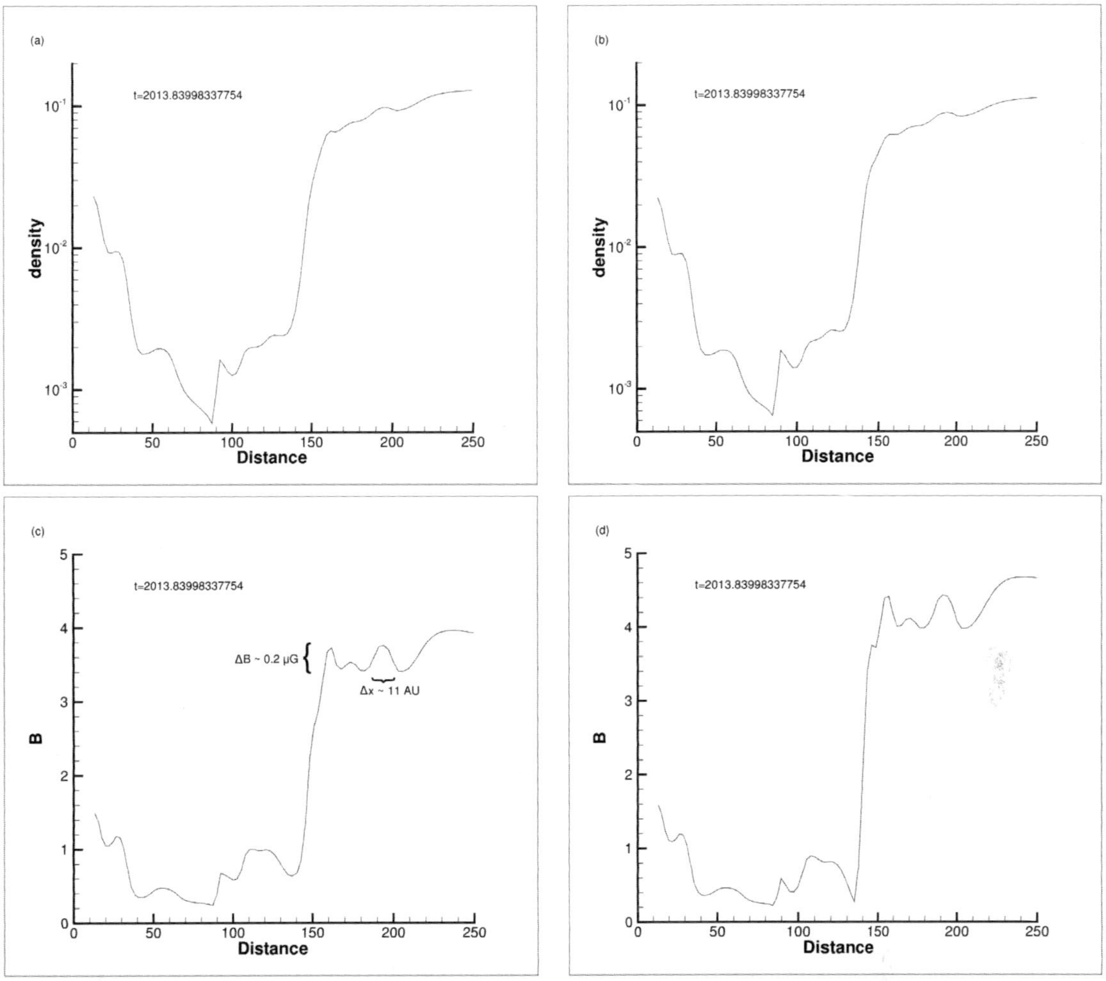

Figure 3. The plasma density (in cm^{-3}) on 29 October 2012 along the trajectories (radial distance in AU) of (a) Voyager 1 and (b) Voyager 2, and the magnetic field strength (in μG) along the trajectories of (c) Voyager 1 and (d) Voyager 2.

for the magnitude of the magnetic field along the Voyager 1 and 2 trajectories on the same date. As before, the outward propagation of these structures along the Voyager 1 trajectory is best observed in the attached multimedia files *V1-logN.mp4* and *V1-logB.mp4* in the online version of this publication. We see that enhancements in the density and magnetic field which can be associated with GMIRs are able to cross the termination shock, slow down and pile up in the heliosheath, and are even able to cross the heliopause into the local interstellar medium.

We compare these to a pair shocks observed by Voyager 1. In Fig. 4, a figure taken from Burlaga et al. (2014) [10], Voyager 1 crossed a transition region T from the solar

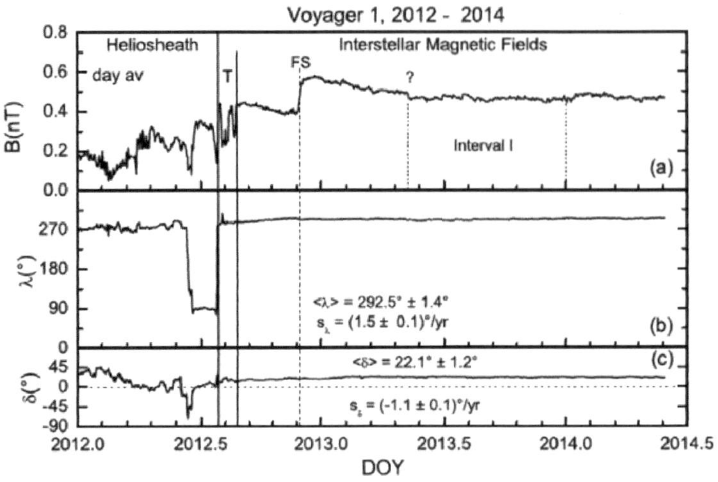

Figure 4. Daily averages from Voyager 1 of the observed (a) magnetic field strength B (nT), (b) azimuthal angle λ (°), and (c) elevation angle δ (°) between 1 January 2012 and 30 May 2014. The current sheet T is now believed to be the heliopause. Two shocks were observed following the heliopause crossing, labeled FS and $?$. Courtesy of Burlaga et al. (2014), Fig. 1 [10].

wind into the LISM. Following that, the spacecraft encountered two shocks, labelled FS and $?$. The former was associated with a jump in the magnetic field strength $\Delta B_{FS} = 0.2$ nT, the latter with a slight decrease of $\Delta B_? = -0.031$ nT, although neither was accompanied by a noticeable rotation in the magnetic field, consistent with the T. The first shock was observed on 30 November 2012, the second on 8 May 2013, for a separation $\Delta t = 160$ days.

Consider now that, based on the local parameters observed by Voyager 1, one may calculate the local Alfvén speed $v_A = 38$ km/s, the local sound speed $v_s = 17$ km/s, and consequently the magnetosonic speed $v_{ma} \approx \sqrt{v_A^2 + v_s^2} \approx 42$ km/s, which we use as a lower limit for the shock speed $v_{shock} \gtrsim v_{ma}$. Using the Voyager 1 speed $v_{V1} = 3.6$ AU/year, we can subsequently calculate a lower limit for the shock separation distance $\Delta x = (v_{shock} - v_{V1})\Delta t \gtrsim 2.1$ AU.

Table 1 shows a direct comparison between the observed shock parameters calculated here and the same observed in the GMIR fluctuations seen in the simulations. In the latter case, the shock jumps ΔB and shock separation Δx are computed as in the intervals shown in Fig. 3(a) for GMIR fluctuations after they have crossed the heliopause. The range of values provided represents an aggregate of GMIR fluctuations over the course of the simulation along the Voyager 1 and 2 trajectories. The magnetic field jumps ΔB observed across the shocks are comparable to those seen in the simulations. The observed shock separation distance Δx is slightly smaller than those seen in the simulations, although it is quite feasible that this gap can be bridged by increasing the resolution in the simulations. The shock speed in the simulations, calculated roughly as the distance traversed by the crest of GMIR fluctuations after having crossed the heliopause divided by the transit time, appears to be slower than the calculated lower limit based on the observed magnetosonic speed v_{ma}. This could perhaps be attributed to the incorrect distance in the simulation of the heliopause, which has been pushed out to $140 - 150$ AU, roughly $20 - 30$ AU farther out than where it was observed by Voyager.

Table 1. Comparison of shock parameters

Shock parameter	Observed by Voyager 1	MHD simulations
Shock jump ΔB	$0.03 - 0.2$ nT	$0.01 - 0.12$ nT
Shock separation Δx	$\gtrsim 2.1$ AU	$3 - 15$ AU
Shock speed v_{shock}	$\gtrsim 42$ km/s	$28 - 33$ km/s

We therefore present GMIRs as a feasible explanation for the shocks observed by Voyager 1 beyond the heliopause. A fully three-dimensional MHD model of the heliosphere shows that the density and magnetic field enhancements associated with GMIR signatures should have observable effects at the location of Voyager 1, fairly consistent with those seen in reality. Further work is needed, however, with higher resolution simulations and a more realistic heliopause location. This will lessen the effects of heliopause and shock broadening and likely increase the correspondence between the simulations and observed shock parameters. Subsequent simulations should better reflect termination shock and heliopause distances, north-south asymmetries, and shock physics. Furthermore, future simulations will enable timing tests to demonstrate when precisely we would expect to see certain solar transients observed at 1 AU to reach the Voyager 1 spacecraft. This was accomplished by Liu et al. (2014) [47] using a one-dimensional MHD model, but a fully three-dimensional model that incorporates the termination shock, heliosheath, and heliopause at their observed locations will increase the predictive capability. Another important question is the nature of these fluctuations in the magnetic field, whether they represent the GMIRs themselves or if the jumps in magnetic field are in fact foreshocks or reverse shocks related to the GMIRs. Further study of the plasma parameters near these magnetic field jumps at higher resolution will help in identifying these structures.

This work was supported, in part, by NASA grants NNX12AB30G, NNX14AJ41G, and NNX14AJ53G, and DOE grant DE-SC00008334. This work was also carried out as part of the IBEX mission, which is part of NASAs Explorer Program. N. Pogorelov highly appreciates discussions at the team meeting Heliosheath Processes and Structure of the Heliopause: Modeling Energetic Particles, Cosmic Rays, and Magnetic Fields supported by the International Space Science Institute in Bern, Switzerland. We also acknowledge NSF PRAC award OCI-1144120 and related computer resources from the BlueWaters sustained petascale computing project. Supercomputer time allocations were also provided on SGI Pleiades by NASA High-End Computing Program award SMD-13-4187 and the Dell Linux cluster Stampede by NSF XSEDE project MCA07S033.

[1] Burlaga L F, Ness N F, Acuña M H, Lepping R P, Connerney J E P, Stone E C and McDonald F B 2005 *Science* **309** 2027

[2] Stone E C, Cummings A C, McDonald F B, Heikkila B C, Lal N and Webber W R 2005 *Science* **309** 2017–2020
[3] Decker R B, Krimigis S M, Roelef E C, Hill M E, Armstrong T P, Gloeckler G, Hamilton D C and Lanzerotti L J 2005 *Science* **309** 2020
[4] Stone E C, Cummings A C, McDonald F B, Heikkila B C, Lal N and Webber W R 2008 *Nature* **454** 71–74
[5] Burlaga L F, Ness N F, Acuña M H, Lepping R P, Connerney J E P and Richardson J D 2008 *Nature* **454** 75
[6] Burlaga L F, Ness N F and Stone E C 2013 *Science* **341** 147
[7] Stone E C, Cummings A C, McDonald F B, Heikkila B C, Lal N and Webber W R 2013 *Science* **341** 150
[8] Gurnett D A, Kurth W S, Burlaga L F and Ness N F 2013 *Science* **341** 1489
[9] Burlaga L F, Ness N F, Gurnett D A and Kurth W S 2013 *Ap. J. Lett.* **778** L3
[10] Burlaga L F and Ness N F 2014 *Ap. J. Lett.* **795** L19
[11] Zhang G and Burlaga L F 1988 *J. Geophys. Res.* **93** 2511
[12] Gosling J T 1993 *J. Geophys. Res.* **98**(A11) 18937–18949
[13] Farrugia C J, Burlaga L F, Osherovich V A, Richardson I G, Freeman M P, Lepping R P and Lazarus A J 1993 *J. Geophys. Res.* **98** 7621–7632
[14] Smith E J and Wolfe J H 1976 *Geophys. Res. Lett.* **3** 137
[15] Gazis P R and Lazarus A J 1983 *Proceedings of JPL Solar Wind Five* vol 228 ed Neugebauer M (NASA Conference Publication) p 509
[16] Gosling J T, Bame S J, McComas D J, Phillips J L, Pizzo V J, Goldstein B E and Neugebauer M 1993 *Geophys. Res. Lett.* **20**(24) 2789
[17] Rouillard A P, Davies J A, Forsyth R J, Rees A, Davis C J, Harrison R A, Lockwood M, Bewsher D, Crothers S R, Eyles C J, Hapgood M and Perry C H 2008 *Geophys. Res. Lett.* **35** L10110
[18] Reames D V 1995 *Rev. Geophys.* **33**(S1) 585–589
[19] Reames D V, Ng C K, Mason G M, Dwyer J R, Mazur J E and von Rosenvinge T T 1997 *Geophys. Res. Lett.* **24** 2917
[20] Roussev I I, Sokolov I V, Forbes T G, Gombosi T I, Lee M A and Sakai J I 2004 *Ap. J. Lett.* **605** L73
[21] Burlaga L F, Schwenn R and Rosenbauer H 1983 *Geophys. Res. Lett.* **10** 413
[22] Burlaga L F 1984 *Space Sci. Rev.* **39** 255–316
[23] Whang Y C 1991 *Space Sci. Rev.* **57** 339
[24] Whang Y C, Lu J Y and Burlaga L F 1999 *J. Geophys. Res.* **104** 19787
[25] Burlaga L F, Ness N F, McDonald F B, Richardson J D and Wang C 2003 *Ap. J.* **582** 540
[26] Burlaga L F, Wang C, Richardson J D and Ness N F 2003 *Ap. J.* **585** 1158
[27] Burlaga L F, Goldstein M L, McDonald F B and Lazarus A J 1985 *J. Geophys. Res.* **90** 12027
[28] Wang C and Richardson J D 2002 *Geophys. Res. Lett.* **29** 1181
[29] Richardson J D, Paularena K I, Wang C and Burlaga L F 2002 *J. Geophys. Res.* **107** 1041
[30] Steinolfson R S 1994 *J. Geophys. Res.* **99** 13307
[31] Whang Y C and Burlaga L F 1994 *J. Geophys. Res.* **99** 21457
[32] Story T R and Zank G P 1995 *J. Geophys. Res.* **100** 9489
[33] Pogorelov N V 2000 *Astrophys. and Space Sci.* **274** 115
[34] Zank G P, Rice W K M, Cairns I H, Bieber J W, Skoug R M and Smith C W 2001 *J. Geophys. Res.* **106** 29363
[35] Zank G P and Müller H R 2003 *J. Geophys. Res.* **108** 1240
[36] Rice W K M and Zank G P 2000 *J. Geophys. Res.* **105** 5157
[37] Zank G P, Müller H R and Wood B E 2001 *Phys. Plasmas* **8** 2385
[38] Washimi H, Zank G P, Hu Q, Tanaka T and Munakata K 2007 *Ap. J* **670** L139
[39] Pogorelov N V, Borovikov S N, Zank G P and Ogin T 2009 *Ap. J.* **696** 1478
[40] Pogorelov N V, Heerikhuisen J, Zank G P, Borovikov S N, Frisch P C and McComas D J 2011 *Ap. J.* **742** 104
[41] Washimi H, Zank G P, Hu Q, Tanaka T and Munakata K 2011 *MNRAS* **416** 1475

[42] Washimi H, Webber W, Zank G P, Hu Q, Florinski V, Adams J and Kubo Y 2012 *Ap. J. Lett.* **757** L2
[43] Borovikov S N, Pogorelov N V and Ebert R W 2012 *Ap. J.* **750** 42
[44] Pogorelov N V, Suess S T, Borovikov S N, Ebert R W, McComas D J and Zank G P 2013 *Ap. J* **772** 2
[45] Pogorelov N V and Zank G P 2005 *Proceedings of the Solar Wind 11 / SOHO 16, Connecting Sun and Heliosphere Conference* vol 592 ed Fleck B, Zurbuchen T H and Lacoste H (ESA Special Publication) p 35
[46] Pogorelov N V, Borovikov S N, Burlaga L F, Ebert R W, Heerikhuisen J, Hu Q, Kryukov I A, Suess S T and Zank G P 2012 *Proceedings of the 10th Annual International Astrophysics Conference* vol 1436 ed Heerikhuisen J, Li G, Pogorelov N and Zank G (American Institute of Physics Conference Series) pp 321–330
[47] Liu Y D, Richardson J D, Wang C and Luhmann J G 2014 *Ap. J. Lett.* **788** L28
[48] Kryukov I A, Borovikov S N, Pogorelov N V and Zank G P 2008 *Proceedings of Numerical Modeling of Space Plasma Flows: Astronum 2007* vol 385 ed Pogorelov N V, Audit E and Zank G P (Astronomical Society of the Pacific Conference Series) pp 265–272
[49] McComas D J, Bzowski M, Frisch P, Fuselier S A, Kubiak M A, Kucharek H, Leonard T, Möbius E, Schwadron N A, SokółJ M, Swaczyna P and Witte M 2015 *Ap. J.* **801** 28

50 years of research on particle acceleration in the heliosphere

L A Fisk

Department of Atmospheric, Oceanic and Space Sciences, University of Michigan
2455 Hayward St., Ann Arbor, MI 48109-2143 USA

Email: lafisk@umich.edu

Abstract. In 1965, and through the late 1960s, the heliosphere was considered to be a passive place, an impediment to the information on the galaxy contained in galactic cosmic ray observations, and on the Sun, from solar energetic particles. All this changed in the early 1970s with the discovery of the Anomalous Cosmic Rays (ACRs), and the subsequent acceptance that the ACRs are ionized interstellar neutral gas that is accelerated in the heliosphere by four orders of magnitude in energy. In the mid-1970s, *Pioneer* 10 & 11 observations provided direct evidence of acceleration. In 1977-78, diffusive shock acceleration was introduced, and subsequently developed in detail, providing compelling explanations for, e.g., the observed acceleration in co-rotating interaction regions, and a likely explanation for the acceleration of ACRs at the termination shock of the solar wind. In 2004 and 2008, the *Voyagers* crossed the termination shock, did not observe the acceleration of the ACRs, but did observe that low-energy particles, up to a few MeV/nucleon, had identical spectra downstream from the termination shock, a distribution function that is a power law in particle speed with a spectral index of -5. When *Voyager* 1 reached ~120 AU, where the high-energy ACRs are at peak intensity, the ACR spectrum is also a -5 spectrum. Moreover, observations of suprathermal tails in the solar wind in the inner solar system have a -5 spectrum, often peaking downstream, but not at shocks. These observations led to the development of a new acceleration mechanism, the pump acceleration mechanism of Fisk & Gloeckler, which can account for all the observed -5 spectra.

1. Introduction

This paper is written in the style of the talk at this conference: It presents the personal reflections and opinions of the author as to the highlights, major events and milestones in research on particle acceleration in the heliosphere during the last 50 years, i.e. starting in 1965. This is not an effort to present a comprehensive review of all pertinent or related topics, or to provide a comprehensive list of references. The author apologizes in advance to anyone who considers that their contribution to this subject is more important than the contributions cited. For readers who would like a more comprehensive and less personal review of particle acceleration in the heliosphere, the Space Science Series of the International Space Science Institute (ISSI) includes *Particle Acceleration in Cosmic Plasmas*, ed. by Balogh, Bykov, Lin, Raymond, and Scholer, which is also published in Space Science Reviews, 172, Issues 1-4, 2012.

2. The situation in 1965

In 1965, the major issue in energetic particle research in the heliosphere was not how particles are accelerated, but rather how they are decelerated. Parker [1] introduced what we refer to today as the

Parker transport equation, for the purpose of describing the modulation of galactic cosmic rays (GCRs):

$$\frac{\partial U}{\partial t}+\nabla\cdot(\mathbf{u}U)-\frac{1}{3}\nabla\cdot(\mathbf{u})\frac{\partial}{\partial T}(\alpha TU)=\nabla\cdot(\vec{\kappa}\cdot\nabla U). \tag{1}$$

Here, U is the differential number density; $\mathbf{u}$ is the solar wind velocity; $\alpha=(T+2T_o)/(T+T_o)$, with T particle kinetic energy and T_o particle rest energy; $\vec{\kappa}$ is the diffusion tensor. Note the second term on the left, which describes the adiabatic deceleration that GCRs suffer while diffusing among magnetic irregularities in the solar wind, which is expanding as it flows radially outward. The adiabatic cooling is substantial, several hundred MeV/nucleon [2].

Gleeson and Axford [3] refined the derivation of equation (1), and introduced the Compton-Getting effect on the streaming, **S**, of energetic particles:

$$\mathbf{S}=\mathbf{u}U-\frac{\mathbf{u}}{3}\frac{\partial}{\partial T}(\alpha TU)-\vec{\kappa}\cdot\nabla U\equiv C\mathbf{u}U-\vec{\kappa}\cdot\nabla U. \tag{2}$$

Here, C is the Compton-Getting coefficient. Equation (2) led to a simple approximate solution to the modulation equation (**S** can be neglected) known as the force field approximation.

In general, then, the attitude toward the heliosphere in the late 1960s and early 1970s was that it was a passive place, an interference for the more important observations of GCRs and what they tell us about the galaxy, and solar energetic particles and what they tell us about the Sun.

3. The discovery of the anomalous component

The situation with regard to particle acceleration in the heliosphere changed dramatically with the discovery of Anomalous Cosmic Rays (ACRs) in the early 1970s, and the subsequent acceptance that this was interstellar neutral gas that is ionized and accelerated in the solar wind, by some four orders of magnitude, as proposed by Fisk et al. [4]. Now the heliosphere was no longer a passive place, but rather a prodigious accelerator of energetic particles.

In the early 1970s, a series of observations began to reveal an unusual component of the energetic particles. The Chicago group [5] noted that the helium flux in the energy range from 20 to 30 MeV/nucleon exceeded the proton flux, which although not terribly dramatic was at least puzzling. Then in rapid succession, the Max-Planck-Maryland group [6], observing at energies of 1-10 MeV/nucleon, and the Goddard group [7], observing above 10 MeV/nucleon, found something that was truly strange. Shown in figure 1 are the observations of Hovestadt et al. [6] and McDonald et al. [7]. The oxygen fluxes are very much higher than the carbon flux and the helium fluxes. Nitrogen was also enhanced, but at that time no other element was observed to be enhanced. This led, in the paper by McDonald et al. [7], to the label “anomalous component,” or now the more common usage of Anomalous Cosmic Rays, or ACRs.

GCR researchers live for discoveries like this. GCRs provide essential information on important objects and processes in the galaxy: supernovae being the classic example—the composition, particularly the isotopic composition, provides information about basic nucleosynthesis. So when ACRs were discovered, the assumption of GCR researchers was that an exotic object had been discovered. In fact, Hoyle and Clayton [8] had a theory for the origin of ACRs based on the composition of white dwarfs in binary star systems. They had a carbon and oxygen rich surface on the dwarf, nuclear burning of the carbon, acceleration in a nova shock and so on. This is the kind of theory that was a thrill to GCR researchers.

When a theory is developed, each contributor approaches the problem from their own background and perspective. I was working on modulation back then, and took one look at the ACR spectra and concluded that they cannot be GCRs. Mel Goldstein, Reuven Ramaty and I had just written a paper about the amount of adiabatic deceleration in the solar wind that is suffered by GCRs, and the energy loss is considerable, several hundred MeV per nucleon, and moreover does not permit spectra as steep as the spectra in figure 1 to survive [2].

So one day in 1973, I was sitting in Reuven Ramaty's office at Goddard discussing, along with Ben-Zion Kozlovsky, the puzzle of the ACR observations. And we had one of those light bulb-goes-off conversations, where you all say – that must be it. We came up with the simple explanation for ACRs that they were interstellar neutral gas that flows into the solar system due to the motion of the Sun through the local interstellar medium; the neutral gas is then ionized by charge-exchange and photoionization, and then picked up by and accelerated in the solar wind to ACR energies.

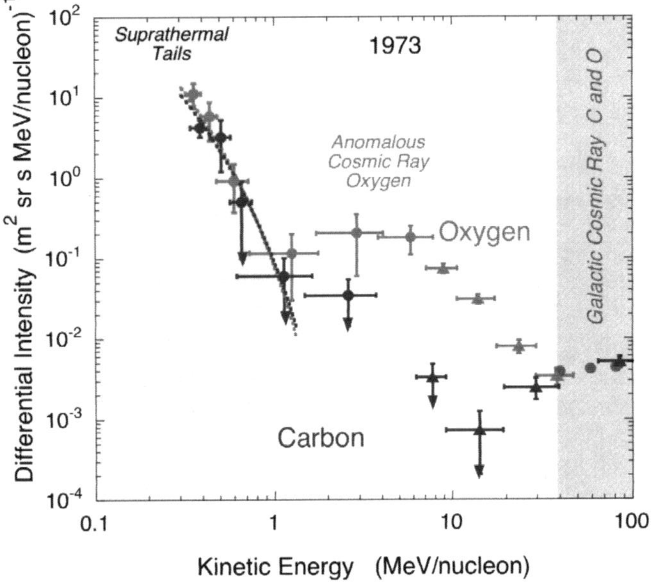

Figure 1. Differential intensity spectra of carbon (blue symbols) and oxygen (red symbols) measured by Hovestadt et al. [6] (filled circles) and McDonald et al. [7] (filled triangles). The oxygen spectrum has an extra component labeled "Anomalous Cosmic Rays" in the energy interval from about 1 to 30 MeV/nuc not seen in the carbon spectrum where only upper limits are shown.

The theory in Fisk et al. [4] provided a natural and easy explanation for ACRs. Interstellar neutral gas is expected to have exactly the correct composition. The elements enhanced in ACRs are all elements with high first ionization potential. They are hard to ionize and therefore are neutral in the interstellar medium and can readily penetrate into the heliosphere. The particles should be ionized only once in the heliosphere. Thus, they would be singly charged and therefore of quite high rigidity, and the concerns over modulation went away. We even predicted that elements like neon would be enhanced in ACRs, an enhancement that was subsequently observed.

I have always been grateful that my name started with 'F', as opposed to 'K' or 'R'. The author list on the Fisk et al. [4] paper is alphabetical, but it is certainly one of the things for which I have been best known.

What the Fisk et al. [4] paper did not do is deal with how ACRs are accelerated. The paper simply said there must be a four-order of magnitude acceleration somewhere beyond the orbit of Earth. The interstellar ions were created primarily in the inner solar system, by photo ionization and charge acceleration, picked up by the solar wind thereby attaining energies of ~1 keV/nucleon, convected outward with the solar wind, and along the way somehow they were to be accelerated to ACR energies.

4. Research on particle acceleration begins to get traction

In the mid-1970s the topic of whether and how particles could be accelerated in the solar wind began to get traction. I was of course interested in the acceleration of ACRs, but this was also when the *Pioneers* were en route outward from Earth. McDonald et al. [9] published a paper comparing *Pioneer* 11 data with *IMP-8* data, and concluded that there was compelling evidence for extensive interplanetary acceleration of low-energy particles. McDonald et al. attributed the acceleration to a basic Fermi mechanism, which Jokipii [10] had written about. In this mechanism particles are stochastically accelerated by interacting with randomly propagating Alfven waves through a cyclotron resonance.

Barnes & Simpson [11] reported acceleration of energetic particles in the solar wind, also from *Pioneer* observations, and they identified co-rotating interaction regions (CIRs) as responsible, noting that the intensities peaked at the interfaces of the CIRs. They only speculated on possible acceleration mechanisms, noting that there were shocks at the interface, but also enhanced turbulence.

I also got in the act in 1976 with a stochastic acceleration mechanism built on transit-time damping, in which particles interact with magnitude fluctuations in the magnetic field through the Landau resonance [12]. This mechanism had an advantage over the cyclotron resonate mechanism cited by McDonald et al. [9] since a cyclotron resonance requires too much pitch angle scattering to be consistent with other observations. The Landau resonance with magnitude fluctuations in the magnetic field did not have that problem.

5. Diffusive shock acceleration

Everything changed again around 1978, with the full development of diffusive shock acceleration. I was working on other problems at the time, but it has always intrigued me that four separate researchers came up with essentially the same theory at the same time. The issue was not acceleration in the solar wind, but rather the acceleration of GCRs at shock waves generated by supernovae. Axford et al. [13] published their theory in the cosmic ray conference that year. Krymsky [14] published in the Soviet literature. Bell [15], and also Blandford and Ostriker [16] published a year later. I suppose the 1977 papers were first, but perhaps it was simply an idea whose time had come.

The concept was quite simple. Particles gain energy at a shock by repeatedly passing back and forth across the shock front, and each time they are compressed between the upstream and downstream magnetic irregularities, which are moving at different speeds, and the particles are accelerated.

The Axford et al. [13] presentation was particularly simple and insightful. If you write down the streaming equation (equation (2)), with the Compton Getting effect of Gleeson and Axford [3], and assume the streaming is continuous across the shock front, with upstream particles reflected off the incoming irregularities, and downstream being simply convected away, you find this simple relationship (for non-relativistic particles):

$$-\frac{u_{up}}{3}\frac{\partial}{\partial T}(2TU) = u_{ds}U - \frac{u_{ds}}{3}\frac{\partial}{\partial T}(2TU), \tag{3}$$

which is expressed in terms of the differential number density, U; u_{up} and u_{ds} are the solar wind speed upstream and downstream of the shock front, respectively. Equation (3) is satisfied by a power law spectrum, $U \propto T^{-\beta}$, where the spectral index depends upon the compression ratio at the shock:

$$\beta = \frac{2(u_{us}/u_{ds})+1}{2(u_{us}/u_{ds})-2} \; . \tag{4}$$

It is straightforward to express these formulae in terms of differential intensity rather than differential number density.

The first formulation of diffusive shock acceleration, contained in the seminal four papers [13-16], has led an enormous body of theoretical research: papers on what happens when cosmic rays have enough energy to influence the incoming flow, cosmic ray mediated shocks. How are particles injected? How can shock acceleration be simulated numerically? And so on; research that continues to this day.

Journal of Physics: Conference Series **642** (2015) 012009 doi:10.1088/1742-6596/642/1/012009

While all this detailed theory work was interesting, what diffusive shock acceleration did for acceleration in the solar wind was to provide a simple and straightforward explanation, one that could readily be understood, for a number of outstanding problems.

As examples, of which there are many, Marty Lee and I wrote a paper in 1981 on shock acceleration in CIRs [17], in which we included both the acceleration at the shocks surrounding the CIRs, as well as deceleration in the upstream solar wind. The model provided a good fit to contemporary observations, particularly at the higher energies, and offered a good explanation for the earlier discovery of acceleration in CIRs by Barnes and Simpson [11]. Pesses et al. [18] wrote a paper on a model that showed that a logical place to accelerate the ACRs was the termination shock of the solar wind. Marty Lee [19] wrote, what I always thought was one of his best papers, on coupling wave excitation and particle acceleration to explain acceleration at the Earth's bow shock. And of course there were countless applications of shock acceleration for propagating interplanetary shocks.

Despite the fact that diffusive shock acceleration had compelling underlying physics and a great deal of convincing theoretical work, nonetheless there were some nagging concerns that observational evidence was lacking that this was all that was happening. Not that diffusive shock acceleration was wrong, just that there was something missing. This was particularly true of propagating interplanetary shocks, which every time they were studied, the simple relationships, like the correlation between the spectral index and the compression ratio, or the expected spatial profile didn't quite fit (e.g., [20]). There was always a possible reason, e.g., time variations. We observe the consequences of the evolution of the shock, not just local conditions. The Earth's bow shock seemed to fit nicely with diffusive shock acceleration, but less so for propagating interplanetary shocks.

Then the *Voyagers* crossed the termination shock of the solar wind, which was the expected location of the acceleration of ACRs, up to 10-100 MeV, and it wasn't. *Voyager* 1 crossed in December 2004 at 94 AU, and saw acceleration of particles up to a few MeV per nucleon, but the real ACRs showed no sign of being accelerated at the location of the *Voyager* 1 crossing [21, 22]. *Voyager* 2 crossed in August of 2007 at 83.7AU, at a distance of more than 100 AU away from the *Voyager* 1 crossing, and also showed only low-energy particle acceleration, with no evidence of the acceleration of the real high-energy ACRs [23, 24].

Of course there could be a reason. McComas and Schwardon [25] presented a model that argued that neither location where the *Voyagers* crossed was expected to accelerate high-energy particles, but rather the real ACRs should be accelerated on the flanks of the termination shock. However, again we are left with needing an excuse why diffusive shock acceleration did not do what had been the expectation for the acceleration of the ACRs for more than 25 years.

Then the *Voyagers* began to make what can be considered to be one of their most significant observations [26, 27]. As *Voyager* 1 penetrated into the heliosheath the spectrum of the low-energy particles that were accelerated at the termination shock jumped around, but then locked onto a power law spectrum with spectral index of -1.5, when expressed as differential intensity, or -5 when expressed as a distribution function. The spectrum is a little steeper, but if you do the Compton Getting correction back into the frame of the solar wind, which is important for the lowest energy particles, it is a remarkably constant spectrum, with spectral index of -1.5 or -5 when expressed as a distribution function, for years. The same spectrum is observed from *Voyager* 2, which took longer to lock onto, but did lock onto a spectrum with a -1.5 spectral index. It appeared as if the termination shock at *Voyager* 1 was traveling inward, and thus *Voyager* 1 penetrated into the heliosheath faster. At *Voyager* 2 the termination shock was not traveling inward and it took longer, but both still locked onto the same spectrum. Even more remarkable, the absolute magnitude of the intensity of these low-energy particles was the same at *Voyagers* 1 and 2, even though the spacecraft were separated by a distance of more than 100 AU.

The particles accelerated at and downstream from the termination shock are not the real, high-energy ACRs. However, when *Voyager* 1 finally reached the location where the ACRs actually peak, at ~120 AU, the spectrum of the real ACRs also has the same shape as the low-energy particles immediately downstream from the termination shock. As shown in figure 2, the spectrum is a power law, expressed here as a distribution function, with spectral index of -5, and an exponential rollover at quite high energies.

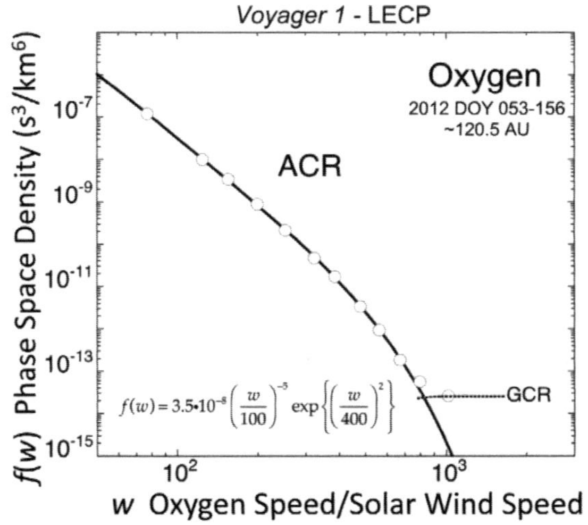

Figure 2. Phase space density of ACR oxygen as a function of oxygen speed/solar wind in the prime acceleration region at ~120 AU. The solar wind speed is taken to be 100 km/s. The data is from the LECP instrument on *Voyager* 1 [28]. The equation shown in the figure is the formula for the common spectrum and provides an excellent fit to the observations (from [29]).

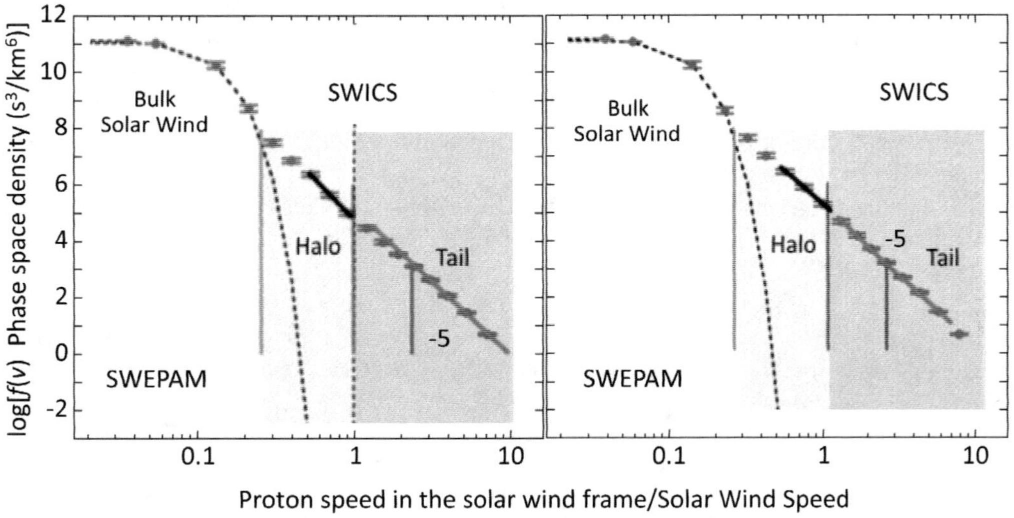

Figure 3. One-hour averaged solar wind frame velocity distribution functions showing the proton bulk solar wind, the halo and the tail segments during hour 11 of August 12, 2001, during which the strong (compression ratio of 3.85 ±0.15) shock passed *ACE* (left panel) and during the hour of peak tail density that was observed one hour downstream of the shock (right panel) (from [29]).

The *Voyager* observations in the heliosheath caused George Gloeckler to ponder in great detail on the suprathermal tails of the solar wind that were being observed in the inner heliosphere by *Ulysses* and particularly by *ACE*. These are tricky observations, because at these low energies, up to 100 keV, the correction to transform into the frame of the solar wind must be done with great care. However, when the correction is done with care you discover that, almost without exception, every time you are in a region where particles are clearly being accelerated, i.e. the intensity is a local maximum, the spectrum, expressed as a distribution function, is also -5, with an exponential rollover.

As is illustrated in figure 3, these prime acceleration sites occur often, but not always, downstream from shocks, and what you actually see at the shock fronts are modulated spectra, as the particles accelerated downstream propagate back to the shock front. It is no wonder that observations of particles accelerated at propagating shocks do not easily fit simple diffusive shock acceleration theory.

As much as you may think that diffusive shock acceleration is the answer to particle acceleration in the heliosphere, you should at least consider that these observations cannot be explained by diffusive shock acceleration. The acceleration does not occur at shocks, and there is no reason that diffusive shock acceleration will yield one particular spectral index, -5.

6. The pump acceleration mechanism

Explaining these observations, spectra with a spectral index that is always -5, for particles accelerated in vastly different conditions in the heliosphere—throughout the heliosheath and in the inner heliosphere—led George Gloeckler and me to develop the pump acceleration mechanism, which we have been publishing about for nine years now [29-35]. As with many new theories, our understanding of the theory and the derivations of the equations evolved over this period. However, from the beginning we knew that the physics of the pump acceleration was just as simple as is the physics of diffusive shock acceleration.

We note first that the -5 spectrum is most pronounced throughout the heliosheath and downstream from shocks in the inner heliosphere. Both are subsonic regions, with extensive compressive turbulence. Further, the regions where the -5 spectrum are most pronounced also have a ready source of particles to accelerate—pickup ions in the case of the heliosheath; the non-thermal distribution of solar wind particles heated crossing the shock in the case of interplanetary shocks in the inner heliosphere. Hence, we need to design an acceleration mechanism that works in the following situation:

- We have a volume of plasma, which is of fixed size, and which contains a population of particles that we are going to accelerate.
- The volume is thermally isolated, which means simply that there is no net inflow or outflow of the particles we want to accelerate, i.e., no large-scale spatial gradients.
- There is also a population of particles that contains the mass, in our case the solar wind, and an embedded magnetic field, and the solar wind and the embedded magnetic field are undergoing compressions and expansions, and are coupled to and cause compressions and expansions of the higher energy particles that we want to accelerate.

Since the volume of our plasma is constant, every compression region must be surrounded by an expansion region, and the compressions and expansions cannot do net work on the particles we want to accelerate. If no work is being done on the particles we want to accelerate, and we have a thermally isolated volume, the energy contained in the particles we want to accelerate must be constant. However, requiring that the total energy in the particles to be accelerated is constant does not restrict us from redistributing the energy from low to high energies through an irreversible process, which is what the pump acceleration does. The combination of cross-field diffusion out of compression regions into the surrounding expansion regions, and flow of particles and energy back into the source particle distribution, all with total energy conserved, is such an irreversible process that will redistribute particles, without changing their total energy, from the initial source population to higher energies.

And now the punch line: If you have a thermally isolated, constant volume system, containing an irreversible process, the system must tend to and achieve equilibrium when it achieves a state of maximum entropy. If you are in a state of maximum entropy and the compressions and expansions still continue, then each compression and expansion must be isentropic, which for an ideal gas requires that each compression and expansion must be adiabatic. If each compression and expansion is adiabatic, then the spectral index of the spectrum of accelerated particles must be -5, as can be seen by considering the Parker transport equation for a compression and expansion with random velocity $\delta\mathbf{u}$ written in terms of a distribution function.

Journal of Physics: Conference Series **642** (2015) 012009 doi:10.1088/1742-6596/642/1/012009

$$\frac{\partial f}{\partial t}+\delta\mathbf{u}\bullet\nabla f-\frac{\nabla\bullet\delta\mathbf{u}}{3}v\frac{\partial f}{\partial v}=0\,. \tag{5}$$

In order for each compression and expansion to be isentropic, or adiabatic, the pressure, P, in particles that are being redistributed to higher energies must satisfy the adiabatic pressure equation, or,

$$\frac{\partial P}{\partial t}+\delta\mathbf{u}\bullet\nabla P+\frac{5}{3}(\nabla\bullet\delta\mathbf{u})P=0\,, \tag{6}$$

which can be achieved only if $f \propto v^{-5}$.

The concept that a -5 spectrum results when we require that energy is conserved in particles undergoing an irreversible process in compressions and expansions was first discovered by Andre Bykov [36] long before anyone worried about the -5 spectrum in the heliosphere. Bykov was doing a numerical calculation of particle acceleration in compressive turbulence in superbubbles in the interstellar medium, and when he applied a nonlinear constraint that the energy imparted to the accelerated particles fed back into the compressions and expansions, and thus energy was conserved in the system, he found a -5 spectrum. This was not noted at the time; the resulting spectrum plotted in Bykov [36] has a spectral index of -3. However, what was plotted was $v^2 f$, i.e., the resulting spectrum for f has a spectral index of -5.

The bottom line is that if you have: (1) a thermally isolated constant volume, containing compressions and expansions, and a low-energy particle population that you want to redistribute to high energies, conditions that, to a reasonable approximation, are readily available in the solar wind, and (2) an irreversible process in your volume, like spatial diffusion, the resulting particle population that gets redistributed to high energies will have a spectrum, expressed as a distribution function, with a spectral index of -5.

The pump acceleration mechanism has been developed beyond simply a concept. The governing equation that describes the time evolution of the distribution function, f_o, the most complete derivation of which is in Fisk & Gloeckler [29], is

$$\frac{\partial f_o}{\partial t}=\frac{1}{v^4}\frac{\partial}{\partial v}\left(\frac{\left\langle(\nabla\bullet\delta\mathbf{u})^2\tau\right\rangle}{9}v\frac{\partial}{\partial v}\left(v^5 f_o\right)\right). \tag{7}$$

Here, τ is the characteristic escape time from a compression region, which is related to the cross-field diffusion coefficient. Equation (7) can be used to determine how rapidly and to what energy the -5 spectrum is established. Also, as might be expected, equation (7) can be used to demonstrate that the pump mechanism is a first-order and not a stochastic second-order acceleration, and thus fully competitive with diffusive shock acceleration.

There has been some confusion about Equation (7) [37]. The expression f_o is not defined as in a stochastic acceleration calculation, or, as you might observe, as the mean distribution function at a given energy. Rather, it is a distribution function that defines the mean pressure of the redistributed particles. Fisk & Gloeckler [29] show, however, that f_o, to any reasonable accuracy that you would care about, is what you observe.

Equation (7) has also been applied, with success, to all the problems we identified as needing an explanation. The curve on the full ACR spectrum observed by *Voyager* 1 at ~120 AU that fits the data so well in figure 2 is the prediction of the pump acceleration mechanism. The pump acceleration mechanism also fully explains George Gloeckler's observations downstream from shocks in the inner heliosphere [29]. Finally, although our subject today is not interstellar physics, we even applied the pump acceleration mechanism to accelerate all galactic cosmic rays in the interstellar medium, essentially Bykov on steroids, and got all the breaks in the spectrum in the right place, with the correct change in spectral index [34].

7. Concluding remarks

We have come a long way since 1965, when we had no indication that there was much acceleration in the heliosphere at all—to where we now have countless examples, and in some cases, like the ACRs, we have examples of acceleration to relatively high energies. Throughout, we have done what we do best in heliophysics. We have used the detailed observations available to us in the heliosphere to constraint and to understand the mechanisms responsible for the acceleration, and in doing so we serve the broader needs of astrophysics, since what we study can occur in other astrophysical settings. In my judgment the explanation of the -5 spectrum, the pump acceleration mechanism, is yet another example of where we can demonstrate the value of heliospheric research to the broader astrophysical community, and we should do so.

Acknowledgments

This work was supported in part by NASA Grant NNX10AF23G, NNX11AP01G, and by NSF Grant AGS-1043012.

References

[1] Parker E N 1965 The passage of energetic charged particles through interplanetary space *Planet. Space Sci.* **13** 9

[2] Goldstein M L, Fisk L A and Ramaty R 1970 Energy loss of cosmic rays in the interplanetary medium *Phys. Rev. Lett.* **25** 832

[3] Gleeson L J and Axford W I 1967 Cosmic rays in the interplanetary medium *Astrophys. J.* **149** L115

[4] Fisk L A, Kozlovsky B and Ramaty R 1974 An interpretation of the observed oxygen and nitrogen enhancements in low-energy cosmic rays *Astrophys. J.* **190** L35

[5] Garcia-Munoz M, Mason G M and Simpson J A 1973 The anomalous 1972 low-energy galactic cosmic ray proton and helium spectrum *Astrophys. J.* **18** L81

[6] Hovestadt D, Vollmer O, Gloeckler G and Fan C Y 1973 Differential energy spectra of low-energy (<8.5 MeV per nucleon) heavy cosmic rays during solar quiet times *Phys. Rev. Letts.* **31** 650

[7] McDonald F B, Teegarden B J, Trainor J H, von Rosenvinge T T and Webber W R 1974 The anomalous abundance of cosmic-ray nitrogen and oxygen at low energies *Astrophys. J.* **187** L105

[8] Hoyle F and Clayton D D 1974 Nucleosynthesis in white-dwarf atmospheres *Astrophys. J.* **191** 705

[9] McDonald F B, Teegarden B J, Trainor J H, von Rosenvinge T T and Webber W R 1976 The interplanetary acceleration of energetic ions nucleons *Astrophys. J.* **203** L149

[10] Jokipii J R 1971 Deceleration and acceleration of cosmic rays in the solar wind *Phys. Rev. Letts.* **26** 666

[11] Barnes C W and Simpson J A 1976 Evidence for particle acceleration of nucleons in co-rotating interaction regions *Astrophys. J.* **210** L91

[12] Fisk L A 1976 The acceleration of energetic particles in the interplanetary medium by transit time damping *J. Geophys. Res.* **81** 4633

[13] Axford W I, Leer E and Skadron G 1977 The acceleration of cosmic rays by shock waves *15th Int Cosmic Ray Conf* vol 11 (Sofia: B'Igorska Academiia na Naukite) pp 132–137

[14] Krymsky G F 1977 A regular mechanism for the acceleration of charged particles on the front of a shock wave *Dokl. Akad. Nauk SSSR* **234** 1306–1308

[15] Bell A R 1978 The acceleration of cosmic rays in shock fronts – I *MNRAS* **182** 147–156

[16] Blandford R D and Ostriker J P 1978 Particle acceleration by astrophysical shocks *Astrophys. J.*

221 L29

[17] Fisk L A and Lee M A 1981 Shock acceleration of energetic particles in corotating interaction regions *Adv. Space Res.* **1** 93

[18] Pesses M E, Eichler D and Jokipii J R 1981 Cosmic ray drift, shock wave acceleration, and the anomalous component of cosmic rays *Astrophys. J.* **246** L85–L88

[19] Lee M A 1982 Coupled hydrodynamic wave excitation and ion acceleration upstream of the Earth's bow shock *J. Geophys Res* **92** 11041

[20] Lario D, Hu Q, Ho G C, Decker R B, Roelof E C and Smith C W 2005 Statistical properties of fast forward transient interplanetary shocks and associated energetic particle events: ACE observations *Proc. Solar Wind 11–SOHO 16 "Connecting Sun and Heliosphere"* ed B Fleck, T H Zurbuchen and H Lacoste (Noordwijk: ESA SP-592) pp 81–86

[21] Decker R B, Krimigis S M, Roelof E C, Hill M E, Armstrong T P, Gloeckler G, Hamilton D E and Lanzerotti L J 2005 Voyager 1 in the foreshock, termination shock and heliosheath *Science* **309** 2020

[22] Stone, E. C., Cummings, A. C., McDonald, F. B., Heikkila, B. C., Lal, N., & Webber, W. R. (2005) Voyager 1 explores the termination shock region and heliosheath beyond, Science, 309, 2017.

[23] Decker R B, Krimigis S M, Roelof E C, Hill M E, Armstrong T P, Gloeckler G, Hamilton D E and Lanzerotti L J 2008 Mediation of the solar wind termination shock by nonthermal ions *Nature* **454** 67

[24] Stone E C, Cummings A C, McDonald F B, Heikkila B C, Lal N and Webber W R 2008 An asymmetric solar wind termination shock *Nature* **454** 71

[25] McComas D J and Schwadron N A 2006 An explanation of the Voyager paradox: Particle acceleration at a blunt termination shock *Geophys. Res. Letts.* **33** L04102

[26] Decker R B, Roelof E C, Krimigis S M and Hill M E 2006 Low-energy ions near the termination shock *Physics of the Inner Heliosheath, AIP Conf. Proc.* vol 858 (Melville: AIP) pp 73–78

[27] Gloeckler G, Fisk L A, Mason G M and Hill M E 2008 Formation of power law tail with spectral index -5 inside and beyond the heliosphere *Particle Acceleration and Transport in the Heliosheath and Beyond, AIP Conf. Proc.* vol 1039 (Melville: AIP) pp 367–374

[28] Krimigis, S. M., C. O. Bostrom, T. P. Armstrong, W. I. Axford, C. Y. Fan, G. Gloeckler, and L. J. Lanzerotti (1997), The Low Energy Charged Particle / LECP / experiment on the Voyager spacecraft, *Space Sci. Rev., 21*, 329–354

[29] Fisk L A and Gloeckler G 2014 The case for a common spectrum of particles accelerated in the heliopshere: Observations and theory *J. Geophys. Res.* **119** 8733

[30] Fisk L A and Gloeckler G 2007 Thermodynamic constraints on stochastic acceleration in compressional turbulence *Proc. Nat. Acad. Sci.* **104** 5749–5754

[31] Fisk L A and Gloeckler G 2008 Acceleration of suprathermal tails in the solar wind *Astrophys. J.* **686** 1466–1473

[32] Fisk L A and Gloeckler G 2009 The acceleration of anomalous cosmic rays by stochastic acceleration in the heliosheath *Advs. Space Res.* **43** 1471–1478

[33] Fisk L A and Gloeckler G 2012a Particle acceleration in the heliosphere: Implications for astrophysics *Space Sci. Rev.* **173** 433–458

[34] Fisk L A and Gloeckler G 2012b Acceleration of galactic cosmic rays in the interstellar medium *Astrophys. J.* **744** doi:10.1088/0004-637X/744/2/127

[35] Fisk L A, Gloeckler G and Schwadron N A 2010 On theories for stochastic acceleration in the solar wind *Astrophys. J.* **720** 533–540

[36] Bykov A M 2001 Particle acceleration and nonthermal phenomena in superbubbles *Space Sci. Rev.* **99** (1/4) 317–326

[37] Jokipii J R and Lee M A 2010 Compression acceleration in astrophysical plasmas and the production of f(v) proportion to v^{-5} spectra *Astrophys. J.* **713** 475–483

Plasma properties at the Voyager 1 crossing of the heliopause

Stephen A Fuselier[1,2] and Iver H Cairns[3]
[1]Southwest Research Institute, San Antonio, TX, USA (sfuselier@swri.edu)
[2]University of Texas at San Antonio, San Antonio, TX, USA
[3]School of Physics, University of Sydney, Sydney, NSW 2006, Australia

E-mail: sfuselier@swri.edu

Abstract. In August 2012, Voyager 1 crossed the heliopause at a distance of 121.5 AU from the Sun. It is argued that the spacecraft entered a region in the outer heliosheath that had the characteristics of a plasma depletion layer. Observed plasma parameters at the heliopause, properties of plasma depletion layers, and some assumptions are used to derive a set of plasma parameters on both sides of the heliopause. Using the density, temperature, and magnetic field magnitude on each side, the corresponding plasma beta and Alfven Mach number (in the outer heliosheath) are derived. These plasma parameters are used to demonstrate that the plasma depletion process is occurring in the outer heliosheath adjacent to the heliopause and these parameters are used to determine if lower hybrid waves are generated locally and if magnetic reconnection is occurring locally at the location of the Voyager 1 crossing. Reconnection may not be an effective source of superthermal electrons at the heliopause, based on the small Alfven speeds there (V_A< 100 km/s) and an empirical connection between electron heating and Alfven speed found in inner solar system studies.

1. Introduction

The heliopause is the location where the total pressures in the inner and outer heliosheath balance. This boundary is considered the outer limit of the solar wind plasma and magnetic field and the beginning of interstellar space. If the heliopause is in motion, then the pressures on the two sides of the boundary may not balance exactly. Recently, there are reports that the Voyager 1 spacecraft crossed this boundary at a distance of 123 AU from the Sun and entered interstellar space [1,2,3].

The density and magnetic field immediately upstream of the heliopause crossing are shown in Table 1. The density is derived from plasma wave instrument data. Specifically, occasional bursts of Langmuir waves in the outer heliosheath near the heliopause allow direct measurement of the density profile near the boundary. The value in Table 1 is determined by extrapolating back to the heliopause location using the density profile observed in the first ~250 days after the crossing on 25 August, 2012 [2]. The magnetic field is measured directly at the heliopause crossing by the Voyager 1 magnetometer. Across the heliopause, there was very little shear in the magnetic field [1].

After crossing the heliopause, the spacecraft entered a region of increasing plasma density and slightly decreasing magnetic field magnitude [1,3]. The density and magnetic field changes in this region are analogous to those observed in the plasma depletion layer upstream from the Earth's magnetopause. A plasma depletion layer [4] is produced by draping of magnetic field lines against a nearly impermeable obstacle, like a magnetosphere or heliosphere. At Earth, the formation and

[1] Southwest Research Institute, e-mail: sfuselier@swri.edu.

properties of the plasma depletion layer are well understood [e.g., 5,6,7], as follows: The magnetic field magnitude increases as the field drapes against the obstacle. This increase drives an increase in the perpendicular temperature of the ions (and electrons) and, according to double adiabatic theory [7,8], a dramatic decrease in the parallel temperature. Above a certain ion temperature anisotropy, the threshold of the electromagnetic ion cyclotron (EMIC) instability is exceeded. The EMIC waves grow rapidly, saturate, and pitch angle scatter ions into the parallel direction. The scattered ions are lost as they propagate along the draped magnetic field lines. Through this loss, the plasma density decreases and the temperature anisotropy increases. Further draping causes the instability to grow again and repeats the process of plasma loss. The temperature anisotropy remains close to the threshold of the EMIC instability throughout the plasma depletion layer. Another important result from this depletion process is a dramatic increase in the Alfven speed and a similar dramatic decrease in the Alfven Mach number.

In addition to temperature anisotropies that depend on plasma beta, opposite gradients in the density and magnetic fields are some of the defining features of a plasma depletion layer. The positive gradient in the density is evident in the Voyager 1 plasma wave data (see Figure 2 in [2]). However, the negative gradient in the magnetic field magnitude is not as evident because of transient increases total magnetic field resulting from shocks propagating past the spacecraft [1].

Table 1 shows the density and magnetic field in the "undisturbed" outer heliosheath upstream from the plasma depletion layer. The density was extrapolated from the plasma wave observations [2] while the magnetic field was estimated from the magnetic field observations [1,9] with the transient increases in the total magnetic field removed. The Alfven speed in Table 1 was computed from these density and magnetic field values. The depletion factor in Table 1 quantifies the change in the Alfven speed from the outer heliosheath to the plasma depletion layer adjacent to the heliopause.

The remainder of this paper uses the observed parameters in Table 1, the properties of the plasma depletion layer, and several assumptions to derive a set of plasma parameters for the local conditions at the Voyager 1 heliopause crossing. These parameters include the temperature and its anisotropy and the parallel and total plasma betas at the Voyager 1 crossing. The plasma conditions are then used to determine if certain plasma processes are active locally at the Voyager 1 heliopause crossing. The plasma processes considered here are the growth of lower hybrid waves in the plasma depletion layer and magnetic reconnection at the heliopause. Finally, the implications for a more extensive depletion layer and non-local effects are discussed.

Table 1: Plasma parameters in the outer heliosheath along the Voyager 1 trajectory

Plasma parameter	Value	Origin
Density in the depletion layer adjacent to the heliopause	$N = 0.047\ cm^{-3}$	Extrapolated from the Langmuir wave profile
Magnetic field magnitude in the depletion layer adjacent to the heliopause	$B = 0.44$ nT	Measured by the magnetometer
Alfven speed in the depletion layer adjacent to the heliopause	$V_A = 44$ km/s	Derived from the density and magnetic field in the depletion layer
Density in the outer heliosheath upstream from the depletion layer	$N = 0.1\ cm^{-3}$	Extrapolated from the Langmuir wave profile
Magnetic field magnitude upstream from the depletion layer	$B = 0.3$ nT	Estimated from the magnetic field measurements
Alfven speed in the outer heliosheath upstream from the depletion layer	$V_A = 21$ km/s	Derived from the density and magnetic field in the outer heliosheath
Degree of plasma depletion =	Depletion Factor = 2.1	Computed from the density and

$B_{PDL}/B_{OHS} * \sqrt{(n_{OH}S/n_{PDL})}$	magnetic field in the depletion layer and outer heliosheath

2. Evidence of Plasma Depletion Layer Processes

The density decrease from the "undisturbed" outer heliosheath to the heliopause is certainly consistent with the plasma depletion process. However, there is additional evidence that the depletion process is active in the outer heliosheath adjacent to the heliopause. This evidence consists of the change in the Alfven speed and Alfven Mach number and an estimate of the changes in the ion temperature anisotropy with and without a plasma depletion layer.

The Alfven speed upstream from the plasma depletion layer is 21 km/s and is determined from the density and magnetic field in the region (Table 1). This speed is lower than the speed of approximately 26 km/s observed for the interstellar neutral Helium [10,11]. Assuming that the interstellar medium is in thermal and collisional equilibrium, the speeds and temperatures of all the neutral and ion species are the same. Thus, if there is no slowing of the ions in the outer heliosheath (i.e., either no shock or a weak shock in the interstellar medium upstream of the heliopause), then the Alfven Mach number is greater than 1 in the region. One of the known effects of a depletion layer is to reduce the Alfven Mach number of the flow past the obstacle [4,5]. The Alfven speed estimated in the depletion layer just beyond the heliopause is 44 km/s (Table 1), which is considerably higher than the bulk plasma flow speed in the outer heliosheath or the interstellar medium. Thus, near the heliopause, the flow is sub-Alfvenic. A similar effect occurs at the Earth's magnetopause. The reduced Alfven Mach number allows magnetic reconnection at the Earth's magnetopause over a wide range of latitudes [12,13], thereby affecting plasma entry into and escape from the magnetopause. This effect is potentially relevant to whether or not magnetic reconnection plays a major role at the heliopause, cf. [14,15].

Further evidence of the depletion process comes from estimates of the temperature anisotropy with and without a depletion layer. Double adiabatic theory [7,8] provides an estimate of the change in the temperatures as the density and magnetic field change, via

$$\frac{d}{dt}\left(T_{||}\frac{B^2}{n^2}\right) = 0 \quad (1)$$

$$\frac{d}{dt}\left(\frac{T_\perp}{B}\right) = 0 \quad (2)$$

where $T_{||}$ is the parallel ion temperature, $T_\perp$ is the perpendicular ion temperature, B is the magnetic field magnitude, and n is the density. To keep the quantities in the parentheses in Equation 1 and Equation 2 constant, $T_{||}$ must decrease dramatically as the magnetic field increases and the density decreases while $T_\perp$ must increase as the magnetic field increases. Using Equation 1 and the change in the magnetic field magnitude and density from the outer heliosheath to the heliopause (see Table 1), $T_{||}$ must decrease by a factor of 7.7 and $T_\perp$ must increase by a factor of 1.5. Therefore, under the assumption that the perpendicular and parallel temperatures are not coupled, $T_\perp / T_{||}$ must increase by a factor of 11.6 from the outer heliosheath to the Voyager 1 location adjacent to the heliopause. This is a very large change in the ion temperature anisotropy and, as will be shown below, such a large change will not occur. Long before the ion distribution becomes this anisotropic, the EMIC instability will grow, saturate, and cause ions to pitch angle scatter from the perpendicular to the parallel direction. This pitch angle scattering effectively couples the parallel and perpendicular temperatures and negates the double adiabatic assumption.

3. Derivation of temperatures and plasma betas

The plasma depletion process is a kinetic process that couples the parallel and perpendicular temperatures through EMIC wave-particle interactions. It is driven by the draping process and the

(3)

plasma throughout the layer maintains marginal stability to EMIC waves. The marginal stability leads to a temperature anisotropy – plasma beta relationship that is maintained throughout the layer [5,6,7].

$$\frac{T_{\perp}}{T_{\parallel}} - 1 \approx \frac{0.715}{\beta_{\parallel}^{0.5}},$$

where $\beta_{\parallel}$ is the ion beta computed from the parallel ion temperature and the numerator and exponent on $\beta_{\parallel}$ are determined by the degree of marginal stability [6]. An exponent of 0.5 is chosen so that Equation 3 is solvable analytically. Solving Equation 3 for $T_{\parallel}$ as a function of $T_{\perp}$ yields a quadratic equation. The first root of the quadratic equation has $T_{\perp} / T_{\parallel} < 1$, which yields a negative $\beta_{\parallel}$, which is unphysical. The second root has $T_{\perp} / T_{\parallel} > 1$, which yields a positive $\beta_{\parallel}$ and so is physical. Table 2 shows the values for $T_{\perp}$, $T_{\parallel}$ (from the second root of the quadratic equation), $T_{\perp} / T_{\parallel}$, the total temperature $T_{Total} = (2\,T_{\perp} + T_{\parallel})/3$, and plasma beta. The perpendicular temperature was estimated by considering charge-exchange of solar wind neutrals with interstellar protons and the resulting convection to the heliopause without energy losses:

$$T_{\perp} = 2 * T_{ISN}(1 - PU\%) + \frac{1}{2}\frac{m_p v_{\perp}^2}{k_B}(PU\%), \tag{4}$$

where T_{ISN} is the (isotropic) average temperature of interstellar neutral Helium measured by IBEX (= 8250 K) [11,16], PU% is the percent of pick up ions relative to the interstellar proton density in the outer heliosheath, and $v_{\perp}$ is the velocity of the pickup ion ring (= slow solar wind velocity of approximately 450 km/s). Equation 4 assumes thermal equilibrium between the bulk ions and neutrals in the outer heliosheath and assumes that the ion temperature increases by a factor of 2 from the interstellar value as the plasma approaches the heliopause. This increase is consistent with global MHD simulations of the heliosphere-interstellar medium interaction [e.g., 17,18].

Equation 4 assumes that pickup ions are generated beyond the heliopause by charge exchange between interstellar protons and secondary neutrals formed earlier by charge-exchange of solar wind protons with interstellar neutrals within the heliosphere. These "secondary" pickup ions form a ring that is perpendicular to the draped magnetic field against the heliopause. Theoretical estimates place the pickup ion percentage relative to the density in the outer heliosheath, PU%, at 0.03% (see, for example, [17,18]), which is the first entry in Table 2.

Table 2: Temperatures and plasma betas derived for various pickup ion fractions, using Equations 3 and 4 and the density and temperature in Table 1, for the depletion layer adjacent to the heliopause at the Voyager 1 crossing.

PU% (percent)	$T_{\perp}$ (K)	$T_{\parallel}$ (K)	$T_{\perp} / T_{\parallel}$	$\beta_{\parallel}$	T_{Total} (K)	β_{Total}
0.03	2.0×10^4	4.2×10^3	4.8	0.04	1.5×10^4	0.13
0.1	2.9×10^4	7.5×10^3	3.9	0.06	2.2×10^4	0.18
2	2.6×10^5	1.6×10^5	1.6	1.4	2.3×10^5	1.9

4. Plasma Processes in the Depletion Layer and at the Heliopause

Table 2 shows that the pickup ions in the outer heliosheath do not contribute significantly to the temperatures and betas unless they are a few percent of the total density. That is, their density needs to be almost 70 times higher than current theoretical predictions. Thus, because the cold interstellar temperature dominates the plasma energy, beta is very low in the plasma depletion layer. Based on

Equation 3, this low-beta plasma supports a very large temperature anisotropy of 4.8. However, this anisotropy is not nearly as large as the ratio $T_{\perp}$ / $T_{\parallel}$ = 11.6 predicted for double adiabatic theory. Therefore, the growth of EMIC waves and the coupling of the parallel and perpendicular temperatures through pitch angle scattering dominates the plasma depletion layer near the Voyager 1 crossing. This is a strong argument that the plasma depletion process is taking place locally along the Voyager 1 trajectory.

In addition to conditions conducive for the growth of EMIC waves, the plasma depletion layer has two other characteristics that relate to plasma processes at the heliopause. The first process is the growth of lower hybrid waves and the second process is magnetic reconnection at the heliopause.

4.1 Lower Hybrid Waves in the Plasma Depletion Layer:

Pickup ions in the outer heliosphere are a free energy source for lower hybrid waves. These waves, with wave vectors almost perpendicular to the local magnetic field, can accelerate electrons parallel to the magnetic field. This ``lower hybrid drive'' process [20] is believed to produce the seed population of suprathermal electrons that are further accelerated in the outer heliosphere by shocks propagating through the region [21,22]. The chain of wave-particle interactions ultimately results in the generation of Langmuir waves, which mode convert to electromagnetic radiation that propagates into the inner heliosheath and heliosphere. This electromagnetic radiation was observed in the 2-3 kHz range by spacecraft in the outer heliosphere beyond about 20 AU [e.g., 23].

Lower hybrid waves start this chain of wave-particle interactions. They are driven unstable primarily when the ring velocity of the pickup ions is less than five times the local Alfven speed [20]. The ring velocity of the pickup ions is fixed at 450 km/s by the slow solar wind speed in the heliosphere. However, the Alfven speed in the outer heliosphere increases in the plasma depletion layer [24]. Table 1 shows that the Alfven speed increases from 20 km/s to 44 km/s near the heliopause. Unfortunately, the increase is only a factor of about 2.1 (the depletion factor in Table 1) and the resulting ratio of the ring velocity to the local Alfven speed is 10.2. This ratio is a factor of two too large for growth of the lower hybrid instability. Therefore, it is concluded that, although the depletion layer improves the conditions for growth of the lower hybrid instability, it is not likely that these waves are driven unstable in the local depletion layer observed by Voyager 1. As a consequence, the seed population for further acceleration by shocks is probably not produced locally. Such electrons could, however, move along the magnetic field to Voyager 1 from a region of the depletion layer that has a stronger reduction in density and increase in magnetic field, where the lower hybrid drive process is able to proceed.

Langmuir waves are observed on two occasions during the ~250 days the Voyager 1 spacecraft spent in the plasma depletion layer [2]. They are associated with shocks propagating in the outer heliosphere [1]. These Langmuir waves may be produced by electron beams that have been accelerated at some distant location in the depletion layer or by the local shock. In particular, if the depletion factor is larger further away from the Voyager 1 location, then the conditions for generation of lower hybrid waves and the associated superthermal electrons would be improved. Petrinec et al. [13] concluded that the depletion factor at the Earth's magnetopause had to be at least 3.6 in order to explain the vast majority (95%) of certain observations in that depletion layer. There are several assumptions that go into this result. In particular, the value of 3.6 is a minimum. Higher depletion factors are possible, but not required for Earth. Thus, the value of 3.6 should be considered as what is generally required at Earth. If the depletion factor was 3.6 at a remote location in the outer heliosheath (i.e., not on the Voyager 1 trajectory), then the Alfven speed would be about 75 km/s and the ratio of the pickup ion ring speed to the Alfven speed would be 6. This ratio is much closer to the required ratio of 5 for the lower hybrid process. Thus, the Langmuir waves observed by Voyager 1 in the depletion layer probably originated from beams produced at the shock, either remotely or locally, from superthermal electrons that were produced at a distant location in the plasma depletion layer where the depletion factor was larger.

4.2 Magnetic reconnection at the Heliopause:

Magnetic reconnection converts magnetic energy into particle energy and changes the topology of field lines in the vicinity of a current sheet like the heliopause. This process has been invoked to explain observations in the inner heliosheath near the Voyager 1 heliopause crossing [e.g., 14]. Magnetic reconnection can occur over a range of shear angles that depends on the plasma betas on either side of the discontinuity. The plasma betas in Table 2, combined with an estimate of the plasma beta in the inner heliosheath at the heliopause and the observed shear angle at the heliopause provide enough information to determine if reconnection is possible locally where Voyager 1 crossed the heliopause.

Table 3 contains the information and assumptions used to compute the plasma beta in the inner heliosheath at the heliopause. By summing over the core ions and pickup ions, the ion temperature was computed as

$$T = T_{SW}(1 - PU\%_{IHS}) + \frac{1}{2}\frac{m_p v_\perp^2}{k_B}(PU\%_{IHS}), \tag{5}$$

where T_{SW} is the temperature of the (shocked) solar wind proton distribution in the inner heliosheath (assumed to be = 2.3 x 10^5 K, or 20 eV), $PU\%_{IHS}$ is the percent of pick up ions relative to the solar wind proton density in the inner heliosheath (assumed to be 16.5% [17,19], and $v_\perp$ is the velocity of the pickup ion ring (= slow solar wind velocity of approximately 450 km/s). In contrast to the outer heliosheath, the ion temperature in the inner heliosheath is dominated by the pickup ion shell because the pickup ion density is so large compared to the cold solar wind proton density.

Table 3: Estimation of the plasma beta in the inner heliosheath just interior to the heliopause and the change in beta across the boundary, for Voyager 1's location.

Quantity	Value	Origin
Magnetic field magnitude in the inner heliosheath at the heliopause	B = 0.25 nT Range: 0.20 – 0.35 nT	Measured by Voyager 1 [Burlaga et al., 2013]
Density in the inner heliosheath at the heliopause	N = 2 × 10^{-3} cm^{-3} Range: (1 – 4) x 10^{-3} cm^{-3}	Assumed, consistent with Voyager 2 observations
Total temperature in the inner heliosheath at the heliopause	T = 2 × 10^6 K	Computed from Equation 5
Plasma beta in the inner heliosheath at the heliopause	β = 2.2 Range: 0.6 – 6.9	Computed from the magnetic field density and temperature in this table
Δβ across the heliopause = $\lvert\beta_{IHS} - \beta_{PDL}\rvert$	Δβ = 2.1 Range: 0.4 – 6.8	β_{HIS} from this table and β_{PDL} from Table 3 with 0.03 PU%

Using the estimate of plasma beta in the inner heliosheath from Table 3 and the estimate of plasma beta in the depletion layer from Table 2 (with the assumption that the pickup ion percentage is 0.03), the change in beta across the heliopause at the Voyager 1 crossing (Δβ) is 2.1 (see the last entry in Table 3). This relatively large change across the heliopause has implications for local reconnection at the boundary. Variations in the density (measured by Voyager 2 in the inner heliosheath) and magnetic field (measured by Voyager 1 in the inner heliosheath near the heliopause) result in a range for Δβ of 0.4 to 6.8.

Swisdak et al. [25] developed a theoretical condition for reconnection onset that depends on the change in beta across a current sheet and the shear angle at the boundary. This condition has been successfully tested at current sheets in the solar wind [26] and has been investigated at the Earth's magnetopause [27] and Saturn's magnetopause [28,29]. Figure 1 shows how the onset of reconnection depends on Δβ and shear angle. Three curves separate Δβ-shear angle regions where reconnection is possible and where it is "suppressed". It is likely that there is no actual suppression of reconnection at the entire heliopause or any other current layer. Instead, what likely happens for high $\Delta\beta$ conditions is that reconnection is initiated at locations on the boundary where the shear angle is large. Thus, Figure 1 provides an estimate of the range of shear angles where reconnection is possible at the heliopause and whether this range includes the shear angle observed at the Voyager 1 crossing. The dark blue bar in Figure 1 shows $\Delta\beta = 2.1$ at the heliopause (from Table 3). Allowing for a range of thicknesses of the current layer from 0.5 to 2 ion inertial lengths (d_i), reconnection can occur at locations where the shear angle is greater than about 55°, with a range of possible minimum shears ranging from 55° to 127°. (For the extrema of the range $\Delta\beta$ = 0.4 - 6.8, the shear angle must exceed 15 and 140°, respectively.) For $\Delta\beta = 2.1$, the range of possible minimum shears is broad, but it does not include the shear angle that was observed at the Voyager 1 crossing. One of the surprises from the Voyager 1 crossing was that the direction of the magnetic field changed very little across the heliopause. Burlaga et al. [1] report that, across the entire transition of the heliopause, the magnetic field deviates from the Parker spiral direction in elevation by about 14° and in azimuth by about 17°. Therefore, Figure 1 shows that $\Delta\beta = 2.1$ is too high to support reconnection locally at the Voyager 1 crossing. Given the extrema of the range of $\Delta\beta$ from Table 3, Figure 1 does indicate the possibility that reconnection could occur locally if the fluctuations in the density and magnetic field were at their limits.

Plasma beta is reduced in a depletion layer. At the Earth and other planets, the lower beta upstream of the magnetopause opens up the range of shear angles where reconnection is possible because the plasma beta inside the magnetosphere is typically less than 1. At the heliopause, the opposite is true. Plasma beta in the inner heliosphere is typically significantly greater than 1 and the plasma betas in the outer heliosphere and in the depletion layer are significantly less than 1. Thus, lowering plasma beta in the depletion layer further (through increased plasma depletion) has little effect on the $\Delta\beta$ across the heliopause. Instead, $\Delta\beta$ is dominated by the plasma beta in the inner heliosphere, which in turn is dominated by the number of ions in the pickup shell. Thus, Figure 1 with $\Delta\beta = 2.1$ is likely a good representation of the conditions across the entire section of the heliopause that faces the interstellar flow. It remains to be determined if there are locations on the heliopause where the shear angle is large enough to support reconnection. Finally, if the solar wind sector was in the opposite direction in the inner heliosphere at the Voyager 1 crossing, then the shear angle would have been close to 180°. For those conditions, reconnection could occur for almost any $\Delta\beta$.

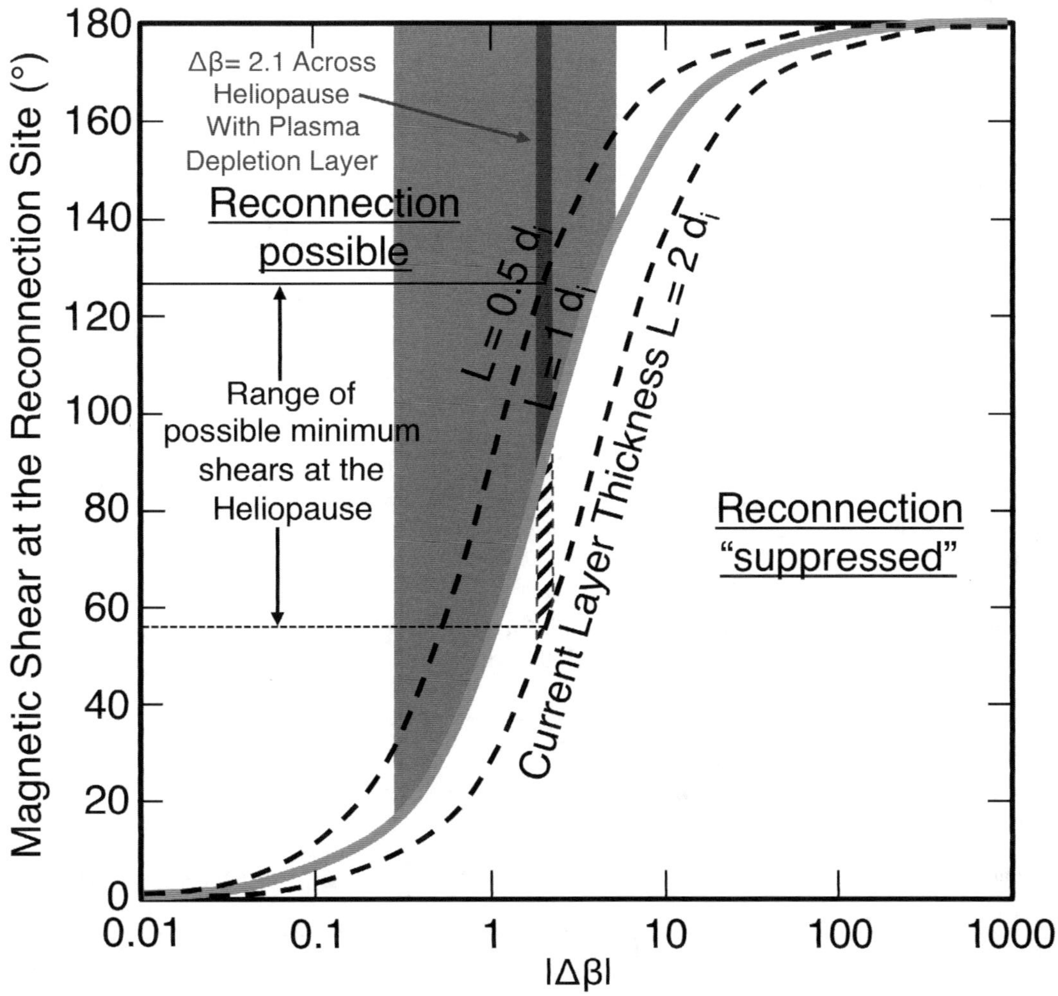

Figure 1: Regions in Δβ-magnetic shear of possible and suppressed reconnection. For the $\Delta\beta = 2.1$ (dark blue line) change across the heliopause at the Voyager 1 crossing, the range of possible magnetic shears that support reconnection is from 55° to 180°. The minimum shear angle of 55° is too large to support reconnection locally at the Voyager 1 crossing (where the shear angle was very small). The extrema of the range of Δβ (light blue shading) could support reconnection down to <20°.

Even if reconnection occurred locally at the heliopause near the Voyager 1 crossing, it is not a source of significant electron heating at the boundary. The small ($V_A < 100$ km/s) Alfven speeds in the plasma depletion layer should result in weak electron heating in the reconnection exhaust [30]. Thus, locally in the vicinity of Voyager 1, an electron seed particle population for further acceleration is created neither by reconnection nor by lower hybrid waves.

5. Conclusions

Using observations from Voyager 1, properties of plasma depletion layers, and some assumptions about the plasma in the outer and inner heliosheath, Tables 1 through 3 contain the plasma parameters near to the heliopause for the Voyager 1 crossing. The plasma in the depletion layer has Alfven Mach number less than 1, is very low beta, is highly anisotropic and is moderately "depleted" compared to the outer heliosheath upstream from the depletion layer (see the depletion factor of 2.1 in Table 1). Locally at the Voyager 1 crossing, the plasma is marginally stable to the Electromagnetic Ion Cyclotron instability, the conditions are not favorable for growth of lower hybrid waves (the ratio of the pickup ring speed to the Alfven speed is too high) and the local shear at the heliopause is too small to support reconnection. Closer to the center of the draping region on the heliopause (away from the Voyager 1 crossing), the plasma depletion is likely greater and the conditions might be favorable for growth of lower hybrid waves. Similarly, away from the Voyager 1 crossing location, the change in plasma beta across the heliopause may be small enough and the local shear at the heliopause large enough to support reconnection. Everywhere on the heliopause, magnetic reconnection, if it occurs, is predicted to be ineffective at heating electrons, based on the small values of Alfven speed ($V_A < 100$ km/s) in the depletion layer and an empirical relation between the Alfven speed and electron heating observed in the solar wind and Earth's magnetosphere [30]. This ineffective electron heating persists even if the asymmetric form of the Alfven speed [31, 32] is considered.

Acknowledgements: The authors would like to thank Gary Zank and the rest of the organizing committee for organizing the 14th Annual International Astrophysics Conference. Also the authors would like to thank the participants for the stimulating discussion at the meeting. Support for this study comes from the IBEX project, part of NASA's Explorer program, and from the Australian Research Council.

References:

[1] Burlaga, L. F., N. F. Ness, and E. C. Stone, 2013, *Science,* **341**, 147.

[2] Gurnett, D. A., W. S. Kurth, L. F. Burlaga, and N. F. Ness, (2013), sciencexpress, 10.1126/science.1241681.

[3] Krimigis, S. M., et al., 2013, Search for the exit: Voyager 1 at heliosphere's border with the galaxy, science, 341, 144.

[4] Zwan, B. J., and R. A. Wolf, 1976, *J. Geophys. Res.*, **81,** 1636.

[5] Anderson, B. J., S. A. Fuselier, S. P. Gary, and R. E. Denton, 1994, *J. Geophys. Res.,* **99,** 5877-5891.

[6] Gary, S. P., B. J. Anderson, R. E. Denton, S. A. Fuselier, and M. E. McKean, 1994, *Phys. of Plasmas, 1,* 1676.

[7] Denton, R. E., B. J. Anderson, S. P. Gary, and S. A. Fuselier, 1994, *J. Geophys. Res.,* **99,** 11,225-11,241, 1994.

[8] Chew, G. F., M. L. Goldberger, and F. E. Low 1956, *Proc. R. Soc. London, Ser. A.,* **236**, 112.

[9] Burlaga, L. F., N. F. Ness, D. A. Gurnett, and W. S. Kurth, 2013b, *Astrophys. J.*, **778:L3**, 1.

[10] Witte, M., 2004, *Astron. and Astrophys.,* **426,** 835.

[11] McComas, D. J., et al. 2015, *Astrophys. J.,* **801:28**, doi:10.1088/0004-637X/801/1/28.

[12] Fuselier, S. A., S. M. Petrinec, and K. J. Trattner, 2000, *Geophys. Res. Lett.,* **27**, 473-476.

[13] Petrinec, S. M., K. J. Trattner, and S. A. Fuselier, 2003, *J. Geophys. Res.,* **108 (A12)**, 1458, doi: 10.1029/2003JA009979.

[14] Swisdak, M., et al., 2010, *Ap. J.,* **710,** 1769.

[15] Swisdak, M., J. F. Drake, and M. Opher, 2013, *Astrophys. J. Lett.,* **774**.

[16] Möbius, E. 2015, *ApJ Suppl. Series*, in press, June,.

[17] Zank, G. P., H. L. Pauls, L. L. Williams, and D. T. Hall 1996, *J. Geophys. Res.,* **101**, A10, 21,639.

[18] Ismodenov, V. V., and D. B. Alexashov, 2015, *Astrophys. J. Suppl. Series*, submitted.

[19] Möbius, E., K. Liu, H. Funsten, S. P. Gary, and D. Winske, 2013, *Astrophys. J.,* **766:129**.

[20] Omelchenko, et al., 1989, Sov. J. Plasma Phys., **15**: 427.

[21] Cairns, I. H., and G. P. Zank, 2002, *Geophys. Res. Lett.,* **29**, 1143, doi:10.1029/2001GL014112.

[22] Mitchell, J. J., I. H. Cairns, and J. Heerikhuisen, 2009, *Geophys. Res. Lett.,* **36**, L12109.

[23] Gurnett, D. A., and W. S. Kurth, 1996, *Space Sci. Rev.,* **78**, 53.
[24] Fuselier, S. A., and I. H. Cairns, 2013, *Astrophys. J.,* **771:83**.
[25] Swisdak, M., et al., 2010,
[26] Phan, T.D., et al. 2010, *Astrophys. J. Lett.,* **719,** L199, doi:10.1088/2041-8205/719/2/L199.
[27] Phan, T. D., et al. 2013, *Geophys. Res. Lett.,* **40**, 11.
[28] Masters, A., et al., 2012, *Geophys. Res. Lett.,* **39**, L08130, doi:10.1029/2012GL051372.
[29] Fuselier, S. A., et al., 2014, *J. Geophys. Res.,* **119**, doi:10.1002/2013JA019684.
[30] Phan, T. D., et al. 2013, *Geophys. Res. Lett.,* **40**, 4475, doi:10.1002/grl.50917.
[31] Cassak, P., and M. A. Shay, 2007, Phys of Plasmas 14, 102114, doi:10.1063/1.2795630.
[32] Swisdak, M. and J. F. Drake, 2007, *Geophys. Res. Lett., 34,* L11106, doi:10.1029/2007GL029815.

IBEX Observations provide strong Evidence that *Voyager 1* is still in the Heliosheath

G Gloeckler[1] **and L A Fisk**
Department of Atmospheric, Oceanic and Space Sciences
University of Michigan, Ann Arbor, Michigan 48109-2143, USA

E-mail: gglo@umich.edu

Abstract. After plasma wave measurements by *Voyager* 1 (V1) revealed a surprisingly high value for the plasma electron density, a value close to that expected in the local interstellar medium, all principal investigators of the *Voyager* mission currently exploring the heliosheath suddenly reversed their position on the location of V1. They concluded unanimously, and NASA announced that V1 has crossed the heliopause and is now in local interstellar space. We have disputed this conclusion, pointing out that to account for all the V1 observations, particularly of the magnetic field direction together with the density, it is necessary to conclude that the higher densities observed by V1 are due to compressed solar wind. In this paper we show that our model for the nose region of the heliosheath can account in detail for the spectral shapes and intensities of Energetic Neutral Hydrogen (ENH) observed by the *Interstellar Boundary Explorer* (*IBEX*) looking in the directions of V1 and *Voyager* 2 (V2). A key feature of our model is the existence of a region, the hot heliosheath, where the outward-moving solar wind is gradually compressed and thus heated, followed by a region, the cold heliosheath, where the solar wind is still compressed but now cold. It is the existence of this cold heliosheath, the region of cold but high-density solar wind, which provides a unique and simple explanation for the low-energy *IBEX* ENH differential intensities. Finally, since this cold heliosheath is the region where V1 must now reside, the low-energy *IBEX* observations provide strong evidence that V1 is still in the heliosphere.

1. Introduction

Differential intensities (spectra) of energetic neutral hydrogen (ENH) measured by the two main instruments on the *Interstellar Boundary Explorer* (*IBEX*) provide invaluable information about the proton velocity distribution throughout, and especially in, the nose region of the heliosheath currently being explored by the two *Voyager* spacecraft. As shown in figure 1, on *Voyager 1* (V1) proton measurements are available only above ~40 keV, and on *Voyager 2* (V2) the core of the solar wind protons (but not any tails) are measured as well. *IBEX* makes line-of-sight measurements, producing average spectra in most directions integrated over time and a small solid angle. The *Voyagers*, on the other hand, measure *in situ* proton spectra at each radial distance in the heliosheath they are exploring. Unfortunately, on V1 there are no proton measurements in the important energy range of pickup protons nor are there solar wind observations, and while on V2 the bulk properties of the core solar wind are available, no information is provided on any tails on the solar wind distributions.

[1] To whom any correspondence should be addressed.

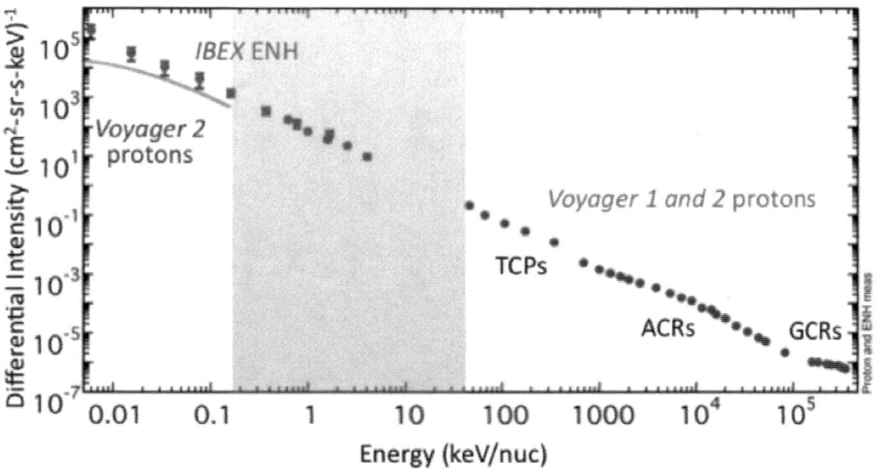

Figure 1. Differential intensity *versus* inertial (sun) frame energy of (1) Energetic Neutral Hydrogen (ENH) measured by *IBEX* in the Voyager 1 direction (red symbols) and (2) greater than 40 keV protons measured in the heliosheath at ~ 115 AU by LECP and CRS on *Voyager 1* (green symbols). No in situ particle are measured by V1 below 40 keV where solar wind protons, transmitted pickup ions and their suprathermal tails reside. On *Voyager 2*, in addition to greater than 40 keV particles, the core of the velocity distribution of solar wind protons is measure (blue curve) still leaving a gap between ~0.1 to 40 keV (shaded region). *IBEX* ENH measurements provide essential information about heliosheath proton populations in the energy ranges not covered by the *Voyagers*.

When at ~122 AU V1 crossed the heliocliff, greater than ~40 keV protons suddenly disappeared and the Galactic Cosmic Ray (GCR) intensity increased. Yet, because the direction of the magnetic showed no change in crossing this sharp boundary, it was generally believed that V1 remained in the heliosheath, although this new region of the heliosheath was clearly very different from that between the termination shock (TS) and the heliocliff. In this new region of the heliosheath energetic particles had easily escape to, and GCRs had easily entered from the local interstellar medium located beyond the heliopause. However, when Gurnett et al. [1] reported exceptionally high (compared to values observed by V2 nearer to TS) plasma densities at ~123 AU just beyond the heliocliff[2], the opinion of many, including all *Voyager* PIs, changed completely. They argued that these high (~0.1 cm^{-3}) density protons had to be of interstellar origin not only because the high values were similar to rough estimates of interstellar proton densitiy, but also because they believed that the heliosheath solar wind could not be compressed. The *Voyager* PIs voted unanimously that V1 has been in the local interstellar medium ever since it crossed the heliocliff, which in retrospect they claimed was really the heliopause. This historically important event was announced to the public in a NASA press conference.

We (Fisk and Gloeckler) have disputed the assertion that V1 has crossed the heliopause, pointing out that to account for all the V1 observations, particularly the lack of any change in the magnetic field direction [2] in crossing the heliocliff, it is necessary to conclude that the high densities observed by Gurnett et al. [1] are due to compressed solar wind. Indeed, by applying the mass conservation law to the solar wind velocity measurements (using energetic particle anisotropies) of Decker et al. [3] and Stone and Cummings [4], that showed a decrease in all three components of the solar wind bulk

[2] Even though the solar wind density is not measured directly on V1, measurements of the density of local thermal electrons are available from plasma wave observation under occasional favorable circumstances.

velocity, it must be concluded that the solar wind becomes compressed beyond ~110 AU along the trajectory of V1. Fisk and Gloeckler [5] developed a model for the nose of the heliosphere (hereinafter referred to as the F&G model) in which a new region (the cold heliosheath) between the heliocliff and heliopause contained a dense but cold (low temperature) solar wind (see figure 2). In the F&G model, sufficiently energetic (e.g. greater than ~ 40 keV) mobile particles could easily escape to the local interstellar medium and GCR could enter the cold heliosheath primarily at the flanks through a porous heliopause[3] provided the heliosheath magnetic field had a favorable polarity. Thus, Gloeckler and Fisk [6] predict that with the higher densities (and hence lower speed) of the compressed solar wind, V1 will encounter another current sheet in the heliospheric magnetic field that results from the change in polarity of the solar magnetic field during the solar cycle. With the then latest estimates of the radial solar wind speed beyond 122 AU, they predicted that V1 should encounter the next current sheet by the end of 2016.

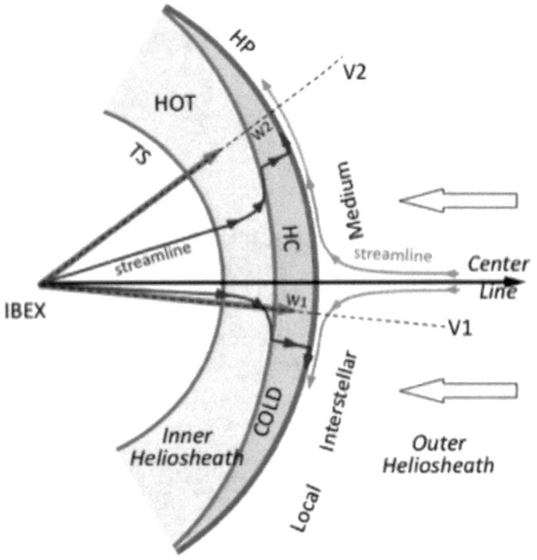

Figure 2. Schematic representation of the nose region of the heliosphere in the planes of constant latitude of V1 (~30° north) and V2 (~30° south). The sharp heliocliff (HC) boundary (red) separates the hot from the cold heliosheath, and is more blunt than either the steady-state termination shock (TS, blue) or the heliopause (HP, green). The brown curves represent two sample streamlines from the F&G model of the nose of the heliosheath. Best fits to the low-energy segment of the *IBEX* ENH spectra in the V1 and V2 directions respectively determine the widths of the cold heliosheath near the flanks (V2 direction) to be $W2 \approx 3$ AU, and near the centerline (V1 direction) to be $W1 \approx 18$ AU, as predicted in the F&G model.

The *Interstellar Boundary Explorer* (*IBEX*) has been observing Energetic Neutral Hydrogen (ENH), created beyond the termination shock (TS) [7]. In particular, *IBEX* measures line-of-sight ENH spectra in nearly all directions including those of V1 and V2 (e.g. [8]). Nearly all ENH seen by *IBEX* is produced in the heliosheath where interstellar neutral hydrogen gas flowing into the heliosphere charge-exchanges with solar wind and more energetic protons (suprathermal solar wind tails and pickup ions with their tails). The ENH spectra (differential intensities as a function of inertial or sun frame energy E), $J_{enh}(E)$, is derived from the proton spectra, $J_p(E,r)$, beyond 100 AU using the standard ENH production equation:

$$J_{enh}(E,\theta,\phi) = \int_{100AU}^{\infty} J_{enh}(E,\theta,\phi,r) N(r) \sigma(v_{rel}(r)) dr \qquad (1)$$

where $N(r)$ is the density of the neutral gas, $\sigma(v_{rel}(r))$ is the H-p charge exchange cross section and $v_{rel}(r)$ is the relative speed between the proton and the interstellar hydrogen gas with which it interacts. Thus, *IBEX* spectra in a given (θ_o,ϕ_o) pixel provide essential information on the proton spectra in the same (θ_o,ϕ_o) direction in the heliosheath. In particular, *IBEX* spectra in the V1 and V2 directions, combined with in situ measurements by V1 of the solar wind velocity and proton spectra above 40 keV from the TS to the heliocliff, and of solar wind bulk parameters as well as >40 keV proton spectra

[3] The heliopause in the F&G model had to be a rotational discontinuity boundary.

by V2, provide, through equation 1, essential information on heliosheath proton spectra in the *IBEX* energy range (~0.005 to ~5 keV) as a function of heliocentric distance *r*.

While most of the attention was focused on explaining the *IBEX* ribbon[4] spectra, Gloeckler and Fisk [10] were among the first to offer an explanation of the then available *IBEX* non-ribbon spectrum in the V1 direction. They considered 4 proton populations in the heliosheth, among them the low density heliosheath solar wind that periodically would be moving radially inward. This inward moving solar wind would then create the ENH that was detected by *IBEX*. However, when Fuselier et al., [8] published updated *IBEX* spectra over the entire energy range of the IBEX-Lo and IBEX-Hi instruments in both the V1 as well as the V2 directions, the Gloeckler and Fisk [10] model predicted much lower fluxes below ~0.3 keV than those of the updated *IBEX* spectra of 2012 (see figure 7 of [8]). Model fits to the non-ribbon *IBEX* spectra by others (e.g. [10,11]), including the most recent one by Zirnstein et al. [12], assumed that pickup ion populations in the heliosheath were the dominant source of the ENH observed by *IBEX*.

In the most recent attempt to explain the *IBEX* non-ribbon spectra, Zirnstein et al. [12] considered the parent populations of *IBEX* ENH to be three forms of pickup ions (transmitted, reflected and injected). Contributions from the solar wind or its suprathermal tail were deemed to be insignificant, because after all, it was generally believed that the solar wind could not be compressed nor develop strong tails anywhere in the heliosheath. Summing up the three pickup proton populations in both the heliosheath (they called the "Inner Heliosheath" or IHS) and the local interstellar medium (they called the "Outer Heliosheath" or OHS) resulted in a complex spectrum that flattens and then bends down below 100 eV (see figure 4 of [12]). In particular, below ~0.15 keV their model flux was at least 10 times lower than fluxes measured by *IBEX*.

IBEX fluxes below ~0.1 keV were never explained. Nobody could find a plasma source for these very low energy ENHs. There is, however, a very simple explanation for the non-ribbon Energetic Neutral Hydrogen (ENH) differential intensities, $J_{enh}(E)$, measured by *IBEX* over its entire energy range. All that is necessary to fit these *IBEX* spectra, is to have a high density hot solar wind just upstream of the heliocliff and a high density but very cold solar wind in the cold heliosheath region between the heliocliff and the heliopause. It is in the cold heliosheath where the lowest energy (< ~0.1 keV) ENH is created.

2. The F&G model for the nose region of the heliosheath

The existence of a cold heliosheath between the heliocliff and the heliopause is an essential feature of the Fisk and Gloeckler [5] model for the nose of the heliosheath. We will briefly describe this model which in the hot heliosheath (between the termination shock and the heliocliff) also allows for the compression and heating of the solar wind and pickup ions, as well as formation of strong suprathermal tails on both the solar wind and pickup proton distributions. In the hot heliosheath the outward moving solar wind is gradually compressed and heated reaching densities of ~ 0.05 cm^{-3} at the heliocliff.

Note in figure 2 that in the F&G model the shape of the heliocliff is more blunt than either the termination shock or the heliopause. This will result in a difference in the widths of the hot and cold heliosheaths in the directions of V1 and V2, which are also noted in figure 2. Figure 2 should be interpreted as an appropriate representation for both the heliographic latitude of V1 (34.6 north) and of

[4]The ribbon is a relatively narrow region of enhanced ENHs, with energies ~1 keV, superimposed on the nose-region spectrum, and stretching like a ribbon across the sky. There have been numerous explanations put forth for the *IBEX* ribbon, many of which are summarized in [Schwadron et al. (2014)], and most of which involve processes occurring beyond the heliopause

V2 (31.1 south). Note that the trajectory of V1 is close to the centerline at its latitude, and thus in this direction the hot heliosheath is relatively narrow and the cold heliosheath relatively wide. In contrast, the trajectory of V2 is closer to the flanks, where the hot heliosheath is relatively wide and the cold heliosheath relatively narrow. The general features of the F&G model [5] are illustrated in figure 2 that shows: (1) A region between the termination shock and the heliocliff (where there is a precipitous decrease in the ACRs). This region is the 'hot heliosheath' (HH) in which the solar wind is gradually compressed and heated as it drifts outward, attaining maximum thermal speed near the heliocliff. (2) A region between the heliocliff and the heliopause, where the solar wind continues to be compressed, but is very cold. This region is the 'cold heliosheath' (CH) from which ACRs, TSPs and the high-energy pickup ions as well as tail particles of the solar wind have escaped[5] into the local interstellar medium. The escape of these mobile particles is primarily along the magnetic field that is lined up with the heliocliff and connects to the heliopause primarily at the flanks. To make this possible, the heliocliff must be more blunt than the heliopause.

The F&G model of the nose region of the heliosheath [5] is based on two *Voyager* observations: (1) In the HH, along a cone of constant heliographic latitude at radial distances along the V1 trajectory beyond ~110 AU the solar wind flow direction turns increasingly more from the radial to a transverse direction. (2) In the cold helioaheath, mobile particles such as pickup ions and ACRs, which contain the pressure, have escaped. Thus, these energetic, mobile particles that are free to move along the magnetic field to balance pressure, can be considered and treated as a separate gas from the solar wind. The solar wind inner core (but not its tail) contains the mass and is responding to local conditions.

The uncoupling from the bulk or inner core of solar wind of the its outer core and tail, and the subsequent escape of the solar wind outer core and tail as well as of the higher energy pickup ions, TSP protons and ACRs, combined with the observation that the solar wind flows primarily on a cone of constant heliographic latitude, makes it quite simple to calculate the streamlines of the solar wind, as is done in [5]. With the streamlines primarily radial and azimuthal downstream from the termination shock, there must be a centerline region, where the streamlines diverge, flowing to the right and to the left of the centerline. Furthermore, the centerline region must occur at all heliographic latitudes, defining a plane that is oriented roughly perpendicular to the heliographic equator. The centerline plane includes the direction of motion of the Sun through the local interstellar medium so that the flow of solar wind into the tail of the heliosheath forms a tail that is opposite to the direction of motion.

A vacuum is created on either side of the centerline by the divergence of the streamlines into which the solar wind can leak. The centerline region thus exerts a drag that reduces the outward flow speed of the solar wind. This decrease in speed results in an increase in solar wind density, converting solar wind ram pressure into thermal pressure. The region where the solar wind is gradually being compressed and heated is the hot heliosheath.

As the solar wind velocity decreases, the azimuthal speed goes to zero before the radial speed and the solar wind continues to flow radially outward to the heliopause, across which the solar wind and the magnetic field have to escape. In the F&G model the heliopause must thus be a rotational discontinuity, which can accommodate flows normal to the discontinuity. In the F&G model the heliopause is at 140 to 145 AU.

The escape of the solar wind across the heliopause will result primarily in the decrease of its pressure since the higher energy particles escape more easily than the lower energy ones. When this escape

[5] See ref [5] for the details of the escape mechanism.

becomes possible[5], the ACRs, the high-energy pickup ions, and the high-energy tail of the solar wind disappear as is observed. This region is thus analogous to a balloon that is cooled. The balloon shrinks in size and the density increases to equal the densities observed by [1]. The region where escape across the heliopause occurs is the cold heliosheath, and the boundary between the hot and cold heliosheath is the heliocliff, marked by the sudden escape of the ACRs.

We emphasize again that in the F&G model the shape of the heliocliff is more blunt than the heliopause resulting in a difference in the widths of the hot and cold heliosheaths in the directions of V1 and V2, as noted in figure 2. Figure 2 is an appropriate representation for both the heliographic latitude of V1 (34.6 north) and of V2 (31.1 south). Note that the trajectory of V1 is close to the centerline at its latitude, and thus in this direction the HH is relatively narrow and the CH relatively wide. In contrast, the trajectory of V2 is closer to the flanks, where the HH is relatively wide and the CH relatively narrow.

3. The ENH Spectra in the Directions of *Voyagers 1* and *2*

Using the general features of the F&G model for the nose region of the heliosheath, we determine the proton velocity distributions as a function of heliocentric distance, *r*, as well as the dimensions of the HH and CH that are required to fit the *IBEX* ENH spectra in the directions of V1 and of V2, respectively. We use the most recent *IBEX* spectra in both the V1 and V2 directions published by Fuselier et al. [8] who have extrapolated their measured spectra back to 100 AU (i.e. to the inner edge of the HH), correcting background and for the extinction that occurs from charge-exchange in the supersonic solar wind between 1 and 100 AU. These V1 and V2 pixel *IBEX* spectra cover the full energy range (~0.006 to ~4 keV) of the *IBEX* instruments and, while the error bars on the lowest four energy points of the spectra are large, they have sufficient accuracy to distinguish among models.

3.1. Proton velocity distributions in the heliosheath

The Energetic Neutral Hydrogen spectra observed by *IBEX* are directly related to the proton spectra in the heliosheath in the *IBEX* energy range through equation 1. In this range reside several important populations of particles. Among them the solar wind with its suprathermal tail, and transmitted pickup protons and their tails make the dominant contributions to the ENH spectra. Unfortunately, as shown in figure 1, energy spectra of protons are not measured on V1 in the *IBEX* energy range and, while on V2 solar wind distributions are measured, they are typically confined to energies below about 0.01 keV, and with V2 still in the hot heliosheath, are currently available only to distances not that far from the termination shock.

Lacking direct measurements of proton velocity distributions in the *IBEX* energy range we use three analytic functions, each with a minimum number of free parameters, that respectively fit velocity distribution of (a) the solar wind with its suprathermal tail, (b) interstellar pickup ions, and (c) suprathermal tails measured directly by ACE and Ulysses in local strong acceleration regions. Making the reasonable assumption that these spectral forms are also representative of the heliospheric proton populations, we use the same analytic function for proton distributions in the heliosheath and then adjust the few free parameters of these functions to match the *IBEX* spectra.

3.1.1. The solar wind proton distribution function in the hot heliosheath

The hot heliosheath (HH) is characterized as a region of high compressive turbulence [2], making it an ideal region for accelerating particles by the Fisk & Gloeckler [13] pump mechanism. Similar regions of high compressive turbulence are often observed downstream of strong propagating (CME) shocks at 1 AU. In figure 3 we show the solar-wind-frame phase space density spectrum of protons measured by ACE downstream of a strong shock [13]. The function (solid curve) that fits the entire spectrum has the simple form:

$$f_{sw} = f_{sw,0}\left[1+\left(\frac{v}{v_{th,sw}}\right)^2\right]^{-2.5}\exp\left[-\left(\frac{v}{v_{roll,sw}}\right)^2\right] \tag{2}$$

With $f_{sw,0} = 4\bullet10^{11}$ s^3/km^6, $v_{th} = 2.3\bullet10^6$ cm/s and $v_{roll,sw} = 4.5\bullet10^8$ cm/s, equation 2 provides an excellent fit to the observed spectrum (red filled circles). The number density derived from this distribution is 18 cm^{-3}. The observed spectrum has a core segment that is reasonably well fit my a maxwellian distribution (dashed curve) with a density of 18 cm^{-3} and thermal of $2.3\bullet10^6$ cm/s, (the same values that were used in equation 2) and a suprathermal -5 power law tail segment above 4 times the thermal speed (~$1\bullet10^7$ cm/s) that rolls over at an e-folding speed of $4.5\bullet10^8$ cm/s.

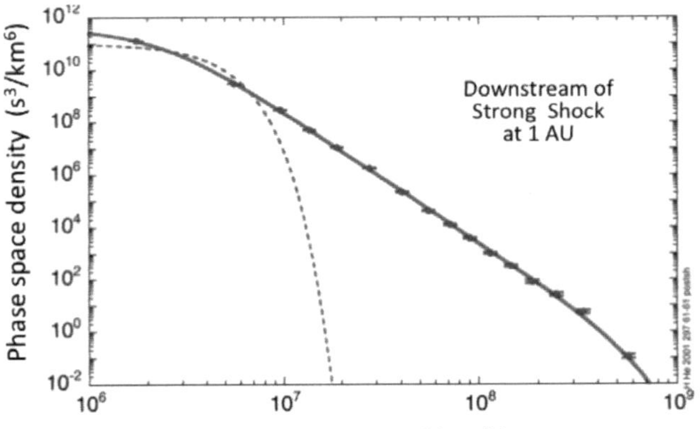

Figure 3. Distribution function of protons (red symbols) observed at 1 AU by ACE SWICS and ULEIS downstream of a strong shock [13]. The blue curve is a fit to the spectrum using equation 2 and parameters listed in the text. The dashed curve is a fit of a maxwellian distribution to the core phase space density (lowest two energies points).

It is reasonable to assume that similar spectra exist in the hot heliosheath, the region downstream of the termination shock. Allowing for the gradual compression of the solar wind as it propagates outward the three parameters in equation 2 will depend on heliocentric distance r. However, the three parameters can be reduced to two by assuming adiabatic compression of the solar wind. The thermal speed is then related to the density by equation 3,

$$v_{th} = v_{th,TS}\left\{\frac{n_{sw}(\langle r_k\rangle)}{n_{TS}}\right\}^{\frac{1}{3}} \tag{3}$$

Thus, equation 2 becomes

$$f_{sw}(r) = f_{sw,0}(n_{sw}(r))\left[1+\left(\left\{\frac{v}{v_{th,TS}}\right\}\left\{\frac{n_{TS}}{n_{sw}(r)}\right\}^{\frac{1}{3}}\right)^2\right]^{-2.5}\exp\left[-\left(\frac{v}{v_{roll,sw}}\right)^2\right] \tag{4}$$

It has only 2 free parameters, the density and the rollover speed $v_{roll,sw}$. The contribution of solar wind protons to the *IBEX* ENH spectra become progressively more important with increasing solar wind density which reaches its highest value (~0.05 cm^{-3}) in the hot heliosheath at the heliocliff.

3.1.2. The transmitted pickup proton and tail distribution functions in the hot heliosheath

Pickup protons (PP) created upstream of the termination shock are compressed and heated as they cross the TS and enter the HH. Convected across the TS, their density and pressure increase is given by the Rankin-Hugonion relations and will depend on the compression ratio of the TS (typically ~3). These pickup protons will make the dominant contribution to the *IBEX* ENH spectra near the TS. Unfortunately, these important populations are not measured by V1. However, Ulysses measured directly [14] pickup proton distributions and their suprathermal tails downstream of Jupiter's bow shock (see figure 4). Because the heliosheath has many similarities to the region downstream of Jupiter's shock we use the same analytic functions that best fit the Jupiter spectra for describing the pickup proton distributions and their tails in the heliosheath downstream of the termination shock.

The function (blue curve in figure 4) that fits to the pickup proton segment of the spectrum downstream of Jupiter's bow shock has the form:

$$f_{PP} = f_{PP,0}\left[1+\left(\frac{v}{4v_{upstrm}}\right)^2\right]^{-15}\exp\left[-\left(\frac{v}{12v_{upstrm}}\right)^2\right]\exp\left[-\frac{3v_{upstrm}}{v}\right] \tag{5}$$

It has only two free parameters: the pickup proton density (that determines $f_{PP,0}$), and the solar wind speed upstream of the shock. With $f_{PP,0} = 2.2{\bullet}10^5$ s^3/km^6 and $v_{upstrm} = 4.2{\bullet}10^7$ cm/s, equation 5 provides an excellent fit to the observed spectrum (red symbols) below $2{\bullet}10^8$ cm/s.

Ulysses HI-SCALE measured the -5 power law suprathermal tail (filled red squares). This segment of the spectrum is well fit by the analytic function:

$$f_{PPtail} = f_{PPtail,0}\left(\frac{v}{4v_{upstrm}}\right)^{-5}\exp\left[-\left(\frac{v}{v_{roll,PPtail}}\right)^5\right]\exp\left[-\left(\frac{2v_{upstrm}}{v}\right)^2\right] \tag{6}$$

It also has only 2 free parameters: the tail density (that determines $f_{PPtail,0}$), and the rollover e-folding speed, $v_{roll,PPtail}$). Setting $f_{PPtail,0} = 2.2{\bullet}10^5$ s^3/km^6, $v_{roll,PPtail} = 10^9$ cm/s, and using for $v_{upstrm} = 4.2{\bullet}10^7$ cm/s (same as in equation 5) yields an excellent fit to the suprathermal tail segment of the measured spectrum. Equation 6 has also only two free parameters, $f_{PPtail,0}$ and $v_{roll,PPtail}$.

It is reasonable to assume that in the hot heliosheath similar transmitted pickup proton distributions and pickup proton tail distributions exist. The free parameters would now be dependent on heliocentric distance r, and we will adjust them to fit the upper energy segment of the *IBEX* ENH spectra.

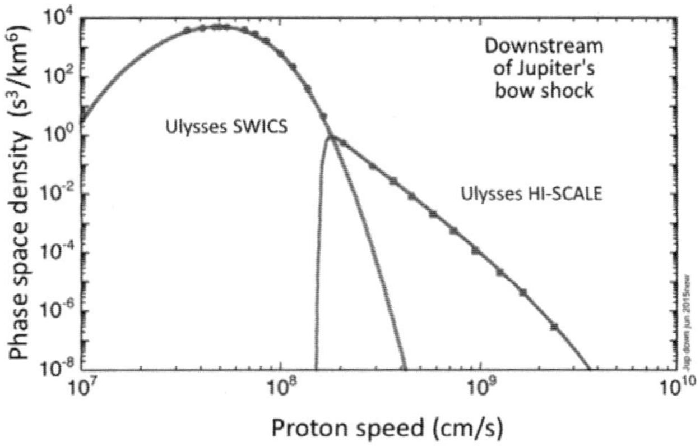

Figure 4. Distribution function of protons (red symbols) observed [14] downstream of Jupiter's bow

shock by Ulysses SWICS and HI-SCALE. The blue and red curves are, respectively, fits to the pickup proton and the suprathermal tail segments of spectrum using equations 5 and 6 and parameters listed in the text.

3.1.3 The proton distribution function in the cold heliosheath

At the heliocliff the solar wind density is most compressed and its pressure has attained its highest value. The density and pressure of the pickup proton tail have also reached their maximum values. However, the pickup proton pressure has to gradually decreasing with increasing r, and at the heliocliff its pressure is at its lowest value. The reason for this is the transfer of energy from pickup protons to the tail. Pickup protons lose the energy they transfer to the tail. The pump acceleration mechanism, with its key assumption that the total energy of the system remains constant, produces the tail. In a system of constant energy the total pressure (sum of the PP and the PPtail pressures) must therefore remaining constant.

Crossing the heliocliff the proton distributions enter the cold heliosheath where the magnetic field lines intersect the heliopause at the flanks. When the heliosheath magnetic field direction is roughly aligned with the direction of the draped local interstellar field (as has been the case since V1 crossed the heliocliff) the highly mobile energetic particles will escape and enter the local interstellar medium. This is why V1 observes of the sudden disappearance of > 40 keV protons and heavier particles.

The velocity distribution of the remaining protons in the cold heliosheath, those that have not escaped, will be

$$f_{cold,remain}(r,v)=\left\{f_{sw}(r,v)\exp\left[-\left(\frac{v}{v_{uncouple}}\right)^4\right]+f_{PP,HC}(v)+f_{PPtail,HC}(v)\right\}g(v) \tag{7}$$

The exponential term represents the speed-dependent uncoupling function of the solar wind tail. The function $g(v) = [1 - h(v)]$ with

$$h(v)=\left\{1-\exp\left[-\left(\frac{v}{v_{cold}}\right)^3\right]\right\} \tag{8}$$

The function $h(v)$ is the particle escape probability. Few low speed particle escape, while hardly any particles having speeds several times the rollover speed, v_{cold}, remaining. The single free parameter in $h(v)$, v_{cold}, is strongly constrained by the disappearance of >40 keV particle observed by V1 after crossing the heliocliff, and by the requirement of pressure balance (some low energy pickup ions must remain in the cold heliosheath to compensate for the pressure lost due to the disappearance of >40 keV particles). The other free parameters in equation 7 (especially $v_{uncouple}$) are determined by matching the lowest energy segment of the *IBEX* spectra.

3.2. Radial dependence of the fit parameters

The *IBEX* V1 and V2 pixel ENH spectra along with *in situ* observations of the energy spectra of protons above 40 keV by V1 allow us to find the radial dependence of the few free fit parameter in equations 4 to 8.

The main free parameter in fitting the time averaged *IBEX* spectra is radial dependence of the time-averaged density of the solar wind in the heliosheath and, to a lesser degree, the density of pickup ions near the termination shock. The radial dependence of the solar wind density that provided the best fit to the middle energy segment of the *IBEX* V1 pixel spectrum is shown in figure 5a. Starting at 0.002 cm^{-3}, the solar wind density gradually ramps up to a value of ~0.05 cm^{-3} at the heliocliff. The density continues to increase rapidly beyond the heliocliff, reaching the local interstellar density about 4 AU downstream of the heliocliff at ~126 AU.

The other essential fit parameter is the time-averaged radial speed, $u_r(r)$ of the solar wind. One needs to know $u_r(r)$ for transforming from the solar wind frame of reference to the sun frame of reference of the *IBEX* measurements. The blue curve in figure 5b shows the dependence of $u_r(r)$ on heliocentric radial distance. Starting near the termination shock, the time-averaged $u_r(r)$ decreased matching the *in situ* measurement[6] of Decker et al. [3]. At about 115 AU $u_r(r)$ reaches its minimum value of only ~ 5 km/s. To fit the lowest energy segment of the *IBEX* V1 spectrum it is necessary to slowly increase $u_r(r)$ to about 13 km/s near the heliopause[7].

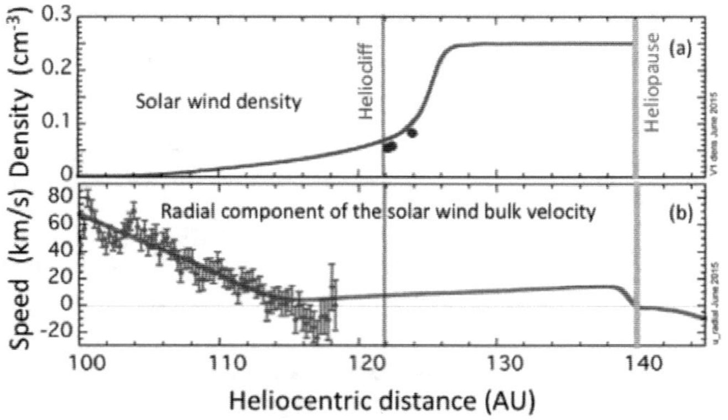

Figure 5. Solar wind number density (upper panel) and radial component of the bulk velocity measured by Decker et al. [3] (lower panel) as a function of heliocentric radial distance. The smooth red and blue curves are respectively the time-averaged density and the radial speed profiles that give best fits to the *IBEX* ENH spectra. Snap-shot measurements of density by Gurnett et al. [1] (blue symbols) and of radial speed by Decker et al. [3] will deviate from the time-averaged values as discussed in the text.

The fit parameters used in equations 3 to 8 give us the radial dependence of the distribution functions of the various proton populations in the heliosheath. From these distribution functions we derive physical parameters (e.g. thermal speeds of the core spectrum and pressures) of each of these populations. In figure 6a we show the variation of the thermal speed of the core solar wind with heliocentric distance. As the solar wind is compressed, its thermal speed increase, reaching 115 km/s at the heliocliff. Upon entering the cold heliosheath the solar wind immediately cools considerably by shedding its strong suprathermal tail. The thermal speed of the solar wind in the CH is reduced to just ~20 km/s[8].

---

[6] *In situ* measurements of the solar wind radial speed and density by the *Voyagers* will fluctuate from the time-averaged values due to the dynamics of the system. These semi-periodic fluctuations over distances of several AU are evident in measurements of $u_r(r)$ and are likely to occur in the solar wind density measurements as well.

[7] The time-averaged speed $u_r(r)$ derived from the *IBEX* measurements beyond the heliocliff is considerably higher than that used by Gloeckler and Fisk [6] to estimate the time of observation of the magnetic sector reversal. This higher radial speed implies that V1 should see the 180° change in the magnetic field direction several years later than estimated in ref [6].

[8] An ideal plasma detector on V1 looking in all directions would easily have detected this solar wind. While V2 has a working (but not an ideal) plasma instrument with its field of view pointing primarily towards the sun, it is not clear that solar wind observations will be made once V2 enters the region of the heliosheath where $u_r(r)$ is less than the speed of the spacecraft (~17 km/s).

Journal of Physics: Conference Series **642** (2015) 012011 doi:10.1088/1742-6596/642/1/012011

IBEX ENH spectra combined with *Voyager* observations of magnetic field strength and energetic particle spectra above 40 keV allow us to obtain a good estimate of the total pressure ($2.5 \bullet 10^{-12}$ dyne/cm^2) in the heliosheath from the sum of pressures of the magnetic field and of the four proton populations (solar wind and its tail, transmitted pickup protons, suprathermal tails and ACRs). In figure 6b we show the radial dependence of the respective pressures. Just downstream of the termination shock transmitted pickup protons (*PP*) account for nearly all the pressure. According to the pump mechanism the tail pressure (*tail*) will gradually increase (due to the increase of the rollover speed of the tail) to its maximum value at the heliocliff. The pressure gained by the tails results in the loss of pressure of the pickup protons. At the heliocliff nearly all of the tail particles and most of the pickup protons vanish, and the pressure of remaining energetic particles drops precipitously.

The pressure resulting from the solar wind and its -5 tail (*SW*) is insignificant in the low-density solar wind. However, as the solar wind is compressed its pressure increased dramatically and at the heliocliff this is by far the dominant pressure (almost $2.5 \bullet 10^{-12}$ dyne/cm^2). When the solar wind enters the cold heliosheath it loses not only its tail but also much of the high-energy portion of its core. These solar wind particles detach from the solar wind inner core and escape. The solar wind in the cold heliosheath is dense because the remaining innermost core particles (which determine the density) have not escaped, but its thermal speed is low because of the loss of outer-core particles. While its pressure drops from $2.5 \bullet 10^{-12}$ dyne/cm^2 at the heliocliff to $1.5 \bullet 10^{-12}$ dyne/cm^2 in the cold heliosheath, it still carries most of the pressure.

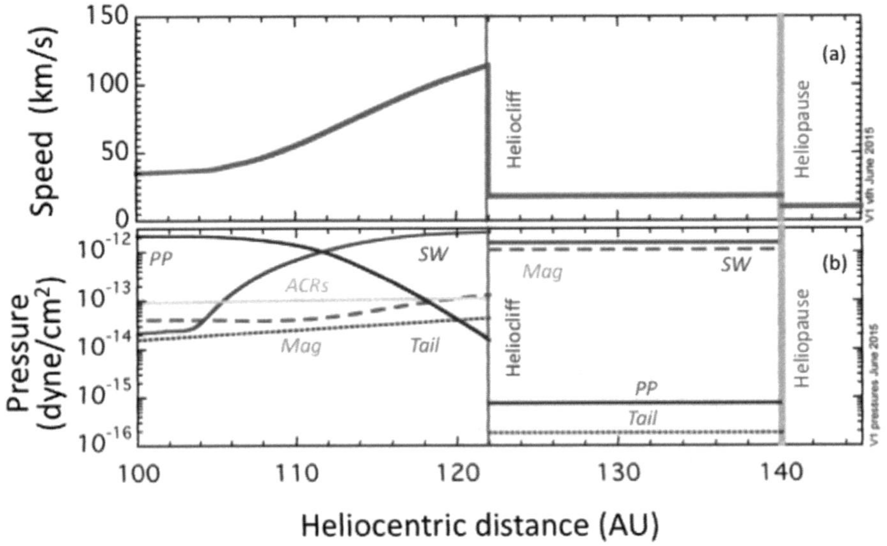

Figure 6. Radial profiles of the thermal speed of the core solar wind (upper panel), and (lower panel) the pressures of the solar wind with its -5 tail (*SW*), pickup protons (*PP*), tail protons (often labeled TSPs), ACR protons and magnetic field (*Mag*). There is an abrupt decrease at the heliopause in the solar wind density and all the particle pressures because of escape of the more energetic, highly mobile particles (described in the text). At the same time there is a large increase in the magnetic pressure required to maintain pressure balance.

The magnetic field pressure is estimated from the magnetic field strength measured by V1 [2]. In the hot heliosheath the magnetic pressure (*Mag*, dashed curve) is small. In the cold heliosheath its pressure jumps to $\sim 10^{-12}$ dyne/cm^2. This large increase in magnetic pressure compensates for the pressure loss from the disappearing ACRs, pickup protons and tails and pressure balance is maintained.

3.3. Fits to the V1 and V2 pixels IBEX Energetic Neutral Hydrogen spectrum

The solar wind frame velocity distributions of the various proton populations in the hot and cold heliosheaths as a function of heliocentric radial distance along either the V1 direction, (θ_{V1},ϕ_{V1}) are computed using equations 3 to 8. The corresponding differential intensities in the sun frame are then given by

$$J_p\left(E,r,\theta_{V1},\phi_{V1}\right)=1.83Ef_p\left(v_{p,radial}-u_{sw,radial},r,\theta_{V1},\phi_{V1}\right) \tag{9}$$

where E, in units of keV, is $[2.283\bullet10^{-8}(v_{p,radial}-u_{sw,radial})]^2$, $v_{p,radial}$ is the radial component of the particle velocity and $u_{sw,radial}$ is the radial component of the solar wind bulk velocity, all in units of cm/s. Note that $u_{sw,radial}$ is a function of r and its variation with r is shown in figure 5b. The ENH differential intensity in the sun frame is then computed using the summation form of equation 1:

$$J_{enh}\left(E,\theta_{V1},\phi_{V1}\right)\doteq\sum_{k=1}^{k=n}J_p\left(E,\langle r_k\rangle,\theta_{V1},\phi_{V1}\right)N\sigma\left(v_{rel}\left(\langle r_k\rangle\right)\right)\Delta r_k \tag{10}$$

We use a constant value of 0.15 cm^{-3} for the neutral hydrogen density and for the p-H charge exchange cross section, σ, we use the formula given by Lindsay and Stebbing [15]. Note that v_{rel}, the relative speed between the interstellar gas and the proton with which it charge-exchanges will also depend on r. Also note that $J_p(E,<r_k>,\theta_{V1},\phi_{V1})$ is the sum of the differential intensities of all proton populations in the *IBEX* energy range (i.e. solar wind, pickup protons and tail particles) at ,$<r_k>$. The summation starts at r_1 = 100 AU and ends at $r_n = r_{heliopause}$ = 140 AU, and the increments, Δr_k, are variable, typically 2 AU when changes in the solar wind density are large and 5 AU when the changes in density are small. Summing over the entire heliosheath from 100 AU to 140 AU (the likely location of the heliopause at V1) produces the heavy black curve in figure 7a. This excellent fit to the V1 pixel ENH spectrum over the entire energy range of IBEX is obtained using the V1 solar wind density profile of figure 6a, the profile of the radial component of the solar wind bulk velocity for V1 shown in figure 6b, a pickup proton density of 0.0008 cm^{-3} and a cutoff speed ('upstream' speed, v_{upstrm}, in equation 5) of 320 km/s just downstream of the termination shock. With increasing distance from the TS, the cutoff speed (v_{upstrm}) was gradually reduced to produce the decrease in the pickup proton pressure shown in figure 6b. Decreasing the PP cutoff speed reduced the pickup proton contribution to highest energy segment of the IBEX spectra. This effect can be seen by examining dashed red curve of figure 7a, obtained by summing equation 9 from 100 AU to 110 AU where the PP cutoff speed remained high. Lowering the cutoff speed will result in a more truncated dashed curve and thus a smaller contribution to the total high-energy segment of the ENH spectrum. Because of the low solar wind density in this co-called low-density hot heliosheath near the termination shock contributions from the solar wind (in the energy range from ~0.01 to 0.1 keV) are small. The primary contribution to the mid energy segment of the *IBEX* spectrum, however, does come from the compressed and hot solar wind near the heliocliff (red curve in figure 7a). The high-density but very cold solar wind from cold heliosheath is the primary source of the lowest energy ENH. The ENH spectrum resulting from this dense but cold solar wind is the blue curve in figure 7a. This spectrum (blue curve) was obtained by summing from 122 to 140 AU, making an 18 AU wide cold heliosheath in the V1 direction.

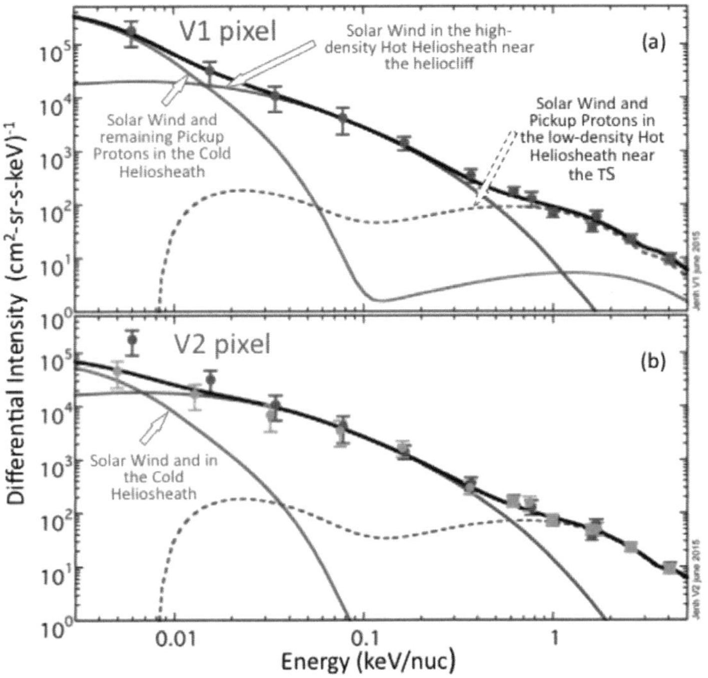

Figure 7. Differential intensity spectra measured by Fuselier et al. [8] in the V1 (panel *a,* red symbols) and V2 (panel *b,* blue symbols) directions, and model spectra described in the text. Main contributions to the *IBEX* spectra come from the cold, high density solar wind (at the lowest energy, blue curve), from the hot, high density solar wind in the hot heliosheath (middle energies, red curve), and from the transmitted pickup protons in the hot, low density heliosheath (highest energies, dashed curve).

The only published V2 pixel ENH spectrum [8] is shown in figure 7b. Above ~0.1 keV the V2 (light blue symbols) and V1 (red symbols) spectra are nearly identical. Below 0.05 keV the V2 ENH fluxes fall more and more below those of V1 and at the lowest energy they are lower by about a factor of 5. Even though the statistical and systematic uncertainties are large, the difference in fluxes is real since the background subtraction and survival probability corrections are independent of the look direction of the *IBEX* instruments. The simplest way to account for the reduced V2 fluxes at the lowest energies is to decrease the size of the cold heliosheath in which these low energy ENHs are created. Indeed decreasing the width of the cold heliosheath in the V2 direction, which is further from the centerline, to ~3 AU while keeping all the other parameters the same as for V1 provides an excellent fit to the V2 pixel ENH spectrum.

4. Closing remarks

We have demonstrated that the *IBEX* ENH observations in the directions of V1 and V2 can be fully accounted for by the model of Fisk and Gloeckler [5] for the nose region of the heliosheath. To fit the ENH observations requires a region of compressed and hot solar wind (the hot heliosheath), with an extensive suprathermal tail, and a region of compressed but cold solar wind (the cold heliosheath), lying between the hot heliosheath and the heliopause.

We have also obtained from fits to the lowest energy segments of the *IBEX* spectra the widths of the cold heliosheaths in both the V1 direction (close to the centerline) and the V2 direction (near the flanks). Assuming that in the V2 direction the heliopause is somewhere between 140 and 145 AU, we predict that V2 will cross the heliocliff at 137 to 143 AU, just 3 AU from the heliopause. But even before reaching the heliocliff, the working V2 plasma detector should begin to see a ramp up in the density and thermal speed before reaching 118 AU.

It is important to emphasize that the cold heliosheath is essential for fitting the *IBEX* ENH observations below ~0.01 keV. These low-energy ENH observations cannot be accounted for by ENHs produced beyond the heliopause, as demonstrated by Zirnstein et al. [12]. Rather to account for these low-energy *IBEX* ENH observations we require the unique features of the cold heliosheath: a dense, cold region with a very slow outward flow in the sun frame of reference. Furthermore, *IBEX* observations indicate that the width of the cold heliosheath near the flanks is much smaller than near the centerline, as is requires in the F&G model. Thus, the low-energy *IBEX* ENH observations provide strong evidence of the existence of the cold heliosheath, which implies that *Voyager* 1 must be in the cold heliosheath, is still in the heliosphere and has not yet crossed the heliopause.

Finally, we note how powerful the *IBEX* observations of ENH spectra have been for determining the properties of the solar wind in the heliosheath. To date, only a handful of such spectra have been published. We are hopeful that in the very near future the *IBEX* team will remedy this by making publically available many more ENH spectra, down to the lowest energies and in many other direction than V1, V2, the ribbon and a few others that have been published. Still, *IBEX*, is a Small Explorer with limited energy resolution, sensitivity, and lifetime. The continuing exploration of the heliosheath, a relatively unexplored but highly scientifically significant region of the heliosphere, requires that we make these higher-resolution, more comprehensive ENH observations especially at the very lowest energies. We call on NASA to implement as soon as possible the high-priority IMAP mission that is designed to make these crucial observations.

Acknowledgments
This work was supported in part by NASA Grant NNX10AF23G and by NSF Grant AGS-1043012. We thank Nathan Schwadron for comments and suggestions.

References

[1] Gurnett, D. A., Kurth, W. S., Burlaga, L. F., & Ness, N. F. 2013, Sci, 341, 1489
[2] Burlaga, L. F., Ness, N. F., & Stone, E. C. 2013, Sci, 341, 147
[3] Decker, R. B., Krimigis, S. M., Roelof, E. C., & Hill, M. E. 2012, Nature, 489, 124
[4] Stone, E. C., and Cummings, A. C. 2011, in Proc. 32nd Int. Cosmic Ray Conf., Vol. 1 (Beijing: IUPAP), 29
[5] Fisk, L. A., and Gloeckler, G. 2014a, ApJ, 789, 41
[6] Gloeckler, G., and Fisk, L. A. 2014, GRL, 41, 5325
[7] McComas, D. J., Dayeh, M. A., Allegrini, F., et al. 2012, ApJS, 203, 1
[8] Fuselier, S. A., Allegrini, F., Bzowski, M., et al. 2012, ApJ, 754, 14
[9] Prested, C., Schwardon, N. A., Passuite, J., et al. (2008), JGR, 113, A06102
[10] Gloeckler, G. and Fisk, L. A. 2010, AIP CP1302, 110
[11] Desai, M. I., Allegrini, F. A., Browski, M., et al. 2014, ApJ, 780, 1
[12] Zirnstein, E. J., Heerikhuisen, J., Zank, G. P., et al. 2014, ApJ, 783, 129
[13] Fisk, L. A., & Gloeckler, G. 2014b, JGR, 119, 8733
[14] Gloeckler, G. and Fisk, L. A. 2007, AIP CP932, 264
[15] Lindsay, B. G., & Stebbings, R. F. 2005, JGR, 110, A12213

Journal of Physics: Conference Series **642** (2015) 012012 doi:10.1088/1742-6596/642/1/012012

High-energy solar particle events in cycle 24

N. Gopalswamy[1], P. Mäkelä[2,1], S. Yashiro[2,1], H. Xie[2,1], S. Akiyama[2,1], and N. Thakur[2,1]

[1]NASA Goddard Space Flight Center, Greenbelt, MD 20771, USA
[2]The Catholic University of America, Washington, DC 20064, USA

nat.gopalswamy@nasa.gov

Abstract. The Sun is already in the declining phase of cycle 24, but the paucity of high-energy solar energetic particle (SEP) events continues with only two ground level enhancement (GLE) events as of March 31, 2015. In an attempt to understand this, we considered all the large SEP events of cycle 24 that occurred until the end of 2014. We compared the properties of the associated CMEs with those in cycle 23. We found that the CME speeds in the sky plane were similar, but almost all those cycle-24 CMEs were halos. A significant fraction of (16%) of the frontside SEP events were associated with eruptive prominence events. CMEs associated with filament eruption events accelerate slowly and attain peak speeds beyond the typical GLE release heights. When we considered only western hemispheric events that had good connectivity to the CME nose, there were only 8 events that could be considered as GLE candidates. One turned out to be the first GLE event of cycle 24 (2012 May 17). In two events, the CMEs were very fast (>2000 km/s) but they were launched into a tenuous medium (high Alfven speed). In the remaining five events, the speeds were well below the typical GLE CME speed (~2000 km/s). Furthermore, the CMEs attained their peak speeds beyond the typical heights where GLE particles are released. We conclude that several factors contribute to the low rate of high-energy SEP events in cycle 24: (i) reduced efficiency of shock acceleration (weak heliospheric magnetic field), (ii) poor latitudinal and longitudinal connectivity), and (iii) variation in local ambient conditions (e.g., high Alfven speed).

1. Introduction

Solar Cycle 24 has been the weakest in the space era as measured by the sunspot number (SSN). The average SSN over the first 73 months of cycle 23 was ~ 46 in cycles 24 compared to 76 over the same epoch in cycle 23. This corresponds to a decrease of ~40% [1]. The activity has already entered into the declining phase, but the number of high-energy solar energetic particle (SEP) events has remained very low. During the first 73 months of cycle 24, there were only two ground level enhancement (GLE) events: 2012 May 17 [2] and 2014 January 6 [3]. Over the same epoch, there were 9 GLE events in cycle 23. Thus the reduction in the number of GLE events is by 78%, much higher than that in SSN. The number of >500 MeV SEP events was also higher in cycle 23 (18 vs. 6 in cycle 24).

The current paradigm for SEPs is that they are accelerated by shocks driven by coronal mass ejections (CMEs) [4]. The majority of CMEs associated with SEPs are halos and the average speed is ~1500 km/s [5] suggesting that the CMEs are very energetic. In cycle 24, the number of fast (≥900 km/s) and wide (≥60 degrees) declined only slightly (~30%), which means the reduction in the number of high-energy SEP events cannot be accounted for by the decline of the number of fast and wide CMEs alone.

It has been suggested that the weakened heliospheric magnetic field reduces the efficiency of particle acceleration leading to the lack of high-energy SEP events [6]. Another study considered CMEs associated with 59 major solar flares of X-ray class ≥M5.0 and found that only 16 of them (or 27%) were associated with large SEP events [7]. Ten other large SEP events were associated with weaker flares (soft X-ray flare size <M5.0). This study showed that the flare size is not a good indicator of SEP events. On the other hand, the CMEs associated with the SEP events were all fast, capable of driving shocks. Furthermore, a large number of fast CMEs (>2000 km/s) associated with SEP events occurred at large ecliptic distances suggesting that the nose of these CMEs were not connected to Earth, so high energy particles accelerated at the nose might not have reached Earth.

In this work, we extend the study to a larger set of SEP events, covering events up to the end of 2014. In addition, we consider a wider range of longitudes (from W00 to W90). We confine ourselves to large SEP events: >10 MeV GOES proton intensity ≥10 particles.$cm^{-2}s^{-1}sr^{-1}$. We compare CME kinematics and source positions over the corresponding epoch in cycle 23 in order to understand the multiple factors involved in the occurrence of GLE events.

2. Data

We make use of the list of large SEP events that are cataloged and maintained at the CDAW Data Center (http://cdaw.gsfc.nasa.gov/CME_list/sepe/) [8]. Since cycle 24 started around 1 December 2008, the study period corresponds to the first 73 months of the cycle. For comparison purposes, we use SEP events from cycle 23 over the corresponding epoch: May 1996 to May 2002. The SEP events were obtained from the NOAA web site, ftp://ftp.swpc.noaa.gov/pub/indices/SPE.txt available until the end February 2014. After this period, we examined the GOES proton data to identify additional events. We also carefully checked the events and separated some compound events. The CME information was obtained from the coronagraphs on board the Solar and Heliospheric Observatory (SOHO) and the Solar Terrestrial Relations Observatory (STEREO). In particular, we use the images from the Large Angle and Spectrometric Coronagraph (LASCO) on board SOHO [9] and the Sun Earth Connection Coronal and Heliospheric Investigation (SECCHI) coronagraphs on board STEREO [10]. For each of the SEP event, we identified the corresponding CME from the SOHO/LASCO catalog [8]. Events to the end of January 2014 were also reported in [7].

There were 37 large SEP events during the study period as listed in Table 1. Column 2 of Table 1 gives the date and time the SEP intensity in the >10 MeV channel increased above the 10 pfu level. Column 3 gives the peak SEP intensity attained. In some cases, the SEP intensity was the highest during the shock arrival at 1 AU marked by the energetic storm particle (ESP) event, but we note only the non-ESP part. For each SEP-associated CME, we identified the solar source as the heliographic coordinates of the associated flare or the centroid of the eruptive filament (column 4). For most of the SEP events, STEREO observations of the associated CMEs are also available, so we were able to fit flux ropes to the CMEs using multiple views from SOHO and STEREO [11]. This gives the flux rope location, which we correct for the solar B0 angle. The resulting final location of the flux ropes is given in column 5. From STEREO EUV images, we identified the source coordinates even for the backside events. We also note the soft X-ray flare size (column 6) and the source NOAA active region (AR) number (column 7) if available. When the source region is a quiescent filament region, we mark it as "EP" for eruptive prominence. For event #37, there were two filament eruptions, one associated with the SEP event, and the other an eastern event. The soft X-ray enhancement from the eastern event was higher, so we do not know the soft X-ray size (marked ????) for the western event we are interested in. Backside source regions are marked as "Back". Column 8 gives the first appearance time of the associated CMEs in the LASCO/C2 FOV. The sky-plane speed (V_{sky}) compiled from the SOHO/LASCO catalog are listed in column 9. The space speeds (V_{sp}) obtained from the flux rope fit are given in column 10. V_{sp} is the peak space speed attained by the CMEs as derived from the flux-rope fit. For event #36, the STEREO data was not useful for fitting. For the last event, there was no STEREO data. Finally, the width (W) of the CMEs is given in column 11, where H denotes halo CMEs. For the non-halo CMEs, the sky-plane width is given in degrees.

Table 1. List of large SEP events from cycle 24

#	SEP Date and UT	Ip pfu	Source Loc.	Final Loc.	Flare Imp.	AR #	CME UT	V_{sky}	V_{sp}	W
1	2010/08/14 12:30	14	N12W56	S20W54	C4.4	11099	08/14 10:12	1205	1658	H
2	2011/03/08 01:05	50	N24W59	N39W58	M3.7	11164	03/07 20:00	2125	2660	H
3	2011/03/21 19:50	14	N23W129	N33W125	Back	11169	03/21 02:24	1341	2022	H
4	2011/06/07 08:20	72	S21W54	S08W51	M2.5	11226	06/07 06:49	1255	1680	H
5	2011/08/04 06:35	96	N16W30	N13W30	M9.3	1 1261	08/04 04:12	1315	2450	H
6	2011/08/09 08:45	26	N17W69	N02W68	X6.9	11263	08/09 08:12	1610	2496	H
7	2011/09/23 22:55	35	N11E74	S02E83	X1.4	11302	09/22 10:48	1905	2474	H
8	2011/11/26 11:25	80	N27W49	N08W47	C1.2	EP	11/26 07:12	933	1187	H
9	2012/01/23 05:30	3000	N28W36	N35W22	M8.7	11402	01/23 04:12	2102	2150	H
10	2012/01/27 19:05	800	N27W71	N33W82	X1.7	11402	01/27 18:27	2508	2625	H
11	2012/03/07 05:10	1500	N17E27	N25E31	X5.4	11429	03/07 00:24	2544	2987	H
12	2012/03/13 18:10	500	N19W59	N28W52	M7.9	11429	03/13 17:36	1898	2333	H
13	2012/05/17 02:10	255	N11W76	S05W76	M5.1	11476	05/17 01:47	1618	1997	H
14	2012/05/27 05:35	14	N10W121	N12W115	Back	11482	05/26 20:57	1966	2623	H
15	2012/06/16 19:55	14	S17E06	S27E02	M1.9	11504	06/14 14:12	981	1626	H
16	2012/07/07 04:00	25	S18W51	S32W62	X1.1	11515	07/06 23:12	1912	2464	H
17	2012/07/09 01:20	18	S17W74	S38W88	M6.9	11515	07/08 16:54	1211	2905	157
18	2012/07/12 18:35	96	S15W01	S23W06	X1.4	11520	07/12 16:48	1360	1415	H
19	2012/07/17 17:15	136	S28W65	S32W79	C9.9	11520	07/17 13:48	814	1881	H
20	2012/07/19 07:45	70	S28W75	S20W88	M7.7	11520	07/19 05:24	1681	2048	H
21	2012/07/23 15:45	12	S17W141	N00W135	Back	11520	07/23 02:36	2142	2621	H
22	2012/09/01 13:35	59	S19E42	S13E40	C8.4	EP	08/31 20:43	1167	1601	H
23	2012/09/28 03:00	28	N06W34	N09W29	C3.7	11577	09/28 00:12	1035	1479	H
24	2013/03/16 19:40	16	N11E12	N17E08	M1.1	11692	03/15 07:12	980	1602	H
25	2013/04/11 10:55	114	N09E12	N14E11	M6.5	11719	04/11 07:24	986	1626	H
26	2013/05/15 13:35	41	N11E51	N15E64	X1.2	11748	05/15 01:48	1511	2294	H
27	2013/05/22 14:20	1660	N15W70	N04W59	M5.0	11745	05/22 13:32	1537	1881	H
28	2013/06/23 20:10	14	S16E73?	????	M2.9	11777	06/21/03:12	1900	1987	>207
29	2013/09/30 05:05	200	N23W25	N16W29	C1.1	EP	09/29 22:12	1025	1864	H
30	2013/12/28 21:50	30	S08W130	N02W127	Back	????	12/28 17:36	1133	1918	H
31	2014/01/06 09:15	40	S15W112	S06W102	Back	11936	01/06 08:00	1402	2287	H
32	2014/01/07 19:55	1026	S12W11	S15W29	X1.2	11944	01/07 18:24	1830	3121	H
33	2014/02/20 08:50	22	S15W73	S07W70	M3.0	11976	02/20 08:00	948	1281	H
34	2014//02/25 13:55	24	S12E82	S10E80	X4.9	11990	02/25 01:25	2147	2218	H
35	2014/04/18 15:25	58	S20W34	S18W18	M7.3	12306	04/18 13:25	1203	1711	H
36	2014/09/11 02:40	29	N14E02	????	X1.6	12158	09/10 18:00	1425	1652	H
37	2014/11/02 21:10	11	N27W79	????	????	EP	11/01 06:12	740	1000	H

The CME identification is unambiguous for all the events except for the 2013 June 23 event (#28 in Table 1). There was a minor particle event associated with the listed CME on June 21, but the origin of the increase on June 23 is uncertain. The increase is well after a shock observed at 1 AU, so it does not seem to be the ESP event associated with the June 21 CME. Because of this, we do not include this event in this study.

3. Analysis and Results

3.1 SEP occurrence rate

There were 37 SEP events in cycle 24 compared to 60 in cycle 23. If we normalize the number of events to the average SSN during the study period, we get 0.80/SSN in cycle 24 very similar to 0.79/SSN in cycle 23. The number of events with >100 pfu is 24 (or 40%) in cycle 23 compared to just 10 (or 27%) in cycle 24. The highest SEP intensity was 3000 pfu in cycle 24 compared to 31700 pfu in cycle 23. There were only 4 SEP events in cycle 24 with intensity >1000 pfu, compared to 10 in cycle 23. The difference is more striking when we consider >500 MeV (6 vs. 18) and GLE (2 vs. 9) events in cycle3 23 and 24. Normalized to SSN, cycle 24 had only 0.04 GLEs/SSN compared to 0.11/SSN in cycle 23.

This is in stark contrast to the >10 MeV events, which are similar when normalized to SSN. We focus on the paucity of GLE events in cycle 24.

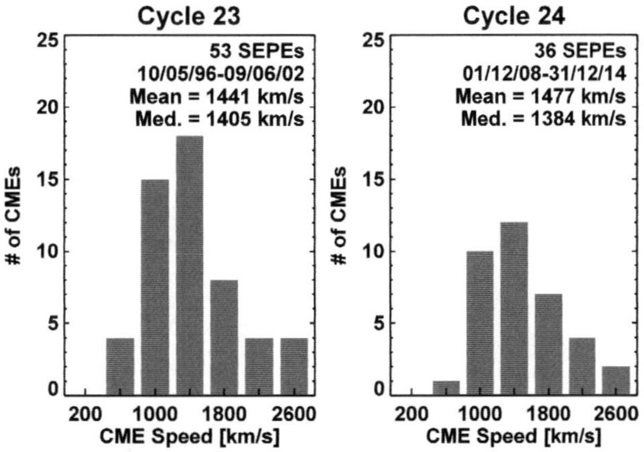

Figure 1. Sky-plane speed distributions of SEP-associated CMEs in cycles 23 (left) and 24 (right). The mean and median values of the distributions are shown on the plot.

3.2. CME kinematics

Figure 1 compares the speed distributions of SEP-associated CMEs in cycles 23 and 24. We are comparing sky-plane speeds because there was no 3-dimensional information in cycle 23. There were actually 60 SEP events during the first 73 months of cycle 23, but only 53 had CME data available (mainly due to SOHO/LASCO data gaps in 1998 and 1999). The distributions are very similar with almost the same mean and median speeds. The most probable value of the speed is in the 1400 km/s bin in both cycles. The range of speeds is also the same. While the speeds are similar, the width distribution is clearly different. We did not plot the width distribution because almost all CMEs are halos (35 out of 36 or 97%) in cycle 24 compared to 75% (40 out of 53) in cycle 23. Such a difference in width distribution was also found for limb CMEs in cycle 24 and was attributed to the anomalous expansion of CMEs due to the reduced total pressure in the heliosphere [6]. It was also recently reported that the occurrence rate of halo CMEs in cycle 24 did not diminish unlike SSN [1]. We note that there were five EP events in Table 1 (16% of the 31 front-side events). CMEs associated with EP events generally have a small initial acceleration and the CMEs attain high speeds at large distances from the Sun [12].

3.3. Source locations

Table 1 shows that the CME sources were located from close to the east limb (E82 for #34) to far behind the west limb (W141 for #21). There were 5 backside events, all of them behind the west limb. The solar sources of the remaining 31 events are plotted in Figure 2. The heliographic grid corresponds to the case of B0=0, so the latitudes are the same as the ecliptic distances (the projection of Earth is at the disk center in this representation). The source locations are the flux rope (FR) latitudes and longitudes obtained from the flux rope fit corrected for the solar B0 angle (see [7] for details). The FR locations are generally different from the flare locations due to non-radial motion of CMEs caused by magnetic pressure gradients around the eruption sites. The latitudes differ by ~10 degrees, while the longitudes differ by ~7 degrees on the average. Both the GLE CMEs have deflections that reduce the ecliptic distance. Only 10 events are in the eastern hemisphere because events from the western hemisphere are magnetically better connected to Earth. The eastern events are generally associated with more energetic CMEs and the intensities are lower. We see that the ecliptic latitudes extend to more than 30 degrees both in the northern and southern hemispheres. The latitudes of the eastern events are within 30-degree latitudes.

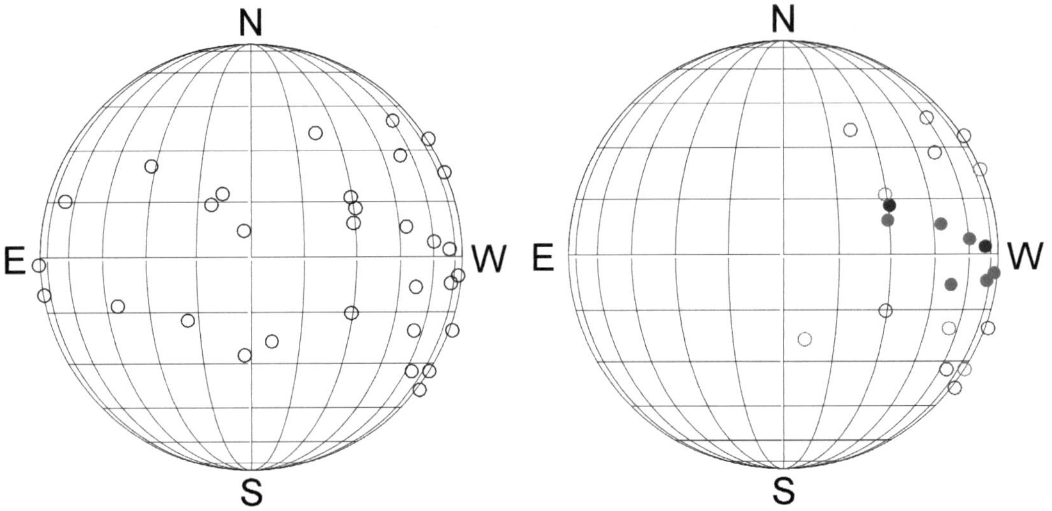

Figure 2. (left) Flux rope locations of 31 frontside SEP events from Table 1. The locations have been corrected for B0 angle (the zero latitude is the ecliptic in this figure). (right) 21 events from the western longitudes (W00 to W90). The red (blue) circles correspond to CMEs with speed < 2000 km/s (≥ 2000 km/s). The filled circles are within the ecliptic latitude range ±13 degrees. The filled circles represent the candidate GLE events discussed in the text.

Table 2. List of candidate GLE events

#	CME Date UT	Imp.	Flr Loc.	FR Loc	B0	Final Loc	Vsp	Max E	Ip
4	2011/06/07 06:16	M2.5	S21W54	S08W51	+0.1	S08W51	1680	330-420	72
5	2011/08/04 03:41	M9.3	N19W36	N19W30	+6.0	N13W30	2450	165-500	96
6	2011/08/09 07:48	X6.9	N17W69	N08W68	+6.3	N02W68	2496	330-420	26
8	2011/11/26 06:09	C1.2	N08W49	N10W47	+1.5	N08W47	1187	40-80	80
13	2012/05/17 01:25	M5.1	N11W76	S07W76	-2.4	S05W76	1997	>700	255
23	2012/09/27 23:24	C3.7	N06W34	N16W29	+6.9	N09W29	1479	80-165	28
27	2013/05/22 13:08	M5.0	N15W70	N02W59	-1.8	N04W59	1881	330-420	1660
33	2014/02/20 07:26	M3.0	S15W73	S14W70	-7.0	S07W70	1281	330-420	22

3.4. Analysis of GLE Candidate Events

For the purposes of this paper, we consider all the western events, as it is well known that GLEs occur preferentially from the western hemisphere [13]. The right-side plot in Figure 2 shows the 21 front-side western events, distinguishing SEP events according to their speeds: <2000 km/s (red) and ≥2000 km/s (blue). We made this distinction based on the fact that GLEs are associated with CMEs with an average speed of ~2000 km/s [14]. We have also isolated SEP events with ecliptic latitudes within ±13 degrees (filled circles). This is based on the observation that the average source latitude of cycle-23 GLE events was 13 degrees [2]. The two GLE events of cycle 24 are well within this 13-degree limit (#13 and #31 in Table 1). Applying all these criteria, we arrive at just 8 events that are in the right latitude range. We have listed the CMEs associated with these 8 events in Table 2 along with the flux rope location, B0 angle, and the final location. We now discuss these events in an attempt to explain why they did not have the GLE component except for the 17 May 2012 event.

3.4.1. The three fastest events. Among the three fastest CMEs in Table 2, only one was associated with a GLE (17 May 2012 event), discussed extensively in [2]. The CME speed (1997 km/s) was just at the average value of GLE-associated CMEs. The two >2000 km/s CMEs did not produce GLEs. The 9 August 2011 event was reported to be associated with narrower CME launched into a tenuous corona with probably higher than usual Alfven speed in the ambient medium (hence weaker shock – see [2]). The 4 August 2011 event was not discussed in [2] because a stricter latitude criterion (±9 degrees) was used based on cycle-23 GLE events that occurred closer to the limb. The 4 August 2011 event occurred at N13W30 just at the latitude limit for GLEs. The W30 longitude in Earth view does not allow us to examine the nature of the corona. Fortunately, the CME was observed by STEREO-Ahead as an east limb event (E70 in that view because the spacecraft was at W100) as can be seen in Figure 3. In the STEREO-Ahead COR2 image taken at 04:24 UT, close to the onset time of the 4 August 2011 SEP event at Earth, we see that the corona overlying the CME is the darkest compared to other parts of the image, suggesting that the corona was tenuous. Although we do not have magnetic field measurement, we speculate that the Alfven speed is higher than usual leading to a weaker shock as in the case of the 9 August 2011 event.

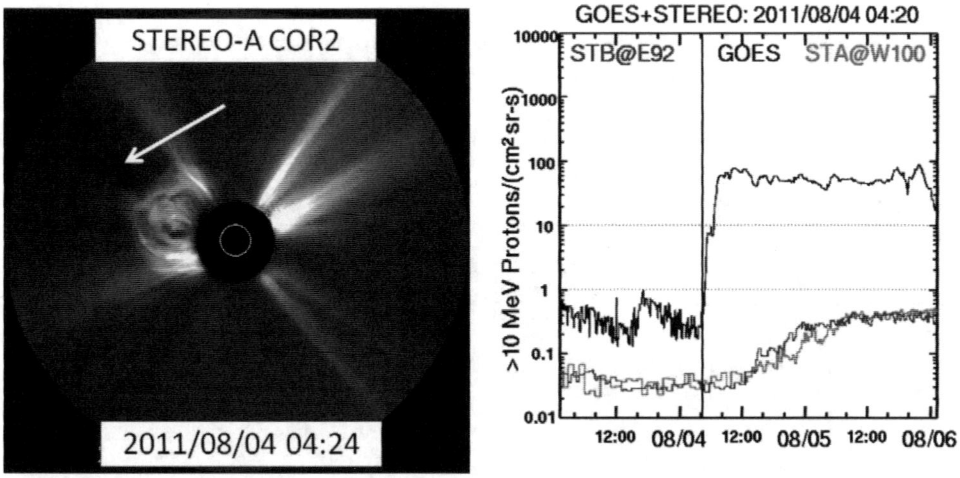

Figure 3. STEREO-Ahead COR2 image (left) showing the 4 August 2011 CME above the east limb. The arrow points to the dim corona above the CME. The >10 MeV proton intensity from GOES and STEREO-Ahead and Behind are shown for comparison. Since the two STEREO spacecraft are poorly connected to the source region, they detected only minor events (<1 pfu).

3.4.2. The two events associated with C-class flares. The 26 November 2011 and the 27 September 2012 CMEs were both associated with C-class flares and discussed in [12]. The 26 November 2011 CME was associated with a filament eruption. Both events were associated with interplanetary type II bursts, but no metric type II bursts. This means the shock forms only in the outer corona, well above the typical height of GLE particle release. From the flux rope fit, we estimate that the shock formed at a height of 4 solar radii (Rs) for the 27 September 2012 CME and 3.25 Rs for the 26 November 2011 CME. In addition, the peak SEP spectra were steep (index = 4.34 for the 26 November 2011 event and 3.25 for the 27 September 2012 event) [15]. The steep spectrum is also evident from the maximum energy (E) of protons detected in GOES differential channels (80 and 165 MeV for the 26 November 2011 and 27 September 2012 CMEs – see Table 2). Furthermore, these CMEs were still accelerating in the inner corona, so the peak speeds listed in Table 2 were not achieved until after the shock formation heights. Therefore, we conclude that in these two events the speed was too low and the shock formed well beyond the typical GLE shock heights (2–3 Rs).

3.4.3 The remaining three events. The 7 June 2011 and 20 February 2014 CMEs had peak speeds well below the typical GLE speeds (1680 and 1281 km/s, respectively). The 22 May 2013 CME was interacting with a preceding CME, but the interaction was taking place at much larger height (~20 Rs) [7]. Even though CME interaction events have high intensity (which is true in this event also) [5], the shocks need to be much closer to the Sun to be relevant for GLE events. Furthermore, the peak speed was only 1881 km/s, not quite the average value of GLE CMEs. The peak speed was also attained when the CME was at ~7 Rs. The metric type II burst started only around 40 MHz, which is a good indication that the shock forms at a larger height (~2 Rs). The typical shock formation heights for GLE events is ~1.5 Rs [2, 16]. The combination of low speed and larger shock formation height seem to have contributed to the lack of GLEs in the 22 May 2013 event. GLEs have been reported to be associated with much lower speeds when preconditioning was present [17]. The 22 May 2013 event seems to be such a case, yet there was no GLE. The overall reduced efficiency of particle acceleration might have also contributed to the paucity of GLEs in almost all the SEP events of cycle 24.

4. Discussion

The above analysis shows that there are three primary factors that affect the acceleration of SEPs to the highest energies: (i) CME speed, (ii) connectivity, and (iii) ambient conditions. The lowest space speed in Table 1 is 1187 km/s, which is definitely high enough to accelerate SEPs. However, such speeds may not be sufficient to make the shock strong enough to accelerate particles to the highest energies. This is compounded by the reduced ambient magnetic field strength in cycle 24 [6,18] that reduces the acceleration efficiency of shocks. For GLEs, the highest CME speed is attained close to the Sun, so the shock is in the high magnetic field region for a longer time. In some cases, the speed was relatively high, but was attained at a larger heliocentric distance. This is clear in filament eruption events and in events where the metric type II bursts start at low frequencies.

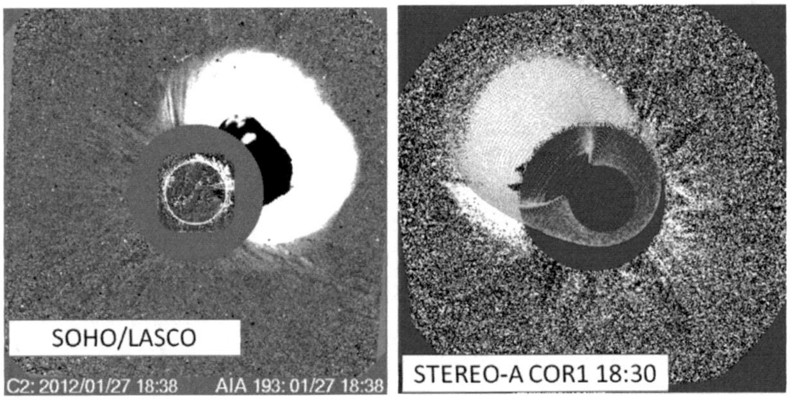

Figure 4. (left) SOHO/LASCO CME on 27 January 2012 at 18:38 UT associated with an intense SEP event. (right) STEREO-A COR1 image at 18:30 UT with the fitted flux rope (green contours) superposed. The CME is heading to the northwest in Earth view and northeast in STEREO-A view because the latter was at W108.

Under connectivity, we considered western events because it is well known that most of the GLE events originate in the western hemisphere. We also considered latitudinal connectivity [19], by which we mean that the shock nose needs to be close to the ecliptic (within 13 degrees based on cycle 23 GLE events. This may suggest that the highest energy particles may be accelerated near the nose, but they do not reach an Earth observer because the nose is too far away from the ecliptic. This is illustrated in Figure 4 for the 27 January 2012 SEP event. The flare location was N27W71 hence longitudinally well connected, but the latitude was very high. The nose of the LASCO CME was at position angle 296 degrees, consistent with the N27 latitude. The flux rope fit in Figure 4 is consistent with the higher latitude of the source region. The fitted longitude was slightly higher (W82) [7] but the latitude was the same as that of the flare location (event # 10 in Table 1). In addition, the B0 angle was -5.6 degrees, making the effective ecliptic latitude as N32.6 and hence worsening the latitudinal connectivity. This is well beyond the average ecliptic distance of 13 degrees for GLE events. From Figure 4 we can infer that only the southern flank of the CME is at the ecliptic. The SEP event was intense (800 pfu) and the GOES differential channel data indicate energetic particles present up to the 510–700 MeV channel [7]. Thus this is an almost GLE event, but not detected by the neutron monitors as a GLE.

In the list of GLE candidates in Table 2, we considered only front-side events. However, we do have the 2014 January 6 GLE event originating from behind the west limb (event #31 in Table 1). Furthermore, there was a GLE in cycle 23 on 2001 April 18 from similar longitudes (W117, see [14]). Therefore, we briefly discuss the possibility of GLEs from behind-the-west-limb events in Table 1. There are 5 such backside events, including the 2014 January 6 GLE. Of these, the 2012 May 27 event is closest in longitude to the limb (W115 or ~ 25 degrees behind the limb) and the ecliptic distance is only N12. This makes it a viable GLE candidate. Other events had longitudes in the range W125 – W135, too far for magnetic connection to Earth. Figure 5 compares the post eruption arcades (PEAs) of the 2012 May 27 and 2014 January 6 events as observed by the STEREO-A EUVI instrument. The PEAs are oriented differently in the two cases (January 2014: east-west and May 2012: north-south). These images suggest that the flux ropes from these source regions have different inclinations. The fitted flux ropes are consistent with this expectation (not shown). The shock surrounding a high inclination flux rope would have a smaller longitudinal extent than that surrounding a low inclination flux rope. We suggest this as the possible reason for not observing GLEs from the 2012 May 27 SEP event because the shock does not extend enough in the longitudinal direction. The different flux rope orientations may also result in different longitudinal extents of SEP events [20].

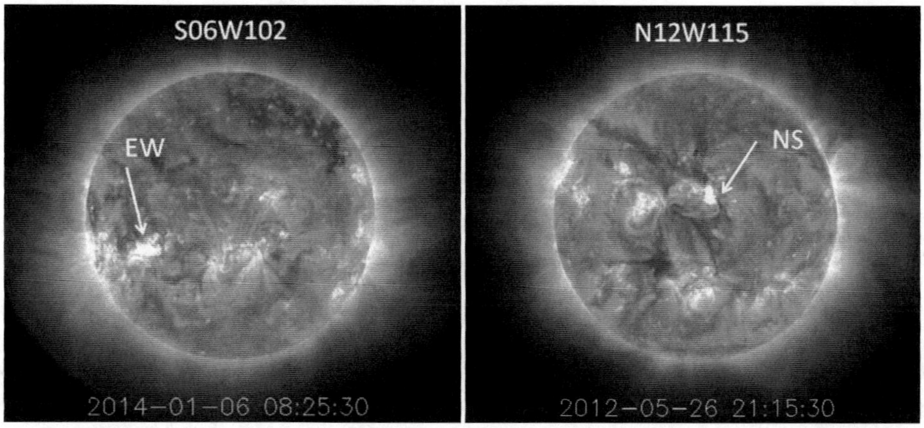

Figure 5. Post eruption arcades (PEAs) associated with the 2014 January 6 (left) and the 2012 May 27 (right) backside SEP events as observed by the Extreme Ultraviolet Imager (EUVI) on board STEREO-A. STEREO-A was located at W151 and W115 when the January 2014 and May 2012 images were obtained. Note the different orientations of the PEAs: east-west (EW) and north-south (NS).

We also considered the possibility of higher Alfven speed in the ambient medium indicated by the tenuous corona. This is qualitative, but can be verified by estimating the coronal magnetic field [21]. Interestingly, most of the >2000 km/s CMEs have higher ecliptic latitudes. It is not clear if this is due to the overall heliospheric magnetic field structure in cycle-24.

5. Summary

We confirm our previous finding that the number of >10 MeV SEP events relative to SSN is similar in cycles 23 and 24, but there were four times more GLE events in cycle 23 than those in cycle 24 during the first 73 months. When normalized to SSN, the rates are 0.04/SSN in cycle 24 compared to 0.11/SSN in cycle 23. The speed distributions of CMEs associated with SEPs are quite similar in the two cycles with little difference in the mean and median speeds. However, the fraction of halos is 97% in cycle 24 compared to 75% in cycle 23. The higher fraction of halos was also reported earlier for a set of limb CMEs associated with soft X-ray flares of size ≥C3.0 [6]. The increased expansion of cycle-24 CMEs seems to have contributed to the increased number of halo CMEs in cycle 23. Since GLE events are a subset of SEP events, we considered front-side SEP events in the western hemisphere as the GLE candidate events. Based on the recent result that the ecliptic distance of CMEs needs to be small, we further narrowed down the events to ecliptic distances ≤13 degrees. This resulted in 8 candidate events including the 17 May 2012 GLE. Among the 7 non-GLE events, two events had very high speed (>2000 km/s), but the ambient medium was not favorable for supporting strong shocks (probably high Alfven speed). Four of the five remaining events had speeds well below the average speed of GLE CMEs (~2000 km/s). The last event (22 May 2013) was an interacting CME event but the interaction distance was ~20 Rs, well beyond the distance where GLE particles are typically released. Furthermore, the peak speed was attained around 7 Rs. Thus we conclude that the CME speed, connectivity of CME nose to Earth, and favorable ambient medium are the key factors that determine whether an SEP event has a GLE component.

Acknowledgments
SOHO is a project of international collaboration between ESA and NASA. STEREO is a mission in NASA's Solar Terrestrial Probes program. The work of NG, SY, SA was supported by NASA/LWS program. PM was partially supported by NSF grant AGS-1358274 and NASA grant NNX15AB77G. HX was partially supported by NASA grant NNX15AB70G.

References
[1] Gopalswamy N, Xie H, Akiyama S, Mäkelä P, Yashiro S, Michalek G 2015 *Astrophys. J.* **804** L23
[2] Gopalswamy N, Xie H, Akiyama S, Yashiro S, Usoskin I, Davila J 2013 *Astrophys. J.* **765** L30
[3] Thakur N, Gopalswamy N, Xie H, Mäkelä P, Yashiro S, Akiyama S, Davila JM 2014 *Astrophys. J.* **790** L13
[4] Reames DV 1999 *SSRv* **90** 413
[5] Gopalswamy N, Yashiro S, Krucker S, Stenborg G, Howard RA 2004 *J. Geophys. Res.* **109**, A12105
[6] Gopalswamy N, Akiyama S, Yashiro S, Xie H, Mäkelä P, Michalek G 2014 *Geophys. Res. Lett.* **41** 2673
[7] Gopalswamy N, Xie H, Akiyama S, Mäkelä P, Yashiro S 2014 *Earth, Planets and Space* **66** 104
[8] Gopalswamy N, Yashiro S, Michalek G, Stenborg G, Vourlidas A, Freeland S, Howard R 2009 *Earth Moon Planets* **104** 295
[9] Brueckner GE et al. 1995 *Solar Phys.* **162** 357
[10] Howard RA et al, 2008 *SSRv* **136** 67
[11] Thernisien A 2011 *Astrophys. J.Suppl* **194** 33
[12] Gopalswamy N, Mäkelä P, Akiyama S, Yashiro S, Xie H, Thakur N, Kahler SW 2015 *Astrophys. J.* **806** 8
[13] Shea MA, Smart DF 2012 *SSRv* **171**161
[14] Gopalswamy N, Xie H, Yashiro S, Akiyama S, Mäkelä P, Usoskin IG 2012 *SSRv* **171** 23
[15] Kahler S W 2001 *J. Geophys. Res.* **106** 20947
[16] Reames DV 2009 *Astrophys. J.* **706** 844
[17] Cliver EW 2006 *Astrophys. J.* **639** 1206
[18] McComas DJ, Angold N, Elliott HA, Livadiotis G, Schwadron NA, Skoug RM, Smith CW 2013 *Astrophys. J.* **770** 2
[19] Gopalswamy N, Mäkelä P 2014 Latitudinal connectivity of ground level enhancement events In: Hu Q, Zank G (eds) Outstanding problems in heliophysics: from coronal heating to the edge of the heliosphere, ASP Conference Series, vol 484., p 63
[20] Lario D, Aran A, Gómez-Herrero R, Dresing N, Heber B, Ho GC, Decker RB, Roelof EC 2013 *Astrophys. J.* **767** 41
[21] Gopalswamy N, Yashiro S 2011 *Astrophys. J.* **736** L17

Journal of Physics: Conference Series **642** (2015) 012013 doi:10.1088/1742-6596/642/1/012013

The collisional relaxation of electrons in hot flaring plasma and inferring the properties of solar flare accelerated electrons from X-ray observations.

N. L. S. Jeffrey[1], E. P. Kontar[1], A. G. Emslie[2], and N. H. Bian[1]

[1] SUPA School of Physics & Astronomy, University of Glasgow, Glasgow, G12 8QQ, UK

[2] Department of Physics & Astronomy, Western Kentucky University, Bowling Green, KY, 42101, USA

E-mail: natasha.jeffrey@glasgow.ac.uk

Abstract. X-ray observations are a direct diagnostic of fast electrons produced in solar flares, energized during the energy release process and directed towards the Sun. Since the properties of accelerated electrons can be substantially changed during their transport and interaction with the background plasma, a model must ultimately be applied to X-ray observations in order to understand the mechanism responsible for their acceleration. A cold thick target model is ubiquitously used for this task, since it provides a simple analytic relationship between the accelerated electron spectrum and the emitting electron spectrum in the X-ray source, with the latter quantity readily obtained from X-ray observations. However, such a model is inappropriate for the majority of solar flares in which the electrons propagate in a hot megaKelvin plasma, because it does not take into account the physics of thermalization of fast electrons. The use of a more realistic model, properly accounting for the properties of the background plasma, and the collisional diffusion and thermalization of electrons, can alleviate or even remove many of the traditional problems associated with the cold thick target model and the deduction of the accelerated electron spectrum from X-ray spectroscopy, such as the number problem and the need to impose an *ad hoc* low energy cut-off.

1. Introduction and the observation of solar flare X-rays

During a solar flare, a very large number of electrons, of the order of 10^{36} electrons s^{-1} are accelerated, but the mechanism responsible for their acceleration remains poorly understood (see [1] for a recent review). The properties of accelerated electrons are mainly deduced by X-ray observations, currently using X-ray imaging and spectroscopy provided by the *Ramaty High Energy Solar Spectroscopic Imager (RHESSI)* [2]. Solar flare X-rays are mainly produced as bremsstrahlung from electron-ion collisions and are optically thin in the solar atmosphere where hydrogen number densities range from around $10^8 - 10^{16}$ cm^{-3}. X-ray observations thus provide an untainted inference of the *emitting* electron distribution, which is generally different from the *accelerated* electron distribution, since the latter can be modified en route by many processes in the solar atmosphere. These processes include Coulomb collisions (mainly with other electrons), and also non-collisional and turbulent processes (for example [3, 4]). Apart from rare cases, such as the *'cold flare'* [5], most solar flare spectra consist of two parts, usually treated as separate entities: a bremsstrahlung power law component dominant above ~25 keV that is produced by the collision and deceleration of fast accelerated electrons, and a thermal component (mostly

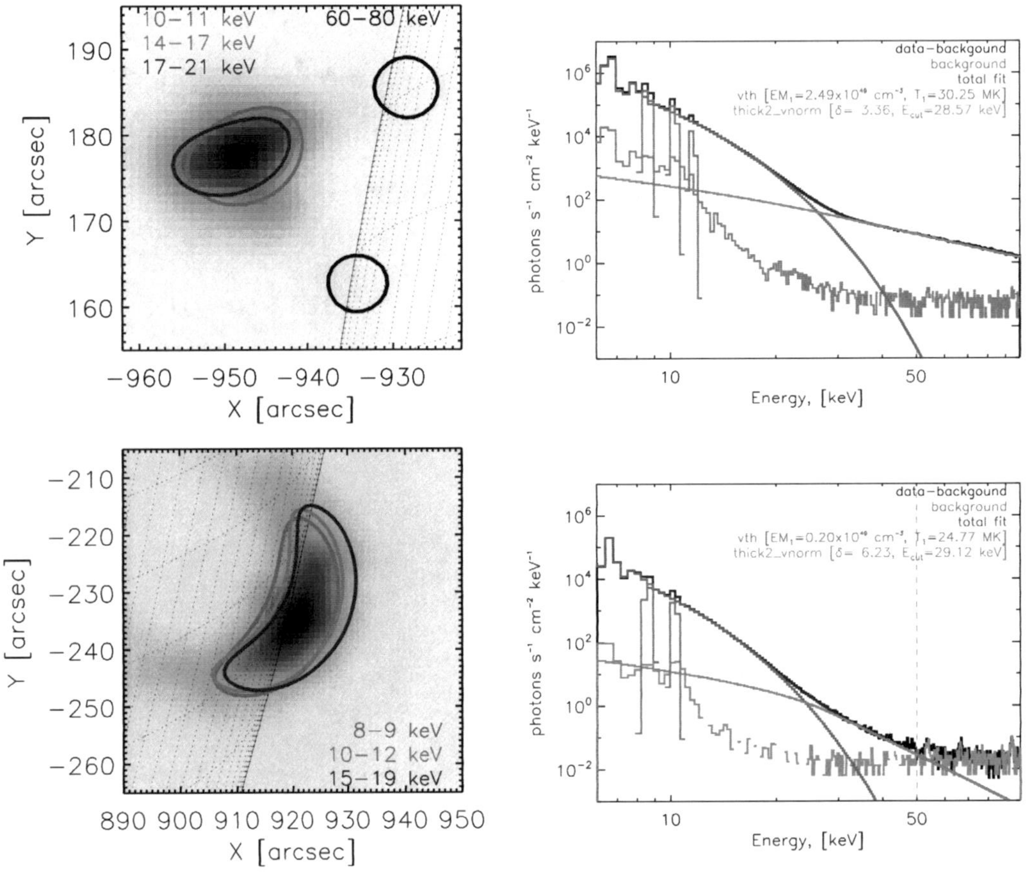

Figure 1. *Left: RHESSI* X-ray images of two solar flares, where the top image shows a 'standard' flare type (SOL2013-05-13T16:04) with a hot coronal source observed at energies of $\sim$ 10 to 25 keV and X-ray footpoints above $\sim$ 30 keV located lower in the atmosphere. The bottom image shows a coronal dominated flare type (SOL2005-08-23T14:30), where both the thermal and nonthermal emission exist cospatially. *Right: RHESSI* X-ray spectra for each flare. Most flares, independent of the flare type have two main components: a hot thermal component dominant at lower energies (red) and a power law nonthermal component at higher energies (blue).

consisting of bremsstrahlung, but also free-bound and line emission), produced by a highly heated plasma (1 keV or more) that completely dominates the X-ray spectrum at energies below $\sim$25 keV (see Figure 1) [6, 7]. *RHESSI* X-ray imaging usually shows that the bulk of hot material $\sim$ 20 MK ($\sim$ 1.7 keV) emanates from a coronal source and the nonthermal X-rays from 'footpoint' sources, located lower in the atmosphere (see Figure 1). There is also some evidence of hot emission of around 10 MK ($\sim$ 0.86 keV) closer to the footpoints [8]. Sometimes the thermal and nonthermal components are cospatial and are both produced in the hot, dense coronal loop, e.g. [9, 10, 11] (see Figure 1).

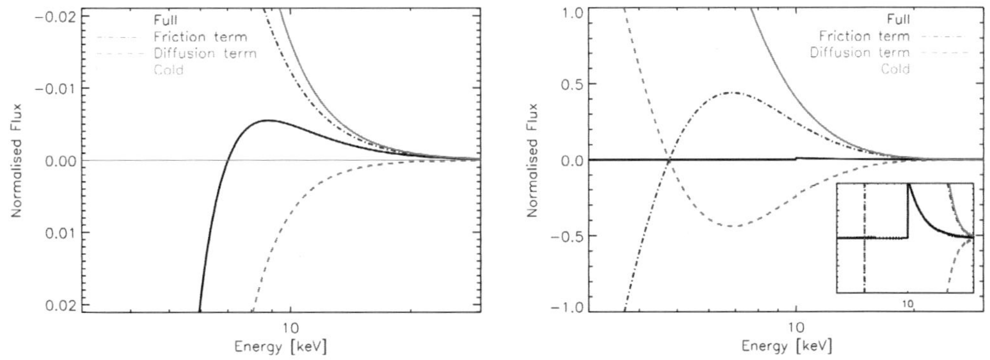

Figure 2. Equations (1) and (2) plotted for *left:* a power law $\langle nVF\rangle(E)$, and *right:* a Maxwellian plus power law $\langle nVF\rangle(E)$, where *blue:* frictional first order term (Equation (2)), *red:* diffusive second order term (Equation (2)) and *black:* both terms (Equation (2)). The green curve represents the cold model (Equation (1)). In the full model, the diffusive term at low energies cancels the frictional term providing a natural cutoff to the accelerated electron distribution, removing the low energy cutoff problem.

2. Finding the flare accelerated electron distribution

Commonly, either by a forward fitting or inversion technique, the cold thick target model e.g. [12, 13, 14] is applied to *RHESSI* X-ray spectra. This model is popular due to its simple analytical form. It assumes that: *1. the accelerated electrons are transported through a 'cold' material, whereby the accelerated electron energy $E \gg T$, where T is the temperature of the background material in keV,* and *2. the observed X-ray bremsstrahlung emission is produced in a region of sufficiently high density that the accelerated electrons lose all of their energy collisionally by friction in this region.* In this model, the dynamics of X-ray emitting electrons is dominated by deterministic collisional energy loss and, hence, the spatially averaged electron spectrum $\langle nVF\rangle$ [15] is related to the accelerated electron spectrum F_0 injected into the source via,

$$F_0(E_0) = -\frac{K}{A}\frac{d}{dE}\left[\frac{\langle nVF\rangle(E)}{E}\right]_{E=E_0}, \tag{1}$$

where E and E_0 are the observed and injected electron energies respectively, K is the collisional parameter and A is the injection area. Equation (1) is plotted in Figure 2 for two different $\langle nVF\rangle$. Solar flare X-ray spectra, in the nonthermal domain, are typically quite steep (Figure 1). Using the above model that only retains the effect of deterministic energy loss requires that the injected electron flux spectra $F_0(E_0)$ are similarly steep. Since the total injected power is a diverging quantity for such steep power-laws, the concept of a 'low-energy cutoff' is frequently used to keep the total number and power of the accelerated electron distribution finite. Such a quantity is also difficult to determine from observation due to the large thermal component dominant at lower energies (Figure 1).

2.1. Hot plasma in the corona

The cold thick target model is not directly applicable to the analysis of hot flaring plasma, since it does not account for the different behaviour and thermalization of electrons at energies close to $E \sim T$. Such electrons are dominated by collisional diffusion and not by systematic energy loss, thermalizing with the background plasma. In order to improve the estimation of the accelerated electron properties, we must take into account the presence of a finite target temperature and

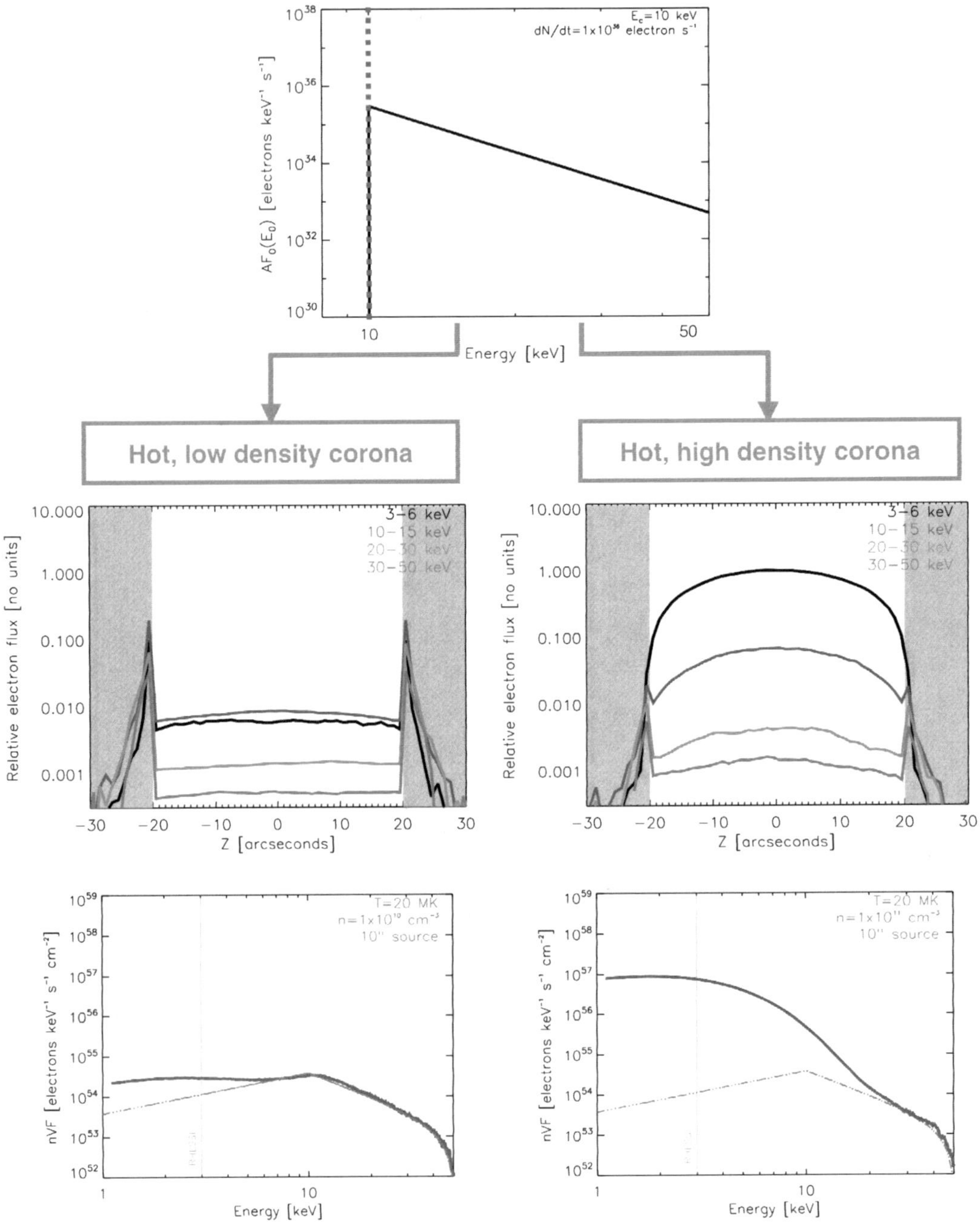

Figure 3. The resulting spatial distributions (parallel to a guiding magnetic field) and spatially integrated spectra for a simulated accelerated electron distribution (top panel) transported through two different solar flare coronal environments (middle panels) of *left*: $T = 20$ MK and $n = 10^{10}$ cm^{-3}, and *right*: $T = 20$ MK and $n = 10^{11}$ cm^{-3}, and injected across $10''$ at the loop top apex ($z = 0''$). The grey regions denote a low temperature, high density 'chromospheric' region. High coronal densities lead to the bulk of electrons stopping in the corona (coronal thick-target flares), but a low density leads to the bulk of electrons stopping in the dense low atmosphere (footpoint dominated flares). *Bottom panels:* the resulting X-ray spectrum (pink) is compared with that of the cold thick target model (blue). The form of the spatial distribution and the integrated X-ray spectrum is dependent on the properties of the background plasma and cannot be ignored if a more correct form for the accelerated electron distribution is to be determined from observation.

the resultant collisional diffusion of electrons. In such a model, the observed electron spectrum $\langle nVF\rangle(E)$ and the accelerated electron spectrum F_0 are related by

$$F_0(E_0) = -\frac{2K}{A}\frac{d}{dE}\left[G\left(\sqrt{\frac{E}{T}}\right)\left\{\frac{d\langle nVF\rangle(E)}{dE} + \frac{1}{E}\left(\frac{E}{T}-1\right)\langle nVF\rangle(E)\right\}\right]_{E=E_0}, \quad (2)$$

where $u = \sqrt{E/T}$, G is known as the Chandrasekhar function (see [16] for more details) and temperature T is in units of keV. Equation (2) reduces to the cold target form, in the high energy, low temperature limit when $E \gg T$, and has two very important distinctions compared to the cold target equation (Equation (1)). First, the first order friction term is modified so that electrons can either lose or gain energy during a collision, which is particularly important at $E \leq T$ (see [17]), and *even more importantly,* there is the inclusion of a second order term that describes energy changes by collisional diffusion, controlling the behaviour of electrons at $E \sim T$, and ultimately leading to thermalization. Equations (1) and (2) are shown in Figure 2 for two different $\langle nVF\rangle(E)$: a power law and a Maxwellian plus power law. Importantly, Figure 2 shows that the diffusive term cancels the deterministic term at low E. For the case of a power law $\langle nVF\rangle(E)$ extending over all E, the injected electron flux $F_0(E_0)$ becomes unphysically negative at low injected energies E_0, producing a natural cutoff to the injected distribution. In the more plausible 'Maxwellian plus power law' case (as might be expected in a hot target), Equation (2) disappears for the Maxwellian component, once again removing the need to apply an arbitrary *low energy cutoff* to the accelerated electron distribution.

2.2. Changes to the observed electron spectrum due to the presence of hot flaring plasma

In order to show how the presence of hot plasma in a flaring solar atmosphere changes the form of the accelerated electron distribution, Figure 3 plots the resulting spatial distributions and spatially averaged spectra for the same injected electron distribution, for the same target temperature $T = 1.7$ keV (20 MK) but with number densities of a) $n = 10^{10}$ cm^{-3} and b) $n = 10^{11}$ cm^{-3}. Unlike in the cold thick target model, the thermalization of fast electrons produces a thermal component at energies <10 keV that increases with the background number density n. Increasing the background temperature, for a given density n, will cause the thermal component to be more pronounced at higher energies (see [16]). Figure 3 exemplifies how the background properties change the 'collisionaly relaxed' electron distribution in the target. From Figure 3, it is easy to recognize the thermalized part in $\langle nVF\rangle(E)$ at low energies. Importantly, we find that the more realistic model with diffusion correctly describes the electron energy loss rate, particularly at $E \sim T$. Hence, fewer electrons are required to produce a given, observed X-ray flux, of an order of a magnitude (see [18]). The number of electrons required to produce a given X-ray flux falls with increasing temperature T and number density n.

3. Discussion and ongoing work

The work summarized demonstrates that a more realistic model that includes the effects of collisional diffusion and thermalization of fast electrons, is preferable for inferring the properties of solar flare accelerated electrons at low energies of 10 to 30 keV from X-ray observations. Accounting for such effects can remove problems such as the low energy cutoff, and reduce the number of electrons required to produce a given observed X-ray flux. Importantly, such a model can be applied to all flare types showing the presence of hot plasma, whether they are 'standard' footpoint-dominated or coronal-source-dominated flares. A full description of the work is provided in [18], and also in [16]. Further, the application of such a model to real X-ray flare data is currently the topic of ongoing work. We hope it will eventually replace the use of the cold thick target model to deduce the accelerated electron distribution from X-ray spectroscopy.

3.1. Acknowledgments
NLSJ was funded by STFC and a SUPA scholarship. EPK and NHB gratefully acknowledge the financial support by an STFC Grant. AGE was supported by NASA Grant NNX14AK56G. NLSJ wishes to thank the organizers for providing her with the opportunity to discuss this work at the conference.

4. References

[1] Zharkova V V, Arzner K, Benz A O, Browning P, Dauphin C, Emslie A G, Fletcher L, Kontar E P, Mann G, Onofri M, Petrosian V, Turkmani R, Vilmer N and Vlahos L 2011 Space Sci. Rev. **159** 357–420 (*Preprint* 1110.2359)
[2] Lin R P, Dennis B R, Hurford G J *et al.* 2002 Solar Phys. **210** 3–32
[3] Kontar E P, Bian N H, Emslie A G and Vilmer N 2014 Astrophys. J. **780** 176 (*Preprint* 1312.0266)
[4] Kontar E P, Ratcliffe H and Bian N H 2012 Astron. Astrophys. **539** A43 (*Preprint* 1112.4448)
[5] Fleishman G D, Kontar E P, Nita G M and Gary D E 2011 Astrophys. J. Lett. **731** L19 (*Preprint* 1103.2705)
[6] Kontar E P, Brown J C, Emslie A G, Hajdas W, Holman G D, Hurford G J, Kašparová J, Mallik P C V, Massone A M, McConnell M L, Piana M, Prato M, Schmahl E J and Suarez-Garcia E 2011 Space Sci. Rev. **159** 301–355 (*Preprint* 1110.1755)
[7] Holman G D, Aschwanden M J, Aurass H, Battaglia M, Grigis P C, Kontar E P, Liu W, Saint-Hilaire P and Zharkova V V 2011 Space Sci. Rev. **159** 107–166 (*Preprint* 1109.6496)
[8] Graham D R, Hannah I G, Fletcher L and Milligan R O 2013 Astrophys. J. **767** 83 (*Preprint* 1302.2514)
[9] Xu Y, Emslie A G and Hurford G J 2008 Astrophys. J. **673** 576–585
[10] Kontar E P, Hannah I G and Bian N H 2011 Astrophys. J. Lett. **730** L22 (*Preprint* 1102.3664)
[11] Guo J, Emslie A G and Piana M 2013 Astrophys. J. **766** 28 (*Preprint* 1303.1077)
[12] Brown J C 1971 Solar Phys. **18** 489–502
[13] Syrovatskii S I and Shmeleva O P 1972 Sov. Astronom. **16** 273
[14] Emslie A G 1978 Astrophys. J. **224** 241–246
[15] Brown J C, Emslie A G and Kontar E P 2003 Astrophys. J. Lett. **595** L115–L117
[16] Jeffrey N L S, Kontar E P, Bian N H and Emslie A G 2014 Astrophys. J. **787** 86 (*Preprint* 1404.1962)
[17] Emslie A G 2003 Astrophys. J. Lett. **595** L119–L121
[18] Kontar E P, Jeffrey N L S, Emslie A G and Bian N H 2015 *ArXiv e-prints* (*Preprint* 1505.03733)

Journal of Physics: Conference Series **642** (2015) 012014 doi:10.1088/1742-6596/642/1/012014

Energetic particle pressure in intense ESP events

D Lario[1], R B Decker[1], E C Roelof[1], A-F Viñas[2],
[1]The Johns Hopkins University. Applied Physics Laboratory, 11100 Johns Hopkins Rd., Laurel, MD 20723, USA
[2]NASA Goddard Space Flight Center. Heliophysics Science Division, Geospace Physics Laboratory, Greenbelt, MD 20771, USA

E-mail: david.lario@jhuapl.edu

Abstract. We study three intense energetic storm particle (ESP) events in which the energetic particle pressure P_{EP} exceeded both the pressure of the background thermal plasma P_{th} and the pressure of the magnetic field P_B. The region upstream of the interplanetary shocks associated with these events was characterized by a depression of the magnetic field strength coincident with the increase of the energetic particle intensities and, when plasma measurements were available, a depleted solar wind density. The general feature of cosmic-ray mediated shocks such as the deceleration of the upstream background medium into which the shock propagates is generally observed. However, for those shocks where plasma parameters are available, pressure balance is not maintained either upstream of or across the shock, which may result from the fact that P_{EP} is not included in the calculation of the shock parameters.

1. Introduction

The highest intensities of solar energetic particles (SEPs) in a solar cycle are usually observed in association with the passage of interplanetary shocks [1]. The particle enhancement associated with the passage of interplanetary shocks is commonly known as an Energetic Storm Particle (ESP) event because of its association with the occurrence of sudden storm commencements [e.g., 2]. If the energy density of SEPs observed around the passage of the shocks becomes comparable to those of the background fields and plasma particles, the effects that such intense energetic particle populations exert on the shocks are not negligible. Such situations are usually described in terms of "cosmic-ray modified" shocks [3, 4, 5, 6, 7, 8, 9] where the effects produced by the energetic particles include the deceleration of the medium where the shock propagates and non-linear effects on the processes of particle acceleration [e.g., 4].

In steady state, the mass and momentum flux conservation relations across the shocks can be written as $\{\varrho_m U_n\}=0$ and $\{\varrho_m U_n^2+P+B_t^2/2\mu_0\}=0$ where {} denotes the difference of the enclosed quantity between the two sides of the shock, ϱ_m is the mass density, U_n is the normal component relative to the shock surface of the flow velocity **U** measured in the shock rest frame, B_t is the component of the magnetic field vector tangential to the shock surface, and P is the particle pressure, which usually is assumed to be isotropic [10]. It is also usually assumed that the only contribution to the particle pressure P comes from the thermal particles computed as $P_{th}=n_pkT_p+ n_ekT_e$ where k is the Boltzmann's constant, n_p and n_e is the number density, and T_p and T_e the temperature of the proton and electron

[1] Correspondence should be addressed to david.lario@jhuapl.edu.

populations forming the background thermal plasma, respectively. However, there are a number of distinct periods and/or regions of interplanetary space when the pressure exerted by the non-thermal energetic particles is comparable to or even exceeds both P_{th} and the magnetic field pressure, $P_B=B^2/2\mu_0$ [11, 12, 13, 14]. Therefore, it is necessary to include in P the non-negligible contribution of the pressure exerted by the non-thermal energetic particles P_{EP} and use $P= P_{EP}+P_{th}$ [15].

In this article we compare three cases where P_{EP} exceeded P_B by more than one order of magnitude and (when plasma observations were available) P_{EP} exceeded also P_{th}. One of these three cases occurred during the most intense SEP event of solar cycle 22 [11] and the remaining cases during the most intense events observed by the spacecraft A of the Solar Terrestrial Relations Observatory (STEREO-A) in solar cycle 24. The highest particle intensities were reached around the passage of interplanetary shocks. The peculiarity common to these three events is the presence of a foreshock region containing decreased magnetic field strength and, when plasma observations are available, depleted solar wind density and gradual increase of solar wind speed. We discuss whether these shocks can be classified as cosmic-ray mediated shocks.

2. Observations

Figure 1 shows hourly averages of (a) 4-6 MeV and (b) 25-53 MeV proton intensities measured by the Low Energy Telescope (LET) [16] and the High Energy Telescope (HET) [17] of the IMPACT (In-Situ Measurements of Particles and CME Transients) investigation [18] on board STEREO-A. The time interval spans from day 1 of year 2009 to day 243 of year 2014. The red circle indicates the highest 25-53 MeV proton intensity over this time interval, whereas the blue circle indicates the highest 4-6 MeV proton intensity. The highest 25-53 MeV proton intensity was observed on day 205 of 2012 [13], while the highest 4-6 MeV proton intensity occurred on day 149 of 2012. These two events were observed in association with the passage of interplanetary shocks and we will show that their properties are comparable to the most intense event observed in solar cycle 22 on day 293 of year 1989 [11].

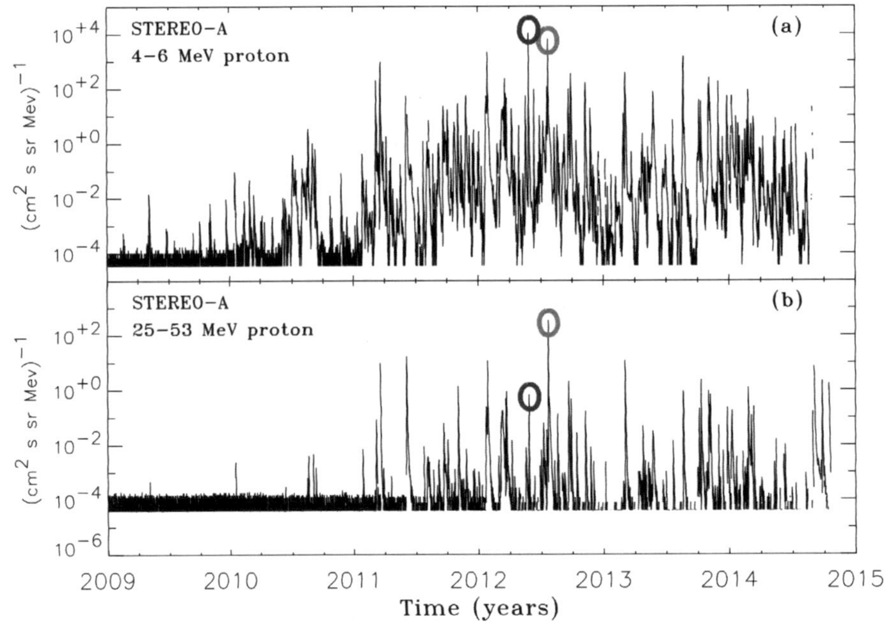

Figure 1. Hourly averages of (a) 4-6 MeV proton intensities measured by STEREO-A/LET [16] and (b) 25-53 MeV proton intensities measured by STEREO-A/HET [17]. The red and blue circles indicate the maximum 25-53 MeV and 4-6 MeV proton intensities measured by STEREO-A in solar cycle 24, respectively.

Figure 2 shows from top to bottom (a) energetic proton intensities observed by (left) the Energetic Particle Sensor (EPS) on board the Geostationary Operational Environmental Satellite-7 (GOES-7) [19] on day 293 of 1989, (center) by STEREO-A/HET [17] on day 205 of 2012 and (right) by STEREO-A/HET on day 149 of 2012, (b) magnetic field magnitude measured by (left) the Goddard Space Flight Center (GSFC) magnetometer on board the Interplanetary Monitor Platfrom-8 (IMP-8) [20] and (center, right) the Magnetic Field Experiment of the IMPACT suite of instruments [21] on board STEREO-A, (c) solar wind proton density, (d) solar wind proton temperature, and (e) solar wind speed as measured by (left) the Massachusetts Institute of Technology (MIT) Faraday Cup on board IMP-8 [20] and (center, right) by the Plasma and Suprathermal Ion Composition sensor (PLASTIC) on board STEREO-A [22]. Figure 2f shows the magnetic field pressure P_B (red symbols), the solar wind thermal pressure P_{th} (blue symbols), and the energetic particle pressure P_{EP} computed over the indicated energy interval (black trace). In the computation of $P_{th}=n_pkT_p+ n_ekT_e$ we have assumed neutrality ($n_p=n_e$) and $T_e=2T_p$ based on statistical surveys of proton and electron temperatures in post-shock plasmas [23]. The computation of P_{EP} and its limitations due to the finite energy range of the energetic particle observations are described in detail in the Appendix.

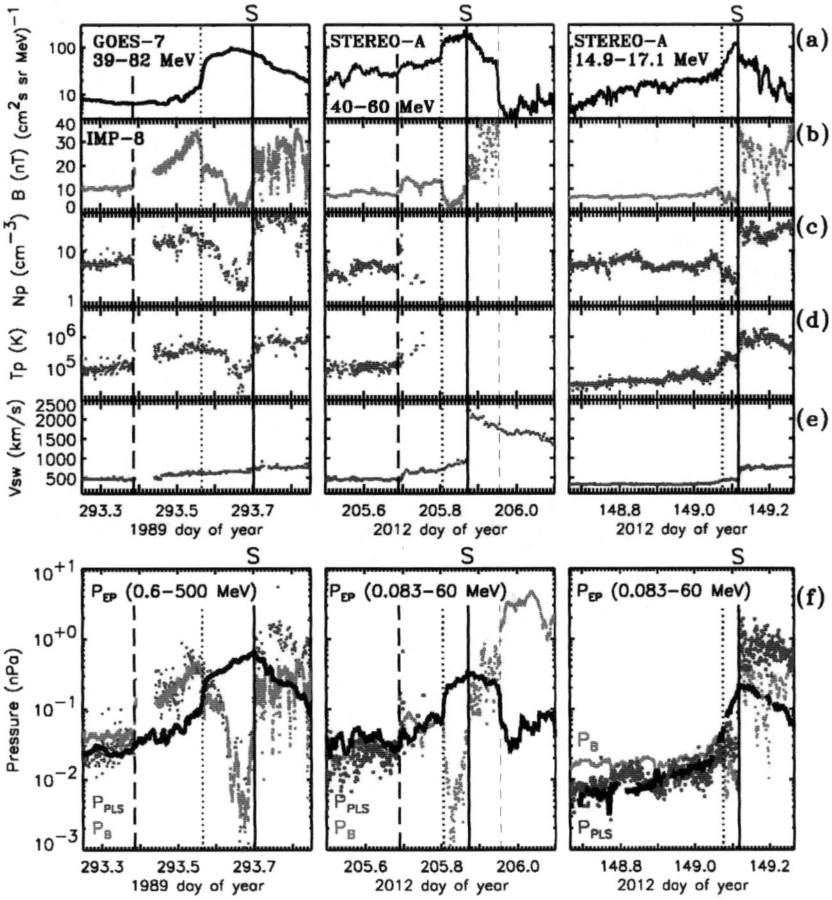

Figure 2. The three most intense ESP events in solar cycle 22 as observed near-Earth (left) and in solar cycle 24 as observed by STEREO-A (center and right panels). From top to bottom we show: (a) proton intensities at the indicated energy, (b) magnetic field strength, (c) solar wind proton density, (d) solar wind proton temperature, (e) solar wind speed, (f) energetic particle pressure (black trace), magnetic field pressure (red symbols) and thermal pressure (blue symbols), for the events on day 293 of 1989 (left), 205 of 2012 (center) and 149 of 2012 (right). See text for details.

The solid vertical lines in Figure 2 (labelled with the symbol S) indicate the passage of interplanetary shocks associated with these intense particle events. The vertical dotted lines in Figure 2 indicate the onset of a depression in magnetic field strength that coincides with an increase of the energetic particle intensities before the arrival of the shock. In the first two events (1989/293 and 2012/205) the decrease of magnetic field and increase of energetic particle intensities occurred abruptly, whereas in the event on 2012/149 the particle increase occurred more gradually, peaking at the arrival of the shock. The periods between the dotted and the solid vertical lines were dominated by P_{EP}, exceeding P_B by two orders of magnitude in the events on 1989/293 and 2012/205 and only by one order of magnitude in the event on 2012/149. The dashed vertical lines in the left and center panels of Figure 2 indicate a prior increase of magnetic field magnitude, solar wind density and solar wind temperature that could be associated with the passage of an earlier weak shock [11, 13].
There are a number of similarities between the three events of Figure 2: (1) the increase in the energetic particle intensities before the shock passage coincides with a decrease of the magnetic field magnitude (indicated by the vertical dotted line), (2) the period before the arrival of the shock is characterized by a steady increase of the solar wind speed V_{sw}, (3) with the exception of the event on 2012/205 where solar wind plasma observations are not available, the period of depleted magnetic field is also characterized by a rarefied solar wind density, (4) the energetic particle pressure P_{EP} exceeded P_B over the period between the dotted and the solid vertical lines, of ~195 minutes during the event on 1989/293, ~96 minutes during the event on 2012/205, and ~60 minutes during the event on 2012/149, (5) with the exception of the event on 2012/205 where P_{th} could not be computed, P_{EP} also exceeded P_{th} in the events on 1989/293 and 2012/149 events. However, whereas the event on 1989/293 showed a decrease of T_p, in the event on 2012/149 T_p increased, and (6) behind the shock P_B and P_{th} dominated P_{EP} throughout the post-shock compressed plasma.
It is possible that the process responsible for the depletion of the magnetic field strength as the energetic particle intensity increased before the arrival of the shock is similar to that proposed for the formation of diamagnetic field cavities observed upstream of the Earth's bow shock [e.g., 24]. Beams of energetic particles propagating away from the shock may generate a diamagnetic current that weakens the background magnetic field. The result is the creation of a crater of tenuous plasma and weak magnetic field bounded by enhanced densities and field strengths [25]. The hot compressed plasma left behind the shock, however, does not allow the creation of these cavities by particles propagating in the downstream region of the shocks. These cavities are therefore only observed in the upstream region of the shocks when the intensity of energetic particles is large enough to create enough pressure to depress the foreshock magnetic field strength and plasma density.

3. Can these shocks be classified as cosmic-ray mediated shocks?

The basic observational features of cosmic-ray mediated shocks are (1) a decrease of the upstream plasma flow U_n in the shock rest frame due to the pressure gradient of the particles ahead of the shock decelerating the background medium, and (2), in steady state, a pressure balance observed in the upstream region of the shock when P_{EP} is included in the estimation of the total pressure [5, 8, 9]. This pressure balance can be also evaluated across the shock by checking if $\{\varrho_m U_n^2+P+B_t^2/2\mu_0\}=0$ [10]. Figure 2e shows that the solar wind speed V_{sw} in the observer's frame steadily increased before the passage of the shock in the three events (especially in the region between the vertical dotted and solid lines). This increase of V_{sw} translates into a decrease of the plasma flow U_n in the shock frame of reference [8, 9] if the shock speed V_{shock} is constant.
In order to precisely estimate U_n, it is necessary to accurately compute both the shock speed and the direction of the normal to the shock. The formulae usually used to compute the shock parameters (i.e. the Rankine-Hugoniot conservation conditions) conventionally neglect the energetic particle contribution to the total pressure. Of the three events depicted in Figure 2, only the one with the complete set of plasma parameters, including the components of the solar wind velocity (the 2012/149 event), allows us to compute the shock parameters.

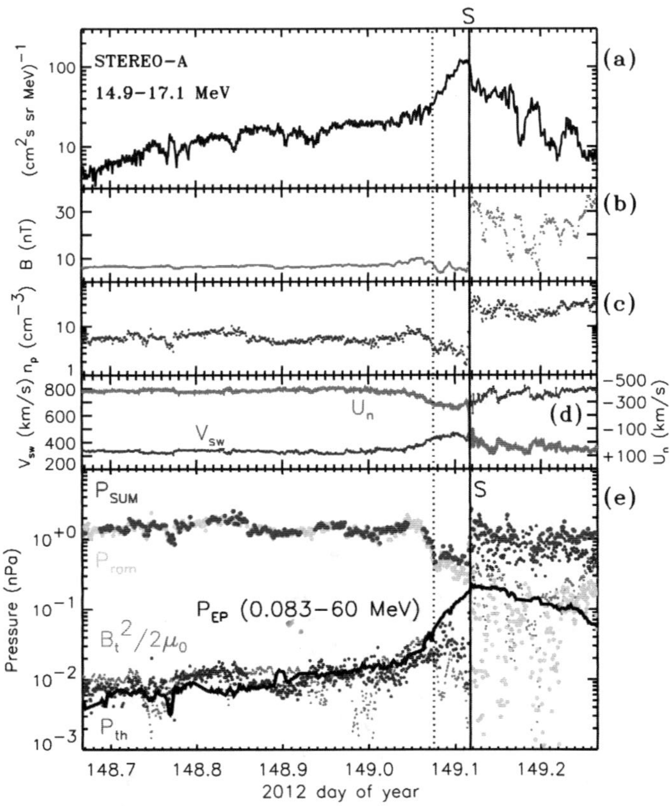

Figure 3. (a) One-minute averages of 14.9-17.1 MeV proton intensities, (b) magnetic field strength, (c) solar wind proton density, (d) solar wind speed Vsw and plasma flow speed in the shock frame of reference, (e) energetic particle pressure P_{EP} over the indicated energy range (black trace), thermal plasma pressure P_{th} (blue symbols), $B_t^2/2\mu_0$ (red symbols), P_{ram} (green symbols) and $P_{SUM}= \varrho_m U_n^2+P_{th}+P_{EP}+B_t^2/2\mu_0$ (purple symbols). Whereas there is a flow deceleration in front of the shock passage, there is no total pressure balance across the shock. See text for details.

We have applied the method described in [26] to compute the shock parameters of the shock on day 149 of 2012 (note that energy conservation equation was not considered in the computation of the shock parameters). The asymptotic solutions of this model provide the following parameters: normal to the shock in the RTN coordinate system **n**=(0.98,-0.17,-0.03), shock speed along the normal V_{shock}=706 ± 38 km s^{-1}, angle between the upstream magnetic field and the normal to the shock θ_{Bn}=76°, Alfvenic Mach number=8.7, magnetic field compression ratio=6.8 and density compression ratio=9.0. Once we know the normal to the shock and the shock speed, we proceed to compute $U_n=\mathbf{V}_{sw}\cdot\mathbf{n}-V_{shock}$, the ram pressure $P_{ram}=\varrho_m U_n^2$ and the quantity $P_{SUM}=\varrho_m U_n^2+P_{th}+P_{EP}+B_t^2/2\mu_0$. Figure 3 shows from top to bottom (a) 14.9-17.1 MeV proton intensities, (b) magnetic field magnitude, (c) solar wind proton density n_p, (d) solar wind speed V_{sw} (blue trace) and the flow speed U_n (gray trace), (e) the thermal pressure P_{th} (blue symbols), the energetic particle pressure P_{EP} over the indicated energy range (black trace), $B_t^2/2\mu_0$ (red symbols), the ram pressure $\varrho_m U_n^2$ (green symbols) and the quantity P_{SUM} (purple symbols). Figure 3d shows that around the time of the particle intensity enhancement (indicated by the vertical dotted line), U_n decreases by about ~100 km s^{-1}. However just before the arrival of the shock, as high-energy proton intensities ceased increasing, U_n stopped decreasing.

Throughout the upstream region of the shock P_{ram} dominated over P_{EP}, P_{th} and P_B, whereas downstream of the shock P_{th} contributed the most to P_{SUM}. As the shock approached the spacecraft and P_{EP} started exceeding both P_{th} and P_B, the deceleration of the plasma flow and the density depletion observed at the time of the vertical dotted line yielded to a decrease of P_{ram} for the whole period comprised between the vertical dotted and solid lines in Figure 3. It is clear that before and across the passage of the shock, P_{SUM} is not constant (in particular at the time of the vertical dotted line). Possible reasons for this lack of balance of the quantity P_{SUM} include the following:

(1) Due to the limited energy coverage of the particle instruments, our calculated P_{EP} is only a lower limit (see Appendix A). If the depleted region of P_{SUM} observed before the shock were due to <83 keV protons, then to fill the depletion would require the intensity of this region to be very high, have a very steep energy spectra, and be localized only in that region (see Appendix A). It would follow that this population, presumably accelerated at the shock, be present also in the downstream region of the shock, but in such a manner that P_{SUM} would have to remain constant across the shock.

(2) The observed depletion in P_{SUM} is a consequence of the depletion in P_{ram}, the dominant pre-shock contribution to P_{SUM}, which derives from the decrease of U_n and low values of n_p. An error in the estimates of any of these terms implies an error in the evaluation of $P_{ram}=\varrho_m U_n^2$. In particular, the evaluation of U_n requires an accurate computation of the shock speed and the shock normal. These shock parameters were derived as an asymptotic solution of the Rankine-Hugoniot conditions that did not include the effects of the energetic particles. However, since we argue that P_{EP} plays a significant role determining the shock parameters, we have an inconsistency that must be addressed in future calculations.

(3) The spatial structure of the shock may not have been steady due to its instability to, for example, sound waves incident from the upstream medium [27], magnetic waves propagating oblique to the average magnetic field [28], and amplification of Alfvén waves by cyclotron resonant interaction with shock accelerated particles [8]. Whereas these fluctuations may lead to the unbalance of P_{SUM}, it is difficult to explain the deep depletion of P_{SUM} before the shock via only fluctuations of U_n and n_p. It is also possible that the high P_{EP} could result from the passage of the spacecraft through a ripple or indentation on the shock surface (~0.01 AU in depth implied by a 1-hour duration and V_{sw}~400 km s^{-1}) that could temporarily trap and enhance intensities of high energy protons able to reflect from two magnetic connection points at the quasi-perpendicular shock while low energy ions were simply transmitted through the shock [e.g., 29].

Therefore, of the basic features of cosmic-ray mediated shocks pointed out by Terasawa et al. [8, 9], we see a relative decrease of U_n in the foreshock region as energetic particle intensities increase, but not a balance of P_{SUM}.

4. Conclusions

The highest energetic particle intensities observed in solar cycle 22 close to Earth and by STEREO-A in solar cycle 24 occur in association with interplanetary shocks. The upstream region of these shocks was characterized by a decrease of the magnetic field strength coinciding with the last increase of energetic particle intensities and, when plasma observations are available, with a depletion of solar wind density. During these periods the pressure exerted by the energetic particles P_{EP} exceeded the magnetic field pressure P_B and the pressure of the background thermal particles P_{th}. These shocks showed also a steady increase of V_{sw} that translates into a decrease of the plasma flow speed in the shock frame of reference as a result of the pressure of the energetic particles reflected ahead of the shock. For the event on day 149 of 2012, when the completeness of plasma parameters allows us to compute shock parameters, we did not observe a total pressure balance. Apparently, the accurate computation of the shock parameters must consider the effects of P_{EP} on the structure of the shocks.

5. References

[1] Lario D, Aran A and Decker R B 2008 *Space Weather* **6** S12001, doi: 10.1002/2008SW000403
[2] Rao U R, McCracken K G and Bukata R P 1967 *J. Geophys. Res.* **72** 4325
[3] Axford W I, Leer E and Skadron G 1977 *Proc. 15th Int. Cosmic Ray Conf.* **11** 132
[4] Axford W I, Leer E and McKenzie J F 1982 *Astron. and Astrophys.* **111**, 317
[5] Drury L O'C and Volk J H 1981 *Astrophys. J.* **248** 344
[6] Bulanov S V and Sokolov I V 1984 *Soviet Astronomy* **28** 515
[7] Blandford R D 1980 *Astrophys J.* **238** 410
[8] Terasawa T, Maezawa K, Hoshino M, Shimada N, Mukai T, Saito Y, Yamamoto T, Kokubun S, Wilken B and Doke T 1999 *Proc. 26th Int. Cosmic Ray Conf.* **6** 528
[9] Terasawa T, Oka M, Nakata K, Keika K, Nose M, McEntire R W, Saito Y and Mukai T 2006 *Adv. Space Res.* **37** 1408
[10] Gurnett D A and Bhattacharjee A 2005 *Introduction to Plasma Physics. With Space and Laboratory Applications.* (Cambridge University Press) p 254
[11] Lario D and Decker R B 2002 *Geophys. Res. Lett.* **29** 1393 doi: 10.1029/2001GL014017
[12] Marhavilas P K and Sarris E T 2011 *IEEE Trans. Plasma Sci.* **39** 1516
[13] Russell C T, Mewaldt R A, Luhmann J G, Mason G M, von Rosenvinge T.T., Cohen C M S, Leske R A, Gomez-Herrero R, Klassen A, Galvin A and Simunac K D C 2013 *Astrophys. J.* **770** 38 doi: 10.1088/0004-637X/770/1/38
[14] Lario D, Decker R B, Roelof E C and Vinas A-F 2015 *submitted to Astrophys. J.*
[15] Roelof E C, Krimigis SM, Mitchell D G, Decker R B, Richardson JD, Gruntman M and Funsten H O 2010 *Pickup Ions Throughout the Heliosphere and Beyond. Proc. 9th Ann. Int. Astrophys. Conf. American Institute of Physics Proceedings* **1302**, 133
[16] Mewaldt R A, Cohen C M S, Cook W R, Cummings A C, Davis A J et al. 2008 *Space Sci. Rev.* **136** 285
[17] von Rosenvinge T T, Reames D V, Baker R, Hawk J, Wessel R, Noal J T, et al. 2008 *Space Sci. Rev.* **136** 391
[18] Luhmann J G, Curtis D W, Schroeder P, McCauley J, Lin R P, et al. 2008 *Space Sci.* **136** 391
[19] Sauer H H 1993 *Proc. 23rd Int. Cosmic Ray Conf.* **3** 250
[20] Scearce C S et al. 1976 *NASA GSFC X Doc.* 695-76-191, Oct. 1976.
[21] Acuña M H, Curtis D, Scheifele J L, Russell C T, Schroeder, Szabo A and Luhmann J G 2008 *Space Sci. Rev.* **136** 203
[22] Galvin A B, Kistler L M, Popecki M A, Farrugia C J, Simunac K D C, et al. 2008 *Space Sci. Rev.* **136** 437
[23] Gosling J T, Baker D N, Bame S J, Feldman W C, Zwickl R D and Smith E J 1987 *J. Geophys. Res.* **92** 8519
[24] Sibeck D, Decker R B, Mitchell D G, Lazarus A J, Lepping R P and Szabo A 2001 *J. Geophys. Res.* **106** 21675
[25] Thomas V A and Brecht S H 1988 *J Geophys. Res.* **93** 11341
[26] Viñas A F and Scudder J D 1986 *J Geophys. Res.* **91** 39
[27] Drury L O and Falle S A E G 1986 *MNRA* **223** 353
[28] Zank G P, Axford W I and McKenzie J F 1990 *Astron Astrophys.* **233** 275
[29] Decker R B 1990 *J. Geophys. Res.* **95** 11993
[30] Muller-Mellin R, Bottcher S, Falenski J, Rode E, Duvet L, Sanderson T R, Butler B, Johlander B and Smith H 2008 *Space Sci. Rev.* **136** 363

Acknowledgments

We acknowledge the STEREO Science Centers for providing the data used in this paper and the STEREO/IMPACT and STEREO/PLASTIC teams for providing these data. DL acknowledges the support from NASA under grants NNX11A083G and NNX15AD03G. ECR, RBD and DL acknowledge NASA support under the ACE grant NNX10AT75G.

Journal of Physics: Conference Series **642** (2015) 012014 doi:10.1088/1742-6596/642/1/012014

Appendix A. Computation of energetic particle pressure P_{EP}

The expression used to compute the pressure exerted by the energetic protons is

$$P_{EP} = (4\pi/3)(2m)^{1/2} \int_{T_1}^{T_2} dT\, T^{1/2} j(T), \qquad (1)$$

where m is the proton mass, $j(T)$ is the differential proton flux at the reference energy T, and T_1 and T_2 are the boundaries over which we evaluate the pressure P_{EP} [11]. Strictly speaking, this expression should be evaluated in the plasma frame [15], but since our lower limit for the proton energies is ~83 keV (corresponding to a proton velocity of v~4000 km s^{-1}), the expression in the spacecraft frame approximates the correct partial pressure of energetic particles to the order of $O(U/v)$<0.10.

Figure A1 shows the computation of P_{EP} for the event on day 149 of 2012. Figure A1a shows the proton intensities measured by the combination of particle instruments on board STEREO-A (i.e. the Solar Electron and Proton Telescope (SEPT) [30], the Low Energy Telescope (LET) [16] and the High Energy Telescope [17]). Figure A1b shows the proton energy spectra at the time of the shock passage (indicated by the solid vertical line in Figure A1a). The differential particle intensities $j(T)$ at each energy channel are indicated by the colored symbols in Figure A1b. We use the mathematical expression $log[j(T)]=a+b\ logT+cT$ (solid black line in Figure A1b) to fit the energy spectra over the energy range 83 keV – 60 MeV.

Owing to the limited energy coverage of the energetic particle instruments, one must wonder about the contribution to P_{EP} of both the most abundant low-energy protons below 83 keV and the protons of energies above 60 MeV. We have assumed several dependences for the energy spectra below 83 keV, from the simple extrapolation of the fitted spectra (indicated by the dashed thick black line in Figure A1b) to power-law functions as T^{0}, $T^{-0.5}$, $T^{-1.0}$, $T^{-1.5}$, $T^{-2.0}$, $T^{-3.0}$ indicated by the orange, purple, cyan, blue, green and red dotted lines in Figure A1b, respectively. Figure A1c shows the integrand of equation (1) at the time of the shock passage. Figure A1d shows how the integrated pressure depends on the maximum energy T_2 of the integration range. The solid thick line considers as lower boundary T_1 the minimum proton energy measured by STEREO-A/SEPT (T_1=83 keV) and the spectra fitted in Figure A1b. The dotted lines and the dashed thick line of Figure A1d assume a lower boundary T_1=4 keV and the extrapolations of the energy spectra to lower energies depicted in Figure A1b. The error in the particle pressure is of the order $O(U/v)$~0.5 near 4 keV, but its contribution to the pressure P_{EP} over the whole energy range is negligible except for the steepest spectrum. Figure A1d also shows that above ~10 MeV, P_{EP} seems to saturate, suggesting that the contribution of the higher energy particles can be considered negligible.

Finally, Figure A1e shows the evolution of P_{EP} around the shock passage considering the individual one-minute spectra throughout the time interval considered in Figure A1a. Whereas the solid black line represents the fitted energy spectra over the energy range T_1=83 keV and T_2=60 MeV, the dotted and dashed lines represent the extrapolation of the energy spectra to lower energies (up to T_1=4 keV) as depicted in Figure A1b. The contribution of the low-energy portion becomes significant for steep energy spectra (e.g., case $T^{-3.0}$) and when the low-energy proton intensities are already elevated (e.g. in the downstream region of the shock).

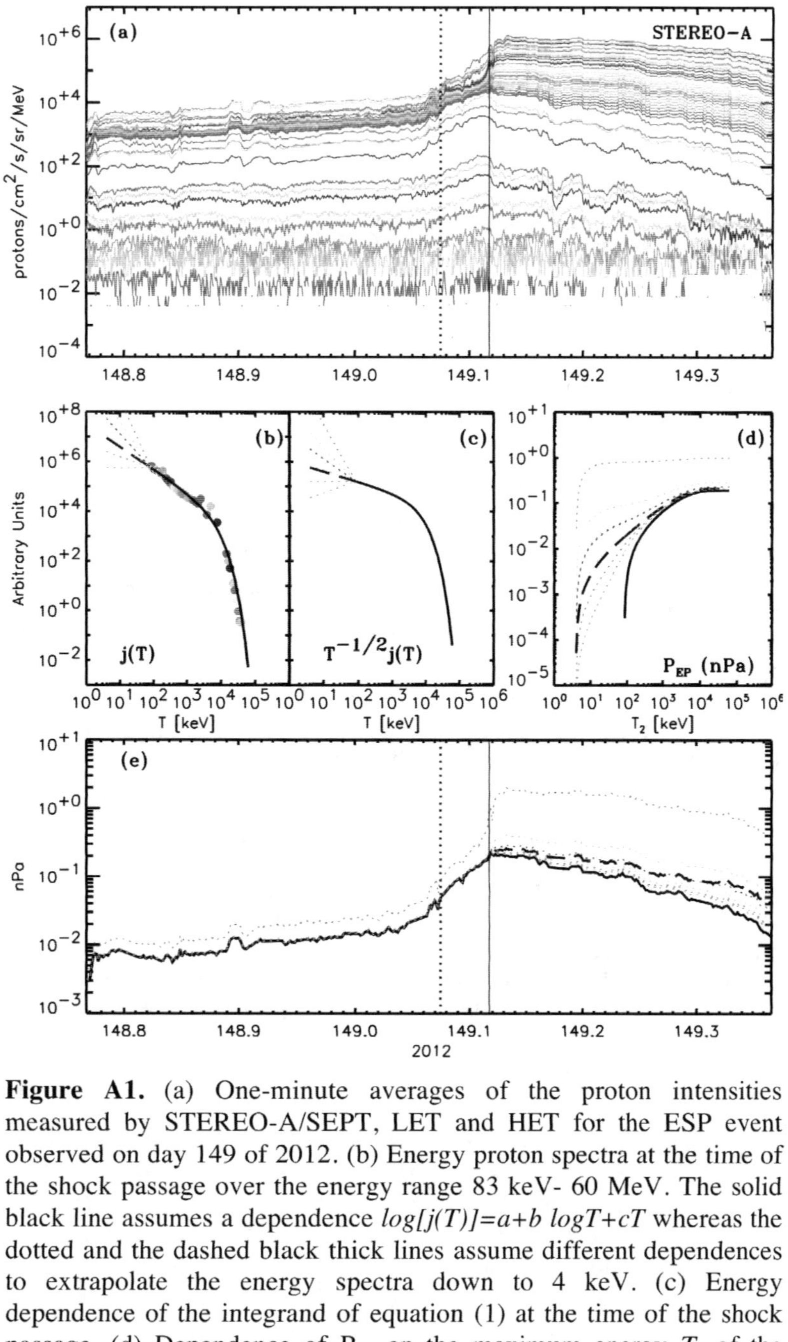

Figure A1. (a) One-minute averages of the proton intensities measured by STEREO-A/SEPT, LET and HET for the ESP event observed on day 149 of 2012. (b) Energy proton spectra at the time of the shock passage over the energy range 83 keV- 60 MeV. The solid black line assumes a dependence *log[j(T)]=a+b logT+cT* whereas the dotted and the dashed black thick lines assume different dependences to extrapolate the energy spectra down to 4 keV. (c) Energy dependence of the integrand of equation (1) at the time of the shock passage. (d) Dependence of P_{EP} on the maximum energy T_2 of the integration ranges. (e) Time evolution of P_{EP} computed over the energy range 83 keV – 60 MeV (solid black line) and over the energy range 4 keV- 60 MeV assuming the extrapolations of the energy spectra to lower energies depicted in panel (b) (dotted and dashed lines). See text for details.

Energetic Ion Acceleration by Small-scale Solar Wind Flux Ropes

Jakobus A. le Roux[1,2], Gary M. Webb[2], Gary P. Zank[1,2], & Olga Khabarova[3]

1. Dept. of Space Science, University of Alabama in Huntsville, Huntsville, USA
2. Center for Space Plasma & Aeronomic Reseach, University of Alabama in Huntsville, Huntsville, USA
3. Heliophysical Laboratory, Pushkov Institute of Terrestrial Magnetism, Ionosphere and Radiowave Propagation RAS (IZMIRAN), Troitsk, Moscow , Russia

E-mail: `jar0013@uah.edu`

Abstract. We consider different limits of our recently developed kinetic transport theory to investigate the potential of supersonic solar wind regions containing several small-scale flux ropes to explain the acceleration of suprathermal ions to power-law spectra as observations show. Particle acceleration is modeled in response to flux-rope activity involving contraction, merging (reconnection), and collisions in the limit where the particle gyoradius is smaller than the characteristic flux-rope scale length. The emphasis is mainly on the statistical variance in the electric fields induced by flux-rope dynamics rather than on the mean electric field induced by multiple flux ropes whose acceleration effects are discussed elsewhere. Our steady-state analytical solutions suggest that ion drift acceleration by flux ropes, irrespective of whether displaying incompressible or compressible behavior, can yield power laws asymptotically at higher energies whereas an exponential spectral rollover results asymptotically when field-aligned guiding center motion acceleration occur by reconnection electric fields from merging flux ropes. This implies that at sufficiently high particle energies, drift acceleration might dominate. We also expect compressive flux ropes to yield harder power-law spectra than incompressible flux ropes.

1. Introduction

Spacecraft observations, magnetohydromagnetic (MHD) turbulence theory and simulations suggest that low-frequency turbulence in the solar wind near Earth consists primarily of a quasi-two-dimensional (quasi-2D) component [1,2,3,4], at least during quiet solar wind conditions. Three-dimensional (3D) compressible MHD turbulence simulations with a strong guide magnetic field support the idea that MHD turbulence is mainly quasi-2D with a strong presence of closely spaced, quasi-2D (helical) small-scale flux ropes (combination of a axial (guide) field and a magnetic island or twisted magnetic field in the plane perpendicular to the guide field) on turbulence inertial range scales that are dynamic (see Dmitruk et al. [4]). Flux-rope dynamics include merging (reconnection) at the current sheets separating neighboring flux ropes, and flux-rope contraction or expansion as well. In the presence of a guide field reconnection can be understood in terms of component reconnection [5], where it is the component of the magnetic field transverse to the guide field that is involved in the reconnection process of two neighboring flux ropes.

Dmitruk et al. [4] also use test particle simulations of particles traversing the electric fields induced by multiple dynamic small-scale flux ropes from the 3D MHD turbulence simulation (with a strong guide field) to show that efficient acceleration and hard power-law distributions result for both ions and electrons. This extended and confirmed earlier work where hard power-law spectra using a 2D MHD turbulence simulation with a guide field [6]. Likewise, MHD simulations of turbulent reconnection at large-scale primary current sheets produce multiple small-scale flux ropes often separated by small-scale current sheets that, through contraction and merging dynamics, can efficiently accelerate test particles to produce power-law spectra [6]. Recent simulations of particle acceleration in merging and contracting small-scale flux ropes of particle-in-the-cell (PIC) codes confirm the formation of power-law spectra [7,8]. This is in contrast to previous PIC modeling that produced exponential spectra [9,10,11], possibly due to computational limitations.

Recent observations show that energetic particle fluxes peak at magnetic field discontinuities including coherent structures separated by current sheets and that the highest energetic particle fluxes occur at the strongest discontinuities most often found behind interplanetary shocks [12]. Observational evidence also exists for small-scale flux rope formation and merging activity at the primary current sheets at leading and trailing edges of interplanetary coronal mass ejection (ICME) structures [10,11], and for energetic particle acceleration in regions of multiple small-scale flux ropes formed near the heliospheric current sheet (HCS) [15]. In summary, a growing body of observational, theoretical and computational evidence suggest that coherent small-scale flux-rope dynamics and dissipation could play an important role in the observed formation of suprathermal ion power-law spectra in the supersonic solar wind. This aspect also needs further investigation in addition to the traditional focus on ion acceleration by interplanetary shocks and random compression and rarefaction regions [16,17,18,19].

Building on previous work [9,10,20], Zank et al. [21] developed a comprehensive kinetic theory framework to model energetic charged particle acceleration by multiple quasi-2D small-scale flux ropes on large spatial scales in the presence of a nonuniform large-scale magnetic field and solar wind flow. By focusing on ion acceleration by the mean electric field induced by multiple flux-rope dynamics of contraction and merging, Zank et al. found power-law spectra with indexes depending on the Alfvénic Mach number and on the ratio of the diffusion over the island contraction time scales. This development was further extended by le Roux et al. [22] using a perturbation analysis augmented with nonlinear effects. This approach enables modeling of the acceleration effects of both the mean and the variance in the electric fields induced by multiple contracting and merging flux ropes in the presence of a strong guide field. In this paper we explore different limits of the extended theory [22] by concentrating mainly on the variance in the flux-rope electric fields to investigate their potential in forming energetic ion power-law spectra in the supersonic solar wind.

2. Solutions of Ion Acceleration by Multiple Contracting and Merging Small-scale Flux Ropes

Our kinetic transport theory was developed using standard guiding center kinetic theory as a basis, resulting in a focused transport equation with Fokker-Planck coefficients for particle diffusion in momentum space due to interaction with several flux ropes. This limits the theory to particle gyroradii $r_g \ll L_I$, where L_I is the characteristic scale size of the small-scale flux ropes [22]. Small-scale flux ropes observed near the heliospheric current sheet at Earth, for example, have a characteristic scale size of $L_I \approx 0.01$ AU [15]. Given that the gyroradius of a 1 keV proton is $\sim 5 \times 10^{-6}$ AU at 1 AU, our theory applies to a wide range of suprathermal particle energies. In our theory we can choose between the limit where the particle parallel scattering mean path $\lambda_\parallel < L_I$ or the opposite limit where $\lambda_\parallel > L_I$. Here we pursue solutions where energetic ions are diffusing along the flux-rope magnetic field ($\lambda_\parallel < L_I$). In this limit the simplifying assumption of a near-isotropic particle distribution is most plausible, enabling us to reduce the basic focused transport equation we derived to a Parker-like transport equation that can be solved analytically [22]. Thus, the solutions discussed below are perhaps best applicable in the enhanced turbulence conditions that one typically find behind interplanetary shocks.

2.1. Drift Acceleration by Incompressible Contracting and Merging Flux Ropes

In this section we discuss briefly ion drift acceleration by the variance in the electric fields induced by multiple contracting and merging flux ropes exhibiting incompressible behavior [9,10]. The basic idea is that ions are energized during flux-rope contraction or merging because they mainly experience curvature drift in the direction of the induced electric fields generated at the strongly curved magnetic field at endpoints of contracting flux ropes, or in the outflow regions of reconnection sites at the center of merging neighboring flux ropes [10,11]. Energy loss is associated with negative betratron acceleration [23] arising from particles conserving their magnetic moment while the magnetic field strength decreases with time during incompressible contraction or merging of flux ropes [10]. Thus, when particles scatter efficiently, as assumed here, they will undergo net stochastic acceleration from experiencing in a random fashion both the mean induced electric field and weakening magnetic field when traversing several contracting and merging flux ropes. If the mean induced electric field and magnetic field of rate of change is zero, or the particles have a purely isotropic distribution, the net energy gain from incompressible flux rope behavior is zero [9,10,22]. However, even during such conditions, the variance in random statistical fluctuations in the induced electric fields and magnetic field time variations of multiple contracting and merging flux ropes results in stochastic acceleration [20,22]. This is the focus of this section.

We assume a spherically-symmetric heliosphere with a radial solar wind outflow from the Sun where spatially parallel diffusing energetic ions feel the effects of large-scale solar wind advection and adiabatic cooling in the supersonic solar wind, and experience momentum diffusion from drift acceleration by incompressible contracting and merging flux ropes. Simplifying our theory accordingly, we solve analytically a steady-state Parker transport equation for the direction-averaged distribution $f_0(r, p')$ as a function of heliocentric distance r in the fixed frame and particle momentum p' in the solar wind flow frame. The equation is

$$U_0 \frac{\partial f_0}{\partial r} - \frac{1}{r^2}\frac{\partial}{\partial r}\left(r^2 \kappa_{rr} \frac{\partial f_0}{\partial r}\right) - \frac{2}{3}\frac{U_0}{r} p' \frac{\partial f_0}{\partial p'} - \frac{1}{p'^2}\frac{\partial}{\partial p'}\left(p'^2 D_{pp}^{INC} \frac{\partial f_0}{\partial p'}\right) = \frac{\dot{N}\delta(r - r_0)\delta(p - p_0)}{16\pi^2 r_0^2 p_0^2}. \quad (1)$$

In this equation U_0 is the solar wind flow speed, $\kappa_{rr} = \kappa_{\parallel}^{eff} \cos^2 \psi$ is the radial diffusion coefficient as a result of projecting the the effective parallel diffusion coefficient $\kappa_{\parallel}^{eff}$ in the radial direction (ψ is the solar wind magnetic field spiral angle), D_{pp}^{INC} is the momentum diffusion coefficient for modeling energetic ion acceleration in response to the variance in the electric fields induced by multiple incompressible contracting and merging flux ropes, and $\dot{N}$ is the constant rate of source particle production at a fixed heliocentric distance r_0 and momentum p_0. In the assumed limit of efficient particle scattering and a strong guide field, parallel diffusion was modeled by extending the usual quasi-linear theory perturbation analysis for undisturbed guiding center motion to include diffusive distortion of particle guiding center trajectories [24]. Effectively, diffusion was modeled as a combination of particle gyro-resonant scattering by parallel propagating Alfvén waves and of scattering in response to the variance encountered in the multi-island end-point magnetic fields that reflect/transmit particles because of the magnetic mirroring force. This generates a nonlinear expression for effective parallel diffusion that needs to be solved. A simplified solution for parallel diffusion was found upon realizing that on large spatial scales scattering by flux ropes can be more efficient than scattering by Alfvén waves in solar wind conditions near 1 AU. The expression for parallel diffusion is

$$\kappa_{\parallel}^{eff} \approx V_A \left(\frac{25}{9\pi} \frac{L_{INC}}{(\langle \delta B_{INC}^2 \rangle / B_0^2)^2} \right), \tag{2}$$

where V_A is the Alfvén speed, L_{INC} is the characteristic length of incompressible flux ropes, and $\langle \delta B_{INC}^2 \rangle / B_0^2$ is the mean fluctuation energy density of the incompressible multi-flux-rope magnetic field normalized to the energy density of the mean magnetic field. Note that parallel diffusion due to random magnetic mirroring and transmission by flux ropes is independent of particle speed. The momentum diffusion coefficient is

$$D_{pp}^{INC} = p'^2 \frac{3\pi}{50} \frac{V_A}{L_{INC}} \left(\frac{\langle \delta B_{INC}^2 \rangle}{B_0^2} \right)^2 \left(\frac{\langle B_{INC} \rangle}{B_0} \right)^4, \tag{3}$$

where $\langle B_{INC} \rangle$ is the mean multi-flux-rope magnetic field strength. Upon conveniently introducing the radial dependencies $\kappa_{rr} = V_{\kappa 0} r$ and $D_{pp}^{INC} = p'^2 D_0 r^{-1}$, whereby both $V_{\kappa 0}$ (a speed determining the rate of diffusive transport) and D_0 (a speed determining the rate of second order Fermi acceleration) are constants proportional to the Alfvén speed, we find a solution to equation (1) given by

$$\begin{aligned} f_0(r, p') &= \frac{\dot{N}}{32\pi^3 r_0^2 p_0^3 V_{\kappa 0}} \sqrt{\frac{V_{\kappa 0}}{D_0}} \left(\frac{r}{r_0} \right)^{-(1-U_0/2V_{\kappa 0})} \left(\frac{p'}{p_0} \right)^{-(3+2U_0/3D_0)/2} \times \\ &\times \ K_0 \left(b\sqrt{\ln^2(r/r_0) + (V_{K_0}/D_0)\ln^2(p'/p_0)} \right), \end{aligned} \tag{4}$$

where b is determined by the expression

$$b^2 = \left(1 - \frac{U_0}{2V_{\kappa 0}} \right)^2 + \frac{1}{4} \frac{D_0}{V_{\kappa 0}} \left(3 + \frac{2}{3} \frac{U_0}{D_0} \right)^2. \tag{5}$$

and K_0 is the modified Bessel function of the second kind (a decaying function with increasing argument size).

The accelerated part of the particle spectra ($p'/p_0 > 1$) converge to power laws with similar power-law indexes for $p' >> p_0$ over a wide range of radial distances at and beyond the particle source position. The power law index can be estimated by assuming $p' \gg p_0$ in the argument of the K_0 Bessel function in equation (4), which leads asymptotically to a power-law particle distribution given by

$$\begin{aligned} f_0(p') &\propto \left(\frac{p'}{p_0} \right)^{-c\left(1+\sqrt{1+(V_{\kappa 0}/D_0)(d/c)^2}\right)}, \\ c &= \frac{1}{2}\left(3 + \frac{2}{3}\frac{U_0}{D_0} \right), \\ d &= 1 - \frac{U_0}{2V_{\kappa 0}}, \end{aligned} \tag{6}$$

weakly modified by a logarithmic function of p' (not shown). When the effects of solar wind advection, adiabatic cooling in the expanding solar wind, and spatial diffusion are ignored by setting $U_0 = 0$ and then $V_{\kappa 0} = 0$ in the expression for the power-law index, we obtain a hard power-law spectrum $f_0(p') \propto p'^{-3}$ because particles only experience second order Fermi acceleration. The accelerated spectrum is steepened when adiabatic cooling partly cancels out stochastic acceleration, and when diffusive particle escape from flux ropes reduces the particle trapping (acceleration) time in individual flux ropes. Advection, on the other hand, tends to harden the spectrum because it effectively reduces diffusion, and particle escape from flux ropes is less efficient. Consequently, the trapping and acceleration time in individual flux ropes increase and the spectrum hardens.

Consider the limit when spatial advection of ions by the solar wind is approximately in balance with spatial diffusion according to the limit $U_0/V_{\kappa 0} \to 2$ (similar to the diffusion-advection limit in standard cosmic-ray modulation theory). Then, diffusive escape of energetic ions from flux ropes is relatively inefficient which limits the steepening of the spectrum. Accordingly,

$$f_0(p') \propto \left(\frac{p'}{p_0}\right)^{-\left(3+\frac{2}{3}\frac{U_0}{D_0}\right)}, \tag{7}$$

so that the power-law index of energetic ions in the supersonic solar wind is determined by a competition between stochastic acceleration by flux ropes and adiabatic cooling in the expanding solar wind. If in addition, stochastic acceleration of energetic ions by flux ropes is weakly dominated by adiabatic cooling, $(2U_0/3D_0 \to 2)$, the ion spectrum becomes a $f_0(p') \propto p'^{-5}$ as is often observed in the solar wind [24]. Such an outcome could be a consequence of a quasi-steady state that forms if one extend our test particle solution to take into account self-consistent interaction between energetic ions and flux ropes [18,25].

In le Roux et al. [22] we also showed that when realistic parameters are incorporated in the full solution of equation (4), acceleration in quiet solar wind conditions near 1 AU is inefficient. Efficient acceleration to hard power-law spectra was found in strong flux-rope conditions expected behind interplanetary shocks. This raises the interesting question of how flux rope acceleration downstream near interplanetary shocks might modify diffusive shock acceleration. Our preliminary results suggest that flux rope acceleration behind shocks can easily result in hardening of the diffusive shock acceleration spectrum. For a discussion of this issue, see Zank et al. [26].

2.2. Field-Aligned Guiding Center Motion Acceleration by Merging Flux Rope Reconnection Electric Fields

In the case of merging neighboring flux ropes with a strong guide field, field-aligned guiding center motion through the reconnection zone occurs mostly parallel or anti-parallel to the direction of the reconnection electric field pointing perpendicular to the magnetic island (twist) part of the flux rope. Therefore, energy gain or energy loss can occur. In the case of multiple merging flux ropes with a finite mean reconnection electric field, energetic ions undergoing efficient scattering will experience 2nd order Fermi acceleration with a net energy gain as long as the particle distribution is not purely isotropic. This case was discussed by Zank et al. [21]. Instead, we focus on how energetic ions undergo stochastic acceleration in response to the statistical variance in the reconnection electric field they encounter when traversing multiple merging flux ropes in the supersonic solar wind, an effect that produces a net energy gain irrespective of whether the particle distribution is isotropic or not [22].

In this limit of our theory, we solve the following steady-state Parker-type transport equation for energetic ions given by

$$U_0\frac{\partial f_0}{\partial x} - \kappa_{xx}\frac{\partial^2 f_0}{\partial x^2} - \frac{1}{p'^2}\frac{\partial}{\partial p'}\left(p'^2 D_{pp}^{REC}\frac{\partial f_0}{\partial p'}\right) = \frac{n_0 U_0 \delta(x)\delta(p-p_0)}{4\pi^2 p_0^2}. \tag{8}$$

In equation (8) it is assumed for simplicity that energetic ions propagate in a one-dimensional heliosphere (Cartesian geometry) where the solar wind flow occurs in the x-direction, and x

is distance relative to the source particles with a momentum p_0 which are injected into the heliosphere at an arbitrary distance labeled $x = 0$ at a fixed rate determined by the flux $n_0 U_0$ (n_0 is the number density of the source particles advected with the solar wind flow at a velocity $\mathbf{U}_0$). A consequence of assuming planar geometry is the neglect of adiabatic cooling of energetic ions. This is reasonable for energetic ions far from the Sun where adiabatic cooling is weak, or in places where stochastic acceleration is relatively efficient, such as in regions of enhanced turbulence behind heliospheric shocks. The spatial diffusion coefficient in the x-direction is $\kappa_{xx} = \kappa_{\|}^{eff} \cos^2 \psi$ where the expression for $\kappa_{\|}^{eff}$ is the same as in equation (2). Therefore, $\kappa_{\|}^{eff}$ is independent of particle momentum. The expression for the momentum diffusion coefficient D_{pp}^{REC} models the effect of the statistical variance in the reconnection electric fields associated with multiple merging flux rope pairs on ion acceleration. Its expression is

$$D_{pp}^{REC} = p'^2 \frac{\pi}{10} \frac{V_A}{L_M} \left(\frac{L_M}{r_g}\right)^2 \left(\frac{\langle \delta B_M^2 \rangle}{B_0^2}\right)^2, \tag{9}$$

where L_M is the characteristic scale size of merging flux ropes, r_g is the maximum particle gyroradius (pitch angle is 90^0), and $\langle \delta B_M^2 \rangle / B_0^2$ is the mean fluctuation energy density of multiple merging flux-rope magnetic fields normalized to the energy density of the mean magnetic field. Assuming non-relativistic particles, inspection of D_{pp}^{REC} reveals that it is approximately independent of p'. This forms a strong contrast with the previous case discussed in Section 2.1 where $D_{pp}^{INC} \propto p'^2$. For simplicity, we assume that both κ_{xx} and D_{pp}^{REC} are independent of distance x. Thus, we specify $D_{pp}^{REC} = D_0$, where D_0 is a constant in units of momentum^2/time instead of speed as in the previous solution. We also set $\kappa_{xx} = \kappa_0$ where κ_0 is a constant in units of speed times length instead of speed as before. We find an analytic solution for the particle distribution which is

$$f_0(x, p') = \frac{\dot{N}}{8\pi^2 p_0^2 D_0} \sqrt{\frac{D_0}{\kappa_0}} e^{\frac{1}{2}\left(\frac{U_0}{\kappa_0}\right)x} \left(\frac{p_0}{p}\right) \times \tag{10}$$
$$\left[K_0\left(\alpha\sqrt{(p'-p_0)^2 + (D_0/\kappa_0)x^2}\right) - K_0\left(\alpha\sqrt{(p'+p_0)^2 + (D_0/\kappa_0)x^2}\right)\right].$$

where $\alpha = (U_0/2\kappa_0)\sqrt{\kappa_0/D_0}$. For $p' \gg p_0$, the accelerated ion spectrum asymptotically becomes

$$f_0(p') \propto \frac{\epsilon(p')}{\sqrt{\alpha p'(1-\epsilon(p'))}} \left[1 - \sqrt{\frac{1-\epsilon(p')}{1+\epsilon(p')}} e^{-2\alpha\epsilon(p')}\right] e^{-\alpha p'(1-\epsilon(p'))}, \tag{11}$$

where $\epsilon(p') = p_0/p' \ll 1$. Note that when $\epsilon(p') \to 0$ $(p'/p_0 \to \infty)$, $f_0(p') \to 0$. Thus,the accelerated ion spectrum rolls over exponentially with increasing momentum and a power law is not found. This can be understood by comparing the transport terms in equation (8) using dimensional analysis. The left-hand-side of the equation then becomes $f_0(p')[1/T_C + 1/T_D + D_0/p'^2]$, where T_C is the spatial advection time scale, T_D is the spatial diffusion time scale, and D_0 is the momentum diffusion coefficient, all independent of particle momentum. This implies that at higher energies the efficiency of spatial transport of ions across flux ropes is maintained. In comparison, transport in momentum space associated with the variance in the reconnection electric fields becomes less efficient at higher energies. In contrast, dimensional analysis of drift acceleration of energetic ions by multiple incompressible flux ropes in Section 2.1 shows that momentum diffusion maintains its efficiency relative to spatial transport at all energies, thus allowing a power-law spectrum to form.

2.3. Drift Acceleration by Compressible Contracting and Colliding Flux Ropes

Here we investigate the possibility that flux ropes can at times also exhibit compressive behavior, thus affecting energetic ion acceleration in a different way. When new elongated magnetic islands form in the plane perpendicular to the guide field and grow initially due to magnetic reconnection as a result of neighboring x-points forming on a current sheet, this process is accompanied by a rapid increase in the density and magnetic field strength as the x-point outflows advect magnetic field lines, causing them to pile-up toward the center of the evolving magnetic island. This suggests that the initial island formation and growth phase is compressible [27,28]. Accordingly, the area surrounded by an individual island magnetic field-line loop decreases while at the same time the path length between the loop endpoints shortens as the loop end points contract [21]. Similar compressible island behavior can be expected when neighboring magnetic islands collide after being attracted to each other. After collision the magnetic islands can either rebound or merge. In addition, Particle-in-cell simulations show that after neighboring islands merge, the newly formed island structures oscillate with alternating expansion and contraction motions [8]. During such pulsations, it is possible magnetic islands are compressive for at least part of the time.

During the early compressive phase of flux-rope contraction, one can think of particle energization occurring due to curvature drift in the direction of the induced electric fields generated at endpoints of contracting flux ropes, as was discussed above for later phase incompressible contraction. However, in the compressive phase particle, positive betatron acceleration also contributes to energization as a consequence of the increase in the flux-rope field strength with time [21]. This is in contrast to later phase incompressible contraction where negative betatron acceleration produce energy loss because the field strength decreases with time [9]. The compressive phase is captured in our Parker-like transport equation in terms of the compression of the flux-rope plasma flow ($\nabla \cdot U_{COM} < 0$).Therefore, when energetic ions traverse multiple compressive flux ropes with a net compression (rarefactions can also occur as discussed above), energetic ions will experience a first order Fermi acceleration effect.

A key difference between incompressible and compressible flux ropes is that the mean contraction effect of multiple incompressible flux ropes only results in a net energy gain for efficiently scattering energetic ions if the particle distribution is not completely isotropic. Thus, a successful 2nd order Fermi acceleration process involving a net mean incompressible contraction effect from several flux ropes is tied to the anisotropic part of the energetic ion distribution. In the case of compressible flux ropes, the mean contraction (compression) from multiple flux ropes is associated with 1st order Fermi acceleration for efficiently scattering ions forming a completely isotropic distribution. When efficiently scattering ions form a nearly, but not completely isotropic distribution, both 1st order Fermi acceleration (associated with the isotropic part of the distribution) and 2nd order Fermi acceleration (associated with the anisotropic part of the distribution) occur in reponse to the net mean compressive contraction of several flux ropes. In addition, 2nd order Fermi acceleration also occurs when energetic ions experience the variance from statistical variations in the divergence of the plasma flow in multiple compressive flux ropes (random compressions and decompressions relative to the assumed mean compression of many flux ropes). This effect occurs irrespective of the anisotropy of the particle distribution.

Below, we investigate the effect of both 1st order Fermi acceleration associated with the mean compression, and 2nd order Fermi acceleration by the variance in the compression of multiple compressible flux ropes on energetic ion acceleration. The 2nd order Fermi acceleration effect from the mean compression associated with the anisotropic part of the ion distribution is ignored because it decays much more strongly with increasing momentum compared to 2nd order Fermi acceleration from the variance in the compression of multiple flux ropes. We solve analytically the steady-state Parker transport equation in spherical geometry for a radial solar wind similar to equation (1). The equation is as follows:

$$U_0\frac{\partial f_0}{\partial r} - \frac{1}{r^2}\frac{\partial}{\partial r}\left(r^2\kappa_{rr}\frac{\partial f_0}{\partial r}\right) + \frac{1}{3}\frac{U_{eff}}{r}p'\frac{\partial f_0}{\partial p'} - \frac{1}{p'^2}\frac{\partial}{\partial p'}\left(p'^2 D_{pp}^{COM}\frac{\partial f_0}{\partial p'}\right) = \frac{\dot{N}\delta(r-r_0)\delta(p-p_0)}{16\pi^2 r_0^2 p_0^2}, \tag{12}$$

where $\kappa_{rr} = \kappa_{\parallel}^{eff}\cos^2\psi$, $U_{eff} = U_{COM} - 2U_0$, and $U_{COM} \approx V_A/\overline{L}_{COM,0}$. The term containing U_{eff} in equation (12) models the competition between particle energization when particles experience a mean compression effect from multiple flux ropes and energy loss when particles undergo adiabatic cooling in the radially expanding solar wind flow. The expression for U_{COM} indicates the mean plasma flow velocity associated with several flux ropes undergoing compression, and $\overline{L}_{COM,0}$ is the characteristic scale size of compressible flux ropes normalized at the energetic ion source distance $r = r_0$ to which it is normalized. In equation (12) it has been assumed that $L_{COM} \propto r$ to model approximately the observed increase in the size of small-scale flux ropes in the solar wind [15,29]. In equation (12), D_{pp}^{COM} is the momentum diffusion coefficient of energetic ions experiencing the effect of the variance in the divergence of the plasma flow when encountering multiple compressible flux ropes. The expression for D_{pp}^{COM} is

$$D_{pp}^{COM} = p'^2\frac{\pi}{25}\frac{V_A}{L_{COM}}\left(\frac{\langle\delta B_{COM}^2\rangle}{B_0^2}\right)^2. \tag{13}$$

The expression for $\kappa_{\parallel}^{eff}$ is the same as in equation (2), except that here we focus on the magnetic field statistics of compressible flux ropes. As in equation (1), for convenience we introduce the radial dependencies $\kappa_{rr} = V_{\kappa 0}r$ and $D_{pp}^{INC} = p'^2 D_0 r^{-1}$, where both $V_{\kappa 0}$ (a speed determining the rate of diffusive transport) and D_0 (a speed determining the rate of second order Fermi acceleration) are constants proportional to the Alfvén speed.

First, we present the solution of equation (12) when $D_{pp}^{COM} = 0$ (1st order Fermi acceleration dominates 2nd Fermi acceleration). The solution is

$$f_0(x,p') = \frac{\dot{N}}{32\pi^2 r_0^2 p_0^3 V_{\kappa 0}}\frac{1}{\sqrt{\pi}b}\left(\frac{r}{r_0}\right)^{-a}\frac{e^{-\frac{\ln^2(r/r_0)}{4\ln(p'/p_0)^{1/b}}}}{\sqrt{\ln(p'/p_0)^{1/b}}}\left(\frac{p}{p_0}\right)^{-a^2/b}, \tag{14}$$

where $a = 1 - U_0/2V_{\kappa 0}$, and $b = U_{eff}/3V_{\kappa 0} = (U_{COM} - 2U_0)/3V_{\kappa 0}$. By taking the limit $p'/p_0 \gg 1$, we find that the accelerated spectrum asymptotically becomes approximately a power law

$$f_0(p') \propto \left(\frac{p'}{p_0}\right)^{-(1-U_0/2V_{\kappa 0})^2 3V_{\kappa 0}/U_{eff}}, \tag{15}$$

weakly modified by a logarithmic function of p' (not shown). Inspection of the power-law index expression shows qualitatively similar effects of transport processes on the spectral slope of the accelerated ion spectrum as discussed for the case of drift acceleration by incompressible flux ropes: (i) Increasing $V_{\kappa 0}$ steepens the spectrum because particle escape from flux ropes is more efficient, thus reducing the contact time with flux rope compressions. (ii) The presence of solar wind advection reduces the net effect of spatial diffusion, thus reducing the steepening effect from diffusive escape on the spectrum because particles have a longer interaction time with compressive flux ropes. (iii) Increasing the compression speed U_{COM} hardens the spectrum because this raises the rate of particle energization. (iv) Adiabatic cooling by the expanding solar wind flow reduces efficiency of particle energization by the mean compression of multiple compressive flux ropes, thus causing steepening of the power-law spectrum. Upon again invoking the diffusion-advection limit $U_0/V_{\kappa 0} \to 2$ (diffusive escape of energetic ions from flux ropes is finite but inefficient), we find that the accelerated spectrum can be quite hard because then approximately, $f_0(p') \propto (p'/p_0)^{\epsilon}$ where $\epsilon \to 0$. If alternatively, $U_{eff} \approx 0$, energization by the mean compression from multiple flux ropes is balanced by adiabatic cooling, and the spectrum is very steep, as expected.

The complete solution of equation (12), which is mathematically similar to the solution of equation (1) for incompressible flux ropes, applies best when 2nd order Fermi acceleration by compressive flux ropes is stronger than 1st order Fermi acceleration, because it does not yield the correct solution in the opposite limit (thus the separate solution presented above when the mean compression effect is dominant). Just as in the incompressible flux-rope case, the asymptotic spectrum for $p'/p_0 \gg 1$ yields a power-law spectrum, but the power-law index is modified by the mean compression of multiple flux ropes, and 2nd order Fermi acceleration is now determined by compressible flux ropes. The expression for the power-law spectrum for $p'/p_0 \gg 1$ is

$$\begin{aligned} f_0(p') &\propto \left(\frac{p'}{p_0}\right)^{-c\left(1+\sqrt{1+(V_{\kappa 0}/D_0(d/c)^2)}\right)}, \\ c &= \frac{1}{2}\left(3-\frac{1}{3}\frac{U_{eff}}{D_0}\right), \\ d &= 1-\frac{U_0}{2V_{\kappa 0}}. \end{aligned} \tag{16}$$

The main differences between the asymptotic energetic ion spectrum for incompressible flux ropes as given by equation (6) and the spectrum compressible flux ropes are that (i) the expression for the stochastic acceleration coefficient D_0 is different (compare the expressions for D_{pp}^{INC} (equation (3)) with D_{pp}^{COM} (equation (13)), (ii) the constant c in equation (16) includes $U_{eff} = U_{COM} - 2U_0$ indicating a three-way competition between particle energization by the mean compression of multiple flux ropes (U_{COM}), adiabatic cooling by the radially expanding solar wind flow (U_0), and energization by the variance in the compression of multiple flux ropes D_0, whereas c in equation (3) includes a two-way competition between adiabatic cooling effect and energization by the variance in the contraction of incompressible flux ropes.

Assuming the diffusion-advection limit $U_0/V_{\kappa 0} \to 2$,

$$f_0(p') \propto \left(\frac{p'}{p_0}\right)^{-\left(3-\frac{1}{3}\frac{U_{COM}-2U_0}{D_0}\right)}. \tag{17}$$

If 2nd order Fermi acceleration is the dominant mechanism for energy change ($D_0 \gg U_{COM}$ and $D_0 \gg U_0$), we get $f_0(p') \propto p'^{-3}$, just as we found when 2nd order Fermi acceleration dominated in the case of drift acceleration by incompressible flux ropes. When 1st order Fermi acceleration by the mean compression of multiple flux ropes is significant and dominates adiabatic cooling $U_{COM} \gg U_0$, we see that the accelerated spectrum will develop a power-law index with a value < 3. Thus, compressible flux ropes have the potential to produce a harder power-law spectrum for accelerated ions than incompressible flux ropes because of the additional contribution of 1st order Fermi acceleration by the mean compression. This situation may occur in the region between an ICME and the HCS, because the expanding ICME pushes against the HCS and compresses the confined magnetic islands near the HCS, which often occur in the vicinity of the HCS. An example of the corresponding energetic particle flux enhancement observed between the HCS and the ICME is shown in [12]. Note that these results apply only for energetic test particles. It remains to be investigated whether this difference in the slope of energetic ions for incompressible and compressible flux ropes will survive when acceleration is modeled self-consistently.

3. Summary

There is growing observational evidence that dynamic, small-scale flux rope structures have a significant presence in the supersonic solar wind, especially near primary current sheet structures such the heliospheric current sheet, and primary current sheets at the leading and trailing edges of the CME structures behind interplanetary shocks, for example [13,14,15]. Observations also increasingly indicate that dissipation of these structures results in the acceleration of suprathermal particles [9,12]. In response we investigated theoretically the possibility that several small-scale flux ropes experiencing contraction, merging, and collisions can accelerate energetic ions to form power-law spectra as simulations suggest [4,11]. This was accomplished by analyzing three distinct limits in the general kinetic transport theory framework that we developed previously [21,22] for energetic ion acceleration by many small-scale flux ropes on inertial scales in the presence of a strong guide magnetic field. The main emphasis was on how the variance in the statistics of flux-rope electric fields induced by flux-rope dynamics affects ion acceleration, because the effect of the mean electric field generated by multiple flux ropes were already discussed before [21]. The three limits considered were (i) energetic ion drift acceleration as a 2nd order Fermi process because of the variance in the induced electric fields from contracting and merging flux-ropes that exhibit incompressible behavior, (ii) energetic ion field-aligned guiding center motion acceleration as a 2nd order Fermi mechanism in response to the variance in the reconnection electric fields generated when neighboring flux ropes merge (reconnect) and, (iii) energetic ion drift acceleration as a 1st order Fermi process due to the mean compression and as a 2nd order Fermi process because of the variance in the compression of several flux ropes exhibiting compressive behavior (e.g., early-time flux-rope contraction or collisions between flux ropes).

Our steady-state analytic solutions, valid for the supersonic solar wind, suggest that both cases of drift acceleration (case (i) for incompressible flux ropes and case (iii)) for compressible flux ropes) can result in accelerated ion spectra that form power laws asymptotically for $p' \gg p_0$. We found asymptotically an exponential rollover for field-aligned guiding center motion acceleration by reconnection electric fields from merging flux ropes (case (ii)). This implies that whereas field-aligned guiding center motion acceleration by reconnection electric fields might be strong at lower energies, at higher energies drift acceleration by incompressible or compressible flux ropes will dominate. This is to be expected because in our theory $D_{pp} \propto p^2$ for the two drift acceleration cases, whereas D_{pp} is approximately independent of particle momentum for the case of field-aligned guiding center motion acceleration. Assuming that compressible flux ropes are just as probable as incompressible flux ropes, the additional 1st order Fermi acceleration effect from the mean compression in the case of compressible flux ropes is expected to yield a harder power-law spectrum compared to the case of incompressible flux ropes where such a 1st order Fermi acceleration effect is absent.

3.1. Acknowledgments

JAlR acknowledges support from NASA grant NNX14AF43G. JAlR and GMW were also supported by NASA grant NNX15AI65G. GPZ was supported by NASA grants NNX08AJ33G, Subaward 37102-2, NNX09AG70G, NNX09AG63G, NNX09AJ79G, NNG05EC85C, Subcontract A991132BT, NNX09AP74A, NNX10AE46G, NNX09AW45G, and NSF grant ATM-0904007 This work was also supported in part by an UAH IIDR award to GPZ. OK was supported by RFBR grant. no. 14-02-00769.

4. References

[1] Matthaeus W H, Goldstein M L and Roberts D A 1990 *J. Geophys. Res.* **95** 20673
[2] Zank G P and Matthaeus W H 1992 *J. Geophys. Res.* **97** 1718
[3] Zank, G P and Matthaeus W H 1993 *Phys. Fluids* **5** 257

[4] Dmitruk P, Matthaeus W H and Seenu N 2004 *Astrophys. J.* **617** 667
[5] Birn J, Hesse M and Schindler K 1989 *J. Geophys. Res.* **94** 241
[6] Gray P C and Matthaeus W H 1992 *AIP Conf. Proc.* **264** 261
[7] Guo F, Liu Y-H, Daughton W and Li H 2015 *Astrophys. J.* **806** 167
[8] Li X, Guo F, Li H and Li G 2015 arXiv:1505.02166
[9] Drake J F, Swisdak M, Che H and Shay M A 2006 *Nature* **443** 553
[10] Drake J F, Swisdak, M and Fermo R 2013 *Astrophys. J.* **763** L5
[11] Oka M., Phan T-D, Krucker S. Fujimoto M and Shinohara I 2010 *Astrophys. J.* **714** 915
[12] Tessein J A, Matthaeus W H, Wan M, Osman K T, Ruffolo D and Giacalone J 2013 *Astrophys. J. Lett.* **776** L8
[13] Song H-Q, Chen Y, Li G, Kong Z-L and Feng S-W 2012 *Phys. Rev. X* **2** 021015
[14] Chian A C-L and Munoz P R 2011 *Astrophys. Lett.* **733** L34
[15] Khabarova O, Zank G P, Li G, le Roux J A, Webb G M, Dosch A, Zharkova V V and Malandraki O E 2015 *Astrophys. J.* http://arxiv.org/abs/1504.06616//
[16] Zank G P, Rice W K M and Wu C C 2000 *J. Geophys. Res.* **105** 25079
[17] Fisk L A and Gloeckler G 2008 *Astrophys. J.* **686** 1466
[18] Zhang M and Lee M A 2013 *Space Sci. Rev.* **176** 133
[19] Ng C K and Reames D V 2008 *Astrophys. J.* **686** L123
[20] Bian N H and Kontar E P 2013 *Phys. Rev. Lett.* **110** 151101
[21] Zank G P, le Roux J A, Webb G. M., Dosch, A and Khabarova O 2014 *Astrophys. J.* **797** 28
[22] le Roux J A, Zank G P, Webb G M and Khabarova O. 2015 *Astrophys J* **A 801** 112
[23] Northrop T G 1963 *The adiabatic motion of charged particles* New York: Interscience p 8
[24] le Roux J A, Zank G P, Li G and Webb G M 2005 *Astrophys. J.* **626**, 1116
[25] Fisk L A and Gloeckler G 2014 *J. Geophys. Res.: Space Physics* 119 8733
[26] Zank G P, Hunana P, Mostafavi P, le Roux J A, Li G, and Khabarova O *J. Phys.: Conf. Series* this volume
[27] Fermo R L, Drake J F, Swisdak M 2010, *Phys. Plas.* **17** 010702
[28] Shen C, Lin J, Murphy N A and Raymond J C 2013 *Phys. Plasmas* **20** 072114
[29] Cartwright M L and Moldwin M B 2010 *J. Geophys. Res.* **115** A08102

Modeling solar wind with boundary conditions from interplanetary scintillations

P Manoharan[1], **T Kim**[1], **N V Pogorelov**[1], **C N Arge**[2] **and P K Manoharan**[3]
[1]Center for Space Plasma and Aeronomic Research, University of Alabama in Huntsville, USA
[2]AFRL/RVBXS, USA
[3]Radio Astronomy Centre, NCRA, Tata Institute of Fundamental Research, India

Abstract. Interplanetary scintillations make it possible to create three-dimensional, time-dependent distributions of the solar wind velocity. Combined with the magnetic field observations in the solar photosphere, they help perform solar wind simulations in a genuinely time-dependent way. Interplanetary scintillation measurements from the Ooty Radio Astronomical Observatory in India provide directions to multiple stars and may assure better resolution of transient processes in the solar wind. In this paper, we present velocity distributions derived from Ooty observations and compare them with those obtained with the Wang-Sheeley-Arge (WSA) model. We also present our simulations of the solar wind flow from 0.1 AU to 1 AU with the boundary conditions based on both Ooty and WSA data.

1. Introduction

The solar wind originates from the corona where the magnetic field in an open configuration, and the flow can accelerate along flux tubes. Depending on the source region of the solar wind (e.g., edge of closed-field region or open-field coronal hole), the solar wind speed can vary between 300 and 800 km/s. Interplanetary Scintillation (IPS) measurements represent one of the many "remote sensing" techniques, which can inform us of the three-dimensional structure of the heliosphere. Here, IPS observations are combined with the Wang-Sheeley-Arge (WSA) semi-empirical solar wind model to simulate the solar wind in a time-dependent manner. The WSA model is based on photospheric magnetic field measurements, and the current version we use employs the Air Force Data Assimilative Photospheric Flux Transport (ADAPT) maps from ground-based observations of photospheric magnetic field to determine the coronal field configuration at the source surface, normally at R = 2.5 R_s (R_s = solar radius), the expansion factor of the magnetic flux, and the coronal hole boundary distance to estimate the distribution of the speed of the solar wind [2, 3, 4]. Approximated outflow at the base of the solar wind with sophisticated three-dimensional magnetohydrodynamic numerical model simulates the resulting flow evolution out to Earth. The model provides radial magnetic field and solar wind speed [17].

2. The Ooty Interplanetary Scintillation Measurements

Interplanetary scintillation (IPS) exploits the scattering of compact radio source by the electron-density irregularities in the solar wind (i.e., IPS is the radio analogy of optical twinkling of star). For example, the plane-wave front traveling through the solar wind is phase modulated by the variations of refractive index of the irregular solar wind density and the random phase modulation leads to a diffraction pattern on the ground. Since the solar wind irregularities drift past the observer's line-of-sight at a velocity V (solar-wind velocity) and the motion of these irregularities converts the diffraction pattern into the temporal intensity fluctuations, which are observed as IPS. The temporal power spectrum of the intensity scintillation observed at the Earth is in fact two-dimensional spatial-intensity spectrum due to strip scans along the line-of sight, transverse to the solar wind velocity vector [13]. From the spectrum of intensity scintillation the speed of the solar wind, shape of density turbulence spectrum, and the structure of the radio source can be derived [14].

The Ooty Radio Telescope (ORT) is a parabolic cylinder 530-m long in the north-south direction and 30-m wide in the east-west direction [20]. The equatorially mounted ORT can track a radio source for ~10 hours in the east-west direction and its electronics system supports a declination coverage of ±60 degrees. The high sensitivity and tracking ability of the ORT allows observation of IPS on a large number of radio sources. The Ooty IPS observations can cover a heliocentric distance range of R ~10 – 250 R_s and at all heliographic latitudes.

3. Wang-Sheeley-Arge Model

The Wang-Sheeley-Arge (WSA) model [8, 6, 5] is a combined empirical and physics based model of the corona and solar wind and an improved version of the original Wang and Sheeley model [21, 22]. The WSA model input is the ground based line-of-sight (LOS) observations of the Sun's magnetic field, in the form of synoptic map. These maps are then used in a magnetostatic potential field source surface (PFSS) model [19, 1, 22], which determines the coronal field out to 2.5 solar radii (Rs). The output of the PFSS model serves as input to the Schatten Current Sheet (SCS) model [18], which provides a more realistic magnetic field topology of the upper corona. Only the innermost portion (i.e., from 2.5 Rs to between 5 and 30 Rs) of the SCS solution, which actually extends out to infinity, is used. (The user sets this outer coronal boundary radius).

An empirical velocity relationship [6, 7] is then used to assign solar wind speed at this outer boundary as a function of two coronal parameters: (1) flux tube expansion factor (fs) and (2) the minimum angular separation (θ_b) at the photosphere between an open field footpoint and the nearest coronal hole boundary. Starting at the center of each of the grid cells on the outer coronal boundary surface and tracing the magnetic field lines down to their footpoints rooted in the photosphere determine these parameters. The flux tube expansion factors are calculated using the traditional definition $fs=(R_{ph}/R_{ss})^2[B_{ph}/B_{ss}]$ [21], where B_{ph} and B_{ss} are the field strengths, along each flux tube, at the photosphere ($R_{ph}=1R_s$) and the source surface ($R_{ss}=2.5R_s$), respectively. The model provides the radial magnetic field and solar wind speed at the outer coronal boundary surface and this may then be fed into an advanced three-dimensional MHD solar wind propagation model.

4. Results

In order to simulate the solar wind outflow to Earth and beyond, the boundary conditions at the source surface are essential. Moreover, the position of the source surface should be close to the Sun and the inner boundary of the heliospheric MHD model is considered in the region where the solar wind can be assumed to be supersonic everywhere. The IPS measurements from Ooty have been traced backward to a surface, from where solar wind propagates steadily. Ooty IPS data measurements are analyzed for a period of about three Carrington rotations from June 1, 2012 to August 31, 2012.

Figure 1 shows the time dependent solar wind velocity distributions at 0.1 AU. Due to the limitations of spatial resolution in the IPS measurements, an average solar wind estimates on 70 sources have been considered for each day in the +90 to -90 degree latitude and 180 degree longitude ranges. To overcome this limitation, the velocity maps are created for each day using data from 13 days prior and after. Thus resulting in a plot with more than 1000 sources, providing a better coverage of the entire solar surface. However, since the accuracy of the solar wind velocity measurements essentially depends on the high signal-to-noise ratio of the power spectrum, the number of IPS estimates used in the velocity distribution plot is limited to about 75 percent of the total number. The velocity distribution has been smoothed by a two-dimensional Gaussian of width ~27 degrees. The velocity distributions are presented for the days June 15, 2012 July 15, 2012 and August 15, 2012. It is evident that there is a significant restructuring of the solar wind velocity structure over the period of two Carrington rotations. This is also observed in the synoptic solar wind velocity map from the IPS observations at the Solar-Terrestrial Environment Laboratory (Nagoya University, Japan). Figure 2 shows the considerable changes occurring in solar wind velocity throughout the year 2012 as observed from STEL.

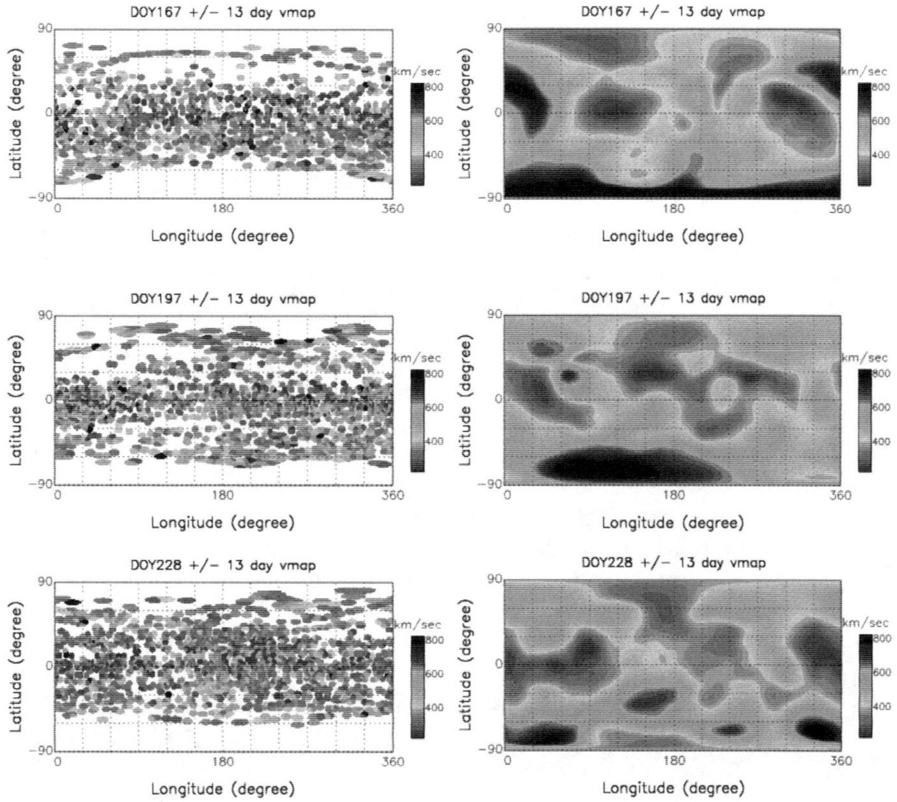

Figure 1. Distributions of solar wind velocity data points at 0.1 AU using IPS measurements (left panel). Time dependent solar wind velocity maps using IPS measurements and Gaussian smoothing (right panel).

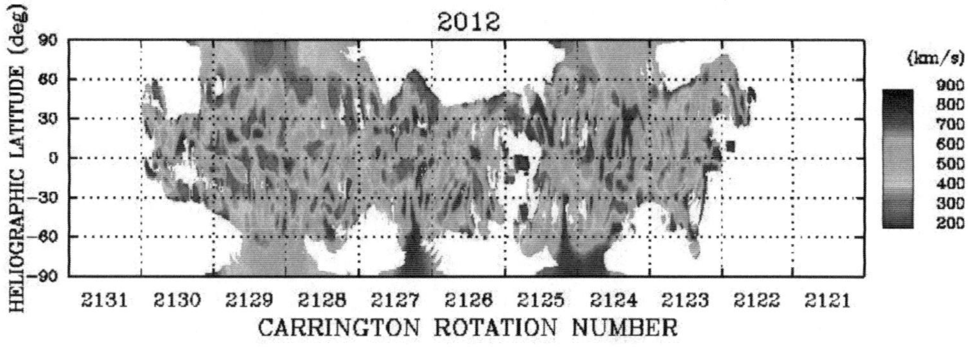

Figure 2. Solar wind velocity map from the STEL using IPS observations for the year 2012

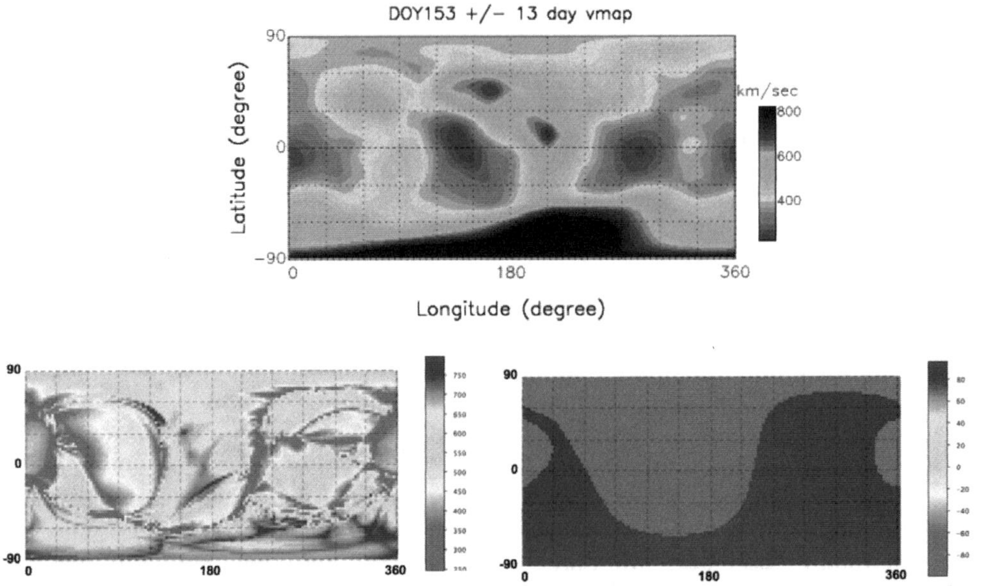

Figure 3. IPS solar wind velocity compared with WSA velocity and magnetic field maps for June 1, 2012, in km/s and nT respectively.

The IPS and WSA velocity maps compare reasonably well in the large scale, though there are some sizable discrepancies where IPS data has spotty coverage. Figure 3 shows the comparison of solar wind velocity distribution from IPS data and magnetic field and velocity data derived from WSA model. It is also seen that towards the northern pole slow solar wind ($V \leq 400$ km/s) is observed and towards the southern pole fast solar wind ($V \geq 400$ km/s) is observed. The slow wind component seen in the WSA velocity map, around the heliospheric current sheet, is also observed in the IPS velocity map. The change in solar wind speed is also seen in the structure of the WSA magnetic field and magnetic field strength.

In the next step of this study, the velocity estimates from IPS measurements, implemented with the WSA model data for the magnetic field, are used as boundary conditions for the Multi-Scale Fluid-Kinetic Simulation Suite (MS-FLUKSS), which has been developed and used by the Center for Space Plasma and Aeronomic Research at the University of Alabama in Huntsville. MS-FLUKSS is a tool for modeling discontinuous flow of partially ionized plasma and plasma-neutral interactions in the heliosphere [15, 16]. Time varying boundary conditions are applied to further investigate the solar wind evolution between 0.1 to 1.5 AU for the period under study.

The ADAPT-WSA model is very good in reproducing the background solar wind speeds at 1 AU [2]. However, the WSA velocities must be reduced substantially at the inner boundary in order to drive a heliospheric MHD model such as Enlil and make reasonable predictions at Earth for the background solar wind [12]. This study was conducted to test whether the IPS data could reproduce the solar wind parameters at 1 AU without having to make such ad hoc velocity adjustments. IPS observations have been incorporated in MHD models to perform various heliospheric measurements [10, 11, 23].

The MHD simulation provides solar wind parameters in the three-dimensional heliosphere. The update rate of the MHD model is typically less than a minute (20-30 seconds per time step). Linear interpolation between the two nearest frames is performed to update the inner boundary at each time step. The solar wind velocities have been extracted at the location of the Earth and are compared with daily averages provided at the OMNI database, as shown in figure 4. Additionally, the simulated radial velocities at the locations of STEREO A and STEREO B are also obtained and compared with the actual observed data sets.

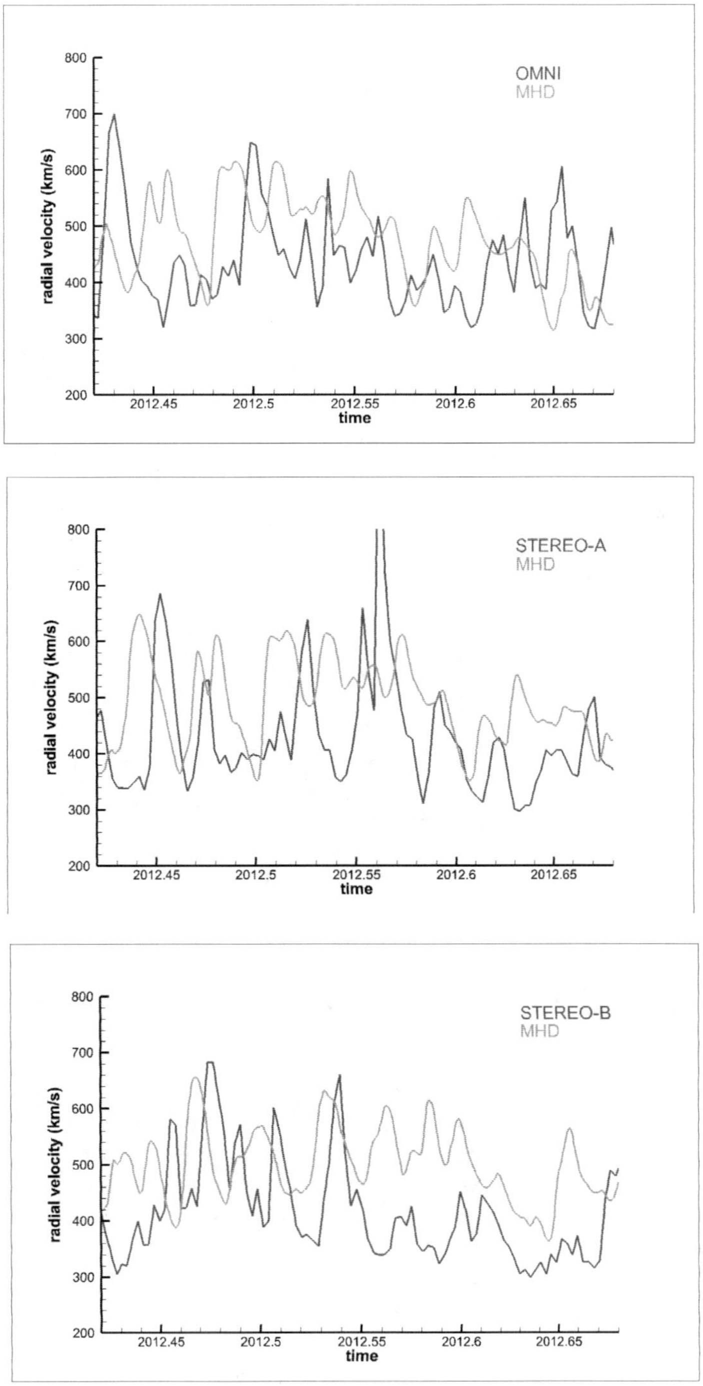

Figure 4. Comparison of radial velocity (km/s) at Earth (top), STEREO-A (middle) and STEREO-B (bottom) with MHD solutions.

It is evident from the above plots that there is a large-scale (i.e., overall) consistency between the in-situ measurements and simulated results. Since the simulated result largely depends on the boundary conditions obtained from the IPS estimates, simulation can be improved by increasing the temporal resolution of the IPS data along with the spatial resolution. Moreover, we can create velocity maps by assigning maximum weightage to estimate obtained on the given day of observation and a decreasing weighting function for measurements taken after and before the day of observation. In this case, the smoothing technique employed by the two-dimensional Gaussian would use varying widths depending on the day of observation. Figure 5 shows velocity distribution maps using varying Gaussian width.

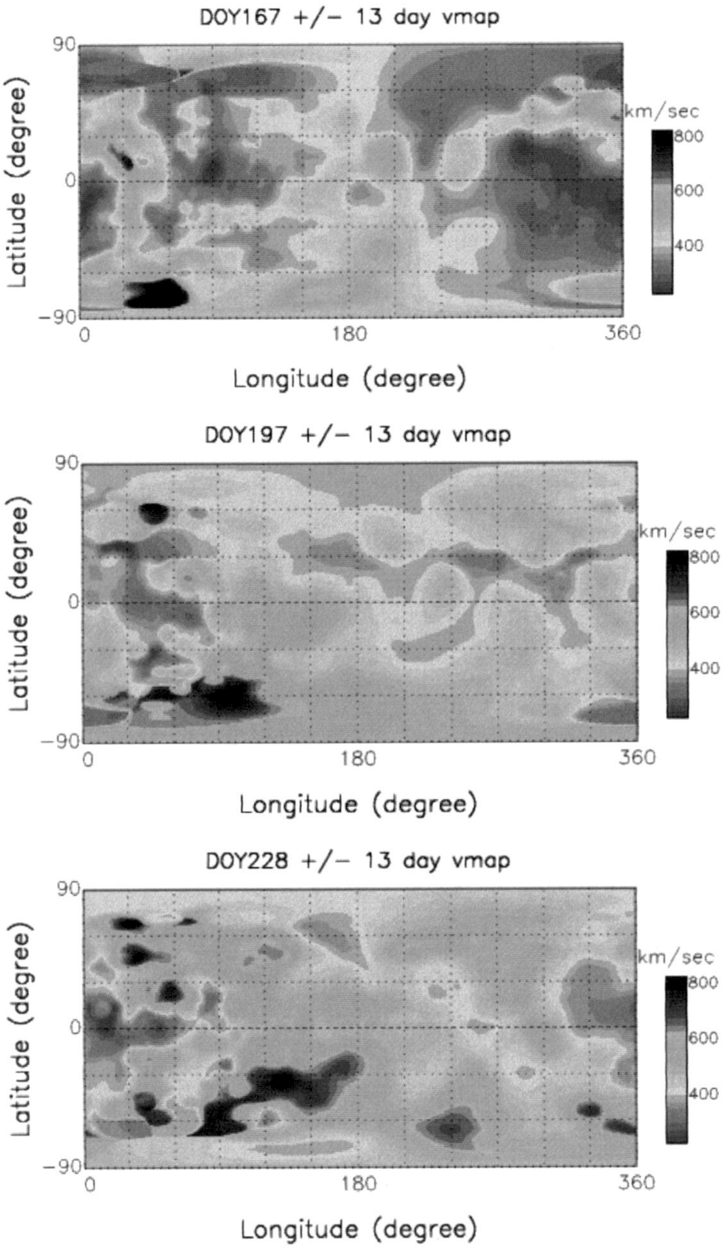

Figure 5. Time dependent IPS velocity maps created using Gaussian smoothing with varying widths.

5. Summary and Conclusion

In this study, time-dependent IPS observations have been used as the time varying inner boundary conditions for the solar wind velocity, together with the magnetic field data derived from the WSA model. This has been used to perform simulations to study the evolution of the solar wind from 0.1 to

1.5 AU. There are discrepancies between in-situ measurements and our MHD simulation results. This could very well be because of the limitations posed from temporal and spatial resolution of the IPS measurements. Further analysis would include using boundary conditions with corrections for the back tracing of the IPS velocity to correct the shift observed in the MHD simulations. Additionally, solar wind heating by turbulence produced by the instability of non-thermal ions in the solar wind should also be analyzed, since it is an observable quantity in a number of spacecraft.

Acknowledgement
This work was supported, in part, by the NSF SHINE project AGS-1358386 and DOE grant DE-SC0008334. We also acknowledge NSF PRAC award OCI-1144120 and related computer resources from the blue-waters sustained-petascale computing project.

References

[1] Altschuler M A and Newkirk G 1969 *Solar Physics* 9 131–149
[2] Arge C N, Henney C J, Hernandez I G, Toussaint W A, Koller J, Godinez H C 2013 *AIP Conf. Proc.* 1539 11-14
[3] Arge C N. Henney C, Koller J, Young S L, Compeau C R, MacKenzie D, Harvey J W 2009 *AGU Fall Meeting Abstract* SM54A-06
[4] Arge C N, Henney C J, Koller, J, Compeau C R, Young S, MacKenzie D, Fay A, Harvey J W 2010 *AIP Conf. Proc.* 1216 343-346
[5] Arge C N, Luhmann J G, Odstrcil D, Schrijver C J, Li Y 2004 *Journal of Atmospheric and Solar-Terrestrial Physics* 66 1295-1309
[6] Arge C N, Odstrcil D, Pizzo V J, Mayer L R 2003 *AIP Conf. Proc.* 679 190-193
[7] Arge C N, Odstrcil D, de Toma G 2005 *AGU Spring Meeting Abstract* SH23C-01
[8] Arge C N and Pizzo V J 2000 *Journal of Geophysical Research* 105 10456-10479
[9] Bisi M M, Jackson B V, Clover J M, Manoharan P K, Tokumaru M, Hick P P, Buffington A 2009 *Ann. Geophysics* 27 4479-4489
[10] Hayashi K, Kojima M, Tokumaru M, Fujiki K 2003 *Journal of Geophysical Research* 108
[11] Kim T K, Pogorelov N V, Borovikov S N, Jackson B V, Yu H S, Tokumaru M 2014 *Journal of Geophysical Research* 119 7981-7997
[12] MacNeice P J, Taktakishvili A, Jackson B V, Clover J M, Bisi M M, Odstrcil D 2011 *AGU Fall Meeting Abstract* SH31C-2026
[13] Manoharan P K 1993 *Solar Physics* 148 153-167
[14] Manoharan P K 2010 *Solar Physics* 265 137-157
[15] Pogorelov N V, Borovikov S N, Florinski V, Heerikhuisen J, Kryukov I, Zank G P, 2009 *Astronomical Society of the Pacific Conf. Ser. 406 Numerical Modeling of Space Plasma Flows: ASTRONUM-2008*
[16] Pogorelov N V, Borovikov S N, Heerikhuisen J, Kim T, Kryukov I, Zank G P 2011, *Astronomical Society of the Pacific Conf. Ser. 444 Numerical Modeling of Space Plasma Flows: ASTRONUM-2010*
[17] Pogorelov N V, Suess S T, Borovikov S N, Ebert R W, McComas D J, Zank, G P 2013 *Astrophysical Journal* 772 2
[18] Schatten K H 1971 *Cosmic Electrodyn.* 2 232-245
[19] Schatten K H, Wilcox J M, Ness, N F, 1969 *Solar Physics* 9 442-455
[20] Swarup G, Sarma N V G, Joshi M N, Kapahi V K, Bagri D S, Damle S H, Ananthakrishnan S, Balasubramanian V, Bhave S S, Sinha R P 1971 *Nature Phys. Sci.* 230 185-188
[21] Wang Y M and Sheeley N R 1992 *Astrophysical Journal* 392 310-319
[22] Wang Y M Sheeley N R 1995 *Astrophysical Journal* 447 L143–L146
[23] Yu H S, Jackson B V, Hick P P, Buffington A, Odstrcil D, Wu C C, Davies J A, Bisi M M, Tokumaru M 2015 *Solar Physics*

Electron Acceleration at a High Beta and Low Mach Number Rippled Shock

S. Matsukiyo[1] and Y. Matsumoto[2]

[1] Faculty of Engineering Sciences, Kyushu University, 6-1 Kasuga-Koen, Kasuga, Fukuoka, 816-8580, Japan

[2] Department of Physics, Chiba University, 1-33 Yayoi-chou, Inage-ku, Chiba, 263-8522, Japan

E-mail: matsukiy@esst.kyushu-u.ac.jp

Abstract. Electron acceleration in a high plasma beta and low Mach number quasi-perpendicular shock is investigated by using two-dimensional full particle-in-cell simulation. Although efficient shock drift acceleration followed by reflection was observed in the previous one-dimensional simulation, no reflected electrons are found due to the effect of shock surface rippling for the particular parameters examined here. Structure of the shock transition region is complex in spite of the high beta and low Mach number situation. In addition to the ion scale fluctuations including the ripple, electron scale fluctuations are also recognized. Among these, downstream fluctuations are dominated by Alfvén ion cyclotron instability, the fluctuations in the foot are due to modified two-stream instability. Electron distribution function in the transition region indicates non-thermal nature. The energy gained by the non-thermal electrons is not explained merely by the shock drift acceleration, implying the importance of local wave-particle interactions.

1. Introduction

Recently, high beta ($\beta > 1$) and relatively low Mach number ($M_{ms} \ll 10$) shocks have been paid special attention as the accelerator of high energy particles. Here, β is the ratio of upstream plasma thermal pressure to magnetic pressure and M_{ms} is magnetosonic Mach number.

Electron acceleration at or near the high beta and low Mach number shocks is actually observed in a variety of astrophysical environments. The heliospheric termination shock is an example. Due to the large abundance of pickup ions, which is a high temperature ion component produced mainly by charge exchange between solar wind ions and interstellar neutral atoms, effective plasma beta upstream of the termination shock is high and magnetosonic Mach number is low. Voyager 2 spacecraft observed flux increase of non-thermal electrons (as well as ions) when it crossed the termination shock [1]. Galaxy cluster mergers often show radio synchrotron emissions from relativistic electrons accelerated near the merger shocks [2, 3, 4, 5, 6]. The intracluster medium usually has rather high temperature, $\sim$ 1-10 keV, so that the corresponding plasma becomes high beta.

One of the widely accepted scenarios of particle acceleration at a collisionless shock is the diffusive shock acceleration (DSA), in which non-thermal particles scattered back and forth across the shock interacting with upstream and downstream converging turbulence gain energy statistically (e.g., [7]). Although the DSA theory assumes pre-existence of mildly accelerated non-thermal seed particles, the mechanism to produce those seed particles has been still an open question, known as the injection problem. In particular, the injection mechanism for electrons has been poorly understood. In terms of magnetized nonrelativistic shocks, the injection model of electrons proposed up to now is only for sufficiently high Mach number shocks at which strong Buneman instability can get excited in the transition region [8, 9]. The instability prefers high Mach number and low beta condition. However, Matsukiyo et al.[10] revisited the so-called shock drift acceleration (SDA) process and showed that SDA prefers high beta and low Mach number situation. They further proposed an electron injection scenario working at high beta and low Mach number shocks based on the SDA as follows. Once some electrons are accelerated through the SDA and reflected at the shock, they form a non-equilibrium velocity distribution function in the foreshock region. The distribution function leads to some instabilities, including electron beam instability and electron temperature anisotropy instability, locally in the foreshock. Some reflected electrons are scattered back toward the shock or its downstream by interacting with the self-generated waves. Then, those electrons are involved in further acceleration processes. Matsukiyo et al.[10] indirectly confirmed the above scenario by performing a series of one-dimensional particle-in-cell (PIC) simulations. Recently, Guo et al.[11] self-consistently reproduced the above scenario by using two-dimensional PIC simulation.

With higher spatial dimensions, in general, structure of a shock changes more than a little from one-dimensional case. In particular the ripple of a shock surface, which is believed to be strongly coupled with ion dynamics in the transition region, is a fundamental structure in a multi-dimensional system. Therefore, it is important to consider how the ripple affects the above injection scenario. In Guo et al.[11] the ripple in the above sense is not seen probably because of their choice of parameters as shown later. In this study we perform a 2D PIC simulation with the same physical parameters used by Matsukiyo et al.[10] in their 1D PIC simulation to discuss the effect of one-dimensional constraint. The simulation setup is briefly represented in section 2 and the results are shown in section 3. The summary and discussions are given in section 4.

2. Simulation Setup

A shock is produced by the so-called injection or reflecting wall method. An upstream magnetized plasma is continuously injected from the right-hand boundary. The plasma is reflected at the left-hand boundary and mixture of the incoming and the reflected plasmas results in a downstream medium. The shock is produced at a boundary between the upstream and the downstream plasmas. Since the simulation frame is the downstream rest frame, the shock propagates in time from left to right. The injection flow is parallel to the shock normal which is along the x-axis. The upstream magnetic field is in the $x-y$ plane. The size of a spatial grid is $\Delta x = \Delta y \approx \lambda_{De}$ where λ_{De} denotes the electron Debye length, and the number of super-particles per cell for the injection plasma is $N_p = 80$ for both electrons and ions. The system size is $L_x \times L_y = 10000\Delta x \times 1024\Delta y$, which corresponds to $67.45c/\omega_{pi} \times 6.91c/\omega_{pi}$. Here, c/ω_{pi} denotes the ion inertial length of the upstream plasma. For the parameters below, L_y is enough to reproduce the ion scale ripple structure which may affect the dynamics of high energy electrons. However, it may not be enough to investigate behaviours of high energy ions. Time resolution is $\Delta t = 0.0578\omega_{pe}^{-1}$, where ω_{pe} is the electron plasma frequency. The Alfvén and the magnetosonic Mach numbers of the shock is $M_A = 7.1$ and $M_{ms} = 2.6$, respectively. The ion to electron mass ratio is realistic ($m_i/m_e = 1836$), the magnetization parameter, or the squared ratio of electron cyclotron to plasma frequencies, is $\sigma = \Omega_e^2/\omega_{pe}^2 = 1/9$. The upstream plasma beta is $\beta = 3(\beta_e = \beta_i = 1.5)$ and the shock angle, the angle between upstream magnetic field and the shock normal, is $\Theta_{Bn} = 85°$.

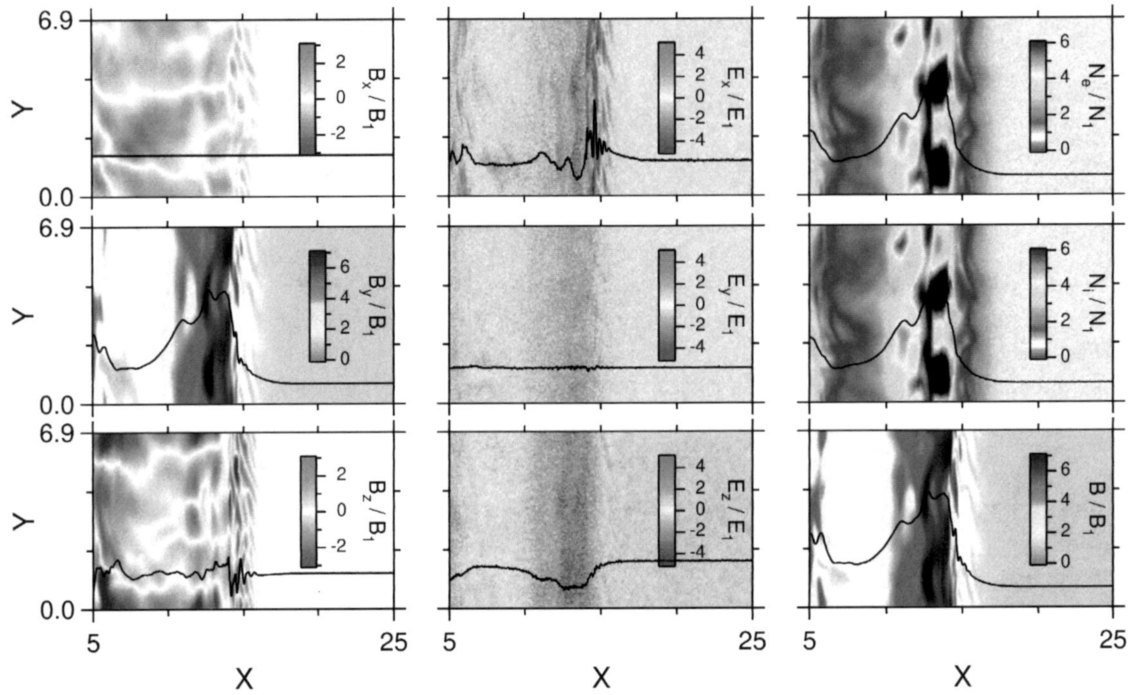

Figure 1. Shock profile at $\Omega_i t = 6.4$. The color scale denotes (left) three components of magnetic field, (middle) electric field, (right) electron and ion densities and magnitude of total magnetic field (right), respectively. In each panel the solid line indicates y-averaged corresponding field profile with arbitrary vertical scale.

3. Results

3.1. Shock Profiles

Fig.1 shows a snapshot of shock profile at $\Omega_i t = 6.4$. Here, Ω_i denotes ion cyclotron frequency defined by upstream magnetic field strength. The left panels indicate three components of magnetic field (B_x, B_y, B_z) and the middle panels show those of electric field (E_x, E_y, E_z). In the right panels from the top the profiles of electron and ion densities (n_e, n_i) and total magnetic field (B) are presented, respectively. The shock is around the center in x. In contrast to the previous one-dimensional simulation, rather complex multi-scale fluctuations are found in the transition region. In each panel y-averaged field profile is indicated by solid line (arbitrary vertical scale). Typical structures of a supercritical quasi-perpendicular shock, i.e., foot, ramp, overshoot and undershoot, are well defined in the right panels, two densities as well as total magnetic field strength. The ripple, the mode 2 structure in the overshoot along the shock surface, is also clearly seen in these panels. The above field structures are more or less time stationary.

In Fig.2 ion (left panels) and electron (right panels) phase space densities along $y = L_y/2$ at $\Omega_i t = 6.4$ are shown. Ion phase space indicates characteristics of a typical supercritical quasi-perpendicular shock where most of the incident ions transmit and a fraction of them reflected at the shock denote large scale gyro motions downstream, which are seen also in the one-dimensional simulation. On the other hand, all electrons transmit and no reflected electrons are produced, which is in contrast to the one-dimensional case.

The origin of rippling in a quasi-perpendicular shock was discussed by a number of authors [12, 13, 14]. Although the issue has not been completely resolved yet, it is considered that downstream temperature anisotropy of ions may play a crucial role. In the particular case here the downstream temperature anisotropy is provided by the ions once reflected by the shock and convected back into the downstream region. We estimated that their temperature anisotropy in the undershoot is $T_{r,\perp}/T_{r,\|} \sim 16$ and local relative density to transmit ions is $n_r/n_{tr} \sim 0.5$. The linear dispersion analysis using these values with assuming a locally homogeneous plasma results in that the maximum growth occurs at $k_{\|}c/\omega_{pi} \sim 1.3$ corresponding mode number in y is 1 or 2. This is what is observed in the undershoot of B_x and B_z in Fig.1. The waves are generated through Alfvén ion cyclotron (AIC) instability [12]. The ripple observed in the overshoot appear to be related strongly with these waves.

The relatively small scale fluctuations seen in the foot is due to the modified two-stream instability [15, 16, 17, 18]. The local distribution function of ions are again divided into two populations, transmitted core ions and reflected ions. The reflected ions are non-gyrotropic so that they are regarded as a cross-field ion beam. Dispersion analysis (not shown) reveals that the generated instability is the modified two-stream instability [19, 16, 17]. Smaller wavelength fluctuations are also present in the foot, ramp, and overshoot, although they are not so clear from Fig.1. They propagate along the local magnetic field so that they appear to be electron temperature anisotropy driven whistler waves.

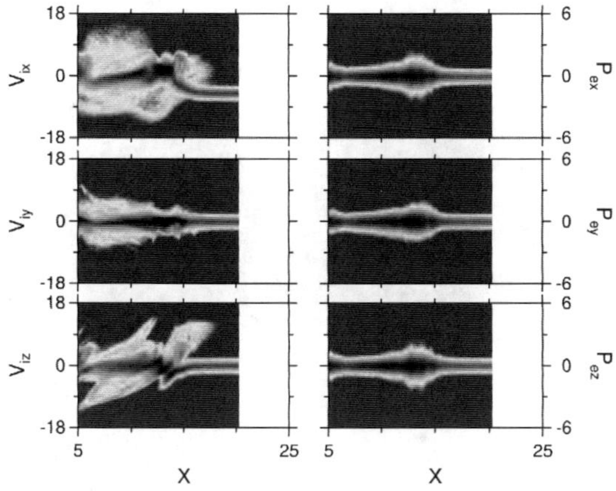

Figure 2. (left) Ion and (right) electron phase space densities along $y = L_y/2$ at $\Omega_i t = 6.4$.

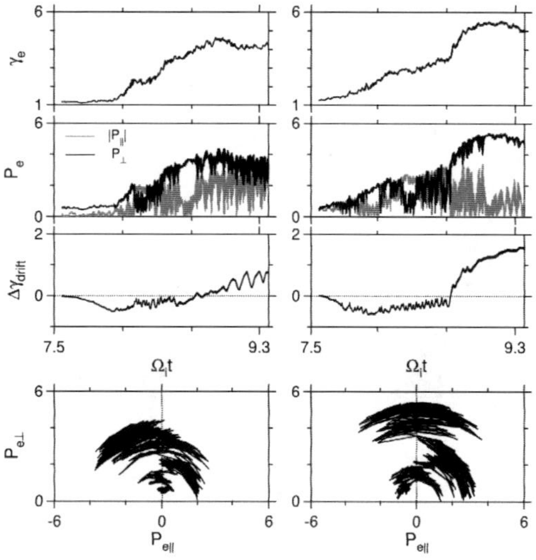

Figure 3. Orbits of two well accelerated electrons. The top three panels show time evolution of energy, momentum (parallel and perpendicular), fraction of energy gained through the shock drift process. The bottom panel denotes the corresponding orbit in $P_{\|} - P_{\perp}$ momentum space.

3.2. Electron Acceleration

Fig.3 represents the trajectories of two electrons which are well energized in the system. When they encounter the overshoot, they pass through the region where the magnetic field is locally weak. This weak magnetic field region in the overshoot appears due to the rippling. The local loss-cone angle in such a region is larger than that defined by average magnetic field in the overshoot. Therefore, if an electron in the process of SDA comes to such a region, it easily enters inside of the loss-cone so that transmits. In the one-dimensional model typical time for an electron to spend in the transition region during the SDA process is roughly about $\sim \Omega_i^{-1}$ [20]. All of the electrons involved in the process of SDA encounter the weak magnetic field region within the above time period by moving along the local magnetic field. As a result, there are no reflected electrons in this particular case.

However, some electrons are locally accelerated to non-thermal energy in the transition region. Fig.4 denotes the local electron energy distribution function obtained in the overshoot. The non-thermal population is clearly seen. The above discussed electrons are the two of them. In Fig.3 it is recognized that there are several acceleration phases during the passage of the transition region. In most of the acceleration phases perpendicular momentum increases. This is in contrast to the SDA in which parallel momentum mainly increases. The trajectories in the $p_{\|} - p_{\perp}$ space show a number of arcs implying that wave-particle interactions play a role. The energy gained through the drift motion along the motional electric field is roughly estimated in Fig.3. In the third panels from the top $\Delta\gamma_{drift} = -eE_{z1}\Delta z/m_e c^2$ is calculated, where E_{z1} denotes the upstream motional electric field and Δz is the distance of drift in the z-direction (anti-)parallel to the motional electric field. Since the motional electric field decreases in the transition region, the actual field at the position of each electron may be weaker than E_{z1}. Therefore, the values of $\Delta\gamma_{drift}$ may be overestimated than the actual energies gained through the shock drift process. Nevertheless, the corresponding energy gain is limited. Hence, at least a part of energy is gained through the process different from the SDA.

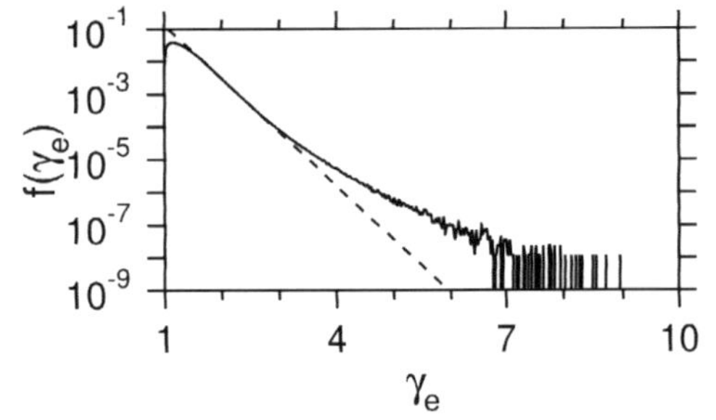

Figure 4. Energy distribution function of electrons in the overshoot.

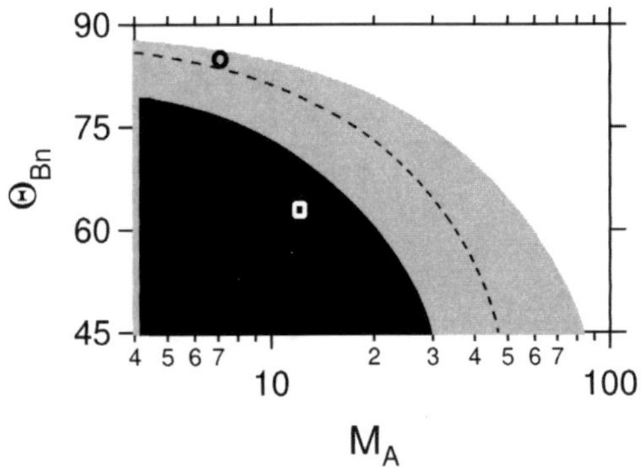

Figure 5. Feasibility of SDA in one-dimensional shocks. The colored region denotes $M_A - \Theta_{Bn}$ parameter space in which some electrons on a velocity shell centered at upstream plasma flow velocity are adiabatically reflected. The gray region corresponds to the shocks with $\beta_e(=\beta_i) = 1.5, \sigma = 1/9, \eta = 3$, while the black region to the shocks with $\beta_e(=\beta_i) = 10, \sigma = 1/29, \eta = 1$, where η is the shell radius in the unit of upstream electron thermal velocity. The solid circle (square) indicates the parameter used in this study and [10] ([11]).

4. Summary and Discussions

In the simulation presented here no incoming electrons are reflected, although the reflection was evident in the previous one-dimensional simulation with the same upstream physical parameters [10]. The reason is the difference of microstructure in the shock transition region between the one and the two-dimensional cases. In contrast to the one-dimensional case multi-scale fluctuations in electromagnetic fields are clearly seen in the two-dimensional simulation (Fig.1). These fluctuations do not alter the y-averaged structures from the one-dimensional simulation. However, the motion of energetic electrons is strongly modified. The ripple plays a crucial role. During its interaction with the shock, a high energy electron encounters the region where the local magnetic field is weak and slips through the region before reflected. This weak magnetic field region is produced as a result of the rippling. In this particular case we believe that the ripple is caused by AIC instability driven by the temperature anisotropy of reflected ions. As McKean et al. [13] pointed out, the local non-gyrotropic reflected ions may drive the instability, although the linear analysis in this region is not straightforward. In the undershoot, however, the local ion distribution function is more gyrotropic and the waves observed in this region may be related with or affected by the ripple in the overshoot. The standard linear analysis assuming the temperature anisotropy of gyrotropic reflected ions gives the maximum growth at the mode number in y between 1 and 2, which is consistent with the simulation result.

When the upstream plasma beta becomes high, the temperature anisotropy becomes small [12]. In Guo et al.[11] using $\beta = 20$ the maximum temperature anisotropy of the whole ions in the transition region is roughly estimated as ~ 3 from their Fig.2. If the reflection ratio is same as our case and the temperature anisotropy is dominantly provided by the reflected ions, the temperature anisotropy of the reflected ions may be roughly half of our case. The maximum growth of the AIC instability in the undershoot in such a case occurs at $k_{\|}c/\omega_{pi} < 0.1$. The corresponding waves can not be actually reproduced in the simulation, because their wavelengths in y are much larger than the system size.

Fig.5 indicates a feasibility diagram of SDA. The colored regions denote $M_A - \Theta_{Bn}$ parameter space where some electrons on the velocity shell having a radius of η times the upstream electron thermal velocity centered at the upstream plasma bulk velocity can be reflected at a shock when two dimensional effects are not taken into account (see details in Matsukiyo et al.[10]). The gray region corresponds to such a parameter space for $\beta_e(=\beta_i) = 1.5$, $\sigma = 1/9$, and $\eta = 3$. Our present simulation is done with the parameters marked by an open circle just below the upper boundary. The black region is similar one but for $\beta_e(=\beta_i) = 10$, $\sigma = 1/29$, and $\eta = 1$. The former two values are used in Guo et al.[11] and the open square denotes the parameters they chose. In contrast to the gray region, $\eta = 1$ is assumed here, because the velocity for $\eta = 3$ exceeds the speed of light in this particular case. For comparison, the boundary for $\eta = 1$ in the first case is indicated by the dashed line. From the figure, it is found that the effect of rippling may push down the boundary. Although we believe that there is a parameter region where the SDA based injection scenario can work, the effect of the ripple should be taken into account.

The present simulation implies that local wave-particle interactions in the transition region appear to play some roles in the injection process. Although there are no reflected electrons in the present case, high energy portion of the electrons in the transition region is clearly non-thermal as seen in Fig.4. The increase of their energy is not provided merely by the drift motion along the shock surface anti-parallel to the motional electric field. Particle trajectories imply the local wave-particle interactions in the transition region. In the simulation at least two electron scale waves are confirmed to be present. One is the oblique whistler waves generated in the foot as a result of modified two-stream instability, which are visible in E_x and all components of $\mathbf{B}$ at $X \sim 15$ in Fig.1. Another one is also whistler waves but propagating almost parallel to the local magnetic field with smaller wavelengths (although not so clear in Fig.1), which appear to be driven by local electron temperature anisotropy. The detailed interaction and energization mechanisms need to be examined elsewhere. Here, we infer that at least in some parameter region injection is achieved by the combined effects of the wave-particle interactions and the SDA. Probably for smaller Θ_{Bn}, some electrons locally energized and pitch angle scattered through the wave-particle interactions during the SDA may be reflected and escape from the shock toward upstream to be involved into further acceleration process. The scenario will be examined in the near future.

In the end we comment on the effect of dimensionality of the simulation. Thomas[21] reported that although there is essentially no qualitative difference in shock structure between two and three dimensional simulations, there are some quantitative differences between them. The amplitude of ripple is smaller in three dimensional case than in two dimensional case. Also the average values at the overshoot of both magnetic field and plasma density are lower in three dimensional simulation. These effects may affect the actual reflection ratio of electrons.

Acknowledgments

We should like to thank H. Washimi, M. Hoshino, T. Amano, T. Kato, Y. Ohira, and R. Yamazaki for useful comments and discussions. The numerical simulation was performed by using the computational resources of the K computer provided by the RIKEN Advanced Institute for Computational Science through the High Performance Computing Infrastructure System Research project (hp120222). This work was supported by Grant-in-Aid for Scientific Research (C) 26400266 (Y. M.), 25400479 (S. M.) and International Space Science Institute (ISSI) (S. M.).

References

[1] Decker, R. B., Krimigis, S. M., Roelof, E. C., Hill, M. E., Armstrong, T. P., Gloeckler, G., Hamilton, D. C., and Lanzerotti, L. J., 2008, *Nature*, **454**, 67
[2] Willson, M. A. G., 1970, *MNRAS*, **151**, 1
[3] Fujita, Y., and Sarazin, C. L., 2001, *Astrophys. J.*, **563**, 660
[4] Govoni, F., and Feretti, L., 2004, *Int'l J Modern Phys. D*, **13**, 1549
[5] van Weeren, R., J., Röttgering, H., J., A., Brüggen, M., Hoeft, M., 2010, *Science*, **330**, 347
[6] Lindner R. R., Baker A. J., Hughes J. P., Battaglia N., Gupta, N., Knowles, K., Marriage, T. A., Menanteau, F., Moodley, K., Reese, E. D., and Srianand, R., 2014, *Astrophys. J.*, **786**, 49
[7] Blandford, R. D., and Eichler, D. 1987, *Phys. Rep.* **154**, 1
[8] Amano, T., and Hoshino, M., 2007, *Astrophys. J.*, **661**, 190
[9] Matsumoto, Y., Amano, T., and Hoshino, M., 2013, *Phys. Rev. Lett.*, **111**, 215003
[10] Matsukiyo, S., Ohira, Y., Yamazaki, R., and Umeda, T., *Astrophys. J.*, **742**, 47
[11] Guo, X., Sironi, L, and Narayan, R., *Astrophys. J.*, **794**, 153
[12] Winske, D., and Quest, K. B., 1988, *J. Geophys. Res.*, **93**, 9681
[13] McKean, M. E., Omidi, N., and Krauss-Varban, D., 1995, *J. Geophys. Res.*, **100**, 3427
[14] Lowe, R. E., and Burgess, D., 2003, *Ann. Geophys.*, **21**, 671
[15] Matsukiyo, S., and Scholer, M., 2006, *J. Geophys. Res.*, **111**, A06104
[16] Umeda, T., Kidani, Y., Matsukiyo, S., and Yamazaki, R., 2012, *J. Geophys. Res.*, **117**, A03206
[17] Umeda, T., Kidani, Y., Matsukiyo, S., and Yamazaki, R., 2012, *Phys. Plasmas*, **19**, 042109
[18] Umeda, T., Kidani, Y., Matsukiyo, S., and Yamazaki, R., 2014, *Phys. Plasmas*, **21**, 022102
[19] Matsukiyo, S., and Scholer, M., 2003, *J. Geophys. Res.*, **108**, 1459
[20] Krauss-Varban, D., Burgess, D., Wu, C., S., 1989, *J. Geophys. Res.*, **94**, 15089
[21] Thomas V. A., 1989, *J. Geophys. Res.*, **94**, 12009

Journal of Physics: Conference Series **642** (2015) 012018 doi:10.1088/1742-6596/642/1/012018

Particle acceleration at shocks propagating in partially ionized plasma

Giovanni Morlino

INFN – Gran Sasso Science Institute, viale F. Crispi 7, 67100 L'Aquila, Italy.

E-mail: giovanni.morlino@gssi.infn.it

Abstract. The shocks of several young supernova remnants (SNR) are often associated with very thin optical filaments dominated by Balmer emission resulting from charge-exchange and collisional excitation between neutral Hydrogen from the interstellar medium and shocked protons and electrons. Optical lines are a direct probe of the conditions at the shock, in particular the width of the narrow and broad components reflect the temperature upstream and downstream of the shock, respectively. When the shock accelerate efficiently non-thermal particles, the shock structure changes producing anomalous Balmer lines and it is possible to use their line shape and their spatial profile to check the efficiency of SNR shocks in accelerating cosmic rays. Here we illustrate the kinetic theory of shock acceleration in presence of neutrals with some applications to young SNRs. We show that in three cases (RCW 86, SNR 0509-67.5 and Tycho) anomalous Balmer lines can be explained assuming that a fraction of $\sim 10\%$ of the total shock kinetic energy is converted into not thermal particles, while in one single case, the northwestern part of SN 1006, there is no evidence of efficient acceleration.

1. Introduction

In the context of supernova remnant (SNR) paradigm for the origin of cosmic rays (CR), particle acceleration takes place at shocks associated with the supernova explosion, and it is described by the non linear theory of diffusive shock acceleration (see [1] for a review). Energy and momentum conservation at the shock in the presence of accelerated particles leads to two straightforward conclusions: 1) since part of the energy is channeled into particle acceleration, the thermal energy (hence the temperature) of the downstream gas is expected to be lower than in the absence of CRs; 2) the dynamical reaction of CRs induces the formation of a precursor upstream of the shock, which results in a deviation of the CR spectrum from a simple power-law behavior. Optical spectra observed from so called Balmer dominated shocks can be used to test these predictions. In fact some young SNR shocks emit optical lines mainly consisting of Balmer Hα emission.

The first detection of bright Hα filaments around the remnants of Kepler, Tycho and the Cygnus Loop was reported by [2]. A peculiarity of this emission is the weakness of forbidden metal lines which implies an high temperature of the emitting region so that radiative cooling and recombination are unimportant. The interpretation of such optical emission remained a mystery up to the seminal works of [3, 4] who proposed that it can be produced by shocks propagating thorough a partially neutral gas. Their model was able to explain the intensity, spectrum and width of the filaments observed in Tycho's SNR, including the weakness of the forbidden metal lines. A peculiarity of Balmer dominated shocks, firstly reported by [4] for the

Tycho's SNR, is that the Hα line is formed by two distinct components, a narrow line with a FWHM of few tens km/s and a broad line with a FWHM of the order of the shock speed. Similar optical profiles are now observed from a bunch of young SNRs both in the Galaxy and in the Large Magellanic Cloud (for a review see [5]).

SNR shocks are collisionless and when they propagates in partially ionized medium, only ions are heated up and slowed down, while neutral atoms are unaffected to first approximation. However, when a velocity difference is established between ions and neutrals in the downstream of the shock, the processes of charge exchange (CE) and ionization are activated and this explain the existence of two distinct lines: the narrow line is emitted by direct excitation of neutral hydrogen after entering the shock front while the broad line results from the excitation of hot hydrogen population produced by CE of cold hydrogen with hot shocked protons. As a consequence, optical lines are a direct probe of the conditions at the shock, in particular the width of the narrow and broad components reflect the temperature upstream and downstream of the shock, respectively. Now, as already pointed out, if the shock accelerate particles efficiently, the shock's structure will be modified, altering the plasma temperature and inducing the formation of a CR precursor. Hence, Balmer emission could be used to provide an indirect measurement of the CR acceleration efficiency and even to gather information on the CR induced precursor.

The first clue that Balmer emission could provide evidence for the presence of accelerated particles was put forward as a possible way to explain the anomalous width of narrow Balmer lines reported for the first time by [6] and [7]: FWHM ranging from 30 to 50 km s^{-1} was detected for four SNRs in the LMC and for the Cygnus Loop, implying a pre-shock temperature around 25,000-50,000 K. Values in the same range have been reported afterwards for other SNRs (see, e.g.[8]). If this were the ISM equilibrium temperature there would be no atomic hydrogen, implying that the pre-shock hydrogen is heated by some form of shock precursor in a region that is sufficiently thin so that collisional ionization equilibrium cannot be established before the shock. Several explanations for this anomaly were proposed but only two of them was considered realistic: 1) the neutral-induced precursor and 2) the CR-induced precursor.

Let us comment the former possibility first. When fast, cold neutrals undergo CE interactions with the slower hot ions downstream of the shock, some fraction of the resulting hot neutrals can cross the shock and move upstream. The relative velocity between these hot neutrals and the upstream ions triggers the onset of CE and ionization interactions that lead to the heating and slowing down of the ionized component of the upstream fluid. The system then tends to develop a *neutral-induced* shock precursor, in which the fluid velocity gradually decreases, and even more important, the temperature of ions increases as a result of the energy and momentum deposition of returning neutrals. A first attempt at investigating the broadening of the narrow line component induced by the neutral precursor was made by [9], using a simplified Boltzmann equation for neutrals, but their calculation does not show any appreciable change of the narrow line width. This conclusion was confirmed by [10, 11], using a fully kinetic approach able to describe the interaction between neutrals and ions in a more accurate way. The physical reason is that the ionization length-scale of returning hot neutrals in the upstream is always smaller than the CE length-scale of incoming cold neutrals. Interestingly enough, [11] showed that the neutral precursor could produce a different signature, namely the presence of a third intermediate Balmer line due to hydrogen atoms that undergone charge exchange with warm protons in the neutral precursor.

The second and more promising possibilities to explain the anomalous width of narrow lines requires efficient particle acceleration which leads to the formation of a *CR-induced* precursor, where ionized plasma is heated before crossing the shock. If the precursor is large enough, CE can occur upstream leading to a broader narrow Balmer line. The first attempt to model this scenario was done by [13] using a two-fluid approach to treat ions and CRs but neglecting the dynamical role of neutrals. A different model was proposed by [14] where momentum and energy transfer between ions and neutrals is included, but the profile of the CR-precursor is assumed *a-priori*. Both works concluded that the observed width of 30-50 km s^{-1} can be explained using a low CR acceleration efficiency.

From the theoretical point of view, the main difficulty in describing the structure of a collisionless shock propagating in a partially ionized medium is that neutrals have no time to reach thermalization and cannot be treated as a fluid. Steps forward in relaxing the fluid assumption have been made by [15] and [16], even if these works neglect the modification induced by neutrals upstream of the shock. A more reliable interpretation of Balmer line profile requires an accurate description of the CR acceleration process where the mutual interplay between CRs, neutrals, ionized plasma and magnetic turbulence is simultaneously taken into account. Such an approach has been developed by [17] using a semi-analytical technique. This work showed that the main physical effect able to broaden the narrow line is the damping of magnetic turbulence in the CR precursor while the adiabatic compression alone is ineffective. Hence the observed widths are compatible also with large acceleration efficiency provided the right level of magnetic damping.

Efficient CR acceleration can also affect the width of broad lines. In fact, when a sizable fraction of the ram pressure is channeled into non-thermal particles, the plasma temperature behind the shock is expected to be lower, and this should reflect in a narrower width of the broad Hα line. Remarkably, there are clues of this phenomenon in two different remnants, RCW 86 [18, 19], and SNR 0509-67.5 in the LMC [20, 21]. In both cases the measured FWHM of the broad lines is compatible with theoretical predictions only assuming fast electron-proton equilibration downstream of the shock, a conclusion which seems to be at odds with both theoretical models and observations [22].

2. The kinetic approach

Here we summarize the kinetic model for shock particle acceleration in presence of neutrals developed in [10, 11, 17]. We consider a stationary system with a plane-parallel shock wave propagating in a partially ionized proton-electron plasma with velocity V_{sh} along the z direction. The fraction of neutral hydrogen is fixed at upstream infinity where ions and neutrals are assumed to be in thermal equilibrium with each other. The shock structure is determined by the interaction of CRs and neutrals with the background plasma. Both CRs and neutrals profoundly change the shock structure, especially upstream where both create a precursor: the CR-induced precursor reflects the diffusion properties of accelerated particles and has a typical spatial scale of the order of the diffusion length of the highest energy particles. The neutral-induced precursor develops on a spatial scale comparable with a few interaction lengths of the dominant process between CE and ionization. The downstream region is also affected by the presence of both CRs and neutrals and the velocity gradients that arise from ionization have a direct influence on the spectrum of accelerated particles. A self consistent description of shock particle acceleration in presence of neutral Hydrogen, requires to consider four mutually interacting components: thermal particles (protons and electrons), neutrals (hydrogen), accelerated protons (CRs) and turbulent magnetic field. We neglect the presence of helium and heavier chemical elements. The interaction terms make the system highly non linear and the solution is found using a iterative scheme similar to the one we introduced in some previous works [23, 24, 10].

Let us start with the description of neutrals. The main difficulty arises from the fact that neutrals cannot be described as a fluid, because in the downstream the collisional ionization length is smaller than the equilibration length. Hence neutrals are described kinetically, using the stationary Boltzmann equation to calculate the evolution of the velocity distribution function, $f_N(\vec{v}, z)$,

$$v_z \frac{\partial f_N(\vec{v}, z)}{\partial z} = \beta_N f_i(\vec{v}, z) - [\beta_i + \beta_e] f_N(\vec{v}, z) , \quad (1)$$

where z is the distance from the shock (which is located at the origin), v_z is the velocity component along the z axis and the electron and proton distribution functions, $f_i(\vec{v}, z)$ and $f_e(\vec{v}, z)$, are assumed to be Maxwellian at each position. The collisional terms, $\beta_k f_l$, describe the interaction (due to CE and/or ionization) between the species k and l. The interaction rate β_k is formally written as

$$\beta_k(\vec{v}, z) = \int d^3w \, v_{rel} \, \sigma(\vec{v}_{rel}) f_k(\vec{w}, z) \,, \tag{2}$$

where $v_{rel} = |\vec{v} - \vec{w}|$ and σ is the cross section for the relevant interaction process. More precisely, β_N is the rate of CE of an ion that becomes a neutral, β_i is the rate of CE plus ionization of a neutral due to collisions with protons, while β_e is the ionization rate of neutrals due to collisions with electrons. A full description of the cross sections used in the calculations can be found in [11].

The isotropic distribution function of CRs satisfies the following transport equation in the reference frame of the shock:

$$\frac{\partial}{\partial z}\left[D(z,p)\frac{\partial f}{\partial z}\right] - u\frac{\partial f}{\partial z} + \frac{1}{3}\frac{du}{dz}p\frac{\partial f}{\partial p} + Q(z,p) = 0 \,, \tag{3}$$

where D is the diffusion coefficient, u the plasma speed and Q the injection term. The z-axis is oriented from upstream infinity ($z = -\infty$) to downstream infinity ($z = +\infty$) with the shock located at $z = 0$. We assume that the injection occurs only at the shock position and is monoenergetic at $p = p_{\rm inj}$, hence the injection term can be written as $Q(z,p) = Q_0(p)\delta(z)$ where $Q_0(p) = \eta_{\rm inj} n_1/(4\pi p_{\rm inj}^2)\delta(p - p_{\rm inj})$. Here n_1 is the number density of ions immediately upstream of the subshock, and $\eta_{\rm inj}$ is the fraction of particles that is going to take part in the acceleration process. $\eta_{\rm inj}$ is related to the subshock compression factor, following the *thermal leakage* model described in [12]:

$$\eta_{\rm inj} = 4/\left(3\sqrt{\pi}\right)\left(R_{\rm sub} - 1\right)\xi_{\rm inj}^3 e^{-\xi_{\rm inj}^2} \,. \tag{4}$$

Here $\xi_{\rm inj}$ is defined by the relation $p_{\rm inj} = \xi_{\rm inj} p_{\rm th,2}$, where $p_{\rm th,2}$ is the momentum of the thermal particles downstream. $\xi_{\rm inj}$ parametrizes the poorly known microphysics of the injection process and is taken as a free parameter with a typical value between 2 and 4. The diffusion properties of particles are described by the diffusion coefficient $D(z,p)$. We assume Bohm diffusion in the local amplified magnetic field:

$$D(z,p) = \frac{1}{3} c r_L[\delta B(z)] \,, \tag{5}$$

where $r_L(\delta B) = pc/[e\delta B(z)]$ is the Larmor radius in the amplified magnetic field. The calculation of δB is described assuming that the only turbulence which scatters particles is the one self-generated by the particles themselves through the resonant streaming instability. These waves are also damped due to several processes. In particular, when the plasma is not fully ionized, the presence of neutrals can damp Alfvèn waves via ion-neutral damping. The equation for transport of waves can be written as:

$$\partial_z F_w = u(z)\,\partial_z P_w + P_w\left[\sigma_{\rm CR}(k,z) - \Gamma_{\rm TH}(k,z)\right] \,, \tag{6}$$

where $F_w(k,z)$ and $P_w(k,z)$ are, respectively, the energy flux and the pressure per unit logarithmic bandwidth of waves with wavenumber k. $\sigma_{\rm CR}$ is the growth rate of magnetic turbulence, while $\Gamma_{\rm TH}$ is the damping rate. For resonant wave amplification the growth rate of Alfvén waves is:

$$\sigma_{\rm CR}(k,x) = \frac{4\pi}{3}\frac{v_A(x)}{P_w(k,x)}\left[p^4 v(p)\frac{\partial f}{\partial x}\right]_{p=\bar{p}(k)} \,, \tag{7}$$

where $p = \bar{p}(k) = eB/km_pc$ is the resonant momentum. The damping of the waves is mainly due to non-linear Landau damping and ion-neutral damping. For the sake of simplicity here we adopt a phenomenological approach in which the damping results in a generic turbulent heating (TH) at a rate $\Gamma_{\rm TH} = \eta_{\rm TH}\sigma_{\rm CR}$. This expression assumes that a fraction $\eta_{\rm TH}$ of the power in amplified waves is locally damped and results in heating of the background plasma.

Finally we need to describe the dynamics of the background plasma which is affected by the presence of accelerated particles and by CE and ionization of neutrals. Protons and electrons in the plasma are assumed to share the same local density, $\rho_i(z) = \rho_e(z)$, but not necessarily the same temperature, i.e., $T_i(z)$ may be different from $T_e(z)$. The equations describing the conservation of mass, momentum and energy taking into account the interactions of the plasma fluid with CRs are:

$$\frac{\partial}{\partial z}\left[\rho_i u_i + \mu_N\right] = 0\,, \tag{8}$$

$$\frac{\partial}{\partial z}\left[\rho_i u_i^2 + P_g + P_c + P_w + P_N\right] = 0\,, \tag{9}$$

$$\frac{\partial}{\partial z}\left[\frac{1}{2}\rho_i u_i^3 + \frac{\gamma_g P_g u_i}{\gamma_g - 1} + F_w + F_N\right] = -u_i\frac{\partial P_c}{\partial z} + \Gamma P_w\,. \tag{10}$$

Here $\mu_N = m_H \int d^3v v_\parallel f_N$, $P_N = m_H \int d^3v v_\parallel^2 f_N$ and $F_N = m_H/2 \int d^3v v_\parallel (v_\parallel^2 + v_\perp^2) f_N$ are respectively the fluxes of mass, momentum and energy of neutrals along the z direction (labelled as $\|$). They can be easily computed once the neutral distribution function is known. P_w and F_w are the pressure and energy flux of waves, while P_c is the CR pressure computed from the CR distribution function:

$$P_c(z) = \frac{4\pi}{3}\int dp\, p^3 v(p) f(z,p)\,. \tag{11}$$

The dynamical role of electrons in the conservation equations is usually neglected due to their small mass. However, collective plasma processes could contribute to equilibrate electron and proton temperatures, at least partially. If the equilibration occurs in a very efficient manner, the electron pressure cannot be neglected and the total gas pressure needs to include both the proton and electron contributions, namely $P_g = P_i + P_e = P_i(1+\beta)$, where $\beta(z) \equiv T_e/T_i$ is the electron to proton temperature ratio and is taken here as a free parameter. While it is well established that electron-ion equilibration in the downstream might be only partial [25, 26], in the presence of a precursor (either induced by the CRs or by the neutrals), also upstream of the shock the level of equilibration becomes an unknown.

In order to solve the set of non-linear equations involving neutrals, ions, CRs and magnetic field, we adopt an iterative method that is fully described in [17]. The input quantities are the values of the shock velocity and all environmental quantities at upstream infinity, where the distribution function of neutrals is assumed to be Maxwellian at the same temperature as that of ions. At the end the calculation provides the distribution function of neutrals as well as the spatial and velocity profile of Balmer emission.

3. CR-precursor vs. neutral-precursor

The formalism presented above represents the first theory of particle acceleration at collisionless shocks in the presence of neutral atoms and allows us to calculate the shock structure, the spectrum of accelerated particles and the Balmer emission from the shock region.

Let us discuss briefly the structure of a shock in presence of neutrals when no particle acceleration occurs. The main effect is the formation of what we call *neutral-induced* precursor in the upstream region: for a wide range of shock velocities, a fraction of the hot neutrals produced downstream by the charge-exchange process can recross the shock toward upstream,

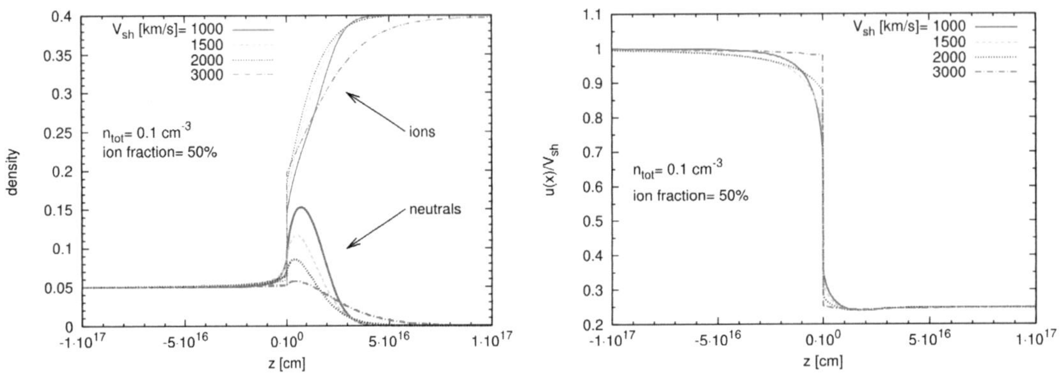

Figure 1. *Left Panel:* Density of neutrals (lower curves) and ions (upper curves) as a function of the distance from the shock upstream (i.e. $z < 0$) and downstream (i.e $z > 0$), for different values of the shock velocity. *Right Panel:* Ion velocity (normalized to the far upstream value) corresponding to the same cases shown in the left panel.

giving rise to a neutral return flux. The subsequent interaction of these neutrals with the upstream ions produces a precursor region where the incoming plasma is heated and slowed down. The role of the return flux is better illustrated by the dynamical quantities associated with ions. In Fig. 1 we show the density (left) and velocity (right) of the ion plasma as a function of the position across the shock ($z = 0$). The different curves refer to different values of the shock velocity as labelled. One can see that the density stays constant upstream until very close to the shock front, where the return flux becomes important. In this region, the fast neutrals from downstream deposit energy into the ion plasma through both ionization and charge exchange, thereby heating the gas. These reactions also deposit momentum in the $-z$ direction, thereby slowing down the ion plasma (right panel). The combination of this induced deceleration and of the ionization process causes the ion density to increase (left panel). One final comment is deserved by the right panel of Fig. 1: here we see that with decreasing shock velocity between 3000 and 1500 km s^{-1} the precursor becomes progressively more pronounced. This trend suddenly changes for V_{sh}=1000 km s^{-1}, when the precursor becomes shorter and steeper: this is due to the fact that in this case the return flux is destroyed by charge-exchange events rather than by ionization, and those occur on a shorter scale.

Now, when also the presence of accelerated particles is included, the shock structure is modified by both neutrals and CRs: neutrals produce a neutral-induced precursor on the typical ionization length scale of fast neutrals in the upstream, while CRs produce a precursor on the typical diffusion length scale of particles at the maximum energy. A good way to visualize the modification induced on the shock by the presence of both neutrals and CRs is to look at the temperature profile of ions in the shock region. In Fig. 2 we compare the temperature profile for a case with and without CRs. We chose a test case with values of parameters typical of a SNR expanding in the cold ISM, at the beginning of the Sedof-Taylor phase, i.e. $V_{\rm sh} = 4000$ km s^{-1}, total ISM density of 0.1 cm^{-3} with ionization fraction $X_{\rm ion} = 0.5$ and temperature $T_{\rm ISM} = 10^4$ K. The black-tick solid line shows the temperature profile without CRs: the temperature increase produced ahead of the shock, on the scale of $\sim 10^{17}$ cm is entirely due to the return flux of neutrals which deposit upstream a fraction of energy subtracted from the downstream. Indeed the length-scale corresponds to the collisional ionization length of returning neutrals with upstream ions. The remaining thin curves show the case when CR acceleration is turned on. For all these curves we assume the same acceleration efficiency of 40% (determined

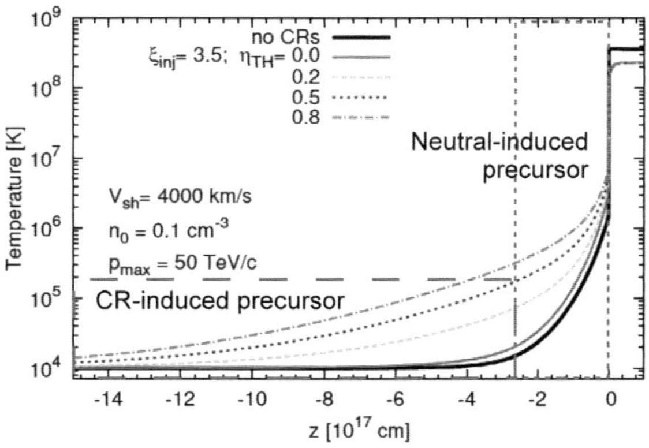

Figure 2. Temperature of the plasma in the shock region for $V_{\rm sh} = 4000\,{\rm km\,s^{-1}}$, $n_0 = 0.1\,{\rm cm^{-3}}$, $X_{\rm ion} = 0.5$ and$B_0 = 10\,\mu$G. When CR acceleration is turned on, we assume $\xi_{\rm inj} = 3.5$ and $p_{\rm max} = 50$ TeV/c. The thick solid (black) line refers to the case with no CRs. The thin (red) solid line refers to the case with CRs but no damping of magnetic turbulence (i.e. $\eta_{\rm TH} = 0$). The remaining thin curves are calculated for increasing values of $\eta_{\rm TH}$, as labelled. The dashed and dotted rectangles mark the CR and the neutral-induced precursor, respectively.

by the injection parameter $\xi_{\rm inj} = 3.5$) and the same maximum proton energy of 50 TeV, but we change the damping efficiency of magnetic turbulence. When the damping is absent, the temperature profile is only slightly different from the case without CRs, the difference being due to the adiabatic compression produced by the CR pressure, while, when the damping is turned on, the temperature increases in the whole CR-precursor in a way proportional to the amount of damping. Notice that the typical CR-precursor length is $D(p_{\rm max})/V_{\rm sh} \approx 4\times 10^{18}\delta B_{\mu{\rm G}}^{-1}$ cm, where δB is the amplitude of Alfvén waves which resonate with particles at $p = p_{\rm max}$. We further notice that all cases with efficient CR acceleration have a temperature in the downstream smaller than $\sim 40\%$ with respect to the case without CRs.

The shape of the spatially-integrated Balmer line emission is plotted in the left panel of Fig. 3, while the right panel shows a zoom in the region of the narrow Balmer lines. The curves refer to the same cases as in Fig. 2, labelled as indicated. The width of the broad component of the Balmer line is appreciably reduced when CR acceleration is turned on, but the results are not sensitive to the amount of TH. On the other hand, the narrow Balmer line becomes broader when CR acceleration is efficient; the effect is more pronounced when non-negligible TH is taken into account. Hence, the distribution function of neutrals becomes broader mainly because of the scattering with a warmer ion distribution in the far precursor (i.e., where TH is more effective), rather than because of the neutral return flux, which operates only within a few CE interaction lengths from the sub-shock.

4. Application to young SNRs

In this paragraph we compare the results of the kinetic theory for Balmer dominated shocks with few noticeable cases of young SNRs. We concentrate only on the two more direct observables from Balmer emission, namely the width of the spatially-integrated broad and narrow lines. Other possible observables, like the spatial profile, the intermediate line or the narrow-to-broad intensity ratio could also be used but, in order to constraint the model, we need more accurate data that the ones available in the literature up to know. In the follow we analyze the broad Balmer emission from three remnants, SN 1006, RCW 86 and SNR 0509-65.7, while for the narrow line we show only one single case, the Tycho's remnant.

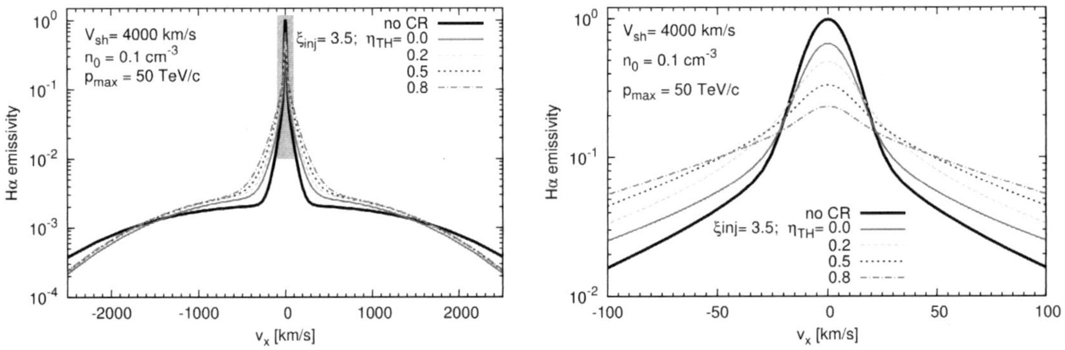

Figure 3. The left panel shows the volume-integrated line profile of Balmer emission for the same cases shown in Fig. 2. The right panel shows a zoom in of the narrow line region, showed with a shadow box in the left panel. The thick solid line shows the case without CRs, while different thin lines are calculated at fixed $\xi_{\rm inj} = 3.5$ and for different values of the TH efficiency as labelled.

The width of the broad Balmer line is probably the most powerful tool to infer the presence of non-thermal particles. If the velocity of the shock is known, the width of the broad line depends mainly on two factor: the electron-to-proton temperature ratio, and the CR acceleration efficiency, while it does not depend on the total ISM density and has a weak dependence on the ionization fraction. In Fig. 4 we show the theoretical prediction of the FWHM of the broad lines as a function of the shock speed and for different value of T_e/T_p, in the absence of CRs. The plot also shows the data for the three SNRs mentioned above. When measured data lie between the top and the bottom lines, they are in principle compatible with the absence of CRs, unless one is able to measure the electron temperature, breaking the degeneracy between different lines. In the following we discuss separately each single remnant.

4.1. SN 1006

The remnant of SN 1006 is a special case in that it presents hints of efficient particle acceleration only in the northeastern (NE) and in the southwestern (SW) regions. Here non-thermal X-ray have been detected, concentrated in thin filaments. As extensively discussed in literature, the thickness of X-ray filaments is interpreted as due to rapid synchrotron losses of electrons in a strong magnetic field ($\sim 100\mu$G), which could be produced by CR-induced amplification (if the CR acceleration is efficient). Moreover the same regions have been detected in γ-rays emission [27]. A further indirect indication supporting efficient CR acceleration is the distance between the contact discontinuity and the forward shock, which is smaller in the NE and SW with respect to the rest of the remnant. In fact, when a non negligible fraction of the shock kinetic energy is converted into CRs, the downstream plasma become more compressible and a reduced distance is expected.

Now, data for the FWHM of the broad Blamer line shown in Fig. 4 (shaded box) are taken from the NW region [28] where none of the above signatures have been detected. Remarkably enough, the detected FWHM is fully compatible with theoretical expectation without CRs and assuming a low level of electron-proton equilibration ($T_e/T_p < 0.2$). This conclusion is further supported by the width of the narrow Balmer line which, in the NW region, is 21 km s^{-1} [8], corresponding to an ISM temperature of 10^4 K, compatible with absence of a CR precursor. It would be extremely interesting to measure the width of Balmer lines also from the NE and the

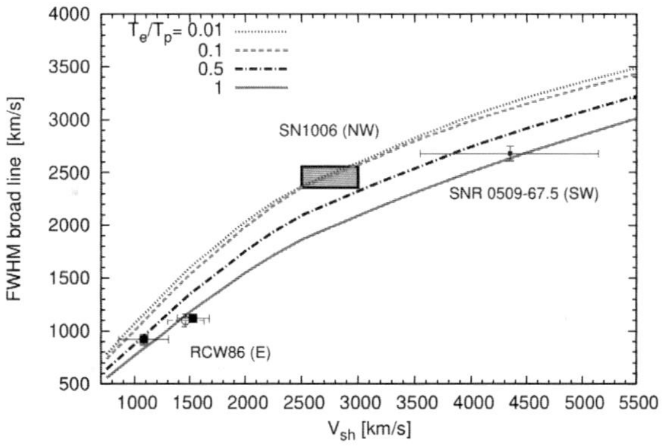

Figure 4. Measured FWHM of the broad Balmer line as a function of the shock speed for three different remnant: RCW 86 (three different locations, one in the northeast (open circle) and two location in the southeast (filled square), data from [18] assuming a distance from the remnant of 2.5 kpc); northwest rim of SN 1006 [28]; southwest rim of SNR 0509-67.5 (the FWHM is taken from [20] while the uncertainty on $V_{\rm sh}$ is taken from the theoretical model in [21]). Lines show the theoretical prediction without CR acceleration for different values of electron to proton temperature ratio in the downstream [17].

SW, but the emission in those regions is so faint that tens of hours of observation are needed to reach the required signal to noise ratio.

4.2. RCW 86

RCW 86 is a shell like SNR and a γ-ray source, detected both in GeV range by FermiLAT and in the TeV range by HESS. The NE part also shows non-thermal X-rays, suggesting a possible efficient production of CRs. The remnant presents many Balmer filaments around the external shell and, recently, the proper motion and the line width of these filaments have been measured [18]. Here we summarize the results presented in [19], where the analysis has been concentrated on the eastern part of the remnant because this is the region showing the highest shock speed, hence the one with a larger probability to accelerate CRs. Because the proper motion is known, the results depends on the assumed distance from us, which is very uncertain, but generally bound between 2 and 3 kpc. The most quoted value in literature is 2.5 kpc, which will be used in the following discussion. The three data points shown in Fig. 5 represent the FWHM measured in three different regions of the eastern part of the remnant by [29]. The error bars in the shock speed reflect the error in the measured proper motion. The three point seem to be marginally compatible with absence of CR acceleration if we assume a full electron-proton equilibration. On the other hand this assumption is at odd with the results by [29], where, using the thermal X-ray emission, the authors were able to infer the electron temperature in the same regions where the Balmer lines have been detected. These values can be used as upper limits for the electron temperature behind the shock, because they are obtained integrating onto a large region in the downstream where Coulomb collisions can enhance the electron temperature. This information allows us to fix an upper limit to the value of T_e/T_p and to break the degeneracy between different theoretical curves. We get $T_e/T_p \lesssim 0.1$ and in Fig. 5 we show, with filled circles, what should be the positions of the measured FWHM if the constrain on the electron temperature is taken into account (always assuming a distance of 2.5 kpc). Hence, to explain the measured value of the FWHM, we conclude that a fraction $\sim 20\%$ of the shock energy is

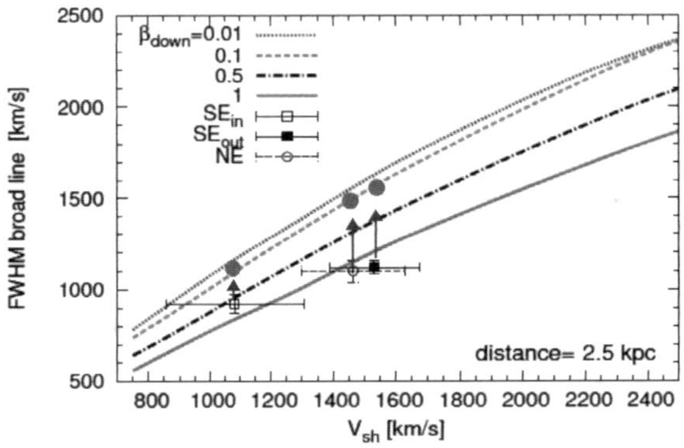

Figure 5. FWHM measured in three different regions of the eastern part of RCW 86 by [29] (see their figure 3). Lines show the theoretical prediction without CRs as in Fig. 4. The filled circles show the value that the FWHM should have in these three locations if one account for the measured electron temperature and if no efficient acceleration were present.

converted into non-thermal particles. On the other hand, if we consider the uncertainty on the distance, the above value ranges between few per cent to 40% for $d = 2$ and 3 kpc, respectively.

An important remark is in order. The shock speed inferred for RCW 86 is relatively low, being < 2000 km s^{-1}, hence we do not expect particles accelerated up to very high energies. Moreover, if the neutral fraction is $> 10\%$, for such low shock speed the neutral return flux could be strong enough to affect the acceleration in the direction to make the spectrum steeper, to the point that the non-thermal energy is dominated by particles with a momentum close to the injection momentum. If this were the case, the γ-ray spectrum observed by FermiLAT and HESS could not be explained by hadronic interaction, but is more probably due to electron inverse-Compton, has suggested in [30] on the basis of the hard spectrum measured by FermiLAT.

4.3. SNR 0509-67.5

The Balmer emission from SNR 0509-67.5 has been studied in [21]. SNR 0509-67.5 is located in the LMC, so that its distance from the Sun is known to be 50 ± 1 kpc. At the moment there are no published measurements for the shock proper motion, but the shock speed can be inferred from evolution models taking into account the measured size of the remnant and its age, which is estimated from the light echo to be 400 ± 120 yr. The point showed in Fig. 4 refers to the FWHM measured in the SW rim of the remnant[20], and the uncertainty in the shock speed correspond to the theoretical estimate done in [21]. Unfortunately the big uncertainty in the determination of the shock speed prevents a firm conclusion on the CR acceleration efficiency. If the shock moves faster than ~ 4500 km s^{-1}, one can conclude that particle acceleration must be taking place with an efficiency of several tens of percent. For lower shock velocity the evidence of particle acceleration becomes less clear because of the uncertainty in the electron-ion equilibrium downstream. We can speculate that if T_e/T_p is as low as the value inferred in SN 1006, than the acceleration efficiency of the SW rim should be $\gtrsim 10\%$. It is worth noting that this speculation is suggested by indirect measurements of the electron-ion equilibration made in several remnants using different techniques, which point towards an inverse relation between the electron-to-ion temperature ratio and the shock speed [22]. On the other hand none of these measurement has been done for SNR 0509-67.5 up to now.

Journal of Physics: Conference Series **642** (2015) 012018 doi:10.1088/1742-6596/642/1/012018

4.4. Tycho

The Tycho's SNR is the best known candidate where to look for efficient CR acceleration mainly because the observed γ-ray spectrum is very steep and is difficult to explain it using leptonic processes, while pion decay produced in hadronic collisions can fit the observations very well, assuming a CR acceleration efficiency $\sim 10\%$ [31]. Moreover many other observations can be simultaneously explained in the framework of non-linear shock acceleration using the same level of efficiency (namely Radio and X-ray spectrum, radio end X-ray morphology, distance between contact discontinuity and forward shock). Unfortunately the Balmer emission in Tycho is faint even if it is present all over around the remnant. In spite of this there is a small region of the shock, called *knot-g* where the Balmer emission is enhanced, probably due to a larger ISM density in that region. These region has been analyzed by several authors and presents two interesting anomalies which could be related to efficient acceleration.

1) A gradual increase of Hα intensity has been measured just ahead of the shock front [32]. This has been interpreted as emission from the thin shock precursor ($\sim 1''$ which implies a thickness of $\sim 3 \times 10^{16}$ cm for a distance of 3 kpc) likely due to CRs. If confirmed, this detection would represent the first direct proof of the existence of a CR precursor. On the other, hand Balmer emission from the upstream can be produced also by the neutral-induced precursor, as showed in [11], and, in order to distinguish between these two possibilities, a careful modeling of the shock is required, taking into account the complex interaction between the CR and the neutral induced precursor. At the moment the most promising technique seems to be the kinetic theory developed in [17].

2) The FWHM of the narrow component has been measured to be 44 ± 4 km s^{-1} [8]. Such values implying a pre-shock temperature between 36,000 and 52,000 K. If this were the ISM equilibrium temperature there would be no atomic hydrogen, implying that the pre-shock hydrogen is heated by some form of shock precursor in a region that is sufficiently thin so as to make collisional ionization equilibrium before the shock unfeasible. The CR precursor is the most plausible candidate to explain such a broadening of the narrow line width. To support this statement in Fig. 6 we compare the value measured from the *knot-g* in Tycho with the theoretical expectation calculated using the kinetic theory. The FWHM of the narrow Balmer line is plotted as a function of the maximum momentum $p_{\rm max}$ and for three different value od the turbulent heating $\eta_{\rm TH} = 0.2, 0.5$ and 0.8 (from top to bottom panel). The three curves in each panel are obtained for $\xi_{\rm inj} = 3.5,\ 3.7,\ 3.8$, which approximately correspond to $\epsilon_{\rm CR} = 0.4, 0.2$ and 0.1 for the chosed value of the shock velocity and total density, i.e. $V_{\rm sh} = 4000\,{\rm km\,s^{-1}}$ and $n_0 = 0.1\,{\rm cm^{-3}}$. The maximum momentum, $p_{\rm max}$ determines the spatial extent of the CR-induced precursor. Larger values of $p_{\rm max}$ imply that there is more time (space) for depositing heat in the upstream, and the width of the narrow Balmer line broadens correspondingly. The effect becomes more pronounced for larger values of the parameter $\eta_{\rm TH}$. The number of free parameters is too large and the FWHM of the narrow line alone cannot be used to constrain none of them. Nevertheless, the FWHM is an increasing function of all the three parameters, $p_{\rm max}$, $\eta_{\rm TH}$ and ϵ_{CR}, hence if the shock speed, the upstream density and neutral fraction are known, we can put lower limit to these parameters. Following this procedure, in the case of Tycho we get: $p_{\rm max} \gtrsim 40$ TeV, $\eta_{\rm TH} \gtrsim 0.2$ and $\epsilon_{CR} \gtrsim 10\%$.

5. Conclusions

The quest for the origin of Galactic CRs remains open. The search for their origin is difficult and all new observational input need to be considered very carefully. Based mainly on the detection of gamma rays from several SNRs and on the morphology of the X-ray emission, which suggests strong magnetic field amplification, SNRs are considered, now more than ever before, as the most likely source of the bulk of CRs. Because the shocks produced by SN explosions often propagate in partially ionized gas, the emission lines of neutral hydrogen have recently been recognized as

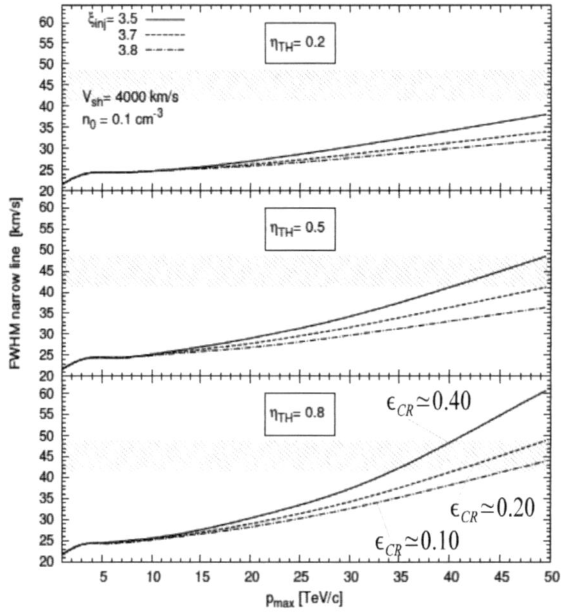

Figure 6. FWHM of the narrow line as a function of the maximum momentum of accelerated protons. The three panels refer to $\eta_{\rm TH} = 0.2, 0.5$ and 0.8. Each panel shows three lines calculated assuming different injection parameter, $\xi_{\rm inj} = 3.5, 3.7$ and 3.8 which approximately correspond to $\epsilon_{CR} = 0.4, 0.2$ and 0.1. The shock speed is 4000 km s^{-1}, the total density 0.1 cm^{-3} with 50% of neutral fraction. The shadow band represent the FWHM measured from the *knot-g* in the Tycho's SNR.

a possible diagnostic tool for CRs. There are three different signatures in the Balmer emission which could reveal the presence of CRs: 1) a shock that is accelerating particles is expected to be less effective in heating the background plasma, which is reflected in a Balmer-line emission with a smaller width than in the absence of CR acceleration; 2) the presence of a CR precursor upstream of the shock could heat the upstream plasma on a scale larger than the charge-exchange length but small than the ionization length, resulting in a larger width for the Balmer narrow line; 3) if the heating of the upstream plasma (especially electrons) is large enough, the collisional excitation between electrons and Hydrogen atoms can produce Balmer emission also from the region ahead of the shock.

Remarkably all these signatures has been observed in some SNRs. Here we summarized the results obtained by the kinetic theory of collisionless shocks in the presence of neutral hydrogen developed in [10, 11, 17] comparing the theoretical predictions with data on the Balmer line emission from four SNRs: SN 1006, RCW 86, SNR 0509-67.5 and Tycho. While the NW region of SN 1006 is compatible with absence of CR acceleration, the other three remnants presents anomalies in the Balmer lies which can be explained assuming an acceleration efficiency around $\sim 10\%$ which is remarkably the amount of efficiency required for SNRs to be the main sources of Galactic CRs. Balmer emission could be used also to gather information on the CR-precursor length, on the magnetic damping in the precursor and on the electron-to-proton temperature equilibration. Unfortunately the quality of data or the lack of important information (like the shock proper motion) prevent a better determination of the CR acceleration efficiency as well as other parameter involved in the theory. However the good news is that the amount and the quality of the data concerning the Balmer emission will increase soon in the next few years. For this reason the analysis of Balmer emission can be regarded as one of the most powerful technique to study the process of shock acceleration and, more generally, to study the physics of collisionless shock.

References

[1] Malkov, M.A. & Drury, L. OC., 2001, Rep. Prog. Phys., **64**, 429
[2] Minkowski R. 1959, IAU, **9**, 315
[3] Chevalier R.A. & Raymond J.C., 1978, ApJ, **225**, L27
[4] Chevalier R.A., Kirshner R.P. & Raymond J.C., 1980, ApJ, **235**, 186
[5] Heng K., 2010, PASA, **27**, 23
[6] Smith R. C., Raymond J. C., Laming J. M., 1994, ApJ, **420**, 643
[7] Hester J.J., Raymond J.C., Blair W.P., 1994, ApJ, **470**, 721
[8] Sollerman, J., Ghavamian, P., Lundqvist, P. & Smith, R. C. 2003, A&A, **407**, 249
[9] Lim A. J., Raga A. C., 1996, MNRAS, **280**, 103
[10] Blasi P., Morlino G., Bandiera R., Amato E., Caprioli D., 2010, ApJ, **755**, 121
[11] Morlino G., Bandiera R., Blasi P., Amato E., 2012, ApJ, **760**, 137
[12] Blasi, P., Gabici, S., Vannoni, G., 2005, MNRAS, **361**, 907
[13] Wagner A. Y., Lee J.-J., Raymond J. C., Hartquist T. W. , Falle S. A. E. G., 2008, ApJ, **690**, 1412
[14] Raymond J. C., Vink J., Helder E. A., de Laat A., 2011, ApJ, **731**, L14
[15] Heng K., McCray R., 2007, ApJ, **654**, 923
[16] van Adelsberg M., Heng K., McCray R., Raymond, J.C., 2008, ApJ, **689**, 1089
[17] Morlino G., Blasi P., Bandiera R., Amato E., Caprioli D., 2013, ApJ, **768**, 148
[18] Helder E. A., Vink J., Bamba A., Bleeker J. A. M., Burows D.N., Ghavamian P., Yamazaki R., 2013, MNRAS, **435**, 910
[19] Morlino G., Blasi P., Bandiera R., Amato E. 2014, A&A, **562**, 141
[20] Helder E. A., Kosenko D., Vink J., 2010, ApJ, **719**, 140
[21] Morlino G., Blasi P., Bandiera R., Amato E., 2013, A&A, **557**, 142
[22] Rakowski C. E., 2005, Advances in Space Research, **35**, 1017
[23] Blasi, P., 2002, Astropart. Phys., **16**, 429
[24] Amato, E. & Blasi, P., 2005, MNRAS, **364**, L76
[25] Ghavamian, P., Raymond, J., Smith, R. C., Hartigan, P., 2001, ApJ, **47**, 995
[26] Ghavamian, P., Laming, J. M., Rakowski, C. E., 2007, ApJ, **654**, 69
[27] Acero, F., Aharonian, F., Akhperjanian, A. G., et al. 2010, A&A, **516**, A62
[28] Nikolić, S. et al., 2013, *Science*, **340**, 45
[29] Helder, E. A., Vink, J., & Bassa, C. G. 2011, ApJ, **737**, 85
[30] Yuan, Q., Huang, X., Liu, S., Zhang, B. 2014, ApJ, **785**, L22
[31] Morlino, G., Caprioli, D. 2012, A&A, **538**, 81
[32] Lee J.-J., Raymond, J. C., Park, S., Blair W. P., Ghavamian P., Winkler P. F., Korreck K., 2010, ApJ, **715**, 146

Particle acceleration and non-thermal emission in Pulsar Wind Nebulae from relativistic MHD simulations

B Olmi[1,2,3], **L Del Zanna**[1,2,3], **E Amato**[2], **N Bucciantini**[2,3] **and R Bandiera**[2]

[1] Dipartimento di Fisica e Astronomia, Università degli Studi di Firenze, Via G. Sansone 1, 50019 Sesto F. no (Firenze), Italy

[2] INAF - Osservatorio Astrofisico di Arcetri, Largo E. Fermi 5, 50125 Firenze, Italy

[3] INFN - Sezione di Firenze, Via G. Sansone 1, 50019 Sesto F. no (Firenze), Italy

E-mail: barbara.olmi@unifi.it

Abstract. Pulsar wind nebulae are among the most powerful particle accelerators in the Galaxy with acceleration efficiencies that reach up to 30% and maximum particle energies in the PeV range. In recent years relativistic axisymmetric MHD models have proven to be excellent tools for describing the physics of such objects, and particularly successful at explaining their high energy morphology, down to very fine details. Nevertheless, some important aspects of the physics of PWNe are still obscure: the mechanism(s) responsible for the acceleration of particles of all energies is (are) still unclear, and the origin of the lowest energy (radio emitting) particles is most mysterious. The correct interpretation of the origin of radio emitting particles is of fundamental importance, as this holds information about the amount of pair production in the pulsar magnetosphere, and hence on the role of pulsars as antimatter factories. On the other hand, the long lifetimes of these particles against synchrotron losses, allows them to travel far from their injection location, making their acceleration site difficult to constrain. As far as the highest energy (X and gamma-ray emitting) particles are concerned, their acceleration is commonly believed to occur at the pulsar wind termination shock. But since the upstream flow is thought to have non-uniform properties along the shock surface, important constraints on the acceleration mechanism(s) could come from exact knowledge of the location and flow properties where particles are being accelerated. We investigate in detail both topics by means of 2D numerical MHD simulations. Different assumptions on the origin of radio particles and more generally on the injection sites of all particles are considered, and the corresponding emission properties are computed. We discuss the physical constraints that can be inferred from comparison of the synthetic emission properties against multiwavelength observations of the PWN class prototype, the Crab Nebula.

1. Introduction

PWNe are a particular class of supernova remnants, born in the violent death of a massive star ($M \gtrsim 8\ M_{\odot}$). The nucleus of the progenitor star gives rise to a highly magnetised, rapidly rotating neutron star, often observed as a pulsar, that emits a magnetised relativistic outflow, the so called Pulsar Wind (PW). Effective confinement of the PW by the slowly expanding debris of the explosion induces the formation of a Termination Shock (TS), at which the flow is slowed down and heated. The Pulsar Wind Nebula arises downstream of this shock as non-thermal emission of the shocked material.

The overall picture can be compared with the solar wind scenario. Even if the nature of the wind is very different in the two cases, both the solar wind and the PW are slowed down at a termination shock, in order to reconnect with the physical conditions of the external medium. The solar wind is a thermally driven outflow, mainly composed of electrons and protons, escaped from the hot solar corona. The PW, instead, is electromagnetically driven and mainly made of electrons and positrons. These are created in the pulsar magnetosphere, where the electrons extracted from the star by the strong electric field present at its surface, interact with photons and develop a pair cascade. Particles and fields leave then the star vicinity in the form of an outflow whose structure is that of a Parker spiral. The region beyond the PW TS, that is the PWN, can be thought analogous to the Sun's helio-sheath. Its outer boundary is the PWN contact discontinuity, where the bubble of relativistic plasma meets the cold supernova ejecta. This is in turn analogous to the heliopause, where the heliospheric plasma meets the interstellar medium.

PWNe are interesting objects from several different points of view. The first obvious reason is that they enclose the most of the pulsar spin-down energy (in the sense that most of the MHD energy released in the pulsar spin down process is deposited in the nebula in the form of Poynting flux and energetic particles), and so they are the perfect place to look at in order to investigate open questions in pulsar physics.

Moreover pulsars have an important role as galactic antimatter factories, and the knowledge of the exact amount of pair production in their magnetospheres can be fundamental in understanding the cosmic ray positron excess measured by PAMELA and AMS02 [1, 2]. The fact that they are so close and bright makes them a perfect place for studying the physics of relativistic magnetized plasmas.

PWNe are also extremely interesting from the point if view of particle acceleration, together with Supernova remnants they are the most powerful accelerators in the Galaxy. While Supernova Remnants are thought to be the primary sources of hadronic cosmic rays in the Galaxy, in PWNe we have direct evidence of the presence of electrons with PeV energies (the highest energy observed in Cosmic rays of Galactic origin). The acceleration of these particles is likely associated with the presence of a magnetised relativistic shock, an environment where acceleration is most difficult to understand, and the acceleration mechanism at work still represents a deep mystery.

2. PWN observations

PWNe show common observational features: a center-filled emission morphology, flat radio spectra and very broad band non thermal emission, usually extending from radio frequencies to X-rays or even γ-rays [3].

The prototype of the class is the Crab Nebula, one of the best studied objects in the sky, and the one from which we have learnt most of what we know about PWNe [4]. The Crab Nebula shows a very broad band spectrum, with primary emission extending from radio frequencies to $\nu \sim 10^{22}$ Hz, due to synchrotron radiation produced by relativistic particles interacting with the intense ($\sim 100 B_{\rm ISM}$) and highly ordered magnetic field of the nebula. At higher energies (up to $\nu \sim 10^{28}$ Hz), the emission is produced via the Inverse Compton scattering process, between electrons and different target photon fields. The most important of these are photons produced by synchrotron emission, photons from the Cosmic Microwave Background and thermal photons emitted by warm dust. Dust, in the form of silicates and graphite at $T \simeq 40-50$ K, is a common left over of SN explosions, and in the Crab Nebula its thermal emission produces a 'bump' in the spectrum at infrared frequencies.

In the last decade, the *Chandra* X-ray observatory has allowed us to observe the high energy emission properties of PWNe with great details. A very puzzling jet-torus morphology was seen in the Crab Nebula and some other PWNe. The presence of polar jets was the most intriguing feature, since they appeared to originate so close to the pulsar location that collimation upstream of the TS seemed to be required. The best candidate mechanism for jet collimation in PWNe appeared to be magnetic collimation, but this can be shown to be very inefficient in ultrarelativistic flows, such as the flow upstream of the TS is.

Thanks to *Chandra* also many different bright and variable features were found, as *knots* and *rings* in the inner region of the nebula [5].

3. The Pulsar Wind

PWNe are powered by the spin-down of young neutron stars that lose their energy in the form of a wind. From pulsar models we know that the wind must be mainly made of electron-positron pairs and magnetically dominated in the close magnetosphere, within the light cylinder of the star. Thus, defining the wind magnetization as the ratio between the Poynting flux and the ram pressure of the wind

$$\sigma = \frac{B^2}{4\pi n m_e \Gamma^2 c^2} \tag{1}$$

with n is the comoving particle density, m_e the electron mass, Γ the wind Lorentz factor and c the speed of light, we expect that $\sigma \gg 1$ for $r \sim r_{\mathrm{LC}}$.

Beyond the light cylinder the streamlines of the outflow become asymptotically radial and the magnetic field becomes predominantly toroidal. This structure is described by the analytical split monopole model [6], which shows an excellent agreement with numerical simulations [7]. The energy flux has an anisotropic distribution in the wind, with $F \propto \sin^2\theta$, where θ is the polar angle, measured from the pulsar rotation axis. In the equatorial plane the split monopole solution implies a sharp change of polarity in the magnetic field, and thus the presence of a current sheet. This current layer would be infinitely thin in the case of an aligned rotator (magnetic axis parallel to rotation axis), but in general pulsars are not aligned, and the wind structure is more complicated. The inclination between the rotational and magnetic axis of the pulsar leads to a 'ballerina skirt' structure: the current sheet oscillates around the equatorial plane within a region of angular extent equal to the inclination angle. Neighbouring stripes of opposite polarities offer the perfect location for magnetic dissipation to occur [8], and hence σ is likely to decrease with increasing distance from the pulsar in some range of latitudes around the equator.

Indeed, a long standing problem in PWN physics is the so called *σ problem*. As we previously mentioned, near the light cylinder the wind is expected to be magnetically dominated ($\sigma \gg 1$). However, effective deceleration of the flow at the TS is only possible if its magnetisation is not too high. In particular, steady state MHD models in 1 or 2 D [9, 10], pointed to $\sigma \sim v_N/c$ at the TS, with $v_N \approx 10^3 \mathrm{km}/s$ the expansion velocity of the nebula. This requirement has somewhat been revised and made less severe by later multi-D MHD simulations [11, 12], but considerable dissipation still appears to be required.

As a distinct contrast there is no *σ problem* in solar wind physics. Since the solar wind is thermally driven, its magnetic energy is nowhere dominant, contrary to what happens at the very basis of the Pulsar Wind.

The split monopole wind structure and its use in multi-D MHD simulations provided a straightforward solution to another puzzle in PWN physics that we already mentioned, namely the jet-torus morphology of the high energy emission from these objects. The anisotropic distribution of the wind energy flux produces an oblate TS, more distant from the pulsar at the equator than along the polar axis, where cusps develop. Therefore, the polar jet, in spite of being seen from very close to the pulsar, originates in fact in the downstream region. Here, as shown in many different simulations [13, 11, 14], magnetic hoop stresses can lead to effective collimation. Even if $\sigma < 1$, compression of the toroidal magnetic field downstream of the TS can bring the field up to the equipartition value. When this happens, the magnetic tension diverts the flow towards the polar axis at intermediate latitudes, collimating the polar outflow into a jet. Of course this process only works if σ is sufficiently large: for the parameters appropriate to describe the crab Nebula, 2D simulations indicate $\sigma > 0.01$.

State of the art 3D simulations seem to indicate that in the Crab Nebula the wind magnetization might be σ = a few, with considerable dissipation taking place downstream of the shock. These findings are however preliminary. 3D simulations only extend to few hundred years after the explosion and extrapolation of their results to the actual age of the nebula seem to indicate that the final strength of the nebular magnetic field might be too low and incompatible with the emission spectrum, suggesting that even higher values of the wind σ might be required. The bottom line is that in order to explain the presence of a jet with the observed properties the pulsar wind magnetization must be likely much larger than 0.001 for most of the flow. As we will discuss later, this lower limit on σ is problematic from the point of view of understanding particle acceleration at the termination shock.

Journal of Physics: Conference Series **642** (2015) 012019 doi:10.1088/1742-6596/642/1/012019

4. Numerical model overview

Many of the open questions in PWNe and pulsar physics appear to be strongly connected with the PW physical properties. Since the wind is cold at its basis (i.e. the particle kinetic energy is much larger than the thermal energy), no radiation is observed to come from the upstream region, that appears to the observer as an under luminous area surrounding the pulsar. As a consequence we do not have the possibility to inspect it directly. What we can do is to use the PWN emission as a diagnostic of the PW properties. To this aim a powerful tool is provided by MHD simulations coupled with non-thermal radiation modelling. One can simulate the nebular dynamics arising from different assumptions on the wind and then compute synthetic emission maps to compare with observations.

We have performed 2D MHD axisymmetric simulations with the shock capturing code ECHO [15]. The physical domain ranges from $r_{min} = 0.05$ ly to $r_{max} = 10$ ly in the radial direction, and from 0 to π in the polar one, with a suitable numerical resolution to balance the computational costs and the needed accuracy (for more details see [16, 17]). We take $v_\phi = B_r = B_\theta \equiv 0$ and the equation of state of an ideal gas with $\gamma = 4/3$, appropriate for the ultra relativistic case.

The initial conditions are chosen in order to best match the Crab Nebula's properties. The PW is injected from the inner boundary of the simulation and expands outwards, impacting the slowly expanding shell of supernova ejecta and giving rise to the relativistically hot and magnetized PWN as the system evolves.

The PW is set up according to the prescriptions of the split monopole model including a 'striped' region (namely a region in which magnetic lines of alternating polarity reconnect and the magnetisation decreases: see [16, 17] for details). In particular we assume an anisotropic energy flux for the PW, which roughly depends on space as predicted by the split monopole model:

$$F(r,\theta) = F_0(r_0/r)^2 f(\theta), \quad f(\theta) = \frac{\alpha + (1-\alpha)\sin^2\theta}{1-(1-\alpha)/3}, \tag{2}$$

where F_0 is a constant referring to the arbitrary distance r_0 from the pulsar (set as 1 ly). The level of anisotropy is governed by the parameter α, which represents the ratio between the energy flux along the polar axis and in the equatorial plane, that must be such that $F(r,\pi/2) \gg F(r,0)$, and thus $\alpha \ll 1$. The magnetic field, taken to be purely toroidal and $\propto r^{-1}$, has a sinusoidal dependence on the polar angle. The width of the striped region is governed by the parameter b: if $b \gg 1$ the striped region disappears, while its extent increases with decreasing b. The magnitude of the field is defined by the initial magnetization at the equator (σ_0).

The three parameters (α, b, σ_0) represent the set of free parameters of the PW model, that must be fixed comparing the computed emission properties with observations.

The broadband Crab Nebula's spectrum can be reproduced only if the contribution of two families of emitting electrons is considered: one responsible for the low energy emission (radio particles) and one for the high energy emission (optical/X-ray particles). We continuously inject particles at the TS with power law spectra for each family, $N_f(E) = K_f E^{-\gamma_f}$. The evolved distribution functions at any place in the nebula are determined by the conservation of particle number along streamlines and by the adiabatic and synchrotron losses that particles experience along their trajectory. The normalization constant K_f and the spectral index γ_f is different for radio and X-ray emitting particles. All the parameters are determined from comparison with the data (see [16, 17] for more details).

In order to compute the non-thermal emission of the nebula, 3 additional tracers have been added to the set of quantities that the code evolves [14, 16, 17]: the maximum particle energy as a function of position, ϵ_∞, namely the energy at a given location that corresponds to an infinite injection energy, the number density of non-thermal particles n and its value immediately downstream of the TS n_0. Once these are known at every point in space, the emission is computed through the standard synchrotron formulae.

5. Simulation results

The flow dynamics corresponding to the best fitting set of PW parameters (σ_0, b, α) [14, 16] allows us to reproduce the oblate shape of the TS, the observed jet-torus morphology and the correct dimensions of the nebula at its present age. In particular we assume an anisotropy parameter of $\alpha = 0.1$, thus the energy flux along the polar axis will be ten times less than at the equator. Larger contrasts are also possible in principle, but we could not investigate them with our current grid: for smaller values of α the TS tends to be very close to the pulsar along the polar axis and it easily ends up crossing the boundary of the region in which the wind is constantly reset; this fact can lead to unphysical results. Larger values of α were instead investigated in the past [11] and found unsuitable to reproduce the expected final dimensions of the nebula and the oblate shape of the TS.

However the imposed axial symmetry leads to excessive compression of the magnetic field around the polar axis, while far from the pole and from the TS the computed field is found to be well below the expected averaged value inferred from observations ($B \simeq 200\,\mu$G). Effects of this behavior are amply discussed in [16].

Our model also compares well with observations of the high energy (optical and X-ray) emission, as discussed in [14, 18].

5.1. On the origin of low energy emitting particles

So far, MHD modelling of PWNe has mainly been concerned with the properties of high energy emission, while ignoring the low energy emission. However knowledge of the origin of low energy emitting particles is of fundamental importance in the attempt to answer to some of the open questions in pulsar physics (such as the role of pulsars as antimatter factories in the Galaxy, the energetic relevance of the hadronic component of the pulsar wind and the acceleration mechanisms working at the TS [19, 20]). In fact, knowing whether radio emitting particles are fossil from previous epochs, or rather currently injected in the nebula as part of the pulsar wind, would allow one to constrain the wind Lorentz factor and the pulsar multiplicity κ, defined as the number of pairs produced in the magnetosphere for each primary electron extracted from the pulsar surface [21]. The question to assess is whether radio particles are part of the pulsar outflow, and accelerated at the TS, or if they are of different origin and accelerated somewhere else.

In the first case the inferred pair multiplicities would be generally larger than current pulsar theories are able to explain [22]: in the case of the Crab Nebula one would obtain $\kappa \sim 10^6$ for the pulsar multiplicity and $\gamma \sim 10^4$ for the wind Lorentz factor, averaged over the solid angle. On the other hand, if radio particles were of different origin, the values of these two parameters would be those inferred from high energy emission alone, namely: $\kappa \sim 10^4$ and $\Gamma \sim 10^6$ [9, 23, 21].

In [16] we studied the problem of the origin of radio particles, considering three different hypotheses. In the first scenario (case A), radio particles are considered as part of the pulsar outflow, continuously injected and accelerated at the TS with the higher energy ones. The second possibility is that they are produced somewhere else in the nebula, for instance in the thermal filaments, and then accelerated by interaction with local turbulence (case B) [24]. Finally, since radio particles are characterized by large number and long life-times against synchrotron losses, they could be of relic origin, born in a primordial outburst of the pulsar soon after the supernova explosion [25] (case C).

The synthetic emission properties at radio frequencies, in the case of the Crab Nebula, are then compared with available data. From observations we expect to find rather uniform radio emission on large spatial scales, with an ellipsoidal morphology [26]. At the same time, we also expect to find variable structures in the inner region (analogous to the optical 'wisps' [27]), changing in brightness on both short ($\Delta t \simeq 2$ months) and long ($\Delta t \simeq 3$ yr) time scales, and moving outward with mildly relativistic velocities ($v \lesssim 0.3c$).

In Fig. 1 emission maps at 1.4 GHz, for case A and B, are shown. In the left panel we show emission maps with small scale features subtracted, as in [26], for a direct comparison with that work: here the expected ellipsoidal morphology is well reproduced by the strong cylindrical symmetry given by the

Journal of Physics: Conference Series **642** (2015) 012019 doi:10.1088/1742-6596/642/1/012019

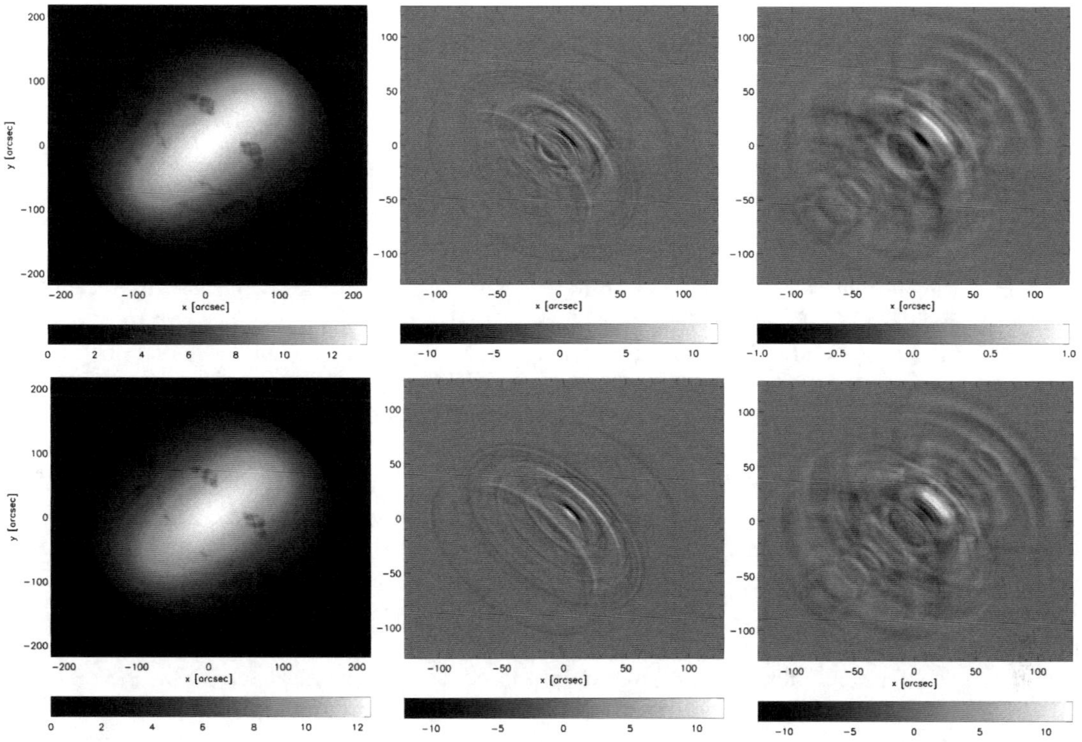

Figure 1. Radio surface brightness maps at 1.4 GHz, in units of mJy/arcsec^2. In the upper row maps for case (A) are shown, while case (B) is in the bottom. In the left panel, emission maps with the small scales subtracted are shown for both cases. In the central and rightmost panels maps of short and long time scales wisp variability are presented.

magnetic field compression around the polar axis.

In the central and right panels we show images obtained by subtracting two maps computed for epochs that are, respectively, 2 months and 3 yr apart. The maps computed under the two assumptions are basically identical. Ring-like structures, very similar to those observed at optical and X-ray frequencies, are clearly seen also at radio wavelengths, in agreement with what observed by [28] in terms of periodicity, locations and outward velocities of the so called 'radio wisps'.

The fact that results for case A and B are almost indistinguishable does not allow us to put strong constraints on the origin of radio particles. On the other hand, the important result we found is that wisps appear with the same properties under the two hypotheses. We can thus conclude that their appearance does not bear any implication on the injection site of the emitting particles, but only on the underlying flow structure: wisps simply trace the local flow and magnetic field properties, as they only arise as a combined effect of Doppler boosting and magnetic field enhancements.

The last case we considered is case C. Here we inject radio particles only at the beginning of the simulation, up to $t = t_b \ll t_{\rm PWN}$, while X-ray and optical particles are treated as in the former cases. This scenario is meant to mimic some sort of primordial outburst in which a very young pulsar would inject a wind with different properties from the current ones, namely higher multiplicity, lower Lorentz factor. What we found is that the resulting emission morphology does not compare well with observations: radio emission would mainly come from the outer region of the nebula (low energy particles advected with the flow would be confined by now to the outskirts of the nebula), while the central region would be much darker than observed. Due to the lack of radio particles in the central region of the nebula, where the magnetic field is highest and the flow is most dynamic and with the fastest speed, no wisp-like features

Journal of Physics: Conference Series **642** (2015) 012019 doi:10.1088/1742-6596/642/1/012019

would be observed in this waveband. In summary, case C can be easily rejected, allowing us to conclude that radio particles cannot be described as a relic population unless a re-acceleration (and re-distribution) mechanism in the nebula is considered for $t > t_b$.

5.2. Investigating the particle acceleration mechanism

In the previous analysis we showed that within a MHD description of the flow wisps are easily accounted for as a result of regions of the nebula in which the magnetic field is strong and flow speed is high and pointing towards the observer. Wisps are seen at radio, optical and X-ray frequencies, and they are not coincident at the different wavelengths, with differences measured both in term of spatial locations and of outward velocities [27, 29]. Within the framework of MHD, the only way to account for the different properties of the wisps at different frequencies is to assume some difference in the spatial distribution of the particles that are responsible for emission in the different wavebands. An obvious difference between particles of different energies (and hence emitting in different wavebands) is the role of energy losses. However, these are not terribly important in any case in those innermost regions of the nebula, from which wisps are observed. Any other difference in the spatial distribution of the particles of different energies suggests differences in the acceleration sites. In other words, if wisps are different at radio and X-ray frequencies, the particles responsible for radio emission must have had a different history than the particles responsible for X-ray emission and the simplest explanation is that they were accelerated in different locations. Here we shall discuss how different injection scenarios reflect on different observable features of the wisps.

Essentially two acceleration mechanisms are proposed to be at work at the TS in the case of the Crab Nebula [21]: Fermi I acceleration (particles gain energy through multiple crossings of the shock front) or driven magnetic reconnection (particles are accelerated by the electric fields associated with magnetic reconnection regions). Typical spectral indices resulting from these two different processes are suitable to describe the distribution of X-ray and radio emitting particles respectively. that can account, respectively, for the high energy and radio spectral indices. The viability of the two mechanisms requires different physical conditions: Fermi I acceleration can be operative at relativistic shocks only where the magnetization is low enough (namely $\sigma < 0.001$) [30, 19], while driven magnetic reconnection requires extremely large multiplicities [31, 32, 19].

In [17] we performed a multiwavelength analysis of the wisps, considering different possibilities for the injection of particles responsible for radio, optical and X-ray wisps. Namely we define three injection scenarios: in case (1) particles are injected uniformly at the TS surface; in case (2) we define a wide equatorial zone, with $\theta \in [20^\circ, 90^\circ]$ and a narrow polar one ($\theta \in [0^\circ, 20^\circ]$); in case (3) we define a narrow equatorial zone, corresponding to the striped region of the wind ($\theta \in [70^\circ, 90^\circ]$), and a wide polar one ($\theta \in [0^\circ, 70^\circ]$). The above angular extents are expressed in the upper hemisphere, but symmetry around the equator is implicit. All the choices are considered for all three ranges of particle energies (radio, optical and X-ray emitting particles), namely, particles in each of those energy ranges can be injected in five different regions: wide or narrow polar cap, wide or narrow equatorial belt, and along the entire shock surface.

Following the analysis performed by [29], wisp profiles are extracted from synthetic intensity maps, during a period of 10 years at the end of the simulation ($t \simeq 950$ yr), and with monthly frequency. Each map is convolved with the appropriate instrumental PSFs, and intensity peaks are extracted from a 3 arcsec wide slice in the upper hemisphere, centered on the polar axis (as shown in Fig. 2). The PSFs employed are those of the instruments used for the observations considered for comparison in each frequency band: in particular we refer to the data analysed in [29, 33, 27]. The relevant observations were obtained with Chandra for the X-rays (with a FWHM=0.5″), Nord Optical Telescope (NOT) for the optical (FWHM=0.75″, as determined based on the average seeing of the period of reference), and VLA for radio frequencies (with a FWHM=1.8″)). For ease of comparison with the data in [29], only peaks with $I \geq I_{max}/3$ are taken into account, with I_{max} the maximum intensity for each map. Results are shown as plots of the radial distance of the local intensity maxima from the pulsar as a function of time

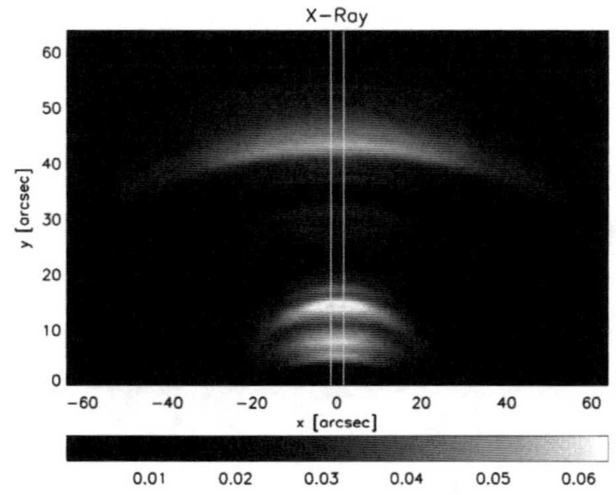

Figure 2. Upper hemisphere of an X-ray intensity map (1 keV) at t = 950 yr. Color scale is linear and in units of mJy/arcsec^2. Vertical lines near the polar axis define the slice used in the analysis of wisp profiles.

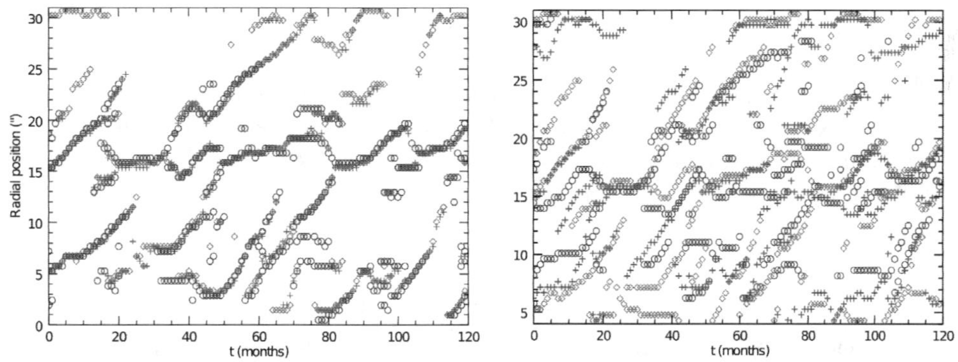

Figure 3. Radial positions of the local intensity maxima (in arcseconds) as a function of time (in months) with red diamonds identifying radio wisps (ν_r = 5 GHz), green crosses optical ones (ν_o = 3.75 × 10^{14} Hz) and blue circles for X-rays (1 keV). On the left case (1) is shown. On the right case (3) is shown, with X-ray particles injected in the equatorial zone and radio ones injected in the complementary sector.

(in months).

Wisps behavior at radio, optical and X-ray wavelengths in the case of uniform injection (case (1)) is shown in the left panel of Fig. 3. As expected wisps are almost coincident at the different energies, with differences only due to the effect of radiation losses.

When the injection location of the different particles' populations is not coincident, the wisps are not coincident anymore at the different wavelengths. Based on our findings, in order to reproduce the X-ray wisps, which are never seen at distances from the pulsar shorter than ~ 6 arcsec, the best scenario is the one in which high energy particles are injected in the narrow equatorial region of case (3) (results are shown in the right panel of Fig. 3). On the other hand, deriving strong constraints on radio particles is much more complicated: again no relevant differences are seen between uniform injection and injection in a wide sector in terms of emission properties. The only scenario that seems to be disfavoured is one in which radio particles are injected in a narrow polar cone: in this case, the variability appears to be strongly reduced.

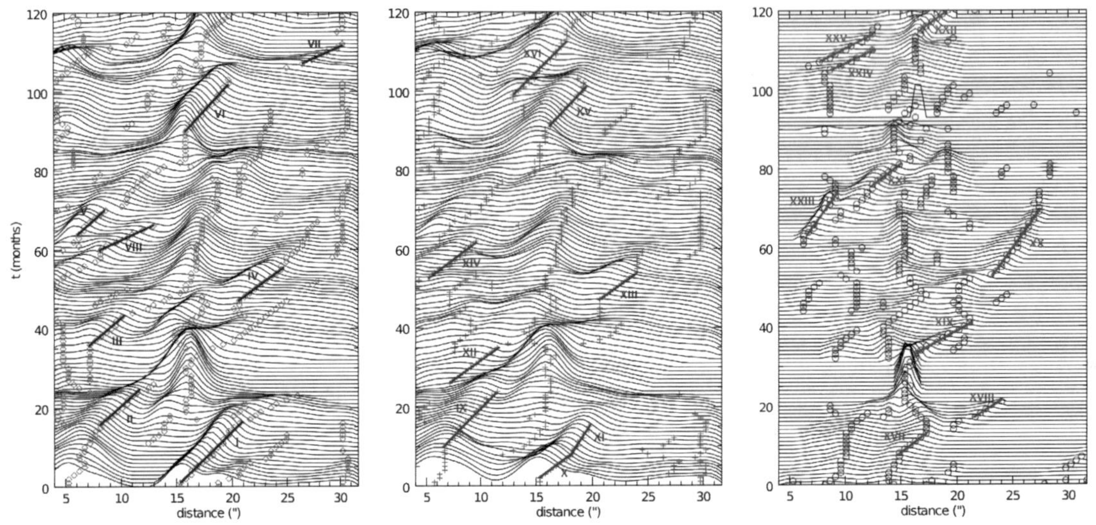

Figure 4. Stack plots with radial profiles of radiation intensity in the radio band (left panel), optical band (middle panel) and X-ray band (right panel). Results refer to case (3), with injection of X-ray particles in the narrow equatorial belt and of radio particles in the complementary wide polar region. Optical emission is contributed by both populations. Lines and Roman numerals are used to highlight the prominent structures used for evaluation of the wisp velocity.

We also computed the outward velocities of the wisps, following the time evolution of the most prominent peaks (an example is shown in Fig. 4). If the prescriptions corresponding to case (3) are followed, with radio particles injected in the wide polar cone and X-ray ones in the narrow equatorial belt, our results are in good agreement with observations: the resulting deprojected velocity is in the range $0.08c \leq v \leq 0.38c$ (to be compared with $0.1c \leq v \leq 0.4c$ from [27, 29]).

6. Summary and conclusions

Pulsar Wind Nebulae can be thought of as a relativistic version of the heliosheath, in which the solar wind is replaced by a magnetised, relativistically hot plasma. Relativistic axisymmetric MHD simulations provide a powerful tool for the investigation of the physics of this plasma, and possibly the best tool to investigate the workings of the underlying stellar source, a fast spinning, highly magnetised neutron star. While many open questions remain, the advent of relativistic MHD simulations has allowed us to answer a number of problems in recent years, and has opened a promising way to answer more.

The MHD interpretation has proven very successful at explaining the high energy morphology of PWNe: within this framework the initially puzzling jet-torus structures observed in the X-ray emission of a number of nebulae have found a straightforward interpretation as the result of the pulsar wind structure, and in particular of the anisotropy of its energy flux.

Nevertheless, this powerful tool had never been used before to investigate lower energy emission, in spite of the fact that clarifying the origin of radio particles is fundamental in order to answer many open questions in pulsar physics. We made a first attempt at a detailed study of the low energy emission properties of PWNe in the framework of 2D MHD. Our main conclusion is that the global radio emission is basically insensitive to the spatial distribution of the particles, as long as they are not pure relics. Whether they are currently part of the pulsar outflow or uniformly distributed in the nebula, the surface brightness maps are almost identical, and also the variability of the inner regions ('radio wisps') is very similar, which proves that wisps, within the MHD framework, naturally arise from the properties of the flow.

Digging more deeply, we performed a multiwavelengths analysis of wisp properties, with the aim of constraining the history, and in particular the acceleration sites, of the emitting particles. The motivation came from observations of the nebula at different frequencies, which showed that wisps are different at radio, optical and X-rays frequencies. In an MHD description of the flow, where wisps arise as a result of the underlying flow properties, this cannot be explained, unless particles responsible for the different contributions have different acceleration sites. We considered different scenarios for the injection of particles in different energy ranges and found that properties of X-ray wisps are best reproduced if injection in a narrow belt around the pulsar rotational equator is considered. The equatorial region is presumably where the flow magnetisation is lower, if effective magnetic dissipation takes place in the striped wind. Then the mechanism responsible for the acceleration of these highest energy particles could be Fermi I, which has been proven to be very effective at relativistic shocks of low enough magnetisation [30, 19], and which naturally provide a particle spectral index close to that inferred from X-ray observations.

Constraining the radio emission properties is much more complicated: the only scenario that radio observations directly exclude is one in which low energy particles are injected in a narrow polar cone, since almost no wisps are found. But emission properties in the case of uniform injection, or injection in a wide polar or equatorial band, are basically indistinguishable.

Based on our findings a possible scenario for the particle acceleration should be the one in which high energy particles are injected in a narrow equatorial cone, where the magnetic dissipation is strong enough to allow Fermi I processes to be viable, while radio particles are injected in a wider region, where physical conditions are such that driven magnetic reconnection would be operative.

References

[1] PAMELA coll 2009 *Nature* **458** 607–609 (*Preprint* 0810.4995)
[2] AMS-02 coll 2013 *Physical Review Letters* **110** 141102
[3] Gaensler B M and Slane P O 2006 *ARA&A* **44** 17–47 (*Preprint* arXiv:astro-ph/0601081)
[4] Hester J J 2008 *ARA&A* **46** 127–155
[5] Weisskopf M C, Hester J J, Tennant A F, Elsner R F, Schulz N S, Marshall H L, Karovska M, Nichols J S, Swartz D A, Kolodziejczak J J and O'Dell S L 2000 *ApJLett* **536** L81–L84 (*Preprint* arXiv:astro-ph/0003216)
[6] Michel F C 1971 *Comments on Astrophysics and Space Physics* **3** 80
[7] Spitkovsky A 2006 *ApJLett* **648** L51–L54 (*Preprint* astro-ph/0603147)
[8] Coroniti F V 1990 *ApJ* **349** 538–545
[9] Kennel C F and Coroniti F V 1984 *ApJ* **283** 694–709
[10] Begelman M C and Li Z Y 1992 *ApJ* **397** 187–195
[11] Del Zanna L, Amato E and Bucciantini N 2004 *A&A* **421** 1063–1073 (*Preprint* astro-ph/0404355)
[12] Porth O, Komissarov S S and Keppens R 2014 *MNRAS* **438** 278–306 (*Preprint* 1310.2531)
[13] Komissarov S S 2004 *MNRAS* **350** 427–448 (*Preprint* astro-ph/0402403)
[14] Del Zanna L, Volpi D, Amato E and Bucciantini N 2006 *A&A* **453** 621–633 (*Preprint* astro-ph/0603080)
[15] Del Zanna L, Zanotti O, Bucciantini N and Londrillo P 2007 *A&A* **473** 11–30 (*Preprint* 0704.3206)
[16] Olmi B, Del Zanna L, Amato E, Bandiera R and Bucciantini N 2014 *MNRAS* **438** 1518–1525 (*Preprint* 1310.8496)
[17] Olmi B, Del Zanna L, Amato E and Bucciantini N 2015 *MNRAS* **449** 3149–3159 (*Preprint* 1502.06394)
[18] Volpi D, Del Zanna L, Amato E and Bucciantini N 2008 *A&A* **485** 337–349 (*Preprint* arXiv:0804.1323)
[19] Sironi L and Spitkovsky A 2011 *ApJ* **741** 39 (*Preprint* 1107.0977)
[20] Arons J 2012 *Space Sci. Rev.* **173** 341–367 (*Preprint* 1208.5787)
[21] Amato E 2014 *International Journal of Modern Physics Conference Series* **28** 1460160 (*Preprint* 1312.5945)
[22] Bucciantini N, Arons J and Amato E 2011 *MNRAS* **410** 381–398
[23] Gaensler B M, Arons J, Kaspi V M, Pivovaroff M J, Kawai N and Tamura K 2002 *ApJ* **569** 878–893 (*Preprint* arXiv:astro-ph/0110454)
[24] Komissarov S S 2013 *MNRAS* **428** 2459–2466 (*Preprint* 1207.3192)
[25] Atoyan A M and Aharonian F A 1996 *MNRAS* **278** 525–541
[26] Bandiera R, Neri R and Cesaroni R 2002 *A&A* **386** 1044–1054 (*Preprint* arXiv:astro-ph/0112459)

[27] Bietenholz M F, Hester J J, Frail D A and Bartel N 2004 *ApJ* **615** 794–804 (*Preprint* arXiv:astro-ph/0408061)
[28] Bietenholz M F, Frail D A and Hester J J 2001 *ApJ* **560** 254–260 (*Preprint* astro-ph/0106339)
[29] Schweizer T, Bucciantini N, Idec W, Nilsson K, Tennant A, Weisskopf M C and Zanin R 2013 *MNRAS* **433** 3325–3335 (*Preprint* 1301.1321)
[30] Spitkovsky A 2008 *ApJLett* **682** L5–L8 (*Preprint* 0802.3216)
[31] Lyubarsky Y E 2003 *MNRAS* **345** 153–160 (*Preprint* astro-ph/0306435)
[32] Lyubarsky Y and Liverts M 2008 *ApJ* **682** 1436–1442 (*Preprint* 0805.0085)
[33] Hester J J, Mori K, Burrows D, Gallagher J S, Graham J R, Halverson M, Kader A, Michel F C and Scowen P 2002 *ApJLett* **577** L49–L52

Evidence for superdiffusive shock acceleration at interplanetary shock waves

Silvia Perri and Gaetano Zimbardo

Dipartimento di Fisica, Università della Calabria, Ponte P. Bucci, I-87036 Rende, Italy

E-mail: silvia.perri@fis.unical.it

Abstract. Recent analysis of time profiles of energetic particles accelerated at interplanetary shocks has shown evidence for superdiffusive transport upstream of the shock fronts, namely for a transport characterized by a particle mean square displacement that grows faster than linearly in time. While for normal, diffusive transport exponential particle time profiles are predicted, a large number of interplanetary shock events, included the termination shock of the solar wind, exhibits energetic particle time profiles that upstream decay as power laws. This power law trend has been derived in the framework of particle superdiffusion. The standard theory of diffusive shock acceleration has been further extended to the case of particle superdiffusive transport (superdiffusive shock acceleration), allowing for the derivation of harder energy spectral indices both for relativistic and non-relativistic particles. Here we test this theory for a couple of interplanetary shock waves that accelerate protons and that have been observed by the ACE spacecraft. We show that power law particle time profiles upstream of the shocks are common and clearly indicate superdiffusive transport. The particle energy spectra in some events are in a very good agreement with the superdiffusive shock acceleration prediction.

1. Introduction

Energetic particles are observed throughout the heliosphere and they originate from various sources: near the boundary of the solar system, the termination shock of the solar wind is a source of energetic particles [1, 2]; particles are observed to be accelerated in planetary magnetospheres [3, 4, 5], at interplanetary traveling shocks [6], and at the forward and reverse shock pairs of corotating interaction regions [7]; violent eruptive events in the solar atmosphere produce solar energetic particles (SEPs) [8, 9]. The mechanism of particle acceleration that has been invoked for more than 30 years is diffusive shock acceleration (DSA), which explained many observations. However, some observations cannot be explained in the classical framework of DSA: hard particle spectra inferred from radio emissions in supernova remnants cannot be reproduced by DSA [10]; particle time profiles at interplanetary shocks do not always follow the exponential profile upstream, in the case of a spatially constant diffusion coefficient D, and a flat profile downstream predicted by DSA [11, 12]; the recent Voyager 2 termination shock crossing has revealed that the spectral slope of energetic ions is compatible with the DSA prediction assuming a compression ratio (i.e., the ratio between the upstream and the downstream plasma speeds) $r = 3$, which is greater than the one observed, $r = 2$, from the spacecraft [2]. This might indicate that we need alternative physical mechanisms to explain particle acceleration. DSA is based on normal (Gaussian) spatial diffusion, but spacecraft observations suggest that the transport of SEPs can vary from normal to scatter-free (ballistic) [13] and the proton parallel mean free path at 1 AU is estimated to range from 0.1 AU [14] to more than 1 AU [15, 16]. Thus, besides normal diffusion, anomalous transport regimes can occur. Anomalous transport regimes are characterized by a mean square displacement for particles $\langle \Delta r^2 \rangle = 2D_\alpha t^\alpha$ with $\alpha \neq 1$. D_α is the anomalous diffusion coefficient having dimensions $[D_\alpha] = L^2/t^\alpha$. For processes slower than normal diffusion $\alpha < 1$ (subdiffusion), for processes faster than normal diffusion $\alpha > 1$ (superdiffusion). Systems exhibiting subdiffusion are characterized by trapping regions, while superdiffusive processes show random walk trajectories characterized by very long displacements. For ballistic transport, $\alpha = 2$. Recently, Refs. [17, 18] using non Gaussian, Lévy, statistics have derived a prediction for particle time profiles far upstream of a shock wave in the case of superdiffusive transport. At variance with DSA, a power law time profile $j(t) \propto t^{-\beta}$ with $\beta < 1$ has been obtained. This has been tested for electrons accelerated at several corotating interaction region shocks [17, 18, 19], for ions accelerated at the termination shock of the solar wind [20], and in a case study of protons accelerated at a coronal mass ejection driven shock [21]. The power law decay fits much better the observed time profiles than the exponential decay predicted by DSA, indicating superdiffusive transport upstream of the shocks.

Further, Refs. [22, 23] have extended the first order Fermi acceleration process to include the case where particles accelerated at shocks propagate superdiffusively. They have developed a new theoretical framework called superdiffusive shock acceleration (SSA), which provides predictions for both the particle energy spectra and the acceleration times (see details in [22, 24]).

In this paper, we will first summarize the main aspects of SSA, then we will test the energy spectra predictions against a couple of interplanetary shocks observed by the *Advanced Composition Explorer* (ACE) spacecraft. In the last Section we will discuss the main results.

2. Theoretical framework

Let an infinite planar shock be the source of accelerated particles, so that the geometry of the system is one dimensional and only variations along the coordinate perpendicular to the shock front (namely, x) are allowed. In such a system, it is possible to reconstruct a particle density profile [17, 18] as

$$n(x, E, t) = \int P(x - x', t - t') S_{\rm sh}(x', E, t') dx' dt', \tag{1}$$

where the source term $S_{\rm sh}(x', E, t') = \Phi_0(E)\delta(x' - V_{\rm sh}t')$ depends on the flux of particles in a given energy channel at the shock, Φ_0, and $V_{\rm sh}$ is the shock speed in the observer frame. $P(x - x', t - t')$ is the probability of finding a particle at the position x at time t if injected at x' at time t' and is called the *propagator* of the process. In the case of a Gaussian propagator, the classical upstream exponential decay predicted by DSA for the density of particles accelerated at a planar shock is recovered, i.e., $n(x, E) \propto \exp(-V|x|/D)$, assuming a spatially constant diffusion coefficient D. However, if we assume anomalous particle propagation, in particular superdiffusive propagation, the one dimensional propagator scales as [25]

$$P(x - x', t - t') = A \frac{t - t'}{(x - x')^\mu}, \tag{2}$$

with $2 < \mu < 3$. This form of the propagator can be derived in the limit of $|x - x'| >> |t - t'|^{1/\mu}$, that is far from the source. Refs. [17, 18] have explicitly derived the particle density under the hypothesis of superdiffusive transport for particles accelerated at a shock wave, that is

$$n(x, E) \propto |x|^{-\beta}, \tag{3}$$

being the exponent β of the power law in Eq. (3) related to the exponent of the superdiffusive particle propagator, $\beta = \mu - 2$. Since superdiffusion is obtained when $2 < \mu < 3$, this implies

$0 < \beta < 1$. Furthermore, a direct relation between μ and the exponent of the mean square displacement α can be obtained, that is $\alpha = 4 - \mu$ [25, 23]. Consequently, from the particle density profiles it is possible to estimate the increase of the mean square displacement using $\alpha = 2 - \beta$. As discussed above, the upstream decay of the particle time profiles has been studied in some shock events in interplanetary space (both for electrons and for protons), and for many of them the superdiffusive prediction reproduces the data much better than the diffusive exponential decay [17, 18, 20, 19, 21].

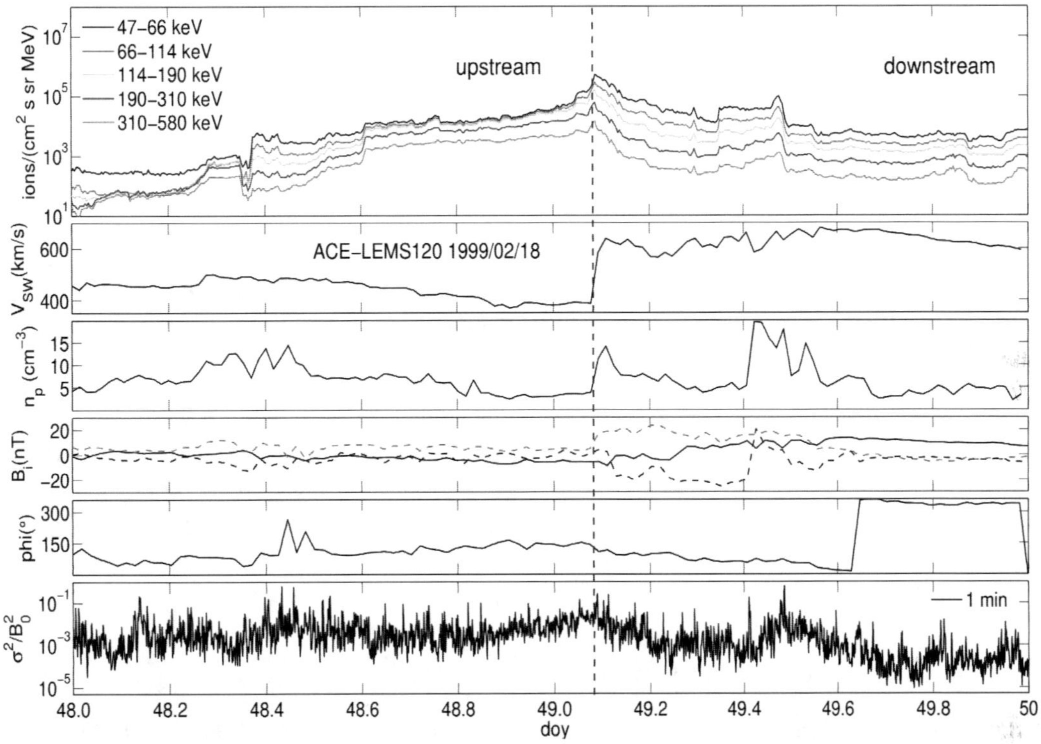

Figure 1. From top to bottom: 5 minute resolution ions fluxes displayed in a log-lin plot in five energy channels (see legend) during an interplanetary shock event detected by the ACE satellite on 1999/02/18; 10 min resolution solar wind bulk speed and proton density; magnetic field components; the azimuthal angle; total magnetic field normalized variance on 1 min time scale. The vertical dashed line indicates the shock time position. Data from http://cdaweb.gsfc.nasa.gov/

Refs. [22, 23] have included superdiffusion within a first order Fermi acceleration process, giving rise to the SSA theory. In particular, they derived the particle density at the shock n_0 using Eq. (1) assuming isotropy and stationary conditions. They computed the probability of escaping from the acceleration region $P_{\rm esc} = \Phi_2/\Phi_1 = n_2V_2/(n_0v/4)$, that is the ratio between the particle flux downstream of the shock and the particle flux from upstream to downstream [12] (Φ_2 depends on the density and on the plasma speed downstream, Φ_1 depends on the particle density at the shock and on the particle velocity). Hereafter, the index 1 indicates the unshocked, upstream region, while 2 refers to the shocked, downstream region. For relativistic particles $P_{\rm esc}$ is directly related to the exponent $\hat{\gamma}$ of the integral energy spectrum (i.e., $N(> E) = N_0(E/E_0)^{-\hat{\gamma}}$), so that determining the escape probability in the superdiffusive

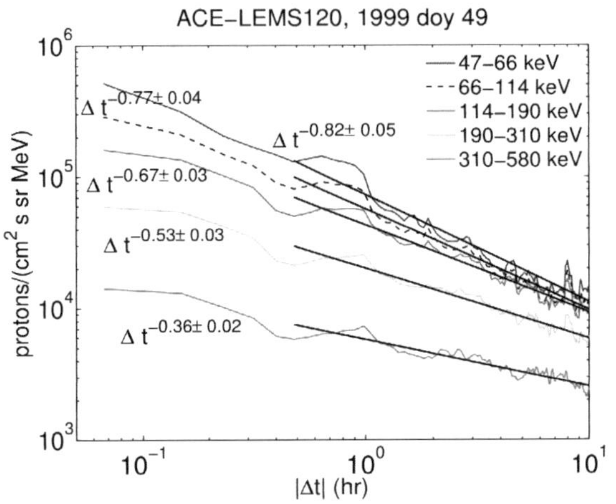

Figure 2. A log-log plot of ion time profiles as a function of the time (distance) from the shock. All the energy channels exhibit a power law trend for more than one decade. The time profiles tend to change in the vicinity of the shock. The best power law fits are indicated by the thick black solid lines. The power law slopes for each energy channel are displayed as well.

framework implies to derive the particle energy spectrum index corrected by the assumption of superdiffusive transport. In the case of non relativistic particles $P_{\rm esc}$ is related to the particle momentum spectrum, so that $\hat{\gamma_p} = P_{\rm esc}/(\Delta p/p)$. Thus, the spectral indices for the differential density predicted by DSA and by SSA [22, 23] can be summarized, respectively, as

$$\begin{aligned} \gamma_p^{\rm DSA} &= \hat{\gamma_p}^{\rm DSA} + 1 = \frac{r+2}{r-1} \\ \gamma_p^{\rm SSA} &= \hat{\gamma_p}^{\rm SSA} + 1 = \frac{6}{r-1}\frac{2-\alpha}{3-\alpha} + 1. \end{aligned} \tag{4}$$

The DSA prediction depends on the compression ratio of the shock $r = n_2/n_1$ only, while in the SSA framework the spectral index also depends on the exponent α. Notice that when $\alpha \to 1$ the classical $\gamma_p^{\rm DSA}$ is recovered. Since $(dN/dE)dE = (dN/dp)dp$, it is straightforward to obtain $\gamma_p = 2\gamma - 1$, which relates the particle momentum spectral index to the particle energy spectral index.

3. Data analysis

We analyze a couple of interplanetary shock events reported in [6] and detected by ACE. Figure 1 shows the 1999/02/18 shock (compression ratio $r \sim 2.1$). Notice in particular from top to bottom: the 5 minute resolution proton time profiles, which tend to peak at the shock front (indicated by a vertical dashed line) in five energy channels, the 10 minute averaged proton bulk speed exhibiting a rapid increases from upstream to downstream, the 10 minutes averaged proton density, the magnetic field components in Geocentric Solar Ecliptic coordinates (notice that the magnetic field tends to fluctuate more in the shocked region), the latitudinal angle that identifies magnetic sectors, and the normalized variance of the total magnetic field at one minute time scale. The variance has been computed as $\sigma^2 = \sigma_x^2 + \sigma_y^2 + \sigma_z^2$, where $\sigma_i^2 = \sum_{j=1}^{n} (B_j - B_0)^2/N$ (the index i runs over the magnetic field components) is evaluated over time windows of N points and B_0 is the mean magnetic field within each window. Those time windows have a one minute length. This value is comparable with the time scale corresponding to the Larmor radius of the energetic protons shown in Figure 1 (i.e., $\tau = 1/f_{\rho_p} = 2\pi\rho_p/V_{sw} \equiv 2\pi m_p v/(q\langle B_1\rangle V_{sw})$, where $\langle B_1\rangle$ is an average magnetic field value upstream). This corresponds to the resonant scale for particles interacting with magnetic field turbulent fluctuations. As addressed in Ref. [26], the normalized variance does not vary significantly upstream as the distance from the shock increases; this is crucial since a constant level of magnetic field variance implies a spatially constant diffusion coefficient which implies an exponential decay of the particle time profiles in the DSA framework.

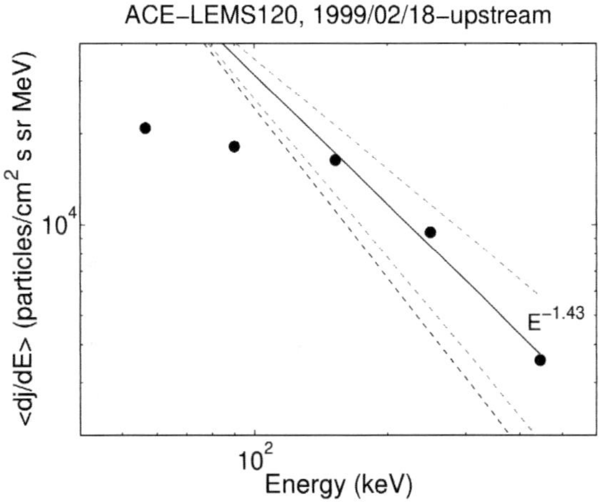

Figure 3. Average differential fluxes as a function of the particle kinetic energy computed upstream of the 1999/02/18 shock from the ion fluxes shown in Figure (1) (filled circles). The best power law fit is indicated by the black solid line. The black dashed line refers to the DSA prediction, while the two red dashed lines indicate the power laws foreseen by the SSA theory for the maximum and minimum values of the particle mean square exponent α derived from the profiles in Figure 2.

Following the data analysis procedure described in Ref. [17], we have plotted in Figure 2 the upstream proton time profiles on log-log axes as a function of the time distance from the shock front (in hours). At about 0.3 hours from the shock the time profiles decay as power laws for all the energy channels considered. Higher energy particle fluxes have not been analyzed since they tend to reach the background level. The best power law fits are shown in Figure 2 as solid black lines and the slopes are also indicated for each energy channel. The power law slope is $\beta < 1$ (see Eq. (3), where $|x| = V_{sh}|\Delta t|$ from single spacecraft observations), thus $\alpha = 2 - \beta > 1$ indicating superdiffusive transport for energetic protons. It is important to remark that close to the shock front ($|\Delta t| < 0.3$ hr) the particle profiles vary and become flatter: this is related to the fact that the propagator form in Eq. (2) does not hold in the vicinity of the source (where a modified Gaussian is obtained for the propagator) [25].

In order to test the SSA prediction for the particle differential spectrum, we have calculated a differential flux $\langle dj/dE\rangle$ averaging the proton time profiles displayed in the upper panel of Figure 1 within the portion of the upstream region from day 48.6 to day 49.0. Notice that the differential flux for non-relativistic particles can be written as $dj/dE = vdN/dE = vE^{-\gamma} \sim E^{-(\gamma-1/2)} \equiv E^{-\gamma_j}$. Thus, we easily get that $\gamma_j = \gamma - (1/2) = \gamma_p/2$, so that obtaining γ_j from the differential flux analysis implies obtaining a γ_p value to be compared with Eq.s (4). Figure 2 reports the observed $\langle dj/dE\rangle$ (filled circles) as a function of the kinetic energy corresponding to the 'central' value of each energy channel. The DSA prediction, based on the observed compression ratio, (see Eq. (4)) is displayed by the black dashed line, while the two red dashed

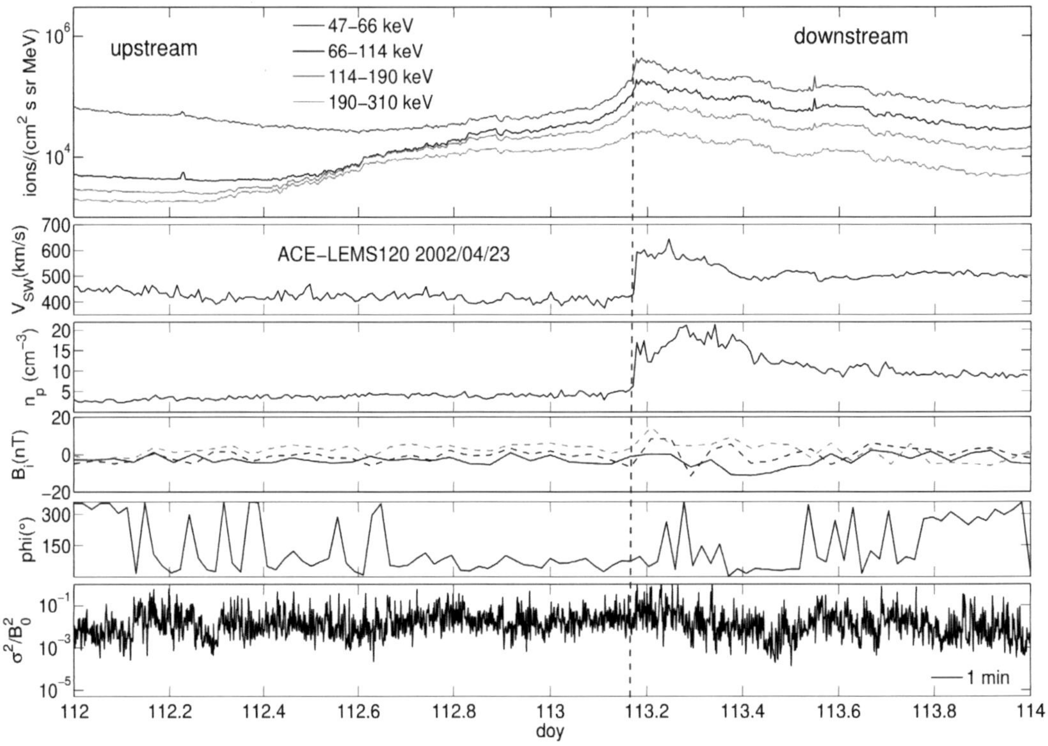

Figure 4. Same as Figure 1 but for the shock event detected by ACE on 2002/04/23.

lines indicate the SSA predictions for the maximum and the minimum β (i.e., α) reported in Figure 2. The latter agree better with the data values. The same analysis on differential fluxes have been reported in Ref. [24] for relativistic electrons accelerated at a corotating interaction region event; the result obtained is in good agreement with the SSA model.

Another interplanetary shock detected by ACE is displayed in Figure 4. It occurred on 2002/04/23 and it has a compression ratio $r \sim 4.0$ due to a large amplitude jump in the proton density from upstream to downstream (see Section 4). Also in this case the proton time profiles analyzed in four energy channels exhibit a power law decay for one decade upstream of the shock front (see Figure 5). The slopes are $\beta < 1$, which agree with superdiffusive transport for energetic protons. In Figure 6 the differential flux (obtained averaging the particle fluxes in each channel from day 112.6 to day 113.0) as a function of the kinetic energy shows again a good agreement with the SSA predictions. Notice that the 'scattering' of the filled circles in Figure 6 is due to the high irregularity in the proton fluxes (see the upper panel in Figure 4), indeed upstream in three energy channels the profiles tend to overlap.

4. Discussions and Conclusions

We have studied the particle differential fluxes upstream of two interplanetary shocks observed at 1 AU by the ACE satellite. The two events indicate a superdiffusive transport for protons in several energy channels, as inferred from the analysis of particle time profiles upstream of the shock front. For the analysis of the spectra we have estimated the compression ratios of the shocks, which have been derived from the proton density measurements upstream and downstream from the front. Since the plasma parameters upstream and downstream can vary

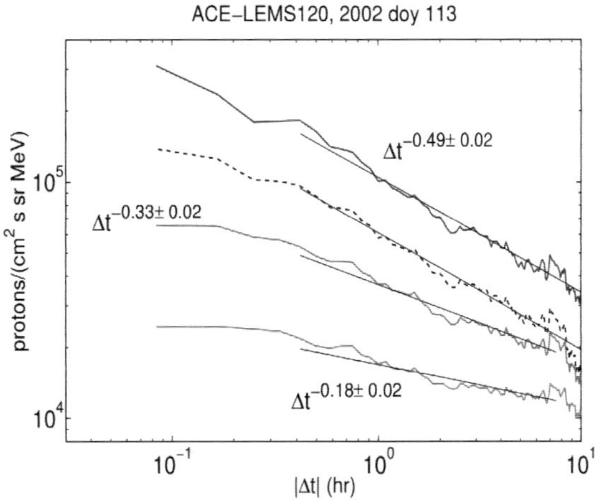

Figure 5. Ion time profiles in log-log axes for the event displayed in Figure 4 as a function of the time distance from the shock. The best power law fits are indicated by the black solid lines and extend for one decade. The power law slopes for each energy channel are reported.

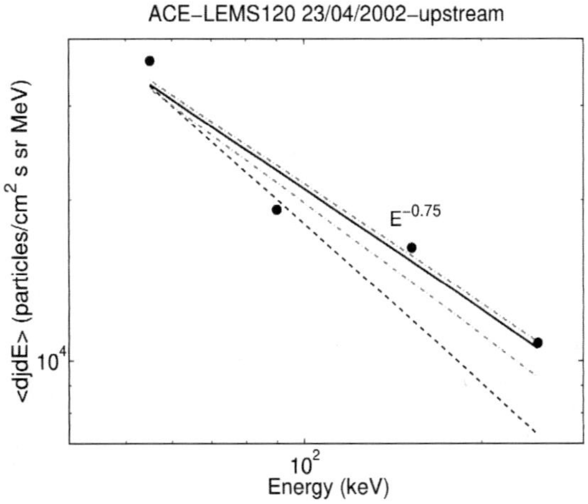

Figure 6. Average differential fluxes as a function of the particle kinetic energy computed upstream of the 2002/04/23 shock (filled circles). The best power law fit is indicated by the black solid line and show a very good agreement with the two power laws predicted by the SSA (red dashed lines). The black dashed line refers to the DSA prediction.

very rapidly, thus giving rise to a high uncertainty in the estimation of the plasma mean values, we have followed the data analysis procedure described in Ref. [24]. More specifically, we considered the plasma data in time intervals in the shocked and in the unshocked regions after a couple of hours from the time of the shock. Thus, we computed plasma density and speed average values within running windows of about 1 hr. The running window can slide over a time period of almost 3 hrs in the two regions, within portions of the signal not affected by large amplitude variations. This procedure tends to minimize the errors associated with the estimation of the plasma mean values, however a certain degree of variation in the plasma measurements upstream and downstream of the shocks gives rise to unavoidable (and sometimes large) errors in their evaluation (see also the discussion in Ref. [27, 6]). Because of both the uncertainty in the derivation of the compression ratio, and the non-regular behavior of the proton time profiles, which in some cases tend to almost overlap in different energy channel (see upper panels of Figures 1 and 4 in the upstream side), the differential fluxes as a function of kinetic energy are not always regular power laws. However, in the two events considered the best power law fit of $\langle dj/dE \rangle$ vs E is in a much better agreement with the power laws predicted by SSA than with the ones from DSA, considering the $\alpha = 2 - \beta$ values deduced from the fits in Figures 2 and 5. It is worth stressing that superdiffusive transport has been found in all the events analyzed from the particle time profiles upstream of the shocks and it has not directly been tested downstream. However, here we assume that particle superdiffusion holds within the downstream region, too. This assumption can be supported by the time trend of the normalized magnetic field variance shown in the bottom panels in Figs. 1 and 4. The variance σ^2/B_0^2 tends indeed to remain constant (or even decrease a bit in the 1999/02/18 event), suggesting that the particle scattering downstream does not increase owing to an increased power in the magnetic field fluctuations.

The analysis reported here, together with the spectrum of relativistic electrons accelerated at a corotating interaction region observed by Ulysses and presented in Ref. [24], is a first step in testing the SSA theory developed in Ref.s [22, 23] and more differential spectra need to be analyzed. In addition, an evaluation of the acceleration time, as presented in Eq. (19) in Ref. [22], directly from spacecraft observations will be presented in a forthcoming paper.

Acknowledgments
We acknowledge discussions at the team meeting "Superdiffusive transport in space plasmas and its influence on energetic particle acceleration and propagation" supported by the International Space Science Institute (ISSI) in Bern, Switzerland.

References

[1] Decker R B et al. 2005 *Nature* **309** 2020
[2] Decker R B et al. 2008 *Nature* **454** 67
[3] Ellison D C, Ḿobius E and Paschmann G 1990 *Astrophys. J.* **352** 376
[4] Fichtner H, Potgieter M, Ferreira S and Burger A 2000 *Geophys. Res. Lett.* **27** 1611
[5] Eraker J H and Simpson J A 2012 *J. Geophys. Res.* **91** 9973
[6] Giacalone J 2012 *Astrophys. J.* **761** 28
[7] Sanderson T R, Marsden R G, Wenzel K P, Balogh A, Forsyth R J, and Goldstein B E 1994 *Geophys. Res. Lett.* **21** 1113
[8] Reames D V 1995 *Rev. Geophys.* **33** 585
[9] Dalla S et al. 2003 *Geophys. Res. Lett.* **30** 8035
[10] Reynolds S P, Gaensler B M and Bocchino F 2012 *Space Sci. Rev.* **166** 231
[11] Lee M A and Fisk L A 1982 *Space Sci. Rev.* **32** 205
[12] Drury L. O'C 1983 *Rep. Prog. Phys.* **46** 973
[13] Lin R P 2005 *Adv. Space Res.* **35** 1857
[14] Bieber J W, Matthaeus W H, Smith C W, Wanner W, Kallenrode M B and Wibberenz G 1994 *Astrophys. J.* **420** 294
[15] Reames D V 1999 *Space Sci. Rev.* **90** 413
[16] Droge W 2003 *Astrophys. J.* **589** 1027
[17] Perri S and Zimbardo G 2007 *Astrophys. J. Lett.* **671** L177
[18] Perri S and Zimbardo G 2008 *J. Geophys.Res.* **113** A03107
[19] Perri S and Zimbardo G 2009b *Adv. Space Res.* **44** 465
[20] Perri S and Zimbardo G 2009a *Astrophys. J. Lett.* **693** L118
[21] Sugiyama T and Shiota D 2011 *Astrophys. J. Lett.* **731** L34
[22] Perri S and Zimbardo G 2012a *Astrophys. J.* **750** 87
[23] Zimbardo G and Perri S 2013 *Astrophys. J.* **778** 35
[24] Perri S, Zimbardo G, Effenberger F and Fichtner H 2015 *Astron. Astrophys. in press*
[25] Zumofen G and Klafter J 1993 *Phys. Rev. E* **47** 851

[26] Perri S and Zimbardo G 2012b *Astrophys. J.* **754** 8
[27] Balogh A, Gonzalez-Esparza J A, Forsyth R J, Burton M E, Goldstein B E, Smith E J and Bame S J 1995 *Space Sci. Rev.* **72** 171

Coronal Heating & Solar Wind Acceleration by Drift Waves

S. Poedts, Ch. Kanella, and G. Lapenta

KU Leuven, Dept. of Mathematics, Center for mathematical Plasma Astrophysics (CmPA), Celestijnenlaan 200 B, 3001 Leuven, Belgium

E-mail: `Stefaan.Poedts@wis.kuleuven.be`

Abstract. An alternative approach to the coronal heating problem, based on the theory of drift waves, has been proposed. The drift mode is the only mode that is able to survive the drastically different (collisional/collisionless) extremes in the different layers of the solar atmosphere. As a matter of fact, this mode is over stable, i.e. able to grow, in both of these extreme situations, and has been called the universally growing mode in the literature. In collisional plasma of the lower layers of the solar atmosphere, the drift mode grows due to the electron collisions and this can be described within the two-fluid model. In the collisionless coronal plasma, however, the mode grows due to a pure kinetic effect, viz. the electron resonance effect in the presence of a density gradient.

It has been shown, with qualitative and quantitative arguments, that the drift waves have the potential to satisfy all coronal heating requirements. The basic ingredient required for the heating is the presence of density gradients in the direction perpendicular to the magnetic flux surfaces. The drift wave theory is well-established and has been explicitly verified experimentally in laboratory (fusion) plasmas, similar (hot, low-beta, highly conductive) to those in the solar atmosphere. In these circumstances, two mechanisms of the energy exchange and heating take place simultaneously: Landau damping in the direction parallel to the magnetic field, and stochastic heating in the perpendicular direction. The latter, in fact, is more effective on ions than on electrons, acts predominantly in the perpendicular direction, and heats heavy ions more efficiently than lighter ions. Moreover, for plasmas at a temperature of 1MK and beyond, the parallel wave field resulting from the drift waves exceeds the Dreicer field so that the bulk plasma species (primarily electrons) can be accelerated/decelerated by the wave in the parallel direction. In addition, this acceleration is more effective on the particles that are already more energetic, resulting in a distribution function considerably different from a Maxwellian, similar to the observed kappa-distribution in the outer solar atmosphere and in the solar wind.

1. Introduction

The solar coronal heating problem, a 75 year old puzzle regarding the observational fact that the temperature in the corona of the Sun is more than two orders of magnitude higher than the surface temperature of the Sun, is still one of the major challenges in solar and astrophysics. As a matter of fact, none of the many proposed models is able to explain all the observational characteristics. This is partly related to the fact almost all the models proposed so far, use magnetohydrodynamics (MHD) as mathematical model. MHD is the simplest model, one-fluid and macroscopic (continuum model) and thus not valid on the length scales at which the actual dissipation takes place. As a matter of fact, in order to be conclusive, the model should take into account the physical conditions in the lower layers of the solar atmosphere. The density and the temperature are not the only physical quantities that change drastically in the solar atmosphere. In fact, the chromosphere plays an important role as an intermediate layer between the photosphere and the corona, where the plasma beta (the ratio of the thermal pressure to the magnetic pressure) changes from above one to much below one ($\leq 1\%$), i.e. where the plasma changes from being thermally dominated in the low atmospheric layers to highly magnetically dominated in the low corona (further out in the corona the plasma β rises again to order one at 1 AU). Moreover, the collisionality of the plasma varies from "strongly collisional" in the photosphere and in the lower part of the chromosphere, over "mildly collisional" in the upper part of the chromosphere, to the "collision-less" plasma regime in most of the corona and beyond. As a result, the ionization degree too varies drastically in the solar atmosphere. It can be as low as 10^{-4} at the temperature minimum (at the top of the photosphere), and is increasing to $\pm 50\%$ at a height of about 2000 km and to nearly 100% in the hot corona. Moreover, Carlsson et al. [1, 2] showed that the chromosphere is in a tricky non-equilibrium ionization and non-LTE (non-local thermodynamic equilibrium) regime.

Hence, the relatively narrow chromosphere is the layer where the plasma jumps from collisional to collisionless, from high beta to very low beta, from neutral to almost fully ionized, from opaque to optically thin (in UV and mm-waves), and increases drastically (by almost 3 orders of magnitude) in temperature. Improved observations by current missions, like Hinode (JAXA/NASA) and IRIS (NASA/ESA), make these complexities more apparent and provide even more evidence for the fact that the chromosphere plays a crucial role in many observed phenomena in the solar atmosphere. This is because the chromosphere is the photosphere/corona interface (cf. above) and the fact that the magnetic field, dominant in the corona, is anchored in the chromosphere and transition region and electric current systems (stressing the coronal field) penetrate the chromosphere.

Vranjes et al. [3, 4] studied the important issue of the energy flux of Alfvén waves bearing in mind the facts that the solar photosphere is very weakly ionized and the dynamics of the plasma particles in this region is heavily influenced by the plasma-neutral collisions. The magnetization of the plasma constituents have been quantitatively examined, showing that the ions and electrons in the photosphere are both un-magnetized, their collision frequency with neutrals is much larger than the gyro-frequency. This implies that eventual Alfvén-type electromagnetic perturbations must involve the neutrals as well. This has important consequences, viz. i) in the presence of perturbations, the whole fluid (plasma + neutrals) moves; ii) the Alfvén velocity includes the total (plasma + neutrals) density and is thus considerably smaller compared to the collision-less case; iii) the perturbed velocity of a unit volume, which now includes both plasma and neutrals, becomes much smaller compared to the ideal (collision-less) case; and iv) the corresponding wave energy flux for the given parameters becomes much smaller compared to the ideal case.

Vranjes et al. [5] then investigated the properties of gas acoustic and ion acoustic modes in a collisional, weakly ionized plasma in the presence of un-magnetized ions and magnetized electrons. In such a plasma, an ion acoustic mode, driven by an electron flow along the magnetic field lines, can propagate almost at any angle with respect to the ambient field lines as long as the electrons are capable of participating in the perturbations by moving only along the field lines. Several effects, including the electron-ion collisions, the perturbations of the neutral gas, and the electromagnetic perturbations, are studied. The neutral sound mode couples to the current driven ion acoustic mode, and these two modes can interchange their identities in certain parameter regimes. The electromagnetic effects, which in the present model imply a bending of the magnetic field lines, result in a further destabilization of an already unstable ion acoustic wave.

2. Drift waves in the solar atmosphere?!

In 2009, Vranjes and Poedts [6] proposed *new paradigm* for the coronal heating problem by applying a multi-component fluid and kinetic description of the drift wave instability to the solar atmospheric plasma. This novel drift wave heating theory appears to be able of satisfying numerous heating requirements of the solar atmosphere. Indeed, a *self-consistent solar coronal heating model* must not only provide an *energy source* for the extremely high temperature in the corona including a reliable and efficient mechanism for the *transfer of energy* from the source to the plasma particles, and this *at a required heating rate* (e.g. $\simeq 10^{-4}$ J/(m^3 s) in active regions) [7]. It should also work *everywhere in the corona* (with different heating requirements in different regions). Moreover, it must also *explain the observed discrepancy between ion and electron temperatures* (typically $T_i > T_e$) and the origin of the large *temperature anisotropy* ($T_\perp > T_\parallel$) with respect to the direction of the magnetic field, particularly for ions [8, 9]. In addition, it should explain the observed *stronger heating of heavier ions* [10]. Clearly, the latter three requirements can never be fulfilled by a single-fluid isotropic pressure model like MHD [11].

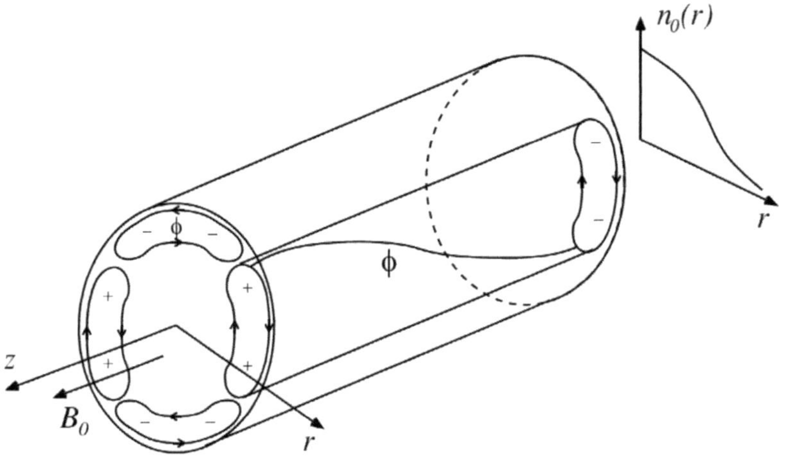

Figure 1. Visualization of a drift wave in cylindric geometry, propagating in the poloidal and axial directions with poloidal wave number $m = 2$, in plasma with a r-dependent equilibrium density. The actual wave fronts are twisted around the axis and have a r-dependent amplitude which reaches a maximum in the area of the largest density gradient. From [12].

The theoretical model proposed by Vranjes and Poedts [6, 12, 13] shows that a) the energy for driving the mentioned drift modes and instabilities, and for the heating of the solar corona is already present in the corona, viz. stored in the ubiquitous density gradients and, b) this energy is naturally transmitted to the different plasma species by well known effects that are, however, beyond the commonly used models and theories. Moreover, it is based on well established, basic plasma theory which has already been verified and confirmed by means of a series of laboratory plasma experiments. It was shown that the basic ingredient necessary for the heating is the presence of gradients of the density (eventually also of the temperature and magnetic field) in the direction perpendicular to the magnetic flux surfaces. Such density gradients are a source of free energy for the excitation of drift waves. The dispersion relation in two-component kinetic theory, under the conditions $k_z v_{Ti} \ll \omega \ll k_z v_{Te}$, $\omega \ll \Omega_i$, and

$$\frac{|k_y|}{|k_z|}\frac{\rho_i}{L_n}\left(\frac{T_e}{T_i}\right)^{1/2} \gg 1,$$

(with L_n the inhomogeneity scale length ($L_n = (n_0'/n_0)^{-1}$) reads [12]:

$$\omega_r = -\frac{\omega_{*i}\Lambda_0(b_i)}{1-\Lambda_0(b_i)+T_i/T_e+k_y^2\lambda_{di}^2}, \qquad \omega_{*i} = k_y \frac{v_{Ti}^2}{\Omega_i}\frac{n_{i0}'}{n_{i0}}, \qquad \nabla_\perp n_{i0} = -\vec{e}_x n_{i0}', \tag{1}$$

for the real part of the frequency and

$$\gamma \simeq -\left(\frac{\pi}{2}\right)^{1/2} \frac{\omega_r^2}{|\omega_{*i}|\Lambda_0(b_i)} \left[\frac{T_i}{T_e}\frac{\omega_r-\omega_{*e}}{|k_z|v_{Te}} \exp[-\omega_r^2/(k_z^2 v_{Te}^2)] + \frac{\omega_r-\omega_{*i}}{|k_z|v_{Ti}} \exp[-\omega_r^2/(k_z^2 v_{Ti}^2)]\right], \tag{2}$$

for the imaginary part. Here

$$\Lambda_0(b_i) = I_0(b_i)\exp(-b_i), \;\; b_i = k_y^2\rho_i^2, \;\; \lambda_{di} = v_{Ti}/\omega_{pi}, \;\; \omega_{*e} = -k_y\frac{v_{Te}^2}{\Omega_e}\frac{n_{e0}'}{n_{e0}}. \tag{3}$$

Note that the *presence of the energy source* is seen already in the real part of the frequency $\omega_r \propto \nabla_\perp n_0$ (compare with Alfvén waves: $\omega = kc_a$). The details of the growth of the wave energy, which is due to the same source, is described by the imaginary part γ.

Two mechanisms of the energy exchange and heating have been shown to take place simultaneously: one due to the Landau effect in the direction parallel to the magnetic field, and another one, *stochastic heating*, in the perpendicular direction. The stochastic heating i) is due to the electrostatic nature of the waves, ii) is more effective on ions than on electrons, iii) acts predominantly in the perpendicular direction, iv) heats heavy ions more efficiently than lighter ions, and v) may easily provide a drift wave heating rate that is several (up to 4) orders of magnitude above the value that is presently believed to be sufficient for the coronal heating [12], i.e., $\simeq 6\cdot 10^{-5}$ J/(m^{3}s) for active regions and $\simeq 8\cdot 10^{-6}$ J/(m^{3}s) for coronal holes. This heating acts naturally through well known effects that are, however, beyond the current standard models and theories.

3. Simple numerical model

As mentioned before, the drift wave theory (both linear and nonlinear) has been tested and confirmed thoroughly via laboratory experiments in the context of controlled thermonuclear fusion research. We here report on the first results of a numerical quantitative study under parameter values that are more relevant for the solar atmosphere. We focussed on lower hybrid drift waves (LHDWs), a subclass of drift waves, that has also been studied and observed in the magnetosphere of the Earth. These waves have an angular frequency ω that satisfies $\omega_{ce} \le \omega \ll \Omega_{ci}$ and $k\, r_{L,e} \le 1 \ll k\, r_{L,i}$, resp., with $r_{L,s}$ the Larmor radius for species s [14]. In this parameter range, the electrons are strongly magnetized and can be described as a fluid, while the ions (protons and heavier ions) are treated as particles. The LHDWs can resonate with ions in the perpendicular direction (w.r.t. the magnetic field) and with electrons in the parallel direction. The latter electron-wave interaction destabilises the modes via the inverse Landau effect [6].

3.1. Set up

We considered a simple Harris current sheet configuration (see Fig. 2 and [15]) which assumes a local Maxwellian velocity distribution, shifted towards the currents direction with a velocity value equal to the diamagnetic drift velocity that arises from the pressure gradients in the system. Our goal is to check how efficient the free energy (in the density gradients) can be transferred to the plasma species, and ultimately, if the plasma species can absorb enough energy to sustain the energy requirements of solar corona. The profiles of the magnetic field and the density (illustrated in Fig. 2) are given by

$$B_z(y) \;=\; B_0\tanh\left(\frac{y}{L}\right)\vec{e}_z, \tag{4}$$

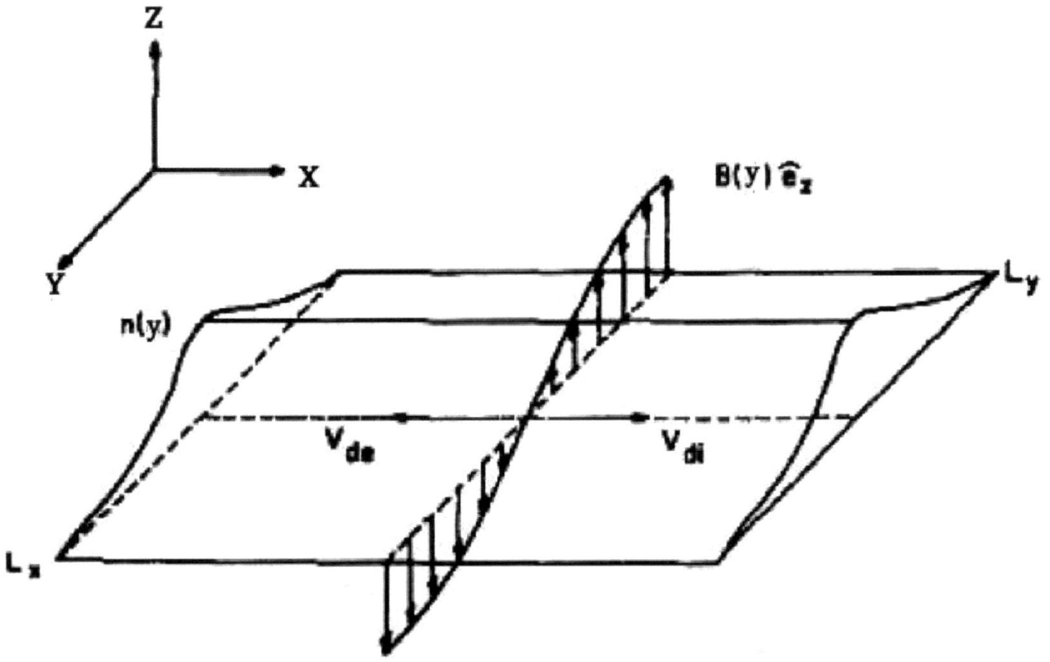

Figure 2. Harris Current Sheet Configuration, illustrating the magnetic field $B_z(y)$ and particle density $n(y)$ profiles considered. The diamagnetic velocities of electrons and ions, generating the current, are denoted by v_{de} and v_{di}, resp. [16].

and

$$n(y) \quad = \quad n_0 \operatorname{sech}^2\left(\frac{y}{L}\right), \tag{5}$$

respectively, where L denotes the thickness of the sheet, and B_0 and n_0 correspond to the values of the magnetic field and density at the centre of the sheet, respectively. The magnetic field profile yields a current density in the x-direction:

$$J_x(y) = \frac{1}{\mu_0}\nabla \times \vec{B} = \frac{B_0}{\mu_0 L} \operatorname{sech}^2\left(\frac{y}{L}\right)\vec{e}_x. \tag{6}$$

We assume quasi-neutrality, i.e. a balance between magnetic and plasma pressure. Initially, $T_i = T_e$, and for two later cases with three species each, $T_i = T_e = T_{O^{+1}}$ and $T_i = T_e = T_{He^{+2}}$.

3.2. Numerics

The simulations are done with IPIC3D, the "implicit Particle-In-Cell 3D" code, which is designed for a multiprocessor architecture and implements implicit methods to solve the equations governing the physics of the particle-heating problem at hand, i.e. the Vlasov equations for the distribution functions of the considered particle species, in combination with the Maxwell equations for the electro-magnetic field. For details and stability conditions etc., we refer to Markidis et al. [17]. For the results discussed in the present paper, the iPIC3D code was applied to a Harris current sheet configuration in which the particle's velocity distribution is Maxwellian. The diamagnetic velocity of each particle species satisfies the relation $v_{*,i}/v_{*,e} = -T_i/T_e$, so that electrons and ions drift to opposite directions and thus generate currents. At the centre of the sheet, the magnetic field has minimal magnitude while the plasma density is maximal there, so that the motion at the central region is dominated by inertia effects. The density gradients are controlled by the parameter L, the thickness of the plasma sheet. The actual value of the ion mass, specified by the parameter m_i/m_e in the code, is not the real one but this does not affect the dimensionless parameters ω_{pe}/Ω_{pe}, $L/r_{L,i}$ and $v_{*,i}/v_{th,i}$, with ω_{pe} the plasma frequency and Ω_{pe} the electron gyro-frequency. We consider a simulation box of $10 \times 5 \times 1/40$ in units of ion skin depth, d_i. The high-resolution grid has $400 \times 200 \times 1$ cells, i.e. $\Delta x = \Delta y = \Delta z = 0.025\ d_i$. Each cell has 9 computational particles and a typical simulation took 20000 cycles.

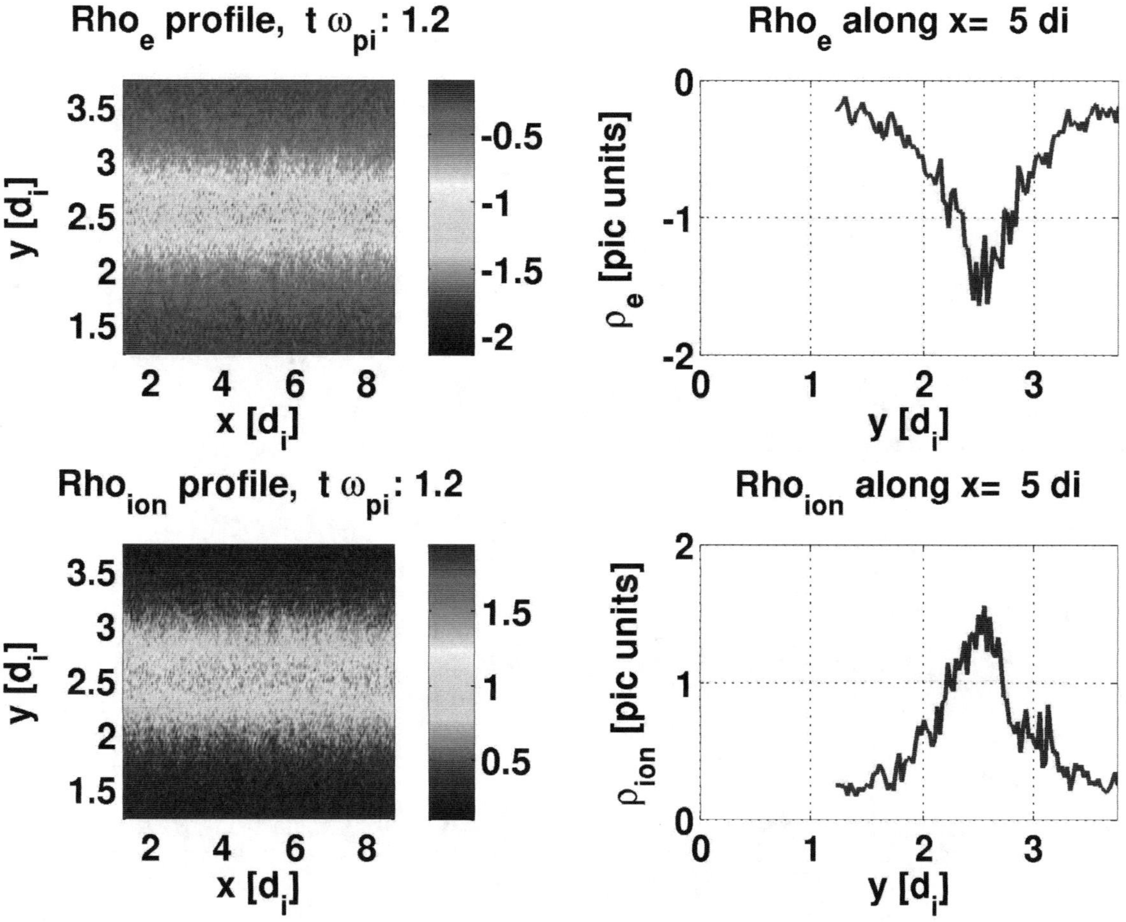

Figure 3. Top left: snapshot of the charge density plotted in the $x - y$ domain. The central region of the current sheet is situated at $y = 2.5\ d_i$. Bottom left: the same quantity for ions that have an artificial mass ratio value $m_i/m_e = 256$. Right: distribution profiles of the charge density of electrons and ions plotted respecting the y-coordinate and evaluated at $x = 5\ d_i$. Distance is in units of ion skin depth d_i and time is normalized by the ion plasma period ω_{pi}^{-1}.

4. Simulation results

4.1. General case with two species only

As a first case, we choose $v_{\text{th,e}} = 0.05\,c$, $T_i/T_e = 1$, and $m_i/m_e = 256$. In Fig. 3, snapshots of the electron and ion charge density are plotted in the plane perpendicular to the magnetic field, i.e. the $x - y$ plane, showing the perturbations that appear at the edges of the current sheet. In order to show how the density profiles change over time, we also plotted the same quantities over the y-direction along $x = 5\ d_i$.

In Fig. 4, shows another snapshot of the same run at a later time, when the non-linear phase has developed. Notice the broader profile in nonlinear phase and the kinking of the current sheet during this phase. Later a kink mode develops due to the Lower Hybrid Drift Instability (not shown here).

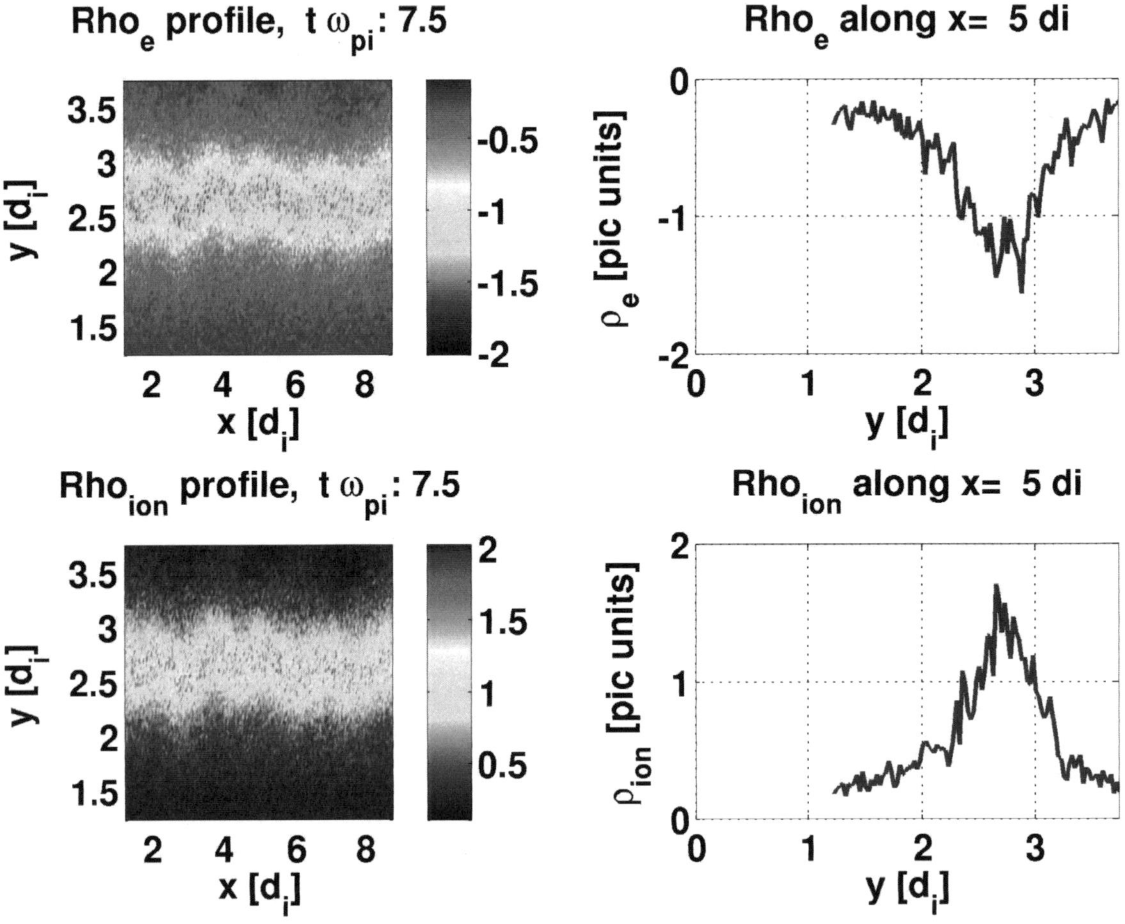

Figure 4. The kinking of the current sheet during the non-linear phase, for the same parameter values and same plot options as Fig. 3.

Next, we calculated the total magnetic and thermal energy density for the whole system by averaging out these energy contributions over the total number of cells and plotting them with respect to time. The results are displayed in Fig. 6, where physical values have been used to calculate the energy density values in SI units. For this case, the initial thermal energy of the plasma species matched the case of the Earth magnetosphere. Notice that the magnetic energy of the system drops. As time elapses, the electrons both give and receive energy ($\mathbf{J}_e \cdot \mathbf{E} > $ or < 0), while the protons only receive energy ($\mathbf{J}_p \cdot \mathbf{E} > 0$), as shown in Fig. 5.

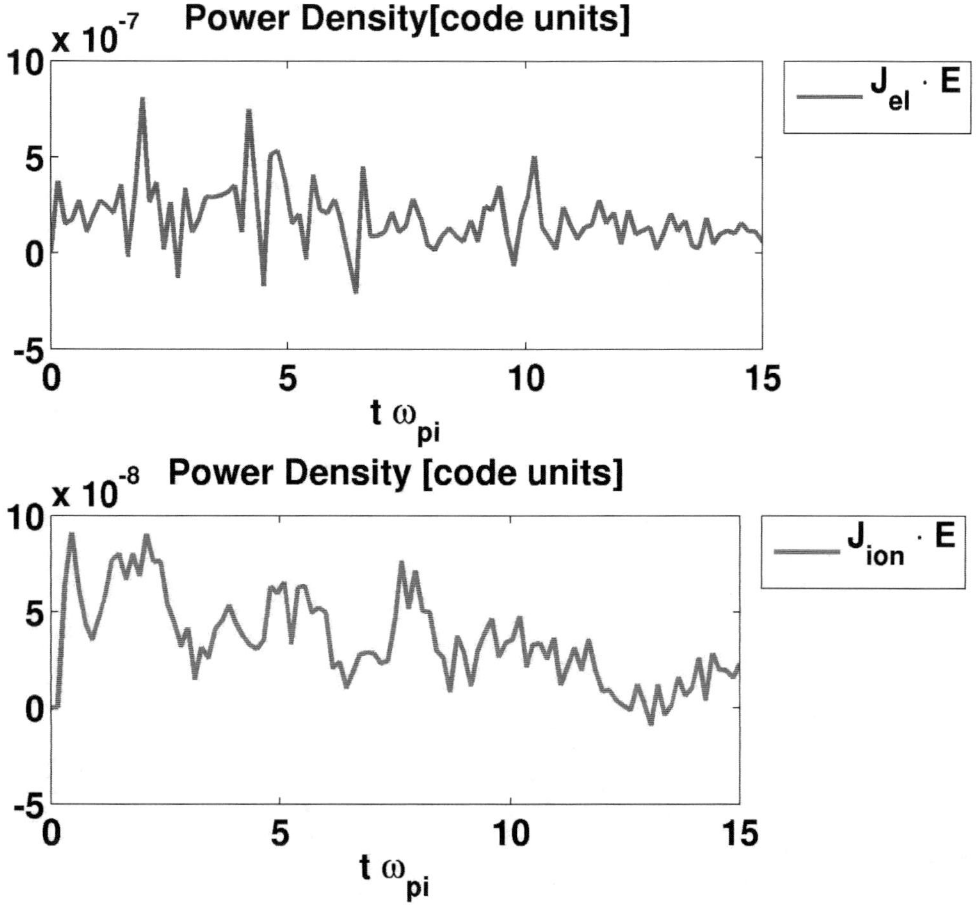

Figure 5. Temporal evolution of the electromagnetic power density equation's terms, represented in code units. The top plot represents the energy exchange between electrons and electrostatic fields and the lower plot corresponds to the energy exchange between ions and electrostatic fields. The time is normalized by the ion plasma period ω_{pi}^{-1}, while the power density of the energy exchange terms is $\mathbf{J}_s \cdot \mathbf{E}\, d_i/(n_0 c^2 m_i^2)$, with s the species.

In Fig. 7, two snapshots are plotted of the distribution of different temperature ratios, for different spatial directions and different species. This demonstrates the obtained temperature anisotropies. As a matter of fact, notice that $T_i > T_e$, i.e. ions are heated more than electrons. Also, $T_{e,\|} > T_{e,\perp}$, so for electrons the parallel direction (w.r.t. the magnetic field) is heated preferably. For ions, on the other hand, $T_{i,\|} \leq T_{i,\perp}$, i.e. the perpendicular direction (w.r.t. the magnetic field) is preferably heated, confirming theory and observations.

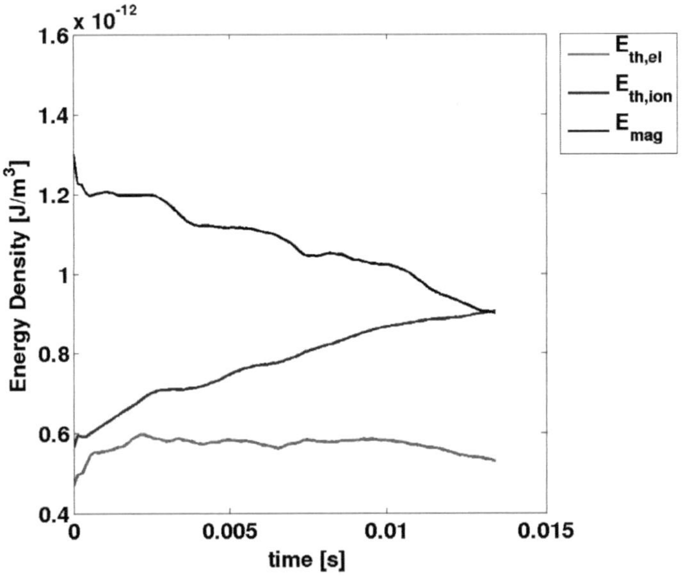

Figure 6. Energy density in SI units for the case of the Earth magnetosphere. The time is normalized by the ion plasma period.

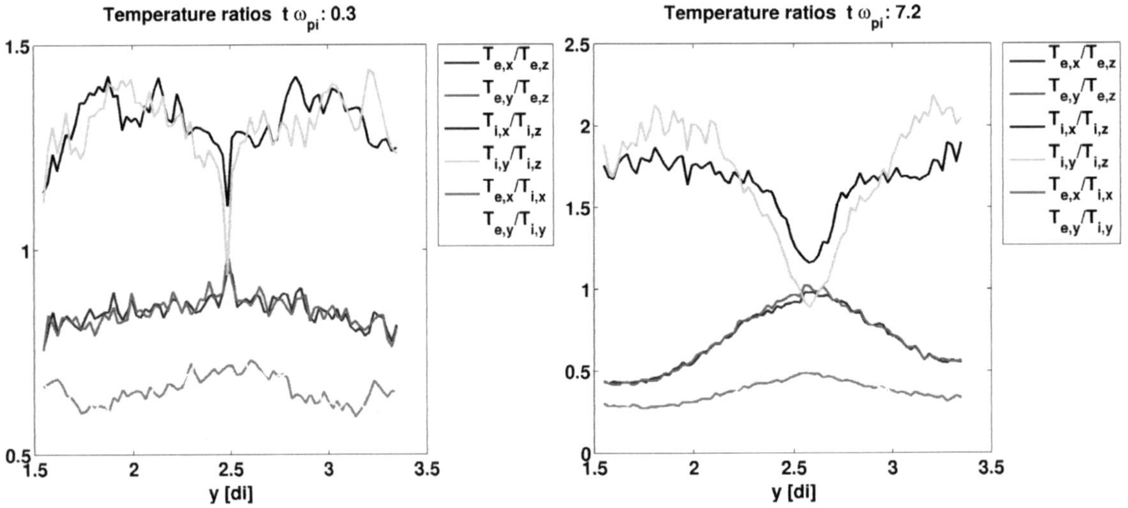

Figure 7. Two snapshots of the temperature ratios, calculated for the different spatial directions and for different plasma species. Left: an initial time-step. Right: a phase during the well developed kink mode. Indices x, y, z denote the spatial coordinates of the Cartesian system in the Harris sheet configuration. The distance is normalized by the ion skin depth d_i and time by the ion plasma period.

4.2. Case $v_{th,e} = 0.05\,c$, $T_i/T_e = 5$, $n_0 = 0.1\,cm^{-3}$ (magnetosphere)

Next, we varied the initial electron thermal velocity and/or the initial temperature ratio between the ions and electrons in different steps in the parameter study.

As a first step, we increased the initial ion temperature by a factor of five, keeping the electron's thermal velocity the same, i.e. $v_{\rm th,e} = 0.05\,c$. In Fig. 8, the conversion to physical units was obtained using a particle density $n_0 = 0.1$ cm^{-3}, typical for the Earth's magnetosphere. As a result, the total energy content is much larger than before. Notice also that the magnetic energy drops faster and more than in the previous case (with $T_i/T_e = 1$). Also, in this case the electrons gain energy during nonlinear phase, in contrast to the case with $T_i/T_e = 1$, and also the protons receive more energy. Notice the fast heating rate (on scale of milliseconds).

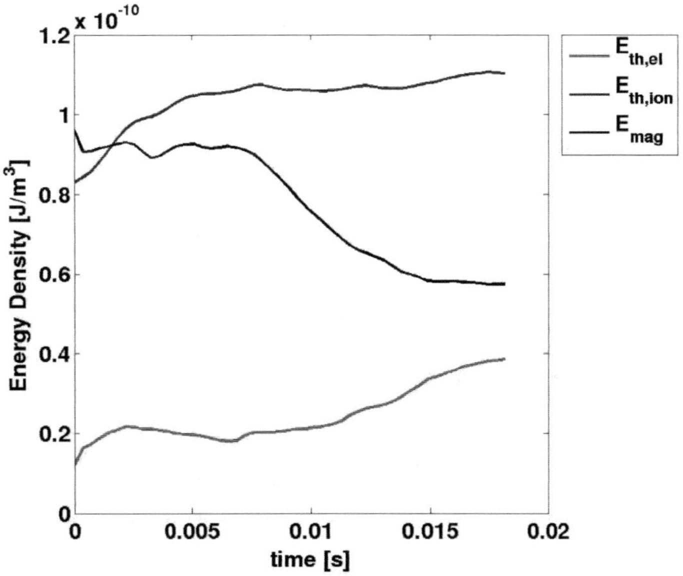

Figure 8. Energy density (SI) for $v_{\rm th,e} = 0.05\,c$, $T_i/T_e = 5$.

4.3. Case $v_{th,e} = 0.035\,c$, $T_i/T_e = 1$, $n_0 = 10^9\ cm^{-3}$ (solar)

Let us now consider the case with the electron's velocity $v_{\rm th,e} = 0.035\,c$. Reducing the only the electron temperature and using $n_0 = 10^9$ cm^{-3} for the conversion to SI, we now have a case more suitable for the solar corona. The resulting energy density evolution is displayed in Fig. 9. Notice the much larger energy content in this case, due to the higher density considered. In this case, the magnetic energy drops faster/more, the electrons gain more energy during the nonlinear phase and the electron energy increases more/faster than in the previous case. Notice that also the protons receive more energy and faster than in the previous cases. The heating occurs extremely fast, in a matter of nanoseconds!

4.4. Cases with three particles

Next, we investigated what happens when there are three different species. First, we consider oxygen atoms as the third species. In the simulation, we specify that the charge density of ions consists of an equal share of protons and Oxygen atoms, while their sum is equal to that of electrons. In other words, the considered oxygen is O^{+1}. Clearly, $m_{O^{+1}}/m_i = 16$ and we specify that, initially, $T_i = T_e = T_{O^{+1}}$. The other parameter values are set as before. This choice of parameters corresponds to the physical situation in the Earths magnetosphere again. For the conversion to SI units, we assumed a particle density $n_0 = 0.1$ cm^{-3}, like before in this case.

Figure 10 displays the energy evolution for this case. It is clear that the oxygen particles are heated more than the protons (cf. observations). Also, the protons heated more than the

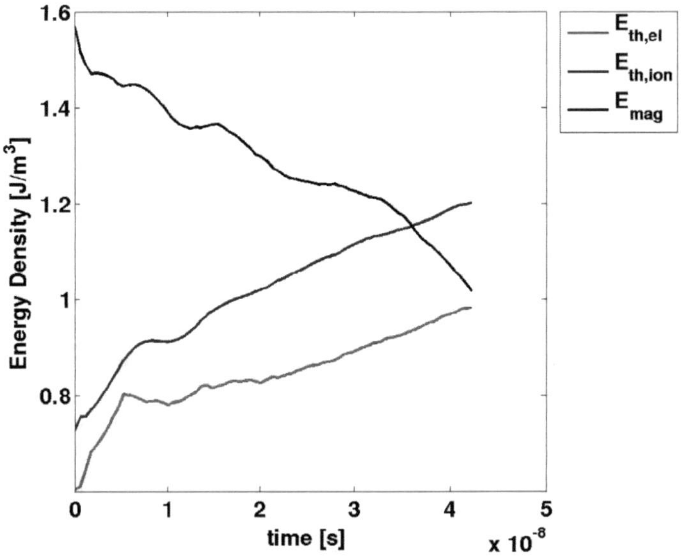

Figure 9. $v_{\mathrm{th,e}} = 0.035\,c$, $T_i/T_e = 1$, energy density (SI)

electrons, but less than in the first case we considered (with only two species), as part of the energy now goes to the heavier ion species. The heating rate is extremely high again. Heating occurs on time scales of milliseconds.

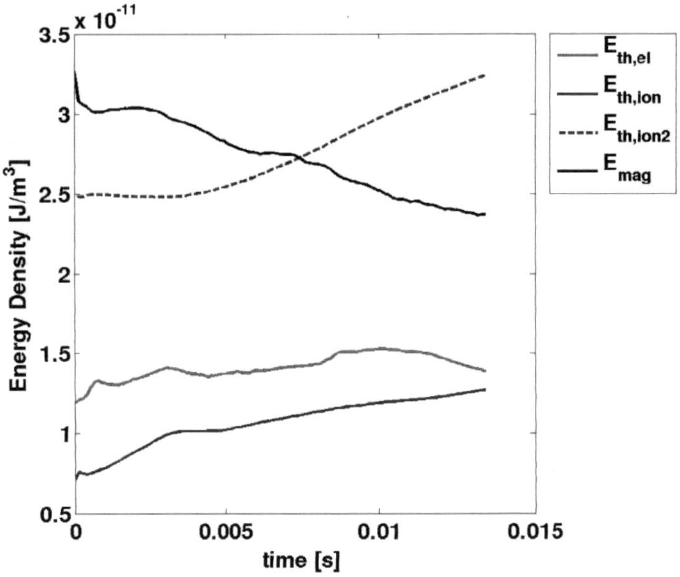

Figure 10. The energy density in SI units for the case of the Earth magnetosphere but now with three species (electrons, protons and oxygen atoms (singly ionised, O^{+1}), with parameter choices $v_{\mathrm{th,e}} = 0.05\,c$, $T_i = T_e = T_{O^{+1}}$, $m_{O^{+1}}/m_i = 16$. Time is normalized to the ion plasma period.

Journal of Physics: Conference Series **642** (2015) 012021 doi:10.1088/1742-6596/642/1/012021

The last case we consider in the present paper, is also a case with three species, but this time we chose Helium ions as the third species. In this case, $v_{\mathrm{th,e}} = 0.05\,c$, and we choose to set $T_i = T_e = T_{He^{+2}}$, initially. Of course, $m_{He^{+2}}/m_i = 4$. For the conversion to SI units, we again assumed a particle density $n_0 = 0.1\ \mathrm{cm}^{-3}$, like in the previous case, for comparison.

Figure 11 displays the energy evolution for this case. It is remarkable that in this case, the (heavier) helium particle population is heated less than the protons, unlike any other heavier ion, just like theory predicts [12]. Note also that the time scales for the heating are extremely short gain, order of milliseconds.

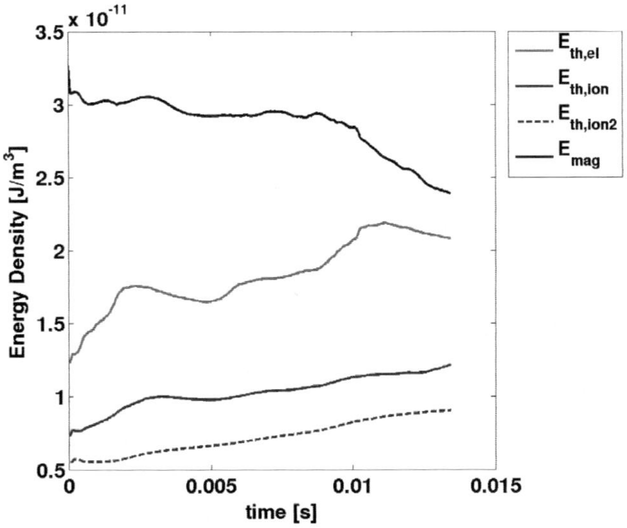

Figure 11. The energy density in SI units for the case of the Earth magnetosphere but now with three species (electrons, protons and helium atoms (doubly ionised, He^{+2}), with parameter choices $v_{\mathrm{th,e}} = 0.05\,c$, $T_i = T_e = T_{He^{+2}}$, $m_{He^{+2}}/m_i = 4$.

5. Conclusions

Vranjes and Poedts [6, 12] proposed a new explanation for the heating of the solar corona based on drift waves. As well-known from basic plasma theory, drift waves are *always present* in (inhomogeneous) plasmas, because these waves are *intrinsically unstable* (in the literature, therefore, this is called the "universal instability"). Any gradient in density (and/or temperature or magnetic field strength) across the magnetic flux surfaces, causes this 'universal' instability that drives these waves to occur. In other words, the drifts waves are 'over stable', i.e. with a frequency with both a real, oscillatory part, and an imaginary (in complex notation at least) growth rate. This distinguishes them clearly from the MHD waves that are often considered in the context of coronal heating, and that need to be driven at the right frequency in order to occur. Given the highly inhomogeneous characteristics of the plasma in the solar atmosphere, drift waves must be always self-generated there, and occur ubiquitously, even though they have not yet been observed (in the literature, any wave mode observed in the solar atmosphere is labeled as one or a combination of the MHD modes).

The alternative heating mechanism proposed by Vranjes and Poedts [6, 12], claiming to have the potential to satisfy all the observational requirements, was tested in a first attempt to quantify its efficiency. As a matter of fact, Vranjes and Poedts only applied linear theory and derived their preliminary conclusions from the large instability windows with high growth rates they obtained for coronal plasma parameter values. Here, a first attempt was made to take into account the non-linear saturation that inevitably occurs after the first (short) linear growth of the mode, to which the previous studies were limited. This was achieved by a relatively simple first numerical set-up and using an advanced particle-in-cell code (iPIC3D). The (first, simple) numerical simulations confirmed the linear theory results on the resulting temperature anisotropy, the fact that ions are heated more than electrons in this mechanism, the fact that heavier ions are heated stronger than lighter ions, except for helium ions, and the extremely fast heating time scales. These first results are, therefore, encouraging and need to be extended in order to be able to draw final conclusions.

As a matter of fact, a full quantitative confirmation of this heating mechanism needs much more advanced numerical simulations and parameter studies to include, e.g., the effect inhomogeneity along field lines, possible geometry effects (on the nonlinear saturation), etc. Clearly, also observational evidence would be welcome for drift waves in the solar atmosphere.

Acknowledgments
These results were obtained in the framework of the projects GOA/2015-014 (KU Leuven) and C 90347 (ESA Prodex). The research leading to these results has also received funding from the European Commission's Seventh Framework Programme (FP7/2007-2013) under the grant agreement SOLSPANET (project n 269299, www.solspanet.eu). This research has been funded by the Interuniversity Attraction Poles Programme initiated by the Belgian Science Policy Office (IAP P7/08 CHARM). For the computations we used the infrastructure of the VSC Flemish Supercomputer Center, funded by the Hercules foundation and the Flemish Government department EWI.

6. References

[1] Carlsson and Stein, *Astrophys. J.* **572**, 626-635 (2002).
[2] Carlsson M., and Leenaarts J., *Astron. Astrophys.* **539**, AA39 (2012).
[3] Vranjes J., Poedts S., Pandey B.P., and De Pontieu B., *Astron. Astrophys.* **478**, 553 (2008).
[4] Vranjes J., Poedts S., and Pandey B.P., 28th ICPIG, Prague, July 15-20, 2007, 5P07-08 (ISBN: 978-80-87026-01-4).
[5] Vranjes J., Pandey B.P., and Poedts S., *Phys. Plasmas* **14**, 032106 (2007).
[6] Vranjes J., and Poedts S., *EPL* **86**, 39001 (2009).
[7] Aschwanden M. J., *Astrophys. J.* 560 (2001) 1035.
[8] Cuseri I., Mullan D. and Poletto G., *Space Sci. Rev.* 87 (1999) 153.
[9] Li X., Habbal S.R., Kohl J.L. and Noci G., *Astrophys. J.* **501**, L133 (1998).
[10] Hansteen V.N., Leer E. and Holtzer T.E., *Astrophys. J.* **482**, 498 (1997).
[11] Klimchuk J., *Sol. Phys.* **234**, 41 (2006).
[12] Vranjes J., and Poedts S., *MNRAS* **398**, 918 (2009).
[13] Vranjes J., and Poedts S., *MNRAS* **400**, 2147 (2009).
[14] Mikhailovskii A.B., and Tsypin V.S., *JETP* **32** (1971).
[15] Harris E., *Il Nuovo Cimento*, **115(23)**, 1955-1966 (1962).
[16] Brackbill J.U., *Comp. Phys. Comm.* , **47(1)**, 1-16 (1987).
[17] Markidis S., Lapenta G., and Uddin R., *Math. and Comp. in Simulation* **80**, 1509-1519 2010.

Journal of Physics: Conference Series **642** (2015) 012022 doi:10.1088/1742-6596/642/1/012022

Plasma and Variability in the Heliosheath

John D. Richardson

Kavli Center for Astrophysics and Space Sciences, Massachusetts Institute of Technology, Cambridge, MA 02139 USA

E-mail: jdr@space.mit.edu

Abstract. Voyager 2 (V2) is making the first direct plasma measurements of the heliosheath and is moving outward about 3 AU/yr. This paper has two foci. It provides an update on plasma conditions in the heliosheath out to 108 AU and discusses the variability of plasma parameters in the heliosheath. The plasma speed continues to be on average constant across the heliosheath and the flow continues to turn tailward, with a flow angle greater than 70° in 2014. The average density and temperature have remained roughly constant since an increase in 2011. The heliosheath continues to be highly variable; this paper shows examples of small-scale variability and shows the changes of the plasma parameters on various scales. Future observations of plasma in the local interstellar medium (LISM) are discussed.

1. Introduction

V2 crossed the termination shock in 2007 [1,2,3,4]. The V2 plasma instrument (PLS) has provided observations in the heliosheath from the crossing at 84 AU to the distance in mid-2015 of 108 AU. The V1 PLS does not provide useful data, so these are the first observations of plasma in the heliosheath. V1 plasma flow speeds have been derived from V1 Low Energy Charged Particle experiment (LECP) and Cosmic Ray Subsystem (CRS) data [5,6,7,8,9], but plasma densities and temperatures cannot be determined.

This paper describes the plasma observations in the heliosheath. The data analysis is described, including how corrections are made in the flow angle to account for the instrument response. The plasma parameters are shown and discussed. The heliosheath is a highly variable region; an attempt is made to quantify these variations. Finally the expected observations in the LISM are illustrated.

2. Data

The V2 PLS instrument observes ions and electron currents with energies/charge from 10-5950 eV/q every 192 seconds using four Faraday Cups [10]. The observed currents are fit with convected isotropic proton Maxwellian distributions to find the plasma velocity, density, and temperature. Three of the cups point towards Earth and the fourth points at right angles to the V2-Earth line. The angular responses of these cups are flat to 45°, then decrease to 0 at 60°. The average plasma flow angle is now over 70° from radial, so plasma is often not detected in the the three cups necessary to determine the velocity. If the flow were steady then flows could never be determined given the high average flow angle. However, the heliosheath plasma velocities, densities and temperatures have remained highly variable throughout the heliosheath. This high variability allows the PLS instrument to observe sufficient currents to determine plasma

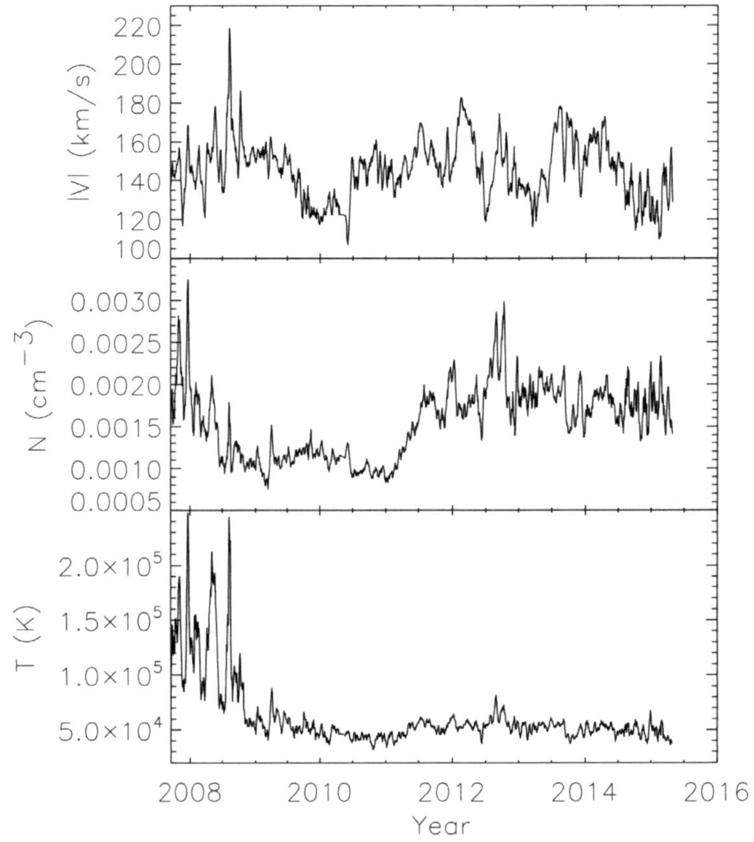

Figure 1. 11-day running averages of the plasma speed, density, and temperature observed by V2 in the heliosheath.

parameters in a fraction of the observed spectra. The combination of large angle flows, tracking gaps, and noise allow only about 5% of possible spectra to be analyzed. Maxwellian distributions fit the observed spectra well when data are available in three of the cups.

Figure 1 shows the heliosheath proton speed V, density N, and thermal speed from the termination shock at 84 AU out to 108 AU. The average speed has remained constant across the heliosheath, although excursions in speed from as low as 120 km/s to as high as 180 km/s have persisted for several months. These observations are in sharp contrast to those at V1, where the speed decreased to near zero five years into the heliosheath.

The density jumped at the termination shock by a factor of 2 to 0.002 cm^{-3}, then decreased by a factor of 2 in 2008, perhaps associated with solar minimum reaching the outer heliosphere. The density increased back to its pre-2008 level in 2011, perhaps marking the end of solar minimum. If these density changes are related to solar minimum, they could be due to V2 at 30° heliolatitude observing more high-speed, low-density wind [11]. However, the speed does not change in this low-density region as one might expect if V2 were in high-speed wind. Factor of 2 changes in the density are observed after 2011 but the average has remained constant.

Figure 2 shows 11-day running averages of the flow angles in the RT and RN planes. The RTN system has R radially outward, T in the plane of the solar equator and positive in the direction of solar rotation, and N completes a right-handed system. Since the observable flow angles are limited by the instrument response, the RT angles must be corrected to account for this limitation. This correction is done with the assumption that the distribution of angles is a Gaussian. This assumption is valid for the plasma density, velocity components, and thermal

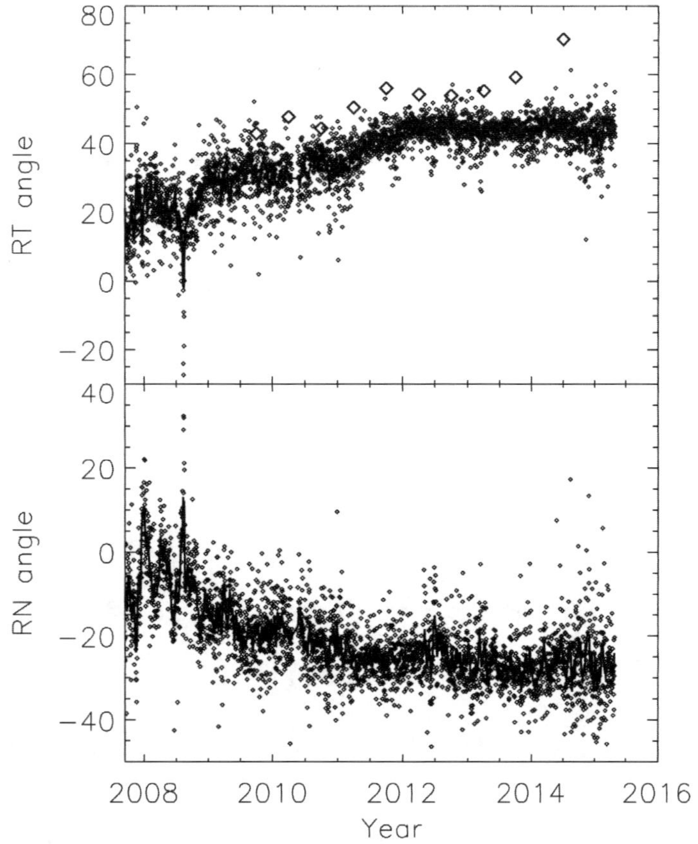

Figure 2. Flow angles in the RT and RN planes observed by V2. The points show individual fits, the line shows 11-day running averages, and the diamonds show values of the RT angle corrected to account for the instrument response.

speeds throughout the heliosheath, and also for the flow angles when they are small [11]. As the RT angle increases, less of the Gaussian distribution is observed and only the tail of the distribution can be used for the fit. Figure 3 shows the distribution of RT angles in 2015. The instrumental cutoff is at about 50°; the data decrease rapidly beyond this cutoff. The Gaussian fit to the data (up to 49°) is very good in the low-angle tail where data are available and gives an average RT angle of 70° in 2015 with a 1-sigma error of 2°. The width of the distribution is 21 km/s with a 1-sigma error of 1 km/s, consistent with the widths in previous years [11].

Figure 2 shows that the flattening of the RT angle in the 11-day averages is an instrumental effect. The corrected RT angle values increase with distance and in 2014 the average RT angle was 70°. Since the RT angle is expected to be near 90° at the heliopause as VR becomes small, this result suggests V2 is getting close to this boundary. The RN angle is much smaller than the RT angle. Flow is moving around the sides of the heliosphere at the V2 position, not over the pole. The RN angle is slowly increasing and was approaching 30° in 2014. McComas and Schwadron [12] suggest that the flow at V2 is not away from the nose of the heliopause but away from the highest pressure observed by IBEX, which is below the nose. Flow away from this direction would give the observed V2 flow directions.

As discussed above, the heliosheath is a region with large fluctuations in plasma parameters. Fig.4 shows an example of three spectra from the PLS B-cup, which looks most directly into the flow, in 2011. These spectra span 35 minutes of time and show a remarkable variation in plasma over this short time scale. In the left panel, the peak currents are at the lowest energies. The

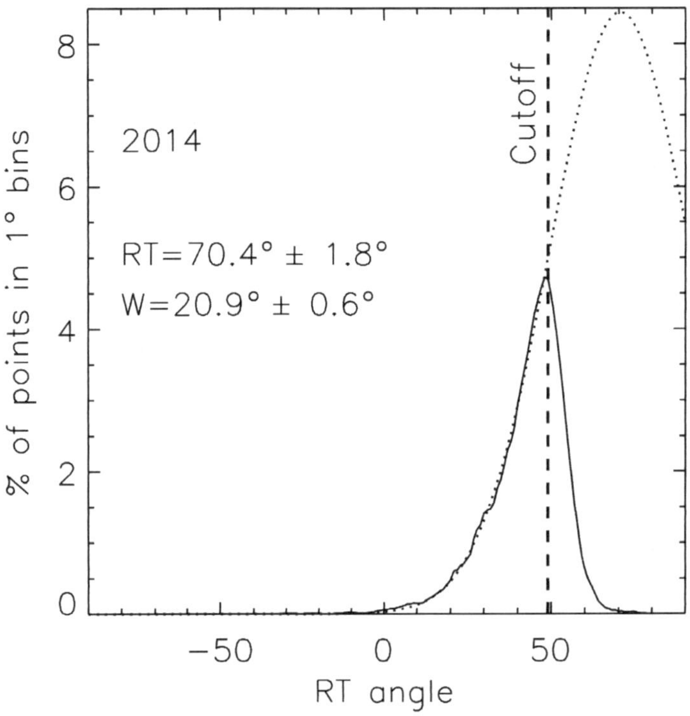

Figure 3. The distribution of RT angles observed in the data (solid line) and the best fit of a Gaussian distribution to these data (dotted line). The dashed line shows the instrument cutoff and the fit parameters are given.

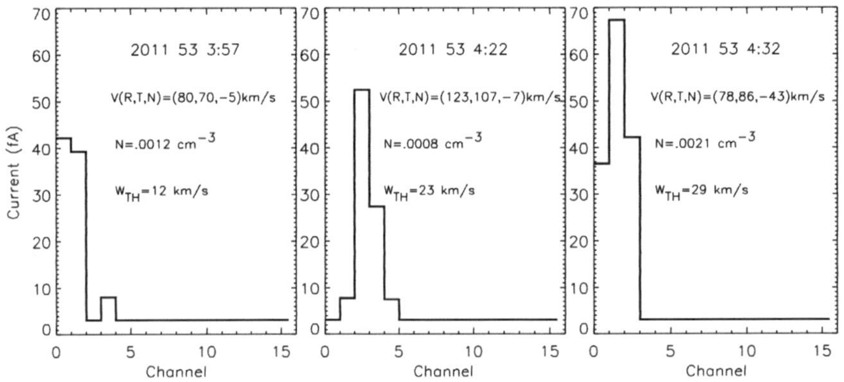

Figure 4. Three V2 PLS B-cup spectra from a 35 minute period in 2011. The best fit parameters are given in each plot. The channels number show a roughly logarithmic energy scale from 10-5950 eV.

parameters derived from fits to all the cups are shown in the figures. The currents in the middle panel peak in the third channel; in 25 minutes the speed increased by 40 km/s, the density decreased by a third, and the thermal speed doubled. Ten minutes later the peak current is in channel 2 and much higher in amplitude. The speed has decreased to the original level but the density is up by roughly a factor of two and the thermal speed is up by 25%. These changes are clear from the spectra; they are not artifacts of the fits.

The variability of the plasma in the heliosheath has not been quantified previously. The next section shows the changes observed in the plasma on a variety on time scales. The data set consists of all the spectra analyzed in the heliosheath. Some of these are undoubtably spurious

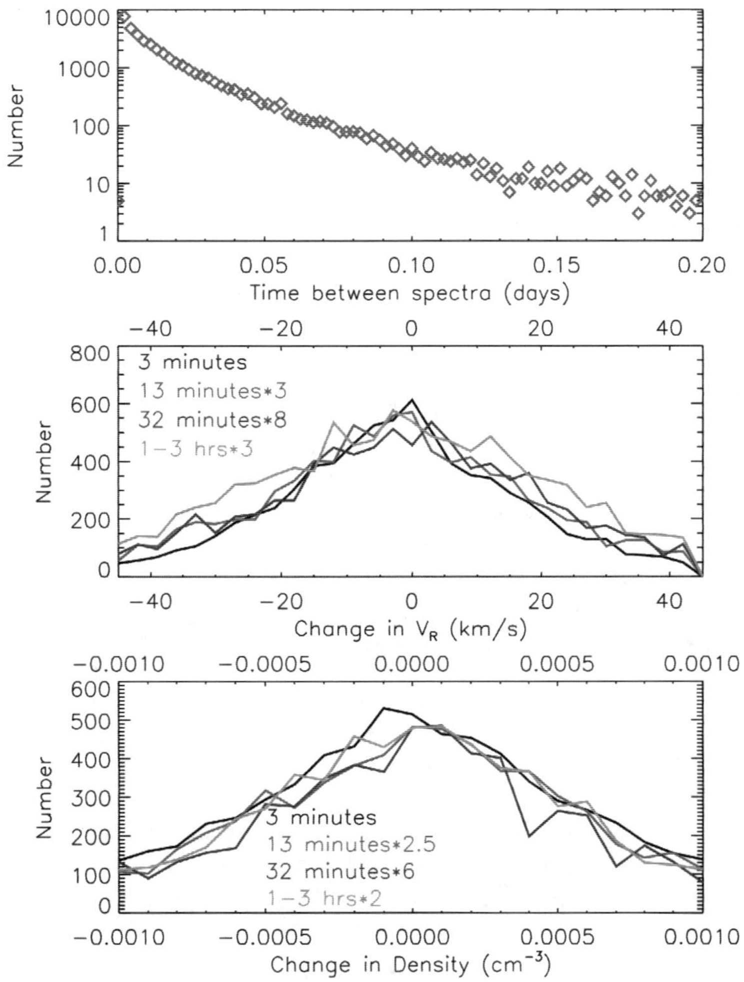

Figure 5. Top: number of data points with the separation given by the x-axis; separations are multiples of 192 sec. Middle, Bottom: the observed changes in V_R and density for four different separations in time as color-coded on the plot.

fits with noise being treated as data. However, since data in three of the four detectors are required to perform these fits, most of the 41,500 fits should be reliable. Previous work shows that the errors in V_R are about 10%, in N about 20%, and in W about 25% [9]. A set of spectra is obtained every 192 sec, so this is the minimum separation. The top panel of Figure 5 shows that about 9,000 points have this minimum separation and about five points have 100 times this separation at the right-hand side of the plot.

The middle and bottom panels show the distributions of the changes in V_R and density for different separations of spectra. Distributions for changes in consecutive spectra (dT = 192 sec) are shown by the black lines, for spectra separated by about 13 minutes, 32 minutes, and 1-3 hours the distributions are shown by red, blue, and green lines, respectively. The distributions are multiplied by the factors shown on the plot to facilitate comparison of the distributions. For V_R the shape of the distribution is similar for the 3-32 minute distributions; the 3 minute distribution is slightly more peaked but this effect is not large. The 1-3 hour distribution is broader, indicating scale lengths in the heliosheath are shorter than a few hours. For the density the profiles do not change shape for time scales up to a few hours.

Figure 6 shows differences in the plasma variation in different regions of the heliosheath. We divide the plasma into three regions: before the decrease in density at 2008.5, during the low density region, and after the density recovery at 2011.5. For each separation in spectra from 192 sec to about 2 hours the plot shows the average relative change between spectra, so for V_R the plot shows (V_R (t+1) - V_R (t))/V_R(t). The red +'s show the data after 2011.5, the black diamonds show data from before 2011.5 and after 2008.5 (the low-density region), and the blue *'s show data from before 2008.5. The average errors in the fit parameters are about 10% for V_R, 27% for N in the low-density region and 22% for N outside the low density region. For both V_R and N the percentage change increases with distance into the sheath for short time separations. For V_R, the change is comparable to the uncertainty for the smallest DT and increases to almost 0.2 at DT 30 minutes, then flattens out. The other two regions also show increases to about 30 minutes, then flatten out, but on average maintain the relative order of larger changes at larger distances. The density profiles are very similar to the V_R profiles, increasing to DT 30 minutes than flattening out. The average ordering of the three regions again remains the same. These data suggest a correlation length in the heliosheath of 30 minutes, which for an average speed of 150 km/s gives a distance of 270,000 km.

3. Plasma observations in the LISM

Voyager 2 may be approaching the heliopause and LISM. It crossed the termination shock 10 AU closer to the Sun than V1. V1 crossed the heliopause at 121 AU; for V2 a 111 AU heliopause crossing would occur within a year. Of course the plasma flows look very different from those observed at V1 so comparison is difficult. The V2 flow direction is changing rapidly and is about 70° from radial. If the flow continues to turn at the current rate it would be at 90°, roughly the angle expected at the heliopause where VR is small, in the next couple of years. Although plasma waves at V1 have given density values at a few places in the LISM, measurements of speed, temperature, and the full density profile must wait for V2. What will V2 observe in the LISM? Since the flows are not directly into the detectors (they are away from the three main sensor cups and likely about 60° from the side sensor) the amount of current observed is very dependent on the thermal speed (temperature). Simulations of the currents in the PLS detectors have been performed for nominal plasma temperatures. The proton flow speed is 26 km/s and the plasma density of order 0.08 cm^{-3}. The LISM temperature has recently been revised upward to 9000K to enable fits to both the Ulysses and IBEX data [13]. As the LISM plasma nears the heliopause it is compressed and heated by a factor of 2-4 [14]. Figure 7 shows the currents expected in the sideways-looking D-cup for T = 18000 K and 36000 K, 2 and 4 times the LISM temperature. In each case currents well above the threshold are observed. For the 36000K case current is also observed in the B-cup, not shown, so a 2D speed, density, and temperature could be derived. For the 18000 K case the speed normal to the detector, density, and temperature could be derived. The right panel shows a prediction of the observed currents using parameters from the Fisk and Gloeckler model [15] of the heliosphere, which suggests that V1 is still inside the heliopause; in this case V2 would observe no currents when it crossed the heliocliff.

4. Discussion and Summary

Voyager 2 continues to see heliosheath flow at a constant average speed which is turning fairly rapidly tailward. Most of the turning is in the RT plane and suggests that the heliopause crossing will occur with a few years. These observations are very different from the flow slowdown into a stagnation region observed at V1. The heliosheath plasma continues to be highly variable.

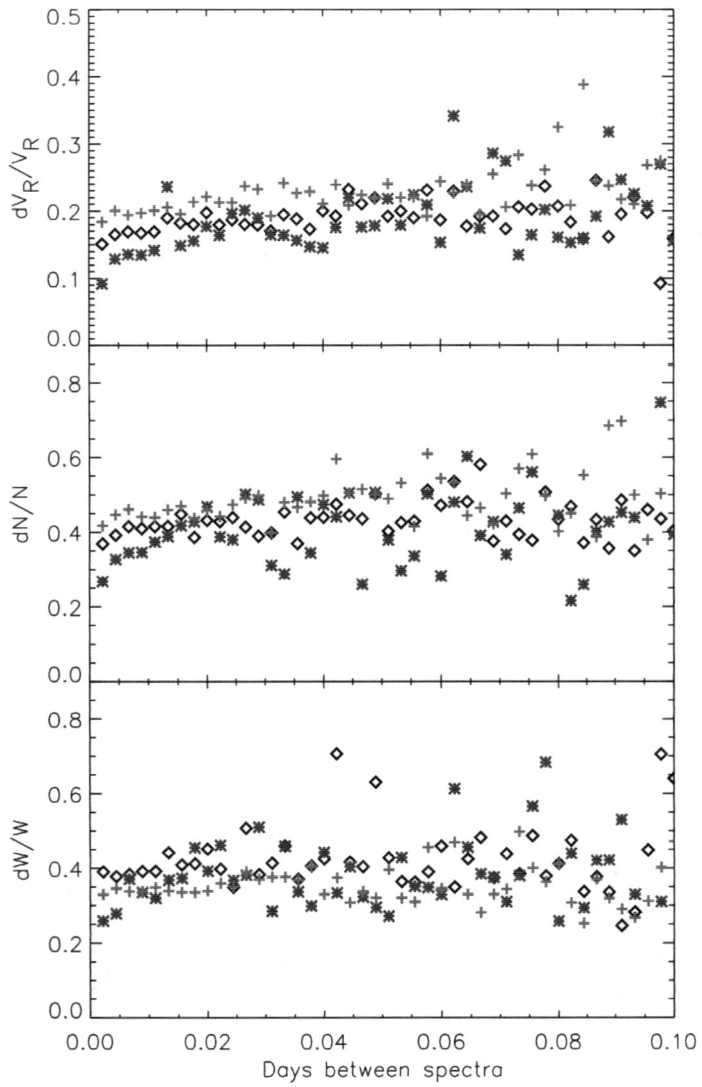

Figure 6. The average change in V_R, density, and thermal speed for each separation in time. The different symbols and colors show different regions of the heliosheath as indicated by the labels.

Changes in plasma parameters on time scales of tens of minutes persist, and have increased in magnitude, across the heliosheath. The plasma is a high-beta plasma, but the pickup ions which likely dominate the plasma pressure are not directly observed, complicating efforts to understand this region. When V2 does enter the LISM the plasma instrument should be able to measure this plasma directly and provide information on the plasma parameters and their variability.

4.1. Acknowledgements

JDR was supported under NASA contract 959203 from the Jet Propulsion Laboratory to the Massachusetts Institute of Technology

[1] Burlaga L F, Ness N F, Acuna M H, Lepping R P, Connerney J E P and Richardson J D 2008 *Nature* **454** 75

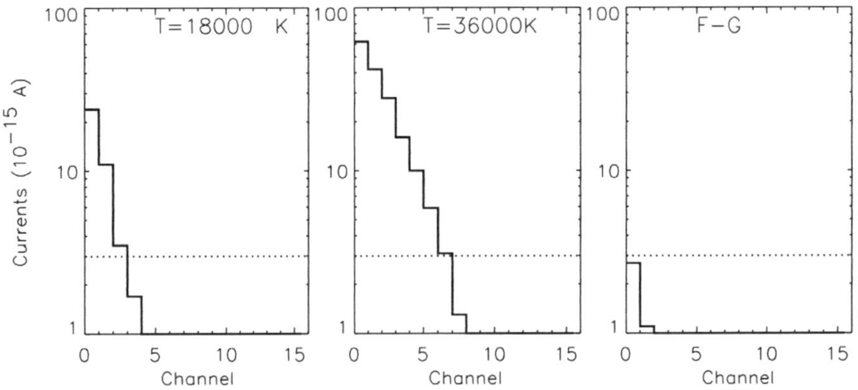

Figure 7. Simulated currents in the V2 PLS D-cup for LISM temperatures at the heliopause of 18000 and 36000 K and for the heliosheath values predicted past the heliocliff [14]. The dashed horizontal lines are the instrument threshold.

[2] Decker R B, Krimigis S M, Roelof E C, Hill M E, Armstrong T P, Gloeckler G, Hamilton D C and Lanzerotti L J 2008 *Nature* **454** 67
[3] Richardson J D, Kasper J C, Wang C, Belcher J W and Lazarus A J 2008 *Nature* **454** 63
[4] Stone E C, Cummings A C, McDonald F B, Heikkila B C, Lal N, Webber W R 2008 *Nature* **454** 71
[5] Decker R B, Krimigis S M, Roelof E C, Hill M E, Armstrong T P, Gloeckler G, Hamilton D C and Lanzerotti L J 2005 *Science* **309** 2020
[6] Decker R B, Krimigis S M, Roelof E C and Hill M E 2010 *AIP Conf. Proc.* **1302** 51
[7] Krimigis S M, Roelof E C, Decker R B, Hill M E 2011 *Nature* **474** 359
[8] Stone E C and Cummings A C 2012 *Proceedings of the 32ND International Cosmic Ray Conference, Beijing 2011* **12** 23
[9] Richardson J D and Decker R B 2014 *Astrophys. J.* 792 196
[10] Bridge H S, Belcher J W, Butler R J, Lazarus A J, Mavretic A M, Sullivan J D, Siscoe G L and Vasyliunas V M 1977 *Space Sci. Rev.* **21** 259
[11] Richardson J D 2011 *Astrophys. J.* **740** 113
[12] McComas D J and Schwadron N A 2014 *Astrophys. J. Lett.* **795** L17
[13] Moebius E et al. 2014 *AGU Fall Meeting* **SH11C-4060**
[14] Zank G P 1999 *Space Sci. Rev.* **89** 413
[15] Gloeckler G and Fisk L A 2014 *Geophys. Res. Lett.* **41** 5325

Charged Particle Energization and Transport in Reservoirs throughout the Heliosphere: 1. Solar Energetic Particles

Edmond C. Roelof
Johns Hopkins University/Applied Physics Laboratory, Laurel MD 20723-6099, USA

E-mail: edmond.roelof@jhuapl.edu

Abstract. "Reservoirs" of energetic charged particles are regions where the particle population is quasi-trapped in large-scale (relative to the gyroradii) magnetic field structures. Reservoirs are found throughout the heliosphere: the huge heliosheath (90<r<110 AU) between the termination shock and the heliopause; the domain of PUIs and ACRs within the solar wind itself; the inner solar system (inside of r<5AU) when filled with SEPs or CIRs; closed structures dragged out from the solar corona by CMEs; and the ring currents of the larger planets (Earth, Jupiter, Saturn, *etc.*) when energized by magnetic storms. Energization (or cooling) of the charged particles within these reservoirs is produced by the interaction when the particle magnetic drifts have a component along the large-scale electric fields produced by plasma convection. The appropriate description of this transport is "weak scattering", in which the particle's first adiabatic invariant (magnetic moment) is approximately conserved while the particle itself moves rather freely along magnetic field lines. Considerable insight into the observed properties of energization processes can be gained from a remarkably simple equation that describes the particle's fractional time-rate-of-change of momentum (d*ln*p/dt) which depends only upon its pitch angle, the divergence of the plasma velocity ($V_\perp$) transverse to the magnetic field), and the inner product of ($V_\perp$) with the curvature vector of the field lines. The possibilities encompassed in this simple (but general) equation are quite rich, so we restrict our application of it in this paper to the compressive acceleration of SEPs within CMEs.

1. Introduction

Here's what this paper on energetic charged particle acceleration *will* and *will not* discuss. It *will* introduce a new expression for the fractional time-rate-of-change of momentum (d*ln*p/dt) for single particle motion in electromagnetic fields in which the first adiabatic invariant, or "*magnetic moment*" $M=p^2(1-\mu^2)/B$, is conserved and parallel electric fields are neglected. This is the familiar "guiding center" approximation for individual particle motion. Consequently, it *will not* deal with the important microphysics of wave/particle interactions, acceleration within the immediate region of shocks, or magnetic reconnection. The results of this paper *will* therefore apply to regions away from such activity, although, if appropriate, those afore-mentioned processes could be also be considered as sources for energetization regions where the guiding center acceleration approximation is valid. It also *will not* attempt to describe the formation of the "seed" populations, *i.e.*, the energetic charged particle populations that exist in the corona prior to the lift-off of the CME and will then be accelerated by the mechanism to be described in this paper. This obviously is an important, but somewhat separate issue.

In this era of sophisticated computational plasma dynamics, it may seem rather naive to invoke a guiding center equation within a large region of space. However, observations over decades have pointed to large regions wherein the wave/particle interactions are either intrinsically weak, or wherein the effects of pitch-angle scattering are dominated by the focusing effects of the large-scale magnetic fields. For instance, we have previously argued (based on observations) that the inner heliosphere (r<1AU) is just such a "weak-scattering" or nearly "scatter-free" region for energetic solar electrons and galactic cosmic rays [1]. The essential condition for applying guiding center theory to particle acceleration is that the magnetic moment is approximately conserved over the *time scale* (τ) of the acceleration *itself*, where $1/\tau = d\ln p/dt$. An *a priori* validation of this assumption would require a simulation based on the actual plasma turbulence within the weak-scattering region. We don't always have such measurements. On the other hand, even if we don't have such plasma measurements, we can establish an *a posteriori* justification of the weak-scattering assumption in remote regions by demonstrating that the *implications* that follow from invoking the assumption are in agreement with observations. We will take the latter approach in this paper with regard to our use of the guiding center expression for $d\ln p/dt$.

2. New form of the guiding center equation for individual particle acceleration

An equation was presented 15 years ago that provides considerable physical insight into the acceleration process for energetic charge particles [2,3]. It is very simple and yet quite general (requiring only the conservation of the magnetic moment with negligible parallel electric fields). It describes the evolution of the particle momentum (p) along a guiding center path. It is valid for both non-relativistic and relativistic energies and time-dependent plasma flows and magnetic fields.

$$d\ln p/dt = (1-\mu^2)/2\ D_\perp \ln B/dt + \mu^2\ \mathbf{V}\cdot\boldsymbol{\kappa} \tag{1}$$

Here ($\boldsymbol{\kappa}$) is the curvature vector of the magnetic field (**B**) and the "transverse convective operator" is defined as

$$D_\perp/Dt = \partial/\partial t + \mathbf{V}_\perp\cdot\nabla \tag{2}$$

Equations (1) and (2) were a re-formulation (in terms of momentum) by Roelof [2,3] of the well-known equation presented by Northrop [4] in terms of energy. That re-formulation is reproduced in the first part of Appendix A. The derivation neglects any effects due to the parallel component of the electric field along the magnetic field ($E_\parallel = \mathbf{b}\cdot\mathbf{E}$). The remaining transverse electric field is taken to be $\mathbf{E}_\perp = -\mathbf{V}\times\mathbf{B}$. This is why only the transverse plasma velocity $\mathbf{V}_\perp = \mathbf{V} - V_\parallel \mathbf{b} = (\boldsymbol{I} - \mathbf{bb})\cdot\mathbf{V}$ appears in the equation for guiding center motion; obviously, $\mathbf{V}\cdot\boldsymbol{\kappa} = \mathbf{V}_\perp\cdot\boldsymbol{\kappa}$. As was pointed out by Northrop [4], the basic physical mechanisms for the acceleration are a combination of the well-known betatron acceleration (in a time-varying magnetic field) plus the work done on a charged particle when its guiding-center drift velocities (due to gradients and curvature of the magnetic field with spatial scales larger than the particle's gyro-radius) have components that lie along the local transverse electric field. Consequently, the latter mechanism will operate unless pitch-angle scattering is so strong and rapid as to degrade the particle's "memory" of the magnetic gradients and curvature. As stated in the Introduction, we therefore assume that this is not the situation in the regions that we are considering in this paper.

Two important points were made in the re-formulation. First, no properties of the particle itself (mass, charge, or energy) appear on the RHS of the momentum equation [2,3]. Therefore, *every* charged particle (protons, ions, or electrons) with the same pitch-cosine (μ) gains (or loses) momentum at the same *fractional* rate, since $d\ln p/dt = (1/p)(dp/dt)$. Second, through Faraday's law, the time-dependence of the magnetic field can be eliminated from the transverse convective operator [3]. However, it was not realized at the time of writing how simple the remaining expression is when this is carried out. We derive the new equation in Appendix A. It contains a rigorous new expression for the transverse convective derivative

$$D_\perp \ln B/Dt = -\nabla\cdot\mathbf{V}_\perp - \mathbf{V}_\perp\cdot\boldsymbol{\kappa} \tag{3}$$

Even though only *spatial* operators appear on the RHS, the new expression for the convective fractional derivative of the magnetic field strength (B) is nonetheless fully valid for *time-dependent* magnetic fields, just as long as the electric field is given by **E**=-**V**×**B**. Simple substitution of Eq. (3) into Eq. (1) yields our new (general) equation for charged particle acceleration.

$$d\,ln\,p/dt = -(1-\mu^2)/2\ (\nabla\cdot\mathbf{V}_\perp+\mathbf{V}_\perp\cdot\boldsymbol{\kappa}) + \mu^2\,\mathbf{V}\cdot\boldsymbol{\kappa} \quad (4)$$

$$= -(1/3)\ \nabla\cdot\mathbf{V}_\perp + (3\mu^2-1)/2\ [(1/3)\nabla\cdot\mathbf{V}_\perp+\mathbf{V}_\perp\cdot\boldsymbol{\kappa}] \quad (4')$$

The coefficient of the 2nd term in Eq. (4') will be recognized as $P_2(\mu)$, the 2nd order Legendre polynomial. Notice the dependence on pitch cosine.

$d\,ln\,p/dt = -(1/2)\ (\nabla\cdot\mathbf{V}_\perp+\mathbf{V}_\perp\cdot\boldsymbol{\kappa})$ $\mu=0$

$d\,ln\,p/dt = -(1/3)\ \nabla\cdot\mathbf{V}_\perp$ $|\mu|=(1/3)^{1/2}$

$d\,ln\,p/dt = \mathbf{V}_\perp\cdot\boldsymbol{\kappa}$ $|\mu|=1$

Thus, the logarithmic acceleration rate behaves like 1st-order Fermi (curvature) acceleration for field-aligned velocities $|\mu|=1$ and like (transverse) compressive deceleration ("cooling") at an intermediate pitch cosine $|\mu|=(1/3)^{1/2}$, the root of $P_2(\mu)$. There is an equal contribution from both terms for mirroring particles $\mu=0$. Otherwise, there is a pitch-angle-weighted mixture of both at all other pitch cosines. Nonetheless, the simple average of d*ln*p/dt over all pitch cosines is identical to the value at $|\mu|=(1/3)^{1/2}$, because the average of $P_2(\mu)$ over pitch cosines (μ) is zero.

The tabulation above reveals that Eq. (4) is very rich in possibilities for charged particle acceleration. For instance, the terms involving the field line curvature vector ($\boldsymbol{\kappa}$) immediately suggest an acceleration in "depolarization" events in planetary magnetospheres (Earth, Jupiter, Saturn) in which a "stretched" magnetic field relaxes to a more dipolar configuration by means of an inward plasma convection. In such a configuration we would have $\boldsymbol{\kappa}\cdot\mathbf{V}>0$, and we would also expect an accompanying compression ($\nabla\cdot\mathbf{V}_\perp<0$), particularly near the plane of the dipole equator. In fact such acceleration may well favor the generation of strong bi-directional field-aligned beams of energetic electrons or ions, because for pitch angles near 0° or 180°, the Fermi acceleration term dominates. This acceleration could operate on time scales less than a bounce time, because it requires only the conservation of the first adiabatic invariant (for the perpendicular energy), rather than the familiar (but possibly slower) conservation of the second adiabatic invariant ($\int ds\,p_\parallel$) that is defined over a bounce time and (by definition) varies slowly.

However, in order to quantify such a suggestion for *net* acceleration, we need to know how long the particle experiences each possible pitch angle because d*ln*p/dt depends strongly on pitch-cosine. This is equivalent to saying that we must have an equation that properly describes the *global* transport of the energetic particles in order to quantify the predictions of our individual-particle acceleration equation from Eq. (4). This is a non-trivial requirement. For instance, in magnetospheres, we must allow for pitch-angle anisotropies that vary strongly along an instantaneous magnetic field line (*e.g.*, the bi-directional beam injection events). Consequently, we will leave the magnetospheric applications to future work. Right now we will concentrate on the simpler case of acceleration in which the curvature term does not come into play, because we have reason to believe that the pitch-angle distribution will be nearly isotropic. Then the contribution of $P_2(\mu)$ averages out in Eq. (4'), leaving only the simple expression

$$d\,ln\,p/dt \approx -(1/3)\ \nabla\cdot\mathbf{V}_\perp \quad (4'')$$

3. Description of general transport within energetic particle "reservoirs"

In a pair of papers, Roelof [5,6] derived a transport equation that describes the post-maximum phase of solar energetic particle events, when the combination of nearly scatter-free propagation along field lines and magnetic mirroring inside 1AU combined with back-scatter beyond 1AU reduces both the field-aligned intensity gradient and field-aligned pitch-angle anisotropy to small values. These conditions produce a particle "reservoir". We will argue that similar conditions exist within the "nose" of an outgoing CME.

$$\partial lnf/\partial t + [\xi v\mathbf{b}+\mathbf{V}_{\perp}+\varepsilon\nabla\times(\mathbf{B}/B^2)]\cdot\nabla lnf - (1/3)(\nabla\cdot\mathbf{V}_{\perp})\partial lnf/\partial lnp = - vB\partial(\xi/B)/\partial s \quad (5)$$

The omnidirectional (p.s.d.) phase-space density $f_0(\mathbf{r},p,t)$ and the parallel anisotropy function $\xi^{\parallel}(\mathbf{r},p,t)$ are written here simply as (f) and (ξ), respectively, in Eq. (5). Since the p.s.d. is by definition non-negative, we have divided the original version (linear homogeneous in f) by (f) itself, converting the dependent variable to *ln*f, thereby leaving a non-homogeneous term on the RHS that is independent of (f). The parallel spatial derivative along a field line $\partial/\partial s=\mathbf{b}\cdot\nabla$ in the non-homogeneous term is defined by the unit vector ($\mathbf{b}=\mathbf{B}/B$) direction of the magnetic field. It describes magnetic focusing of the pitch angle.

The physical content of the RHS of Eq. (5) is clear. The coefficient of the term describing particle acceleration ($\partial lnf/\partial lnp$) is precisely the compressive acceleration term given by Eq. (4") that we derived from guiding center motion, thus validating its intuitive meaning for general transport as well. We easily recognize that the vector coefficient of the spatial transport (∇lnf) describes: first, the parallel transport velocity along field lines (which is a small effect if the parallel anisotropy ξ is small, as it is in a reservoir); next the $\mathbf{E}\times\mathbf{B}$ drift velocity ($\mathbf{V}_{\perp}$) due to field line motion; and finally the magnetic field drifts that are actually producing the energization via their components along the transverse electric field. In this last transport term, the drift transport is characterized solely by the independent variable $\varepsilon=pv/3Qe$. Thus the global transport within the reservoir should be mainly a competition between acceleration in compression regions (or cooling in expansion regions) and the tendency for particles to drift out of those regions. The acceleration rate is the same for all species, while the drift velocities depend on the parameter (ε), which is 2/3 of the energy-per-charge for non-relativistic particles. We will consider these two competing processes for SEPs from the viewpoint of guiding center motion in the following two Sections, bearing in mind that their interaction is most rigorously combined in the reservoir transport described by Eq. (5).

4. SEP acceleration: Observational characteristics

Certain simple patterns in the behavior of solar energetic particles (SEPs) emerge when only magnetically "well-connected" events are selected for study. These are the events in which the spacecraft happens to be on interplanetary field lines that are connected at the tune of the first-arriving SEPs to the location marker on the Sun that is customarily taken as the site of the largest flare (observed in Hα, EUV, xrays, *etc.*) having the closest onset time association with the SEP onset. The solar longitude of the foot-point of the magnetic field line (Φ_0) is usually estimated using the Parker relation $\Phi_0=\Phi+\Omega R/V$, where (Φ) is the longitude of the spacecraft, (R) is its radial distance, and V is the solar wind velocity at the time of the observation. To good approximation (for observations at 1 AU), R is the distance from the center of the Sun, and it matter little whether ecliptic or heliospheric longitudes are used. As pointed out explicitly in the papers that first justified this connection relation [7,8], it does *not* require that the field line be a Parker spiral all the way back to the corona, but only that the plasma element (into which the field is "frozen") did not undergo significant acceleration (or deceleration) during its transit to the spacecraft.

Customarily, "well-connected" is taken to mean that the longitude of the foot-point of the field line is within ~30° of the flare longitude. On such field lines, the pitch-angle distribution of fast SEPs (v>0.1c) during the rise-to-maximum phase is usually strongly anisotropic, with a beam-like intensity distribution that can be characterized by an exponential in pitch cosine $j\approx j_0\exp(a\mu)$, where the parameter (a~1) is a measure of the anisotropy. Such a pitch-angle distribution was anticipated as a consequence of the "focusing" effect of the $\sim 1/r^2$ Parker field overcoming any pitch-angle scattering while the particle is en route to 1 AU [9]. The implication of such a p.a.d. is that most of the first-arriving particles are essentially traveling along the field line ($v_{||}=\mu v \approx v$), and so they must have been emitted at a time earlier by $\Delta t \approx s/v$ than their arrival time. For 250 keV electrons (with v=0.7c), this is about 10 minutes at 1 AU (where s≈1.2 AU). Thus, during the portion of the rise-to-maximum phase of a 250 keV electron event while these beam-like anisotropies are observed, we can construct the time histories of the injection process on a time scale considerably less than 10 minutes. Analysis of such events led to the discovery that these fast electrons commenced their interplanetary injection with delays of as much as 20 minutes after the type III radio emission from the associated flares [10]. An interpretation offered (and quantitatively supported by comparing SEP electron injection times with SOHO/LASCO observations of the velocity histories of the associated CMEs), was that the injection did not commence until the leading edge of the CME had attained an altitude of several solar radii into the corona [11].

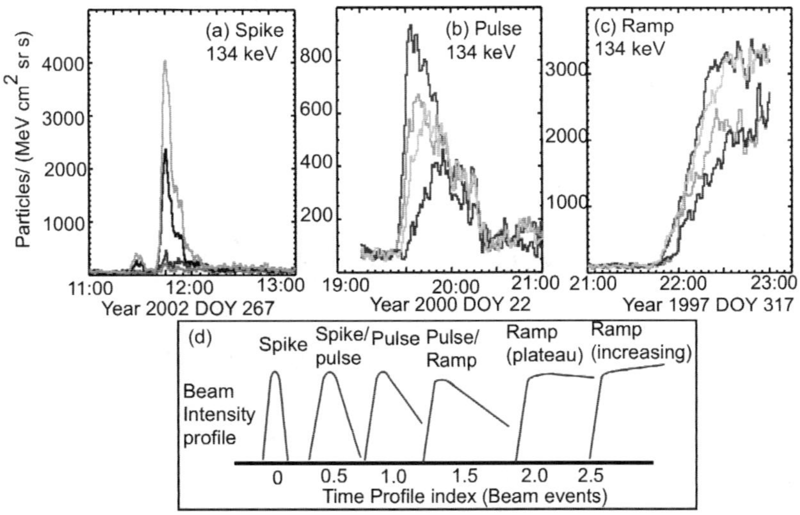

Figure 1. A continuum of beam-like injection profiles observed for the ~200 magnetically well-connected 40-300 keV ACE/EPAM electron events during Solar Cycle 23. From [12].

Haggerty and Roelof [12] made a comprehensive study of the patterns that emerge from the compilation the properties of ~200 magnetically well-connected 40-300 keV electron events during Solar Cycle 23. Examples of electron beam-like pitch-angle anisotropies are presented in the upper panels of Figure 1, in which measured ACE/EPAM intensities are plotted for four different directions around a 30° cone in the Sunward hemisphere. Small intensity events often appeared as "spikes", symmetric in their rise and fall and lasting <20 minutes in all. Moderate intensity events had a "pulse" shape, with a decay much slower than the rise. Large intensity events, although somewhat slower to rise than pulses, had a "ramp" shape, with a plateau or even a continued increase lasting at least an hour. When ~200 such well-connected beam-like electron events were analyzed over the complete Solar Cycle 23 [12], we found a continuum of such shapes, orderly enough for us to assign a "time profile index" ranging from 0 for a spike to 1 for a pulse and >2 for a ramp (as shown in the lower panel of Fig. 1). Figure 2 compares the distributions of peak intensity spectral index, peak intensity itself, injection

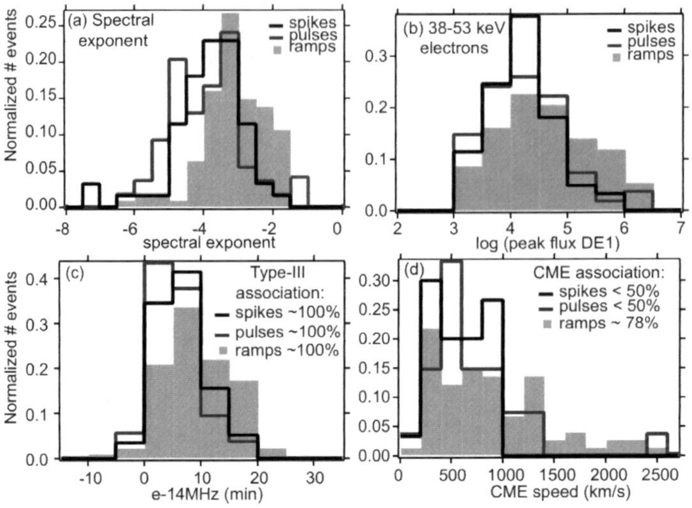

Figure 2. Essential properties of beam-like injection profiles (see Fig. 1). From [12]

delays (at the Sun) from 14 MHz type III radio bursts, and plane-of-sky speed of the associated CME. Although spikes, pulses and ramps exhibit somewhat similar distribution of all these parameters, the ramps have harder spectra, higher peak intensities, longer injection delays, and are associated with the fastest CMEs.

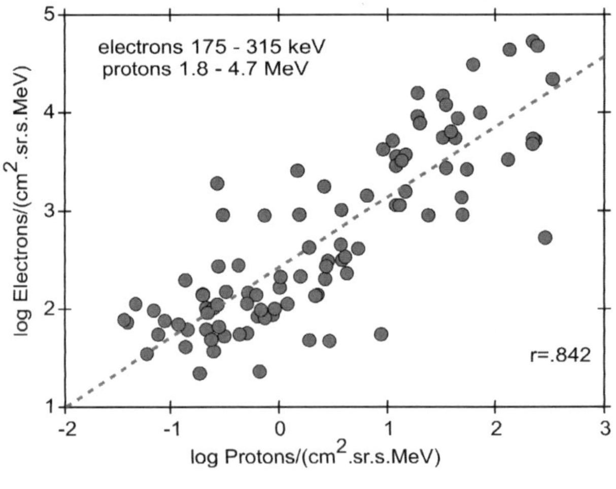

Figure 3. Maximum intensities of near-relativistic electrons *vs.* those of 1.8-4.7 MeV protons in magnetically well-connected beam-like electron events. From [12].

How can we explain consistently such an over-all orderly pattern in this continuum of well-connected electron events? Interplanetary diffusion cannot have a significant effect on the event histories, because there are beam-like pitch-angle anisotropies throughout the rise-to-maximum phase of the well-connected events. Magnetic reconnection in the corona might be a viable explanation for the spikes, but because it is inherently a sporadic process, one would have to invoke a well-behaved "ensemble" of many such reconnection events over several hours (lasting while the CME was in the high corona) in order to produce the relatively smooth profile of the ramps. On the other hand, CME-driven shocks might explain the ramps, but in order to produce the spikes, one would have to invoke a shock that "switched" on and off in only a few minutes. Finally, if one claimed that the sources of the spikes were magnetic connection while that of the ramps were high coronal CME-driven shocks, then why did we find a smooth continuum of time profiles from shocks through pulses to ramps throughout Solar Cycle 23?

Many aspects of the pattern found for the near-relativistic electrons were also found during Solar Cycle 23 in ~20 SEP well-connected beam-like proton events (20-100 MeV) within about the same velocity range (0.2c<v<0.5c) as the ACE/EPAM electrons. These strong proton pitch-angle anisotropies were measured by the two-dimension detector SOHO/Erne [12]. Detailed fitting of the rise-to-maximum of both the proton and electron intensities near 0° pitch angle (in the same ~20 events) revealed similar (although not identical) injection profiles, along with correlated injection delays. Even for lower energy ions (1.8-4.7 MeV, 0.006<v<0.010c), the peak intensities in well-connected events with beam-like near-relativistic electron anisotropies are correlated (r=0.842) to within an order of magnitude [12]. See Figure 3.

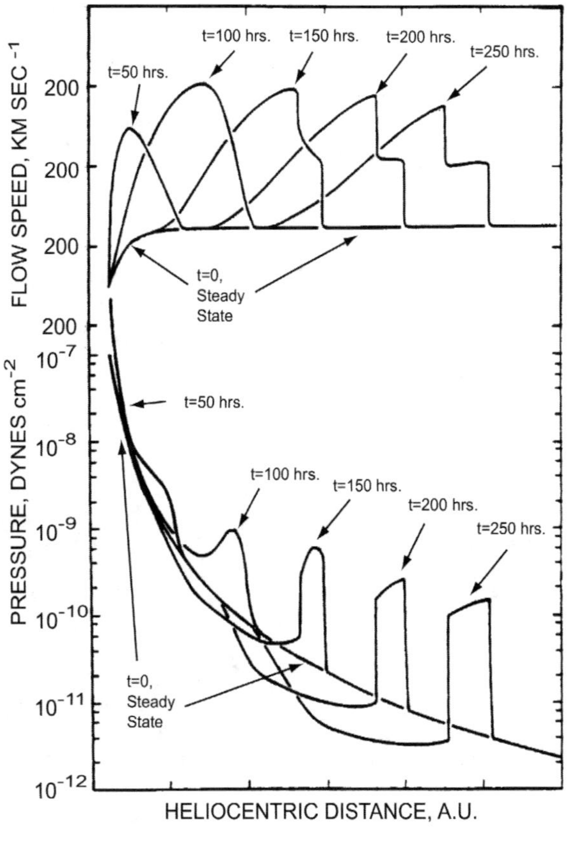

Figure 4. Snapshots of solar wind flow speed and pressure as functions of heliocentric distance at different times during the outward evolution of a high-speed stream driven by a slowly-increasing pressure pulse. Calculated from a simple 1-D gas dynamic code. Adapted from [13] and [14].

5. Suggested mechanism: Acceleration *within* coronal mass ejections

The implication of the previous Section is that the simple patterns observed in the injection histories for *both* near-relativistic electron and proton events demand a correspondingly simple and general acceleration/injection mechanism that must produce similar injection histories (and delays) for *both* protons and electrons. The rest of this paper will describe a mechanism that we believe can satisfy these nearly all-encompassing observational properties.

There is a fundamental property of the launch of CMEs through the corona out into the solar wind: they are driven by over-pressure that produces a radial plasma velocity profile that steadily steepens until (if it is fast enough) it drives a shock. Figure 4, adapted from [13] and [14], portrays a simple 1-D calculation of a relatively modest CME by Hundhausen [13] in which the temperature (and thereby the pressure) was increased and decreased by a factor of four at the inner boundary over 100 hours. The significance for us lies in the general shape if the radial velocity profile.

Consider a generic CME while it is only a solar radius or so above the solar surface. After it is launched, there will be a steady steepening of the velocity-radius profile. For fast CMEs, the steepening will eventually form a shock. What is important to us is that there is always a region inside the leading edge of the disturbance where $\partial V_r/\partial r<0$. We will assume in this region that: (i) the SEP pitch-angle distribution is nearly *isotropic*; and (ii) the swept-up magnetic field is *draped* across the evolving front, *i.e.*, it is strongly *non-radial*. Now consider the implications of our acceleration Eq. (4") and follow along with the aid of Figures 5 and 6.

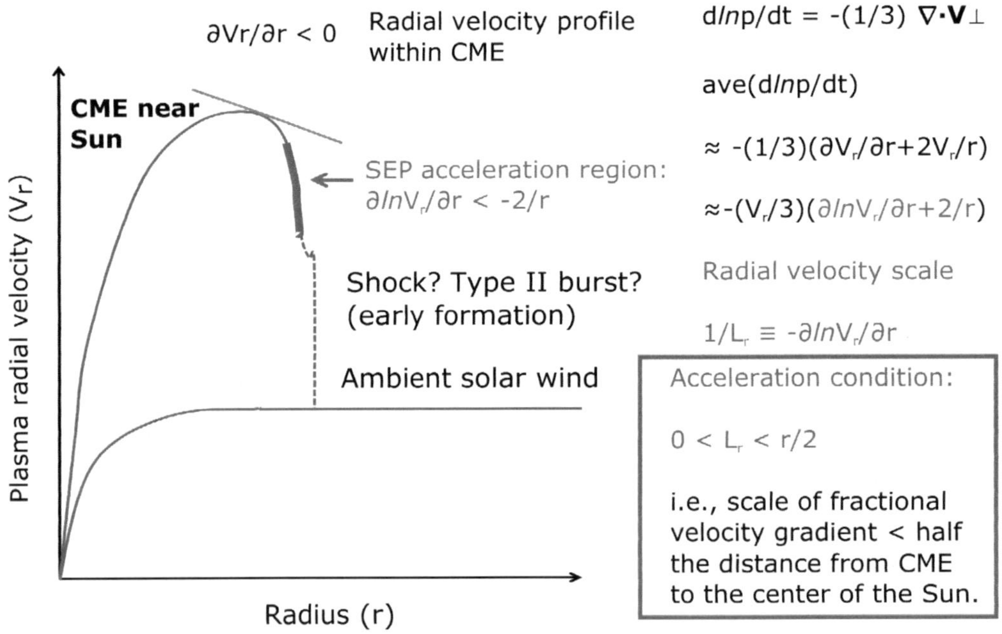

Figure 5. The basic conditions for compressive acceleration of SEPs within the driver plasma of CMEs.

5.1 The acceleration time

Since $\mathbf{V}_\perp=\mathbf{V}-V_\parallel\mathbf{b}$, we can write

$$\nabla\cdot\mathbf{V}_\perp = \nabla\cdot\mathbf{V} - B\partial(V_\parallel/B)$$

so that

$$d\ln p/dt \approx -(1/3)\ \nabla\cdot\mathbf{V}_\perp \approx -(1/3)[\ (1/r^2)\partial/\partial r(r^2V_r) - B\partial(V_{||}/B)/\partial s]$$

In the nose region of the CME, the plasma velocity may be taken to be predominantly radial. Because of assumption (ii), we can assume that $V_\perp \approx V_r$ and also that $|V_{||}| << V_r$.

$$\nabla\cdot\mathbf{V}_\perp \approx (1/r^2)\partial/\partial r(r^2V_r) - B\partial(V_{||}/B)/\partial s$$

$$= \partial V_r/\partial r + 2V_r/r - \partial V_{||}/\partial s + V_{||}\partial \ln B/\partial s \qquad (6)$$

Since the nose region is broad, neither ($V_{||}$) nor *ln*B should vary strongly along the field, and $V_{||}$ is already a small quantity compared to V_r. So we can neglect those terms in this important region.

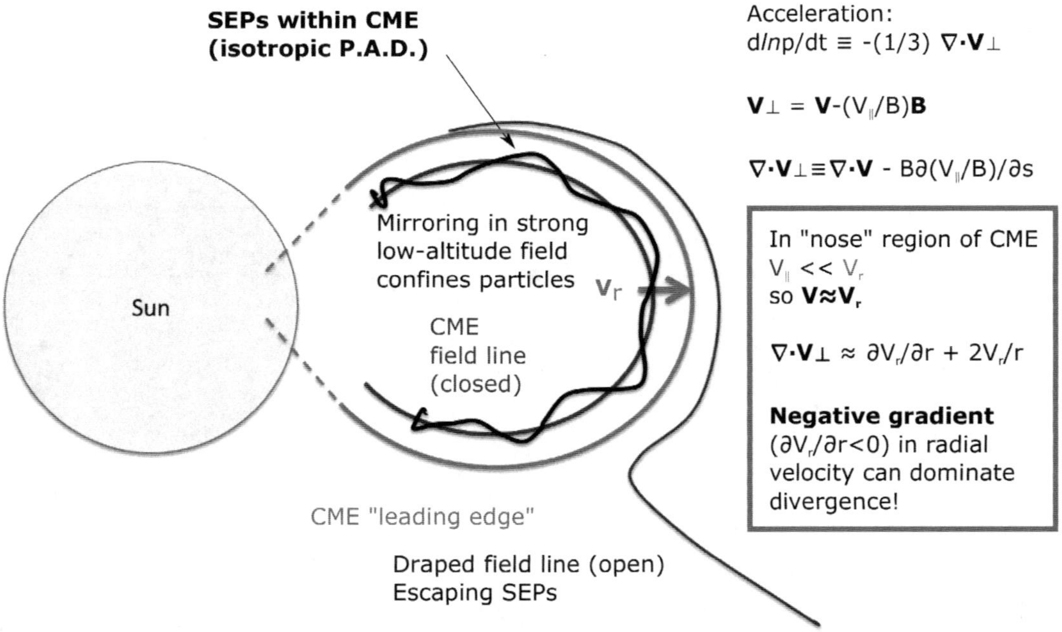

Figure 6. The global topology of CMEs that favors compressive acceleration of SEP near the nose.

The term ($\nabla\cdot\mathbf{V}_\perp$) will therefore be *negative* (implying acceleration, not deceleration) if $\partial V_r/\partial r < -2V_r/r$, so if we define a gradient scale length (L_r) for the *decreasing* radial velocity by $\partial \ln V_r/\partial r = -1/L_r$, then

$$\nabla\cdot\mathbf{V}_\perp \approx -V_r(1/L_r - 2/r) \qquad (7)$$

Thus the SEPs will be *accelerated* (because $\nabla\cdot\mathbf{V}_\perp<0$) in the nose region of the shock whenever $L_r<r/2$ (half the radial distance from the center of the Sun).

We can get an estimate of the characteristic time for energization from the guiding center equation for $d\ln p/dt$ (averaged over all pitch cosines for consistency with the assumed isotropy in the nose region). Defining this *growth* time as $1/\tau$, we see that

$$(-1/3)\nabla\cdot\mathbf{V}_\perp = 1/\tau \approx (-V_r/3)(-1/L_r + 2/r) = (V_r/3R_s)(R_S/L_r - 2R_S/r) \qquad (8)$$

To pick some numbers, let's take V_r=1000 km/s, so that $V_r/3R_s$=(1/2100s). Then in a nose region where the radial velocity gradient scale was $L_r=(1/10)R_s$ at $r=2R_S$, we obtain the e-folding time for momentum increase:

$$\tau \approx (2100s)/(10-1) = 3.9 \text{ minutes}$$

The e-folding time for *energy* increase alone (non-relativistic) will be half of this (2 minutes), or the same (4 minutes) for fully relativistic particles. Of course, this guiding center estimate assumes that the particle has remained in the nose region for the entire e-folding time. The very magnetic drifts that energize (or de-energize) the particle will also transport the particle (transversely) within the energization region. We now turn to that important issue.

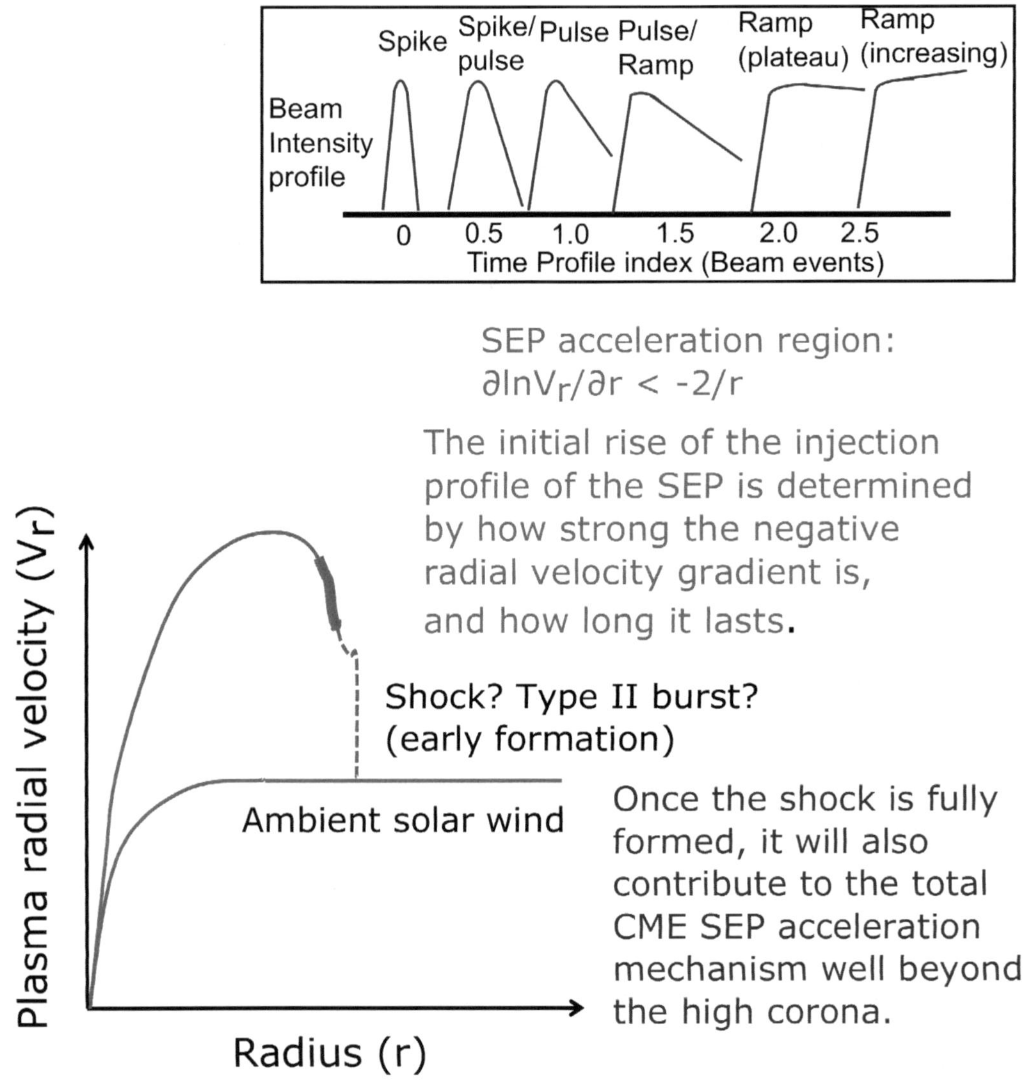

Figure 7. The basic properties of compressive SEP acceleration within CMEs that explain the simple patterns of magnetically well-connected SEP events (spikes, pulses and ramps) over the past Solar Cycle. See Figs 1, 2, and 3.

5.2 The "residence" times

The question of whether the net energization of the SEPs that escape upstream of the CME is positive or negative would be most properly answered by solving the time-dependent reservoir transport equation that includes processes such as escape from the region due to parallel anisotropy (ξ) or transverse motion across field lines due to magnetic curvature and gradient drifts. The $\mathbf{E}\times\mathbf{B}/B^2$ drift is, in this transverse (draped) magnetic field geometry, essentially the radial plasma velocity (V_r) that moves with the CME front. However, we can make some rough estimates of whether the SEPs are contained long enough to be significantly accelerated just from the guiding center equations. The usual expressions for the gradient and curvature drifts are

$$\mathbf{V}_D = (pv/QeB)\,\mathbf{b}\times[(1-\mu^2)/2\,\nabla \ln B + \mu^2\,\boldsymbol{\kappa}] \tag{9}$$

We can express $pv/QeB=(A/Q)\gamma\beta^2c^2/\Omega_0$, where $\Omega_0=eB/m_0$ is the non-relativistic gyro-frequency for protons (for electrons, multiply x 1836). We'll do the calculation for protons (or electrons), because the ion factor (A/Q) can be folded in at the end. To good approximation we have the "rule-of-thumb" $\Omega_0(s^{-1}) \approx B(nT)/10$, so $c^2/\Omega_0 \approx 9\times10^{11}\ km^2s^{-1}/B(nT)$. We make the rough estimates $|\nabla \ln B|\sim 1/L_B$ and $|\boldsymbol{\kappa}|=\kappa$, and we take an isotropic average over the pitch cosine. The application of the inequality $|\mathbf{b}\times(\mathbf{g}+\mathbf{h})|< |(\mathbf{g}+\mathbf{h})|<|\mathbf{g}|+|\mathbf{h}|$ to the vector cross-product yields

$$V_D \le (\gamma\beta^2c^2/\Omega_0)(1/L_B+\kappa) \tag{10}$$

Then the distance a particle drifts due to magnetic field gradient and curvature during the (isotropic) acceleration time (τ) estimated in the last section is bounded from below by

$$|\mathbf{V}_D|\tau \le (\gamma\beta^2c^2/\Omega_0)\,(1/3)(1/L_B + \kappa)/[(V_r/3)(1/L_r - 2/r)]$$

$$= (\gamma\beta^2c^2/\Omega_0V_r)\,(L_r/L_B + \kappa L_r)/(1-2L_r/r) \tag{11}$$

We have assumed again that the scale of the radial velocity gradient is much smaller than the radial distance of the acceleration region ($L_r<<r$). Scaling the radial plasma velocity (V_r) by 1000 km/s (as in the previous section) and $V_D\,\tau$ by 1 $R_S=7\times10^5$ km, we obtain the lower limit

$$(V_D\tau)/R_S \le (\gamma\beta^2)/B(nT)(9\times10^{10}\ km^2s^{-1})/[(1000\ kms^{-1})(7\times10^5\ km)]\ (1000\ km/s/V_r)\ (L_r/L_B + \kappa L_r)$$

$$= (\gamma\beta^2)\ (130)/B(nT)\ (1000\ kms^{-1}/V_r)\ (L_r/L_B + \kappa L_r) \tag{12}$$

If the leading edge of the CME is inside of r<0.1 AU, then B(nT)>500 nT and the factor 130/B(nT)<0.3. If we assume that all of the magnetic inhomogeneity scales (L_B and $1/\kappa$) are comparable to the gradient scale of the radial velocity (L_r), then for a CME with $V_r=1000\ kms^{-1}$ well inside r<<0.1 AU≈20R_S we conclude that the transverse magnetic drift distance ($V_D\tau$) during one momentum e-folding time (τ) is

$$(V_D\tau)/R_S << \gamma\beta^2 \tag{13}$$

Consequently even a relativistic proton ($\gamma>1$ and $\beta\approx1$) should drift only a small fraction of a solar radius (R_S) in a momentum e-folding time. Non-relativistic particles will drift only a tiny fraction of a solar radius. For multiply-charged ions we just include the factor (A/Q) on the RHS of Eq. (13). Therefore the only other confinement constraint involves the parallel (presumably weak-scattering) transport. However, the global topology of the leading portion of the CME is generally regarded as magnetically closed (*e.g.,* "light-bulb", "ice-cream cone", lifting flux rope). Moreover, when the particle leaves the acceleration region in the leading portion of the CME (where $\partial ln V_r/\partial r < -2/r$), it will encounter the much stronger magnetic fields at lower altitudes. Outside the nose region, this is a mirror confinement regime and is completely consistent with the reservoir requirement of weak field-aligned intensity gradients. Consequently, it would appear from the foregoing discussions that an efficient acceleration region for SEPs exists within the structure of outgoing CMEs.

6. Compression acceleration signatures in coronal images

We mention the intriguing possibility that the SEP acceleration time (τ) could be extracted *directly* from coronal white light imagers. This possibility is justified in a companion paper from this Conference [15], in which we argue that an estimate of the plasma *compression* ($-\nabla\cdot\mathbf{V}$) can be extracted from outward-propagating features *within* the CME, *i.e.,* within the driver of the CME shock. The extraction follows simply from the integration of the equation for the electron density (N), *i.e.,* for the conservation of electrons. Upon expansion and re-arrangement,

$$\partial N/\partial t + \nabla\cdot(n\mathbf{V}) = 0 \quad \text{becomes} \quad D ln N/Dt = -\nabla\cdot\mathbf{V} \tag{14}$$

where $D/Dt=\partial/\partial t+\mathbf{V}\cdot\nabla$ is the Lagrange convective derivative for a function co-moving (but evolving) with the (full) plasma velocity (**V**). Carrying out the implied integration, the average value of the compression along the trajectory of the centroid of the feature is

$$\langle -\nabla\cdot\mathbf{V}\rangle = (1/\Delta t)\ln\{N[\mathbf{r}(t+\Delta t),t+\Delta t]/N[\mathbf{r}(t),t]\} \tag{15}$$

In the companion paper, the local electron densities are related to the brightness in the white light *coronagraph* pixels. We shall therefore proceed in this paper assuming that we have a measurement of the compression $\langle -\nabla\cdot\mathbf{V}\rangle$ for the outward-propagating density feature.

As we just showed in the previous Section, we have the identity

$$\nabla\cdot\mathbf{V} = \nabla\cdot\mathbf{V}_\perp + \nabla\cdot(\mathbf{B}V_{||}/B) = \nabla\cdot\mathbf{V}_\perp + B\partial(V_{||}/B)/\partial s$$

This equation becomes really useful if we can make an estimate of the divergence of the parallel plasma velocity, or even more simply (as in the geometry of the nose region of the CME), we can argue that $V_{||}$ is negligible compared to the radial velocity: $V_\perp \approx V_r$ so that $V_{||} \ll V_r$. Then we can approximate the *average* value of the acceleration rate along the path of the density maximum in the brightest pixels of images taken over a time interval $\Delta t = t_2 - t_1$.

$$\langle 1/\tau\rangle = (1/\Delta t) \int_{t1}^{t2} dt/\tau[\mathbf{r}(t),t] \approx (1/3\Delta t) \int_{t1}^{t2} dt(-\nabla\cdot\mathbf{V}) = -(1/3\Delta t) ln(N_2/N_1) \tag{16}$$

In the companion paper [15], we suggest how the electron densities $N_1=N(\mathbf{r}_1,t_1)$, $N_2=N(\mathbf{r}_2,t_2)$ can be extracted from the normalized pixel brightnesses in sequential coronagraph images. Note that any *linear* instrument calibration and response factors will cancel out of the density ratios. Also note that we do *not* have to estimate: either the plasma velocity (**V**): or the time derivative ($\partial N/\partial t$): or the spatial gradient (∇N) of the electron density! All we have to do is track the feature's normalized *brightness* in the images. Of course, we must still estimate of the maximum electron density in the feature. A method is suggested in [15]. If this approach proves feasible, it will give us observational "closure" for well-connected SEP events -- when we have concurrent coronagraph images that capture the radial brightness histories of density features in the associated CME.

7. An observation of compressive acceleration of SEPs at 1AU

We have been discussing compressive SEP acceleration within CMEs escaping through the corona, but we now believe that there was a fortuitous observation by the ISEE-3 spacecraft of an atypical SEP event (which happened to be among the largest of Solar Cycle 21) that was undergoing compressive acceleration within an ICME (interplanetary CME) at 1AU [16]. The report was titled "A Major Solar Shock-Associated Energetic Storm Particle Event Wherein the Shock Plays a Minor Role", and the unusual configuration of the event is summarized in Figure 9. Because of the excellent angular coverage of the Low Energy Telescope, we could deduce from the beam-like pitch-angle distributions of the ~0.10-1.0 MeV ions that they were passing *adiabatically* through the shock, from the downstream region into the upstream region, with no signature of local shock acceleration. In addition, for the first quarter hour (~0.003 AU) behind the shock, the "slipped" layers of non-Parker-oriented magnetic field also contained ion beams that had to be moving along field lines that would eventually cross the shock. Further behind, the pitch-angle distributions were nearly isotropic, suggesting that this was the region that was the immediate source of the particle beams observed behind (and passing through) the shock.

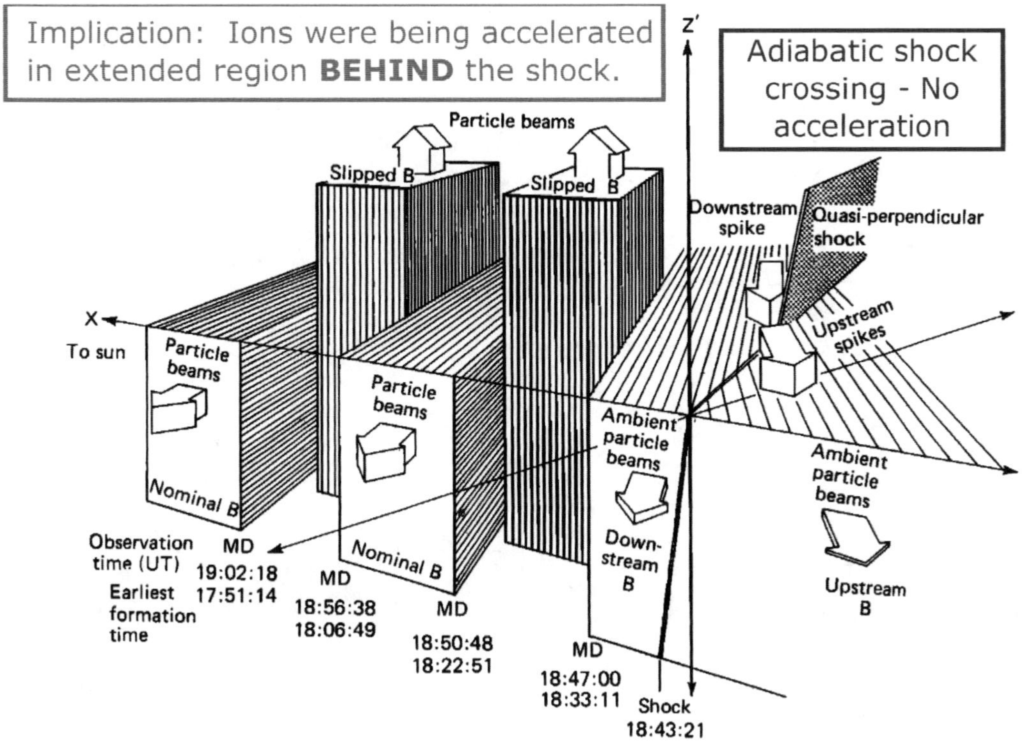

Figure 8. Stylized visualization of the shock/multiple-magnetic-discontinuity ensemble that controls the SEPs. All structures are populated with ambient particle fluxes that are strongly anisotropic, while intense particle beams are observed both upstream and downstream of the shock. The measured pitch-angle distributions imply that the particles are passing adiabatically through the shock (with no acceleration). From [16].

We now (30 years later) know what to look for, and we find it in Figure 9 (taken from the original paper) [16]. The key is the solar wind velocity history (V_{SW}). It jumps from ~300 km/s (upstream) to only ~400 km/s immediately behind the shock. This small jump is consistent with the observed lack of measureable shock acceleration of energetic ions. However, the downstream velocity continues to rise during the next 2 hours (~0.030 AU *behind* the shock at an average velocity of 600 km/s) to as much as 750 km/s. This implies a strongly *negative* fractional velocity gradient $\partial ln V_r/\partial r \approx [ln(750/400)]/0.03$ AU, yielding (in the notation of the previous Section) a scale length for the negative gradient of $L_r \approx 0.05$, so that $L_r << r/2 = 0.50$ AU. We thus estimate that the SEPs were being accelerated in this *interplanetary compression region* with average characteristic time $\tau \approx L_r/V_r \sim (7.5 \times 10^6$ km/s)/(600 km/s) ~3.5 hr. Since this is very short compared to the transit time from the Sun for the ICME, we believe that it indicates efficient compressive acceleration in this ICME at 1AU, and we therefore take it as *in situ* evidence in support of the proposed mechanism of compressive acceleration in CMEs back at the Sun.

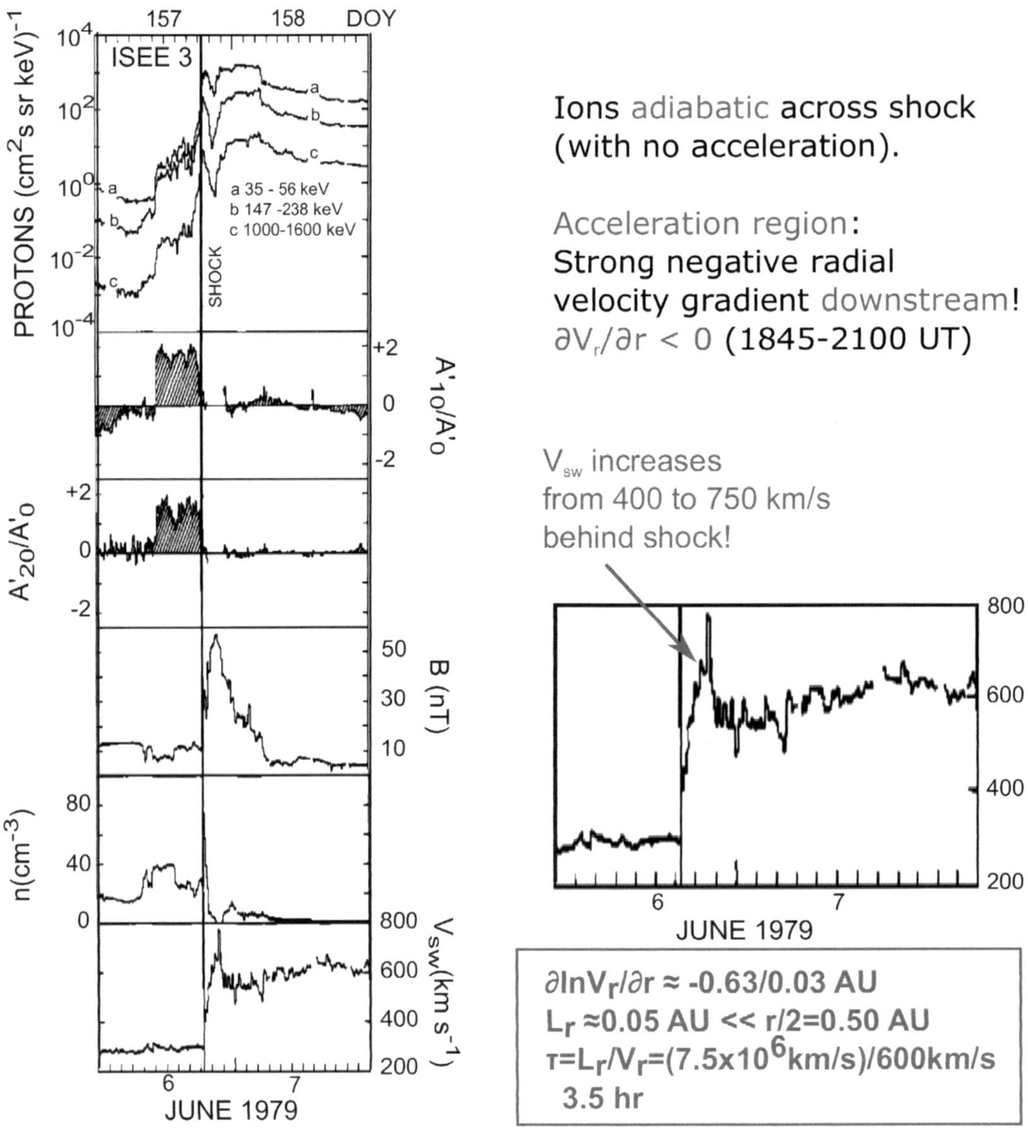

Figure 9. Energetic particle, magnetic field and plasma data averaged over 10-minute intervals from ISEE-3 during 6-7 June 1979. The estimate of the acceleration time (τ) is explained in the text. From [16].

8. Discussion and Summary

We have introduced a general guiding center equation for fractional momentum acceleration ($d\ln p/dt$) that: is independent of particle identity (mass, charge, energy); is valid for non-relativistic and relativistic particles; and holds for time-dependent electric and magnetic fields. The acceleration is produced by a combination of betatron acceleration and work done on the particle when its drift velocity in the gradient and curvature of the magnetic field has a component along the transverse electric field. Under the conditions within a CME (*e.g.*, pitch-angle isotropy), it takes the form $d\ln p/dt=-(1/3)\nabla\cdot\mathbf{V}_\perp$, so that acceleration occurs as a result of the compression of the plasma velocity component transverse to the magnetic field. For reasonable CME parameters, we obtain acceleration rates $1/\tau=d\ln p/dt$ on the order of a few minutes. We have presented corroboration for this mechanism from the unusual circumstance in which it was observed *in situ* by ISEE-3 during an SEP event at 1AU in June 1979.

Compressive acceleration, in its general concept, has appeared in many forms in the literature, *e.g.*, [17] and [18], but it usually also invokes meandering of field lines or turbulent motions of the plasma and/or pitch-angle scattering (none of which are required for our proposed mechanism). Although they may indeed be operative in a CME, none of these manifestations makes the very simple prediction that the acceleration itself is *independent* of all particle properties, depending only on the divergence of the (transverse) plasma velocity. In brief, we suggest that our new Eq. (4') has the very general properties required to explain the rather simple but broadly observed patterns in the rise-to-maximum phase of SEP events. The main variation in their time histories (spikes, pulses, ramps) is the result of how long the compression region lasts in the outgoing CME (see Fig.7), while the secondary variations in those profiles among different particle species may well be the consequence of their different "residence times" in the accelerating region which may be determined by their magnetic field drift velocities (which depend only on energy/charge). Finally, we should bear in mind the point made in the Introduction, that the final SEP spectrum may also depend on the properties of the "seed particle" population upon which the proposed compressive acceleration mechanism acts.

9. Acknowledgments.

The efforts of E. C. R. were supported by the NASA ACE and Voyager Interstellar missions under contracts NNX10AT75G and NNN06AA01C, respectively. This synthesis of observations was made possible by the many papers published in this century by Robert P. Lin and the University of California/Berkeley group, George M. Simnett and the University of Birmingham group, and our own group at the Johns Hopkins University/Applied Physics Laboratory.

10. Appendix. Derivation of the new equation for individual particle motion

Northrop's formulation [4] of the guiding center equation for energy (W) changes began with a non-relativistic equation

$$dW/dt = q\,d\mathbf{R}/dt\cdot\mathbf{E} + M_0\partial B/\partial t$$

where the non-relativistic magnetic moment is $M_0=m_0v_\perp^2/2B=m_0v^2(1-\mu^2)/2B$ and the perpendicular drifts are

$$d\mathbf{R}_\perp/dt = \mathbf{E}\times\mathbf{B}/B^2 + (m_0/qB)\mathbf{b}\times(v_\perp^2/2\,\nabla\ln B + v_\parallel^2\,\partial\mathbf{b}/\partial s)$$

$$= \mathbf{E}\times\mathbf{B}/B^2 + (m_0v^2/qB)\mathbf{b}\times[(1-\mu^2)/2\,\nabla\ln B + \mu^2\,\partial\mathbf{b}/\partial s]$$

Relativistically, the gradient and curvature drifts are increased by the relativistic factor (γ) because of the corresponding increase in the particle's gyroradius. In the limit of $\mathbf{E}_\perp$ "small", *i.e.*, V<<c, the dominant terms are still just the relativistic generalizations of the non-relativistic expressions.

$$d\mathbf{R}_\perp/dt = \mathbf{E}\times\mathbf{B}/B^2 + (\gamma m_0/qB)\mathbf{b}\times(v_\perp^2/2\,\nabla\ln B + v_\parallel^2\,\partial\mathbf{b}/\partial s)$$

$$= \mathbf{E}\times\mathbf{B}/B^2 + (\gamma m_0 v^2/qB)\, \mathbf{b}\times[(1-\mu^2)/2\, \nabla lnB + \mu^2\, \partial\mathbf{b}/\partial s]$$

leading to

$$dW/dt = qd\mathbf{R}/dt\cdot\mathbf{E} + (M_r/\gamma)\partial B/\partial t$$

where the relativistic expression for the adiabatic invariant is stated to be

$$M_r=\gamma^2 M_0=\gamma^2 v_\perp^2/2m_0 B$$

In the re-formulation [2,3], we simply set dW=vdp (which holds both relativistically and non-relativistically), and we obtained (still assuming that $E_\parallel=0$)

$$vdp/dt = q\mathbf{E}\cdot(\gamma m_0 v^2/qB)\mathbf{b}\times[(1-\mu^2)/2\, \nabla lnB + \mu^2\, \partial\mathbf{b}/\partial s] + (\gamma m_0 v^2/2B)(1-\mu^2)\partial B/\partial t$$

This simple change of dependent variable (from W to p) introduces a common factor on the RHS of $\gamma m_0 v^2$=pv, so our general result (relativistic or non-relativistic) is

$$d\,lnp/dt = \mathbf{E}\cdot(1/B)\mathbf{b}\times[(1-\mu^2)/2\, \nabla lnB + \mu^2\, \partial\mathbf{b}/\partial s] + (1/2B)(1-\mu^2)\partial B/\partial t$$

Please note carefully that since we have assumed that $E_\parallel=0$, then we have $\mathbf{E}=-\mathbf{V}\times\mathbf{B}$ for the total electric field, and therefore *only* $\mathbf{V}_\perp$ should appear explicitly in our equations.

All that remains to be done is to complete the inner vector product $(1/B)\mathbf{E}\cdot\mathbf{b}\times\mathbf{A}$ for some vector ($\mathbf{A}$) with either $\mathbf{A}=\nabla lnB$ or $\mathbf{A}=\mathbf{b}/\partial s=\boldsymbol{\kappa}$ (the curvature vector of the magnetic field lines). The "box product" identity is

$$(1/B)\mathbf{E}\cdot\mathbf{b}\times\mathbf{A} = (\mathbf{b}\times\mathbf{V})\cdot(\mathbf{b}\times\mathbf{A}) = \mathbf{V}\cdot\mathbf{A}-(\mathbf{b}\cdot\mathbf{V})(\mathbf{b}\cdot\mathbf{A}) = \mathbf{V}_\perp\cdot\mathbf{A}_\perp$$

After a trivial rearrangement of terms, we obtain the form of the guiding center equation for the fractional time-rate-of-change of the particle momentum.

$$d\,lnp/dt = (1/2)(1-\mu^2)(\partial lnB/\partial t + \mathbf{V}_\perp\cdot\nabla lnB) + \mu^2\, \mathbf{V}_\perp\cdot\boldsymbol{\kappa}$$

Note that, although the perpendicular subscript ($\mathbf{V}_\perp$) is redundant in the term ($\mathbf{V}_\perp\cdot\boldsymbol{\kappa}$) because the curvature is always transverse to (**B**), it is necessary to indicate it explicitly in the term ($\mathbf{V}_\perp\cdot\nabla lnB$), because the gradient of (B) may have a parallel component.

The direct path to our new equation follows from realizing that with $\mathbf{E}=\mathbf{E}_\perp=-\mathbf{V}\times\mathbf{B}$ we can immediately write $\mathbf{E}_\perp=-\mathbf{V}_\perp\times\mathbf{B}$ when we substitute the electric field into Faraday's law.

$$\partial\mathbf{B}/\partial t = -\nabla\times\mathbf{E}_\perp = \nabla\times(\mathbf{V}_\perp\times\mathbf{B})$$

$$= \mathbf{V}_\perp(\nabla\cdot\mathbf{B}) - \mathbf{B}(\nabla\cdot\mathbf{V}_\perp)+\mathbf{B}\cdot\nabla\mathbf{V}_\perp - \mathbf{V}_\perp\cdot\nabla\mathbf{B}$$

$$\partial\mathbf{B}/\partial t+\mathbf{V}_\perp\cdot\nabla\mathbf{B} = D_\perp\mathbf{B}/dt = -\mathbf{B}(\nabla\cdot\mathbf{V}_\perp)+\mathbf{B}\cdot\nabla\mathbf{V}_\perp$$

Since $D_\perp/dt$ is a linear differential operator, we can write

$$B\, D_\perp B/dt = (1/2)\, D_\perp B^2/dt = (1/2)\, D_\perp(\mathbf{B}\cdot\mathbf{B})/dt = \mathbf{B}\cdot D_\perp\mathbf{B}/dt$$

$$\mathrm{B}\,\mathrm{D}_\perp\mathrm{B}/\mathrm{dt} = \mathrm{B}^2\,(\boldsymbol{\nabla}\cdot\mathbf{V}_\perp) + \mathbf{B}\cdot(\mathbf{B}\cdot\boldsymbol{\nabla}\mathbf{V}_\perp)$$

If (s) is the distance along the field line, then

$$\mathbf{B}\cdot(\mathbf{B}\cdot\boldsymbol{\nabla}\mathbf{V}_\perp) = \mathbf{B}\cdot(\mathrm{B}\partial\mathbf{V}_\perp/\partial\mathrm{s}) = \mathrm{B}^2\mathbf{b}\cdot(\partial\mathbf{V}_\perp/\partial\mathrm{s})$$

$$\mathbf{b}\cdot(\partial\mathbf{V}_\perp/\partial\mathrm{s}) = \partial(\mathbf{b}\cdot\mathbf{V}_\perp)/\partial\mathrm{s} - \mathbf{V}_\perp\cdot\partial\mathbf{b}/\partial\mathrm{s} = -\,\mathbf{V}_\perp\cdot\boldsymbol{\kappa}$$

This leaves only two terms

$$\mathrm{D}_\perp ln\mathrm{B}/\mathrm{dt} = -\boldsymbol{\nabla}\cdot\mathbf{V}_\perp - \mathbf{V}_\perp\cdot\boldsymbol{\kappa}$$

Thus, using Faraday's law to eliminate the explicit time dependence in the magnetic field, the guiding center equation finally takes the remarkably simple form, which is nonetheless valid for time-dependent variations in both **B** and $\mathbf{V}_\perp$.

$$\mathrm{d}ln\mathrm{p}/\mathrm{dt} = (1-\mu^2)/2\ \mathrm{D}_\perp ln\mathrm{B}/\mathrm{dt} + \mu^2\,\mathbf{V}\cdot\boldsymbol{\kappa}$$

$$= -(1-\mu^2)/2\ (\boldsymbol{\nabla}\cdot\mathbf{V}_\perp + \mathbf{V}_\perp\cdot\boldsymbol{\kappa}) + \mu^2\,\mathbf{V}\cdot\boldsymbol{\kappa}$$

11. References

[1] Roelof, E. C. (2008) Scatter-free propagation of energetic solar electrons and galactic cosmic rays in the inner heliosphere, *Proc. 7th Ann. Int'l. Astrophys. Conf., AIP Conf. Proc.,* **1039**, 174-183.

[2] Kunow, H., *et al.* (including E. C. Roelof) (1999), Co-rotating Interaction Regions at High Latitudes, *Space Sci. Rev.,* **91**, 221-268.

[3] Roelof, E. C. (2000), High Latitude Observations of Corotating Interaction Regions: Remote Sensing Using Energetic Particles, 242-249 *American Institute of Physics* (New York), *AIP Conf. Proc. No.* **528**, 242-249.

[4] Northrop, T. G. (1963) *The Adiabatic Motion of Charged Particles,* (Interscience, New York).

[5] Roelof, E. C. (2012a) Energetic particle reservoirs: 1. Derivation of the transport equation, *Proc. 11th Ann. Int'l. Astrophys. Conf., AIP Conf. Proc.,* **1500**, 174-179.

[6] Roelof, E. C. (2012b) Energetic particle reservoirs: 2. Solutions of the transport equation and comparison with data, *Proc. 11th Ann. Int'l. Astrophys. Conf., AIP Conf. Proc.,* **1500**, 180-185.

[7] Nolte, J. T. and E. C. Roelof (1973a) Large-scale structure of the interplanetary medium: 1, High coronal source longitude of the quiet-time solar wind, *Solar Physics*, *33*, 241.

[8] Nolte, J. T. and E. C. Roelof (1973b) Large-scale structure of the interplanetary medium: 2, Evolving magnetic configurations deduced from multi-spacecraft observations, *Solar Physics*, *33*, 483.

[9] Roelof, E. C. (1969) Propagation of solar cosmic rays in the interplanetary magnetic field, Ch. VII of *Lectures in High Energy Astrophysics*., H. Ogelman and J. R. Wayland, ed. NASA SP-199.

[10] Krucker, S., D. E. Larson, R. P. Lin, and B. J. Thompson (1999) On the origin of impulsive electron events observed at 1 AU ,*Astrophys. J.,* **519**, 864-875.

[11] Simnett, G. M., E. C. Roelof, and D. K. Haggerty (2002) Near-relativistic electrons accelerated by shocks associated with coronal mass ejections (CMEs), *Astrophys. J.,* **579**, 854-862.

[12] Haggerty, D. K. and E.C. Roelof (2009) Probing SEP acceleration processes with near-relativistic electrons, *Proc. 8th Ann. Int'l Astrophys. Conf.* AIP Conference Proceedings, **1183**, 3-10.

[13] Hundhausen, A. J. (1973), Non-linear model of high speed solar wind streams, *J. Geophys. Res.,* **78**,1528-1542.

[14] Gosling, J. T. and V. J. Pizzo (1999) Formation and evolution of corotating interaction regions and their three-dimension structure, *Space Sci. Rev.,* **89**, 21-52.

[15] Roelof, E. C. and Vourlidas, A. (2015), On extracting plasma compression signatures from white light coronal images, *This Conference.*

[16] van Nes, P., E. C. Roelof, R. Reinhard, T. R. Sanderson, and K.-P. Wenzel, (1985) A major shock-associated energetic storm particle event wherein the shock plays a minor role, *J. Geophys. Res.*, **90**, 3981-3994.

[17] Giacalone, J. and J. R. Jokipii (2007), Adiabatic compression acceleration of fast charged particles, *Astrophys. J.*, **660**, 336-340.

[18] Fisk, L. A., and G. Gloeckler (2014), The case for a common spectrum of particles accelerated in the heliosphere: Observations and Theory, *J. Geophys. Res.*, ***119**, 8733-8749.*

On extracting plasma compression signatures from white light coronal images

E C Roelof[1], **A Vourlidas**[1],
[1]The Johns Hopkins University. Applied Physics Laboratory, 11100 Johns Hopkins Rd., Laurel, MD 20723, USA

E-mail: edmond.roelof@jhuapl.edu

Abstract. We offer the intriguing possibility that the divergence of the plasma velocity ($\nabla\cdot\mathbf{V}$) in outward-moving density features can be extracted *directly* from coronal and heliospheric white-light images of coronal mass ejections (CMEs). The relevant empirical formula is quite simple to apply. It follows directly from the conservation of electrons. We derive an approximate relation between the measured white-light intensities in the brightest image pixels and the dominant electron densities along the pixel line of sight by invoking the generalized mean value theorem for integrals. If the formula holds with acceptable accuracy, we then may have a measure of not only ($\nabla\cdot\mathbf{V}$) itself, but also the related compressive acceleration rate of solar energetic particles within the CME.

1. Introduction

The quantity ($-\nabla\cdot\mathbf{V}$) measures the *compression* of the plasma within the feature. This is a fundamental measure of plasma dynamics in a CME. In particular, the nature of the compression in the sheath (or driver) region of the CME should vary strongly with the altitude of the driver. For instance, the velocities of Type II shock-associated CMEs can exhibit a variety of acceleration signatures (~±10 m/s^2) out to several solar radii (R_S), e.g., [1]. Those CMEs with constant velocities can be characterized by approximately steady *expansion* in which $\nabla\cdot\mathbf{V}\approx 2V_r/r>0$. However, very low in the corona during their launch and acceleration phase, the driver region should contain a volume of strong over-pressure with the consequent *compression* of the mass ahead of it, *i.e.*, $\nabla\cdot\mathbf{V}<0$.

We are proposing an observational method for extracting the quantity $\nabla\cdot\mathbf{V}$ *directly* from white-light images of the solar corona. The method should be valid in either the compression ($\nabla\cdot\mathbf{V}<0$) or expansion ($\nabla\cdot\mathbf{V}>0$) regimes, but, for the reasons given above, we expect that compression (particularly strong compression) is most likely to be observed at the lowest altitudes in the corona, *i.e.*, close to the innermost edges of coronagraph images. This poses an observational challenge, but we believe it will be well worth the effort.

There is an additional and closely related scientific motivation for being able to detect compression regions in the low corona. It should not be too surprising that the plasma compression can be related to the *acceleration* of solar energetic particles (SEPs). A simple and direct quantitative relation between the characteristic e-folding time for SEP acceleration (τ) and the plasma "transverse" *compression* ($-\nabla\cdot\mathbf{V}_\perp$) was derived in the companion paper [2]. The inverse ($1/\tau$) of the particle

[1] Correspondence should be addressed to edmond.roelof@jhuapl.edu.

acceleration time is defined as the fractional time-rate-of-change of the particle's momentum (p) along its guiding center trajectory, and the relationship to plasma compression is

$$1/\tau = d\ln p/dt \approx (1/3)(-\nabla \cdot V_{\perp}) \quad (0)$$

The relationship holds even if the plasma flow and the related electric and magnetic fields are time-dependent. It also holds for both non-relativistic and relativistic charged particles, and (remarkably) for *all* charged particles (since neither mass, energy, nor charge appears on the RHS).

2. Observations of CMEs associated with SEP events

We first present some observations that suggest to us that there are compressional regions within outgoing CMEs that will accelerate SEPs. Kahler and Vourlidas [3] made a comparison of the properties of 116 CMEs observed by LASCO and the peak intensities of 20 MeV protons measured by Wind/EPACT during the associated western hemisphere SEP events (1998-2002). The proton peak intensities generally increased with increasing CME velocity, but had a wide spread (factor>10^3) about this trend. Calling most intense 15 events matching the trend "SEP-Rich" and the least intense 16 events "SEP-Poor", the authors concluded that the SEP-Rich events were "clearly larger events in the low corona".

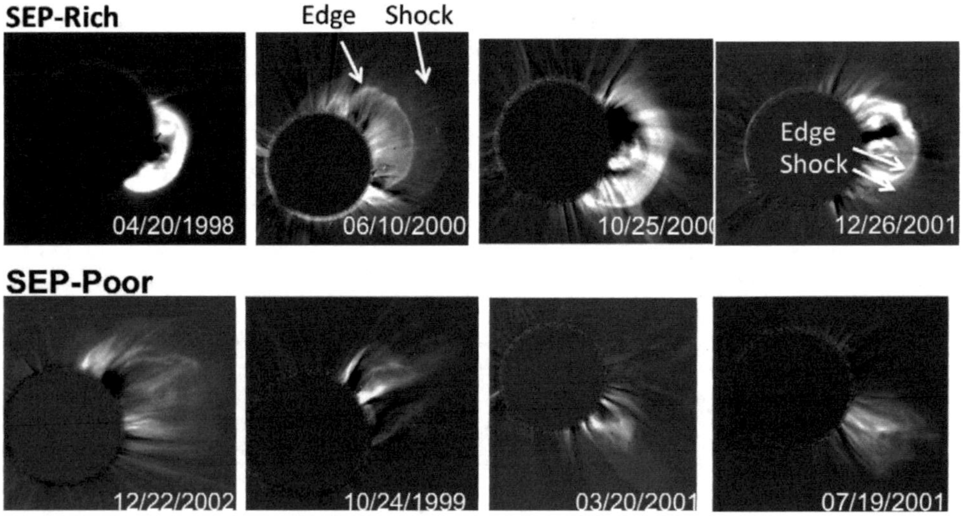

Figure 1. *Upper row*: Examples of 'SEP-Rich' CMEs as defined in Kahler and Vourlidas [3]. They are characterized by a bright front ('edge') behind the fainter shock front. In some cases, there's a clear separation between the shock and driver edge (marked by arrows) while in other cases they appear to overlap. *Lower Row*: Examples of 'SEP-Poor' CMEs. Dark disk is inner edge of the SOHO/LASCO coronograph at $r=2R_S$. The SEP-Rich CMEs are clearly larger events that are brighter lower in the corona.

Eight CME images drawn from this joint data set are displayed in Figure 1. The dark central disc is the occultation by the SOHO/LASCO coronagraph for $r<2R_S$. The SOHO/LASCO images have been calibrated (excess mass) and scaled identically; for more details of the image preparation see [4] and [5]. The upper row displays CME images associated with "SEP-rich" events. The shocks (at $\sim 5R_S$) are easily identifiable and are well separated by $\sim 1R_S$ from the bright leading edges of the CMEs (at $\sim 4R_S$). The lower row displays images associated with "SEP-poor" events. The lack of sharp and bright fronts suggests that a shock is unlikely to have developed, but the CME did in fact produce an SEP event (albeit a relatively small one). There would still be regions of plasma compression within the CME (too weak to drive a strong shock) that nonetheless could have accelerated the observed SEPs. Although the strongest brightening may occur when the leading edge of the CME is closer to the Sun than $r=2R_S$ (the inner limit of the images), we believe that a careful measurement of such images will yield a useful estimate of the compression ($-\nabla\cdot\mathbf{V}$) for $r>2R_S$, sufficient to discriminate between positive values (implying compression) and negative values (implying expansion).

3. Basic relation between electron density and plasma compression

The extraction procedure for obtaining estimates of the plasma compression ($-\nabla\cdot\mathbf{V}$) from white-light images of the corona follows from a version of the familiar conservation equation for electron density (N). As explained in the companion paper [2], expansion and re-arrangement of the terms yields

$$\partial N/\partial t + \nabla\cdot(n\mathbf{V}) = 0 \quad \text{becomes} \quad D\,ln N/Dt = -\nabla\cdot\mathbf{V} \qquad (1)$$

where $D/Dt=\partial/\partial t+\mathbf{V}\cdot\nabla$ is the Lagrange convective derivative for a function co-moving (but evolving) with the plasma velocity (**V**).

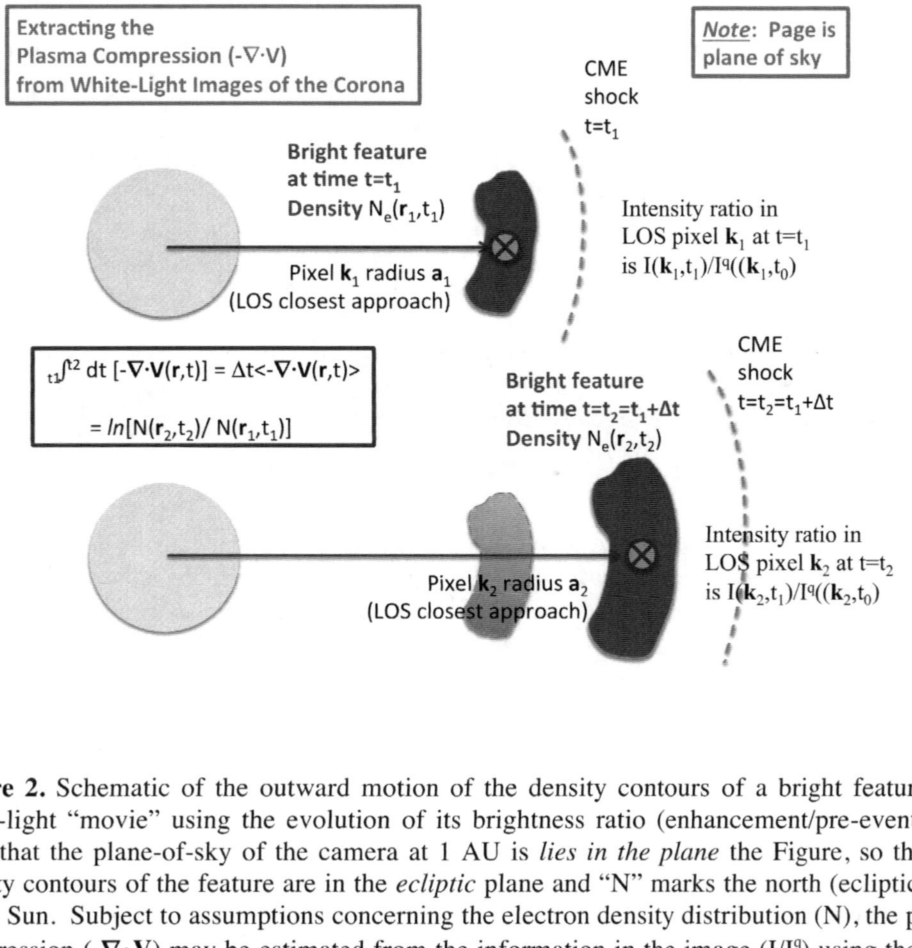

Figure 2. Schematic of the outward motion of the density contours of a bright feature in a white-light "movie" using the evolution of its brightness ratio (enhancement/pre-event=I/I^q). Note that the plane-of-sky of the camera at 1 AU is *lies in the plane* the Figure, so the blue density contours of the feature are in the *ecliptic* plane and "N" marks the north (ecliptic) pole of the Sun. Subject to assumptions concerning the electron density distribution (N), the plasma compression ($-\nabla\cdot\mathbf{V}$) may be estimated from the information in the image (I/I^q) using the basic relationship among them (written in the box).

Then we can formally integrate the convective derivative along the characteristic plasma path $d\mathbf{r}=\mathbf{V}(\mathbf{r},t)dt$ over the time range $t_1<t<t_2$, thus obtaining the *local* path-averaged plasma *compression* $<-\nabla\cdot\mathbf{V}>$, which has the units of inverse time.

$$\int_{t1}^{t2} dt\,[-\nabla\cdot\mathbf{V}(\mathbf{r},t)] = \Delta t<-\nabla\cdot\mathbf{V}(\mathbf{r},t)> = ln[N(\mathbf{r}_2,t_2)/\,N(\mathbf{r}_1,t_1)] \quad (2)$$

Here $N(\mathbf{r}_1,t_1)$ and $N(\mathbf{r}_2,t_2)$, are the electron densities at heights $\mathbf{r}_1$, $\mathbf{r}_2$ at times t_1,t_2, respectively and $\Delta t=t_2-t_1$. The geometry is sketched in Figure 2.

4. Extraction of compression from coronal white light images

Now suppose an outward moving brightness feature within the CME itself, not necessarily at the shock , can be identified in successive coronal or heliospheric images. See Figure 2. Note that the brightest pixel of the feature need not necessarily be propagating radially. Since the corona is optically thin in white light, the (calibrated) image brightness $I(\mathbf{k},t)$ in the pixel direction specified by the unit vector ($\mathbf{k}$) is proportional to the line-of-sight (LOS) integral (over $d\mathbf{r}=\mathbf{k}dl$) of the electron density $N(\mathbf{r},t)$

$$I(\mathbf{k},t) = \int dl\, N(\mathbf{k}l,t) \int d^2\omega\, \sigma(\boldsymbol{\omega}\cdot\mathbf{k})I_0(\mathbf{k}l,\boldsymbol{\omega}) = I_0 \int dl\, N(\mathbf{k}l,t)\, \Delta\omega(\mathbf{k}l)\, \sigma_{ave}(\mathbf{k}l) \quad (3)$$

where I_0 is the photospheric brightness in white light and $\sigma_{ave}(\mathbf{k}l)$ is the value of the Thomson scattering cross-section of a photospheric photon (unit direction $\boldsymbol{\omega}$) averaged over $\Delta\omega(\mathbf{k}l)$, the solid angle subtended by the Sun at LOS position $\mathbf{r}=\mathbf{k}l$. Consistent with the usual method of analyzing coronagraph images of moving structures, we compare the *ratio* $\Gamma(\mathbf{k};t,t_0)$ of the LOS intensity $I(\mathbf{k},t)$ in the pixel direction ($\mathbf{k}$) within the CME to the pre-event intensity of the "quiet" corona, $I^q(\mathbf{k},t_0)$ in the same pixel ($\mathbf{k}$) at an earlier time $t_0<t$.

$$\Gamma(\mathbf{k};t,t_0) = I(\mathbf{k},t)/I^q(\mathbf{k},t_0) \quad (4)$$

It is this pixel brightness ratio (Γ) that is actually depicted in coronograph movies of outgoing CMEs. It is therefore a directly *observable* quantity, and it is particularly useful because many of the coronograph calibration factors required for estimating the absolute intensities (I and I^q) cancel out of the ratio.

4.1. Simplification of the LOS integral of the intensity

A more useful representation of the LOS position $\mathbf{r}=\mathbf{k}l$ involves the angle (ψ) from the center of the Sun between the point of closest approach ($\mathbf{a}$) and the point ($\mathbf{r}$) on the LOS. Please refer to Figure 3 and note that the plane of sky is out of the page. The coordinate transformation is $\mathbf{r}=\mathbf{a}+\mathbf{a}\mathbf{k}\tan\psi$ so that $r=a\sec\psi$ and $dl=a\sec^2\psi$ with $-\pi/2<\psi<\pi/2$. Then a simple calculation will reveal that the solid angle subtended by the photospheric limb from position ($\mathbf{r}$) is

$$\Delta\omega = 2\pi(1-\cos\alpha) = 2\pi[1-(1-\cos^2\psi\, R_S^2/a^2)^{1/2}] \quad (5)$$

because $\sin\alpha=R_S/r=(R_S/a)\cos\psi$. We ignore limb darkening (so the illumination is uniform across the apparent solar disk), because it occupies a negligible fraction disk as viewed along the LOS. When the LOS is tangent to the photosphere ($a=R_S$) we have $\Delta\omega=2\pi(1-\sin\psi)$, *i.e.*, the solid angle encompasses the whole Sunward hemisphere at closest approach itself, when $\psi=0$. For $a>R_S$ we can apply the binomial expansion to $\Delta\omega$ with $x=(R_S/a)\cos\psi$ that converges for all $x^2<1$.

$$\Delta\omega = \pi x^2(1+x^2/4+x^4/8+5x^6/64+\ldots) \quad (5')$$

For LOSs with points of closest approach $a>2R_S$, we need therefore retain only the leading term in Eq. (5'), $\Delta\omega \approx \pi(R_S/a)^2\cos^2\psi$, because the next order correction term $(x^2/4)$ is never more than ~6% above the true value. On the other hand, for LOSs that pass within a few R_S of the Sun, the incident-direction-integrated Thomson (polarization) scattering term (σ_{ave}) will depend only weakly on the photon scattering direction ($\boldsymbol{\omega}$) through $\cos\psi=\mathbf{k}\cdot\boldsymbol{\omega}$. From here on, we will both approximate $\Delta\omega$ using only its leading term in Eq. (5') and also pull σ_{ave} out from under the LOS integral.

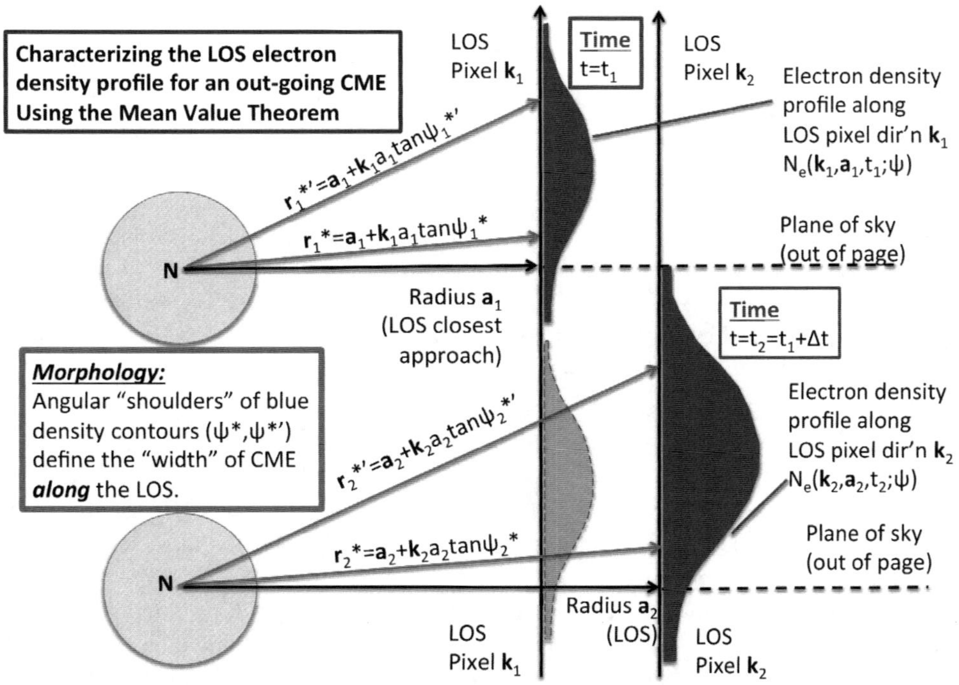

Figure 3. The generalized mean value theorem for integrals delimits the electron density profile (blue) as a function of time by the angular "shoulders" (ψ^*,$\psi^{*\prime}$) of the density function along pixel LOS. The plane of sky is out of the page while the LOS pixel directions ($\mathbf{k}_1$ and $\mathbf{k}_2$) are in the page, so that we are looking down on the north polar region (N) of the Sun.

Thus the equation we will use for the LOS intensity in pixel ($\mathbf{k}$) becomes

$$\begin{aligned} I(\mathbf{k},t) &= \pi a I_0 \int_{-\pi/2}^{\pi/2} d\psi\, \sec^2\psi\, [1-(1-\cos^2\psi\, R_S^2/a^2)^{1/2}]\, \sigma_{ave}(\mathbf{a}+\mathbf{a}\mathbf{k}\tan\psi)\, N(\mathbf{a}+\mathbf{a}\mathbf{k}\tan\psi,t) \\ &\approx (\pi R_S^2 I_0 \sigma_{ave}/a) \int_{-\pi/2}^{\pi/2} d\psi\, N(\mathbf{a}+\mathbf{a}\mathbf{k}\tan\psi,t) \end{aligned} \quad (6)$$

For the reasons that will soon become apparent, we re-express the electron density $N(\mathbf{r},t)$ at time (t) in terms of its ratio (γ) to the density $N^q(\mathbf{r},t_0)$ at the same position prior to the launch of the CME at time $t_0<t$.

$$\gamma(\mathbf{a}+a\mathbf{k}\tan\psi;t,t_0) = N(\mathbf{a}+a\mathbf{k}\tan\psi,t)/N^q(\mathbf{a}+a\mathbf{k}\tan\psi,t_0) \quad (7)$$

We then re-group these factors under the integral in Eq. (6) to obtain our basic imaging relation

$$I(\mathbf{k},t) = (\pi R_S^2 I_0 \sigma_{ave}/a) \int_{-\pi/2}^{\pi/2} d\psi\, \gamma(\mathbf{a}+a\mathbf{k}\tan\psi;t,t_0)\, N(\mathbf{a}+a\mathbf{k}\tan\psi,t_0) \quad (8)$$

4.2. Application of the generalized mean-value theorem for integrals (GMVTI)

In order to estimate the plasma compression using Eq. (2), we need an estimate of the actual electron density ratio $N(\mathbf{r}_2,t_2)/ N(\mathbf{r}_1,t_1)]$. We obviously need to make further assumptions and approximations in order to use Eq. (8) to estimate the distribution of the electron density along different LOSs, *e.g.*, see [3] for a detailed discussion of extracting mass densities (or electron densities) from coronagraph images of CMEs. We are assisted in this by the generalized mean value theorem for integrals (GMVTI), a very useful form of which is $\int dx f(x)g(x) = f(x^*)\int dx g(x)$. The theorem states that if $g(x)>0$ is a continuous function and $\int dx g(x)$ is bounded by some value $\Delta y<\infty$ (even if the range of x is infinite), then there must be at least one value of x^* (and there may be more) at which the theorem is satisfied. The theorem follows trivially from the ordinary theory of the mean after a change of variable of integration to $dy=dx g(x)$ so that $\int dy=\Delta y$, because the equation transforms to $\int dy f[x(y)]=f[x(y^*)]\Delta y$, revealing that $f(x^*)=f[x(y^*)]$ is the mean value of $f[x(y)]$ over the interval Δy. For brevity, we can write $f^*=f(x^*)$ where we also write $x^*=x(y^*)$.

We apply the GMVTI directly to Eq. (8).

$$\begin{aligned} I(\mathbf{k},t) &= (\pi R_S^2 I_0 \sigma_{ave}/a)\, \gamma^* \int_{-\pi/2}^{\pi/2} d\psi\, N^q(\mathbf{a}+a\mathbf{k}\tan\psi,t_0) \\ &= \gamma(\mathbf{a}+a\mathbf{k}\tan\psi^*;t,t_0)\, I^q(\mathbf{k},t_0) \end{aligned} \quad (9)$$

The quantity γ^* is the *mean value* of the density enhancement ratio ($\gamma=N/N^q$) for the LOS integral weighted by the pre-event density (N^q), in a manner that we will describe in the next Section. By the GMVTI, (γ) *must* take on its mean value (γ^*) at some point(s) along the LOS, which we designate by $\mathbf{r}^*=\mathbf{a}+a\mathbf{k}\tan\psi^*$. This is how we know that there is at least one angle (ψ^*) such that $\gamma^*=\gamma(\mathbf{a}+a\mathbf{k}\tan\psi^*;t,t_0)$. We have also recognized that the remaining integral in Eq. (9) is just the pre-event LOS intensity $I^q(\mathbf{k},t_0)$, as can be seen immediately from Eq. (6). Then Eq. (9) reduces to an equality between the intensity ratio of Eq. (4) and the density ratio of Eq. (7), evaluated at the angle(s) (ψ^*) defined by the GMVTI.

$$\Gamma(\mathbf{k};t,t_0)= \gamma(\mathbf{a}+a\mathbf{k}\tan\psi^*;t,t_0) \quad (10)$$

We have therefore obtained a relation between the ratio ($\Gamma=I/I^q$) of the enhanced intensity at time (t) to the intensity in the same pixel at the pre-event time ($t_0<t$) and the ratio ($\gamma=N/N^q$) of the corresponding electron densities at the *same* (unknown) angle (ψ^*) along the LOS for pixel (**k**) which has a distance of closest approach (**a**). This equation between intensity and density ratios is our fundamental analytical tool for extracting the electron densities, as can be seen by re-writing it explicitly in terms of the densities.

$$N(\mathbf{k},\mathbf{a},\psi^*,t) =N^q(\mathbf{k},\mathbf{a},\psi^*,t_0)\, \Gamma(\mathbf{k};t,t_0) \quad (11)$$

The utility of Eq. (11) depends upon two things: 1) our knowledge of the pre-event electron density distribution $N^q(\mathbf{k},\mathbf{a},\psi^*,t_0)$, and 2) the properties of the angle(s) (ψ^*) that come out of our application of the generalized mean theorem.

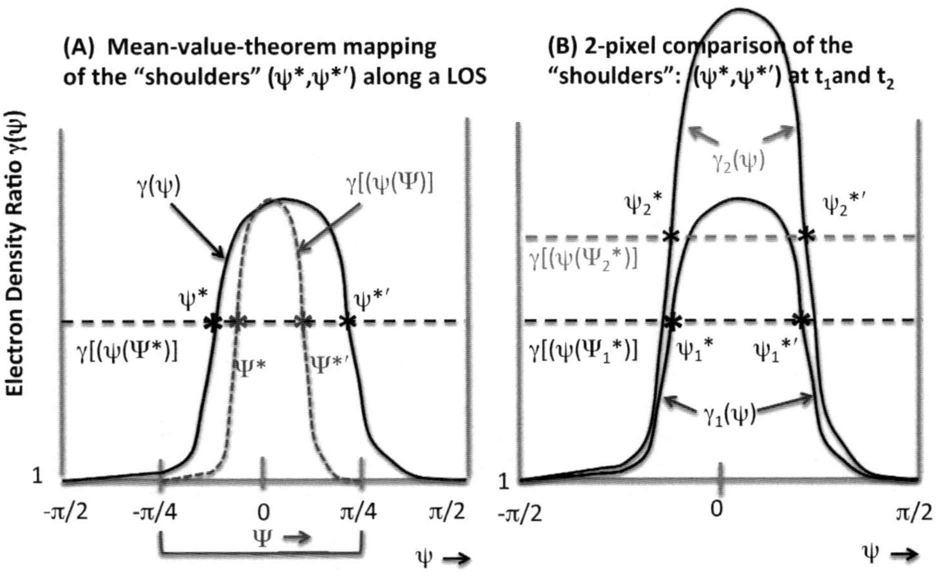

Figure 4. Functional relationships in the application of the generalized mean value theorem for integral (GMVTI) to the LOS density ratio profiles $\gamma(\psi)$ given by the black curves. (A) The mapping of $\gamma(\psi)$ in black to $\gamma[\psi(\Psi)]$ in blue in terms of the pre-event-density-weighted angle (Ψ). The GMVTI implies the existence of the two "shoulders" ($\Psi^*, \Psi^{*\prime}$) where the density ratio equals its average value $\gamma^*=\gamma[\psi(\Psi^*)]$ over the LOS integral. The inverse mapping ($\Psi \rightarrow \psi$) then gives the "shoulders" in terms of the actual LOS angles ($\psi^*, \psi^{*\prime}$). (B) If the shape of the CME is self-similar in two successive images (t_1, t_2) of the white-light "movie", then the pairs of values of corresponding the "shoulder" angles ($\psi_1^*, \psi_1^{*\prime}$) and ($\psi_2^*, \psi_2^{*\prime}$) should be approximately equal in the corresponding brightest pixels ($\mathbf{k}_1, \mathbf{k}_2$).

4.3. Interpreting the GMVTI

Eq. (11) cannot be interpreted until we can understand the behavior of the values of the angle (ψ^*) implied by the GMVTI. We can do this, because we actually possess a considerable amount of information, once we assume that our LOS pixel direction (**k**) of the (relatively) brightest feature tracks the maximum of the (relative) electron density with time across the plane of sky. Many CMEs appear to have a single dominant density maximum along the LOS, and this is the way we drew our sketch in Fig. 3. This means that there will be two angles ($\psi^*, \psi^{*\prime}$) that will bracket that dominant density maximum. The mean GMVTI is rather robust in this sense, because secondary density maxima (as long as they are significantly smaller than the dominant peak), will have little effect on the values of ($\psi^*, \psi^{*\prime}$). We will say that this pair of angles delimits the "shoulders" of the dominant density feature along the LOS. In other words, these "shoulders" provide a rather robust means of *localizing* the density peak somewhere between (ψ^*) and ($\psi^{*\prime}$), without having to specify the exact mathematical shape of the density peak itself.

Another simplifying assumption – one commonly made in white-light imaging – is that the pre-event electron distribution 1) depends *locally* predominantly on the radius (r) and 2) can be characterized by a *local* power-law dependence $N^q(\mathbf{r}) \approx N^q(a)(a/r)^{b(a)}$, where (a) is some normalizing radius. The formulas we will need are derived in the Appendix. In particular, when we set (a) equal to the LOS radius of closest approach and use the geometric reation $r = a\sec(\psi)$, the power-law relation becomes

$$N^q(\mathbf{k},\mathbf{a},\psi) = N^q(\mathbf{k},\mathbf{a})\cos^{b(a)}\psi \quad \text{where} \quad N^q(\mathbf{k},\mathbf{a}) = N^q(\mathbf{k},\mathbf{a},\psi=0) \tag{12}$$

We can now go back to the basic LOS integral in Eq. (9) and insert this locally spherically symmetric pre-event density distribution.

$$\int_{-\pi/2}^{\pi/2} d\psi\ \gamma(\mathbf{k},\mathbf{a},\psi;t,t_0)\ N^q(\mathbf{k},\mathbf{a},\psi) \approx N^q(\mathbf{k},\mathbf{a}) \int_{-\pi/2}^{\pi/2} d\psi\ \gamma(\mathbf{k},\mathbf{a},\psi;t,t_0)\cos^b\psi$$

where $\gamma(\mathbf{k},\mathbf{a},\psi;t,t_0) = N(\mathbf{a}+a\mathbf{k}\tan\psi,t)/N(\mathbf{a}+a\mathbf{k}\tan\psi,t_0)$ is the event-to-pre-event density ratio. This allows us to introduce the new variable of integration (Ψ) and transform the integral over (ψ) to one over (Ψ).

$$d\Psi = \cos^{b(a)} d\psi \tag{13}$$

This angular function $\Psi(\psi)$ and its implied inverse $\psi(\Psi)$ are essential to the GMVTI. For mathematical convenience in our illustrative example, we will take b=2 exactly, because we saw that there was a relatively small variation in the power-law index $2.0<b<2.5$ for $2R_S<r<8R_S$. Then, with the condition $\Psi(0)=0$, the new variable of integration is $\Psi(\psi)=(\psi/2)+(1/4)\sin 2\psi$. The original integration range $-\pi/2 < \psi < \pi/2$ maps into $\pi/4<\Psi<\pi/4$ in the transformed integral. The inverse function $\psi=\psi(\Psi)$ is transcendental, but it always exists. The transformed integral in Eq. (9) now simply equals $(\pi/2)\gamma^*$.

$$\int_{-\pi/2}^{\pi/2} d\psi\ \gamma(\mathbf{k},\mathbf{a},\psi;t,t_0)\cos^2\psi = \int_{-\pi/4}^{\pi/4} d\Psi\ \gamma[\mathbf{k},\mathbf{a},\psi(\Psi);t,t_0] = (\pi/2)\ \gamma[\mathbf{k},\mathbf{a},\psi(\Psi^*);t,t_0] \tag{14}$$

The effect is to shrink the angular range by a factor of two, but with little distortion $\Psi=\psi+O(\psi^3)$ near zero values for both variables. These relations are sketched in Figure 4A. Note that the mean value (γ^*) of the density ratio is not affected by the mapping of the independent variable: $\gamma^*=\gamma[\psi(\Psi^*)]=\gamma(\psi^*)$.

4.4. Tracking the brightest pixels in the outgoing CME

Although we cannot know the location ($\mathbf{r}=\mathbf{a}+a\mathbf{k}\tan\psi^*$) precisely (only because we do not precisely know the mean-value angle ψ^* precisely), we are justified in assuming that it is somewhere along the LOS of the pixel within the bright feature, because that is how we selected the pixel **k** at time $t>t_0$. The feature cannot be bright unless it contains a *localized* enhancement of electron density. This simple inference from our assumption, that the density has a single (dominant) maximum along the LOS, allows us to extract significant quantitative information on the plasma compression without detailed knowledge of the precise geometry of the enhancement.

We do this by tracking the bright feature in two successive images. From Eq. (11) we have the ratio

$$N(\mathbf{k}_2,\mathbf{a}_2,\psi_2^*,t_2)/N(\mathbf{k}_1,\mathbf{a}_1,\psi_1^*,t_1) = N^q(\mathbf{k}_2,\mathbf{a}_2,\psi_2^*,t_0)/N^q(\mathbf{k}_1,\mathbf{a}_1,\psi_1^*,t_0)\ \Gamma(\mathbf{k}_2;t_2,t_0)/\ \Gamma(\mathbf{k}_1;t_1,t_0) \tag{15}$$

It is worth noting at this stage of the argument that if we are following pixels from a spacecraft at 1AU of an outgoing feature ($\mathbf{k}_1,\mathbf{k}_2,\ldots$), then the successive LOS vectors are very nearly parallel. If the change $|\ \mathbf{a}(\mathbf{k}_2)-\mathbf{a}(\mathbf{k}_1)| < 1R_S$, then they are parallel to within $<1/214.3$ rad $= 0.3°$.

The actual plasma "track" of *shoulder* of bright feature that originates at $\mathbf{r}_1{}^*=\mathbf{a}_1+a_1\mathbf{k}_1\tan\psi_1{}^*$ moves to a position $\mathbf{r}_2=\mathbf{a}_2+\mathbf{k}_2 a_2\tan\psi_2$ during time interval $t_1<t<t_2$ that is determined by the actual plasma velocity $\mathbf{V}[\mathbf{r}(t),t)]$ so that

$$\mathbf{r}_2 = \mathbf{a}_2+\mathbf{k}_2 a_2\tan\psi_2 = \mathbf{a}_1+a_1\mathbf{k}_1\tan\psi_1{}^* + {}_{t1}\!\int^{t2} dt\,\mathbf{V}[\mathbf{r}(t),t)] \quad (16)$$

Because the pixel directions $\mathbf{k}_1$ and $\mathbf{k}_2$ actually "track" the brightest region, we know the LOS radii of closest approach $\mathbf{a}_1$ and $\mathbf{a}_2$ in the plane of sky. We have just shown that we can interpret the angle pairs $(\psi_1{}^*,\psi_1{}^{*\prime})$ and $(\psi_2{}^*,\psi_2{}^{*\prime})$ as specifying the angular width (the "shoulders") of the density enhancement along the LOSs. Many observational studies over the past decades have indicated a nearly-constant angular width with radius for individual CMEs (the "ice-cream-cone" model), so we may set $\psi_2\approx\psi_2{}^*$ and $\psi_2{}'\approx\psi_2{}^{*\prime}$. That will be our basic "tracking" assumption for deriving the compression of the plasma.
Therefore, we can take the shoulder track defined by Eq. (16) as going approximately from $\mathbf{r}_1{}^*=\mathbf{a}_1+a_1\mathbf{k}_1\tan\psi_1{}^*$ to $\mathbf{r}_2{}^*=\mathbf{a}_2+a_2\mathbf{k}_2\tan\psi_2{}^*$, or from $\mathbf{r}_1{}^{*\prime}=\mathbf{a}_1+a_1\mathbf{k}_1\tan\psi_1{}^{*\prime}$ to $\mathbf{r}_2{}^{*\prime}=\mathbf{a}_2+a_2\mathbf{k}_2\tan\psi_2{}^{*\prime}$, on the other shoulder. In other words, the density enhancement profile along the two LOSs is taken to be *self-similar* between to LOSs not too far separated in distances of closest approach (a_1 and a_2). See Fig. 4B. By the same reasoning, the tracks of the pair of shoulders must *bracket* the actual track of the *maximum* of the density enhancement. Since we are assuming that all LOS density profiles are self-similar between the shoulders, then the plasma compression (which only depends on density *ratios*) should be comparable along all *three* tracks. This assumption will indeed be justified if the compression is mainly caused by the *radial* to the divergence

$$\nabla\cdot\mathbf{V} = (1/r^2)\partial(r^2V_r)/\partial r+(1/r\sin\theta)\partial(\sin\theta V_\theta)/\partial\theta+(1/r\sin\theta)\partial V_\phi/\partial\phi \approx (1/r^2)\partial(r^2V_r)/\partial r \quad (17)$$

with a relatively weaker contribution *transverse* to the radial axis of the CME. Another way of saying this is that the self-similarity of the shapes of expanding CMEs implies that the transverse expansion is weak. Thus, even though we don't know the locations of the "shoulder" tracks at angles $(\psi^*,\psi^{*\prime})$ implied by the MVTI, we have some justification that the self-similarity implies that we can assign the average value of the compression $<-\nabla\cdot\mathbf{V}>$ to that within the density peak itself.

5. Observational formula for the plasma compression

The relations from the previous Section allow us to "extract" the time-averaged plasma compression $<-\nabla\cdot\mathbf{V}>$ from successive disturbed images using the exact relation of Eq. (2) from the conservation of electrons and our derived relations in Eq. (9) and Eq. (11) from the GMVTI from either shoulder (ψ^* or $\psi^{*\prime}$).

$$(1/\Delta t)\,{}_{t1}\!\int^{t2} dt\,[-\nabla\cdot\mathbf{V}(\mathbf{r},t)] = ln[N(\mathbf{r}_2,t_2),N(\mathbf{r}_1,t_1)] \approx ln[N(\mathbf{k}_2,\mathbf{a}_2,\psi_2{}^*,t_2)/N(\mathbf{k}_1,\mathbf{a}_1,\psi_1{}^*,t_1)]$$
$$= ln[N^q(\mathbf{k}_2,\mathbf{a}_2,\psi_2{}^*,t_0)/N^q(\mathbf{k}_1,\mathbf{a}_1,\psi_1{}^*,t_o)\,\Gamma(\mathbf{k}_2;t_2,t_0)/\,\Gamma(\mathbf{k}_1;t_1,t_0)] \quad (18)$$

We can immediately apply the result from our discussion of the pre-event local spherical density model (see the Appendix) to the estimate of the pre-event ratio

$$N^q(\mathbf{k}_2,\mathbf{a}_2,\psi_2{}^*,t_0)/N^q(\mathbf{k}_1,\mathbf{a}_1,\psi_1{}^*,t_o) \approx (r_1{}^*/r_2{}^*)^{b12} = |a_1\cos,\psi_1{}^*/\,a_2\cos\psi_2{}^*|^{b12} \approx (a_1/a_2)^{b12} \quad (19)$$

because self-similarity implies that $\psi_1{}^*\approx\psi_2{}^*$ which in turn implies (to even better approximation) that $\cos\psi_1{}^*\approx\cos\psi_2{}^*$.
This gives us our final formula for plasma compression, in the approximation of a spherically symmetric model for the pre-event coronal electron density distribution

$$(\Delta t) <[-\nabla\cdot\mathbf{V}(\mathbf{r},t)> \approx -b_{12}\, ln(a_2/a_1) + ln[\Gamma(\mathbf{k}_2;t_2,t_0)/\Gamma(\mathbf{k}_1;t_1,t_0)] \quad (20)$$

The first logarithm in the square brackets stems from the radial dependence of the pre-event density. It will always be negative, because $a_1<a_2$ for an outgoing CME. Consequently, if there is to be plasma compression, the density feature will have to *brighten* sufficiently as it moves outward so that the second logarithmic term (Γ_2/Γ_1) can overcome the pre-event density ratio term containing (a_2/a_1), if there is to be time-averaged plasma *compression* ($\nabla\cdot\mathbf{V}<0$). This condition can be stated as an inequality:

$$[\Gamma(\mathbf{k}_2;t_2,t_0)/\Gamma(\mathbf{k}_1;t_1,t_0)] > (a_2/a_1)^{b12} \text{ implies that } \nabla\cdot\mathbf{V}<0 \quad (21)$$

The only *un-observed* quantity is the position-averaged power-law index (b_{12}) of the pre-event density. It may be taken from published studies like [6] or [7], but perhaps it may be possible in some events to derive a pre-event value b_{12} from "quiet" images over the several days prior to the SEP event. In any case, the remaining logarithm involves only the *ratio* of the brightness ratios at the two times (Γ_2/Γ_1) These are precisely the measured pre-event-normalized pixel brightnesses that are used to make movies of CMEs. Note that any *linear* instrument calibration and response factors will cancel out of the brightness ratio, and we have argued that any Thomson scattering corrections will be negligible in the low corona.

6. Summary

Under the assumptions made here, the normalized intensity ratio (Γ), corresponds directly to the density compression ratio Γ_{CR} introduced in [4] and [5] to estimate the strength of CME shocks in white light images. Also note that we do *not* have to estimate the plasma velocity (**V**) itself, nor the time derivative ($\partial N/\partial t$) nor the spatial gradient (∇N) of the electron density! All we have to do is track the feature's normalized white-light *brightness*: if it is *increasing sufficiently* to satisfy the inequality in Eq. (21), then we have a measure of the *compression* of the plasma within the region occupied by that feature, and conversely, an *insufficient increase* implies plasma *expansion*.
We suggest that the strongest brightening is much more likely to occur when the leading edge is closer to the Sun, perhaps inside of the radius of the occulatation disk (~1.5-2 R_S) on contemporary spacecraft coronagraphs. Nonetheless, even with this observational limitation, we believe that a careful measurement of such images will yield a useful estimate of the quantity $<(-\nabla\cdot\mathbf{V})>$, according to Eq. (4), regardless of whether its value is positive (implying compression) or negative (implying expansion). When there is compression, Eq. (0) -- taken from the related paper [2] – may be used to estimate the rate of compressive acceleration of solar energetic particles occurring within the observed CME, thus providing observational "closure" for the proposed mechanism for SEP acceleration.

7. References

[1] Michalek G, Gopalswamy N, Yashiro S, and Bronarska K 2015 "Dynamics of CMEs in the LASCO Field of View." *Solar Physics* **290**, 903–17. doi:10.1007/s11207-015-0653-8
[2] Roelof E C 2015 "Charged particle energization and transport in reservoirs throughout the heliosphere: 1. Solar energetic particles", *This Conference*
[3] Kahler S W and Vourlidas A 2005, Fast coronal mass ejection environments and the production of solar energetic particle events, *J. Geophys. Res.* **110**, A12S01, doi:10.1029/ 2005JA011073
[4] Vourlidas A, Howard R A, Esfandiari E, Patsourakos S, Yashiro S and Michalek G 2010 "Comprehensive Analysis of Coronal Mass Ejection Mass and Energy Properties over a Full Solar Cycle." *Astrophys. J.* **722**, 1522

[5] Ontiveros V, and Vourlidas A 2009, "Quantitative Measurements of Coronal Mass Ejection-Driven Shocks from LASCO Observations." *Astrophys. J.* **693**, 267–75
[6] Leblanc Y, Dulk G A and Bourgeret J-L 1998, "Tracing the electron density from the corona to 1AU", *Solar Phys.*, **183**, 165-180
[7] Saito K, Poland A J and Munro R H 1977, "A study of the background corona near solar minimum", *Solar Phys.*, **55**, 121-134

Acknowledgments
The efforts of E. C. R. were supported by the NASA ACE and Voyager Interstellar missions under contracts NNX10AT75G and NNN06AA01C, respectively, while those of A.V. were supported by NASA and JHU/APL internal funds.

Appendix A. Simple model for a spherically symmetric electron density distribution
We will treat a simple example in this paper, that of a spherically symmetric "quiet" corona $N^q(r)$ within which the electron densities obey a "local" power-law dependence on radius. The statistical analysis of electron densities using dekameter solar radio Type III bursts observed by ISEE-3 from $\sim 2R_S$ to 1 AU by Leblanc *et al.* [6] found an average density $N \propto 1/r^2$ from 1 AU into $r=8R_S$. The power law gradually steepened to $N \propto 1/r^{2.5}$ from $8R_S$ to $2R_S$, in good agreement with the oft-quoted estimates of Saito *et al.* [7]. We can model this behavior as a "local" power law in radius with a radius-dependent index b(r).

$$d\,ln N^q/d\,ln r = -b(r) \qquad ln[N^q(r_1)/N^q(r_2)] = -\int_{r_1}^{r_2} d\,ln(r)\, b(r)$$

If we expand $\gamma(r)$ in a Taylor series about either the radius $b(r)=b_1+ln(r/r_1)db/d\,ln r_1+\ldots$ or the outer radius $b(r)=b_2+ln(r/r_2)db/d\,ln r_2+\ldots$, integrate the two expressions as above, and then average the results, we obtain

$$ln[N(r_1)/N(r_2)] = b_{12}\, ln(r_2/r_1) + (1/4)\, ln^2(r_2/r_1)(db/d\,ln r_1 - db/d\,ln r_2) + \ldots$$

where $b_{12}=(1/2)(b_1+b_2)$. The coefficient of second-order term in the already small quantity $ln(r_2/r_1)$ involves the *difference* of the slope derivatives, so it can clearly be neglected for two radii relatively close together such that $ln(r_2/r_1)<<1$ for $r_2>r_1$. In our LOS coordinates (with the pixel direction **k** held constant), we have $r^*=a\sec\psi^*$, so our estimate of the pre-event ratio finally becomes

$$N^q{}_2/N^q{}_1 \approx (r_2/r_1)^{b12} = (a_2/a_1)^{b12}\,(\cos\psi_2/\cos\psi_1)^{b12}$$

Broken Power-law Distributions from Low Coronal Compression Regions or Shocks

N. A. Schwadron[1], M. A. Lee[1], M. Gorby[1], N. Lugaz[1], H. E. Spence[1], M. Desai[2], T. Török[3], C. Downs[3], J. Linker[3], R. Lionello[3], Z. Mikić[3], P. Riley[3], J. Giacalone[4], J. R. Jokipii[4], J. Kota[4] and K. Kozarev[5]

[1] University of New Hampshire, Durham, AH, 03824
[2] Southwest Research Institute, San Antonio, TX 78228, USA
[3] Predictive Science Inc., San Diego, CA 92121
[4] University of Arizona, Tucson, AZ 85721
[5] Harvard-Smithsonean Center for Astrophysics, Cambridge, Mass 02138

E-mail: n.schwadron@unh.edu

Abstract. Coronal Mass Ejection (CME) expansion regions low in the corona ($< 2 - 3$ R$_s$) are highly efficient for the acceleration of energetic particles. Because the acceleration occurs over a finite spatial region, there is a regime where particles diffuse away and escape from the acceleration sites, leading to the formation of broken power-law distributions. This paper highlights recent results indicating that CME expansion and acceleration in the low corona may cause rapid particle acceleration and create large solar energetic particle events with broken power-law distributions.

1. Introduction

The low corona creates the conditions necessary for rapid particle acceleration. Sophisticated CME simulations have shown that diffusive shock acceleration at driven shocks may be responsible for accelerating solar energetic protons up to very high energies of 10 GeV for very strong shocks [1–4].

[5] showed results of the Energetic Particle Radiation Environment Model [EPREM, 6] for describing particle acceleration in the corona. The model includes an energetic particle solver describing the evolution of the particle distribution using the focused transport equation. The particle acceleration and propagation model requires information about the evolution of the plasma based on simulated magnetohydrodynamic (MHD) fields.

The MHD fields are solved using CORHEL (Corona-Heliosphere) for the ambient solar corona, solar wind [7–9] and CMEs. CORHEL is based on the MAS (Magnetohydrodynamic Algorithm outside a Sphere) code [e.g., 10–12] that solves standard viscous and resistive MHD equations in spherical coordinates. The model incorporates an empirical coronal heating function, thermal conduction parallel to the magnetic field, and radiative losses.

Rapid particle acceleration in the low corona after fast CME expansion is often traced back to large-scale compression regions or shocks near the expanding CME legs [5; 13; 14]. In the cases of particle acceleration from compression regions [see 15; 16] and from shocks [e.g., 17; 18], particles rapidly scatter across the structure and gain energy after each crossing of the compression gradient or shock. The total energy gain for shocks depends on the number of shock crossings [19; 20]. In the case of compressions, the situation is more complex since different scattering events sample different parts of the compression and therefore lead to varying gains in energy. Diffusive acceleration often requires a quasi-perpendicular configuration [21–23] or a short scattering mean free path [24–26] for efficient particle acceleration.

The common occurrence of broken power-law distributions [27–30] is another important observational factor in the development of our understanding large SEP events, particularly ground level events. [31] find that the power-law break has a strong rigidity dependence, suggesting strong wave-particle interactions. **The effects of shock geometry on spectral breaks has been modeled by [32] based on temporally limited particle acceleration.**

The formation of the broken power law remains an important problem. In this paper, we show how broken-power law distributions can result from size-limited shocks or compressions that often form in the low corona. **We note that the basic elements of this physical process have parallel analogs in Earth's magnetotail. Though in a different part of parameter space (subsonic flow versus shocks) charged particles are energized in the magnetotail through adiabatic compression and through energy-dependent gradient drifts along electric fields, which are controlled fundamentally by the finite dimensions of the system. These factors in turn produce non-Maxwellian particle distributions with spectral breaks [33–35], demonstrating the ubiquity of these same governing processes that are the focus of this work for which the complexity is much greater.**

2. Diffusive Particle Acceleration at Low Coronal Shocks and Compressions

Our analysis is motivated, in part, by recent results of CME simulations, which reveal how a CME rapidly accelerated low in the corona causes extremely strong compression regions [5]. Consider the configuration in Figure 1. A compression or shock region is driven by a CME expansion and acceleration. Notably the compression is strongest where the CME acceleration or expansion is most rapid [36].

Diffusive acceleration at shocks is solved for using the Parker transport equation:

$$\frac{\partial f}{\partial t} + \mathbf{u} \cdot \nabla f - \nabla \cdot (\mathsf{K} \cdot \nabla f) - \frac{\nabla \cdot \mathbf{u}}{3} p \frac{\partial f}{\partial p} = Q_0 \delta(x) \delta(z) \delta(p - p_{inj}) \quad (1)$$

where

$$\mathsf{K} \;=\; \kappa_{\|} \hat{e}_b \hat{e}_b + \kappa_{\perp} (\mathsf{I} - \hat{e}_b \hat{e}_b) \quad (2)$$

and $\kappa_{\|}$ and $\kappa_{\perp}$ are coefficients of diffusion parallel and perpendicular to the magnetic field, respectively. The distribution function is given by f, the plasma velocity is given by $\mathbf{u}$, and the **particle momentum in the plasma frame is given by** p. The diffusion tensor indicated here is diagonal and neglects an off-diagonal term, κ_A, due to drift is sometimes included in the diffusion tensor.

The source term represents a modification of the standard approach to diffusive shock acceleration theory. In the standard treatment, particles are injected at a fixed energy, the injection energy, and at the location of the shock ($x = 0$). The shock or compression is within the y-z plane centered on $x = 0$. Magnetic field lines lie within plane of coplanarity x-z. **Particles injected at a fixed location** $z = 0$ **are accelerated while the field connection point sweeps across the shock structure.** A similar concept was used to model particle acceleration at the blunt termination shock [37].

To develop a quantitative description, we solve for a Green's function for injection at a fixed location, at $z = 0$. We then integrate over the region along the shock, from $z = 0$ to $z = L$. The

Journal of Physics: Conference Series **642** (2015) 012025 doi:10.1088/1742-6596/642/1/012025

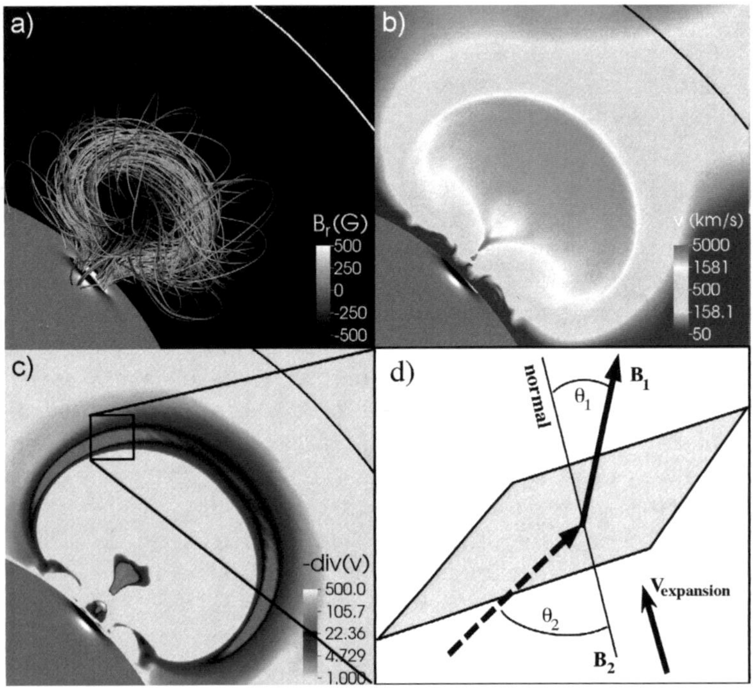

Figure 1. (From [36]). Strong compression from the expansion of a CME in the low corona leads to rapid particle acceleration [5]. The erupting magnetic flux rope (Panel a) is shown along with the photospheric field strength B_r (greyscale on the solar surface). The CME accelerates rapidly (Panel b). Strong compressions or shocks are formed, **showing strong velocity divergence, $\nabla \cdot \mathbf{u}$ (Panel c) expressed in code units corresponding to 7×10^{-4} s^{-1}** . The blow out (Panel d) shows a magnetic flux bundle piercing the shock.

source-term scaling constant Q_0 is

$$Q_0 = u_1 \epsilon f_{\text{inj}} L p_{\text{inj}} \tag{3}$$

where u_1 is the upstream flow speed in the frame of the shock and the injection efficiency is given by ϵ. The distribution function at the injection energy is f_{inj}, and the size-scale of the shock is given by L.

We track $\nabla \cdot \mathbf{V}$ in Figure 2 along a magnetic flux bundle. The acceleration region is most pronounced where the CME expansion and acceleration is strongest. The outward CME expansion speed is < 600 km/s at 1.5 solar radii. By the time this expansion reaches 1.8 solar radii, the speed reaches ~ 1500 km/s. Above 1.8 solar radii the acceleration of the CME slows and the speed levels off at ~ 1400 km/s. The divergence shown in Figure 2 (right panels) provides an excellent indicator for the pronounced effects of the CME expansion and acceleration. This divergence is largest between 1.5 and 2 solar radii and then levels off at larger distances.

The Parker transport equation (1) includes the divergence in the plasma velocity (4th term on left-hand-side). This term leads to the acceleration of energetic particles. The CME acceleration is prompt and localized (between 1.5 and 2 solar radii) leading to prompt energetic particle acceleration inside of 2 solar radii.

The configuration of the accelerator provides a connection point between the shock or compression and a magnetic field line (or flux bundle) that moves along the structure. This motion occurs as particles are accelerated. On average, the higher the energy that particles gain,

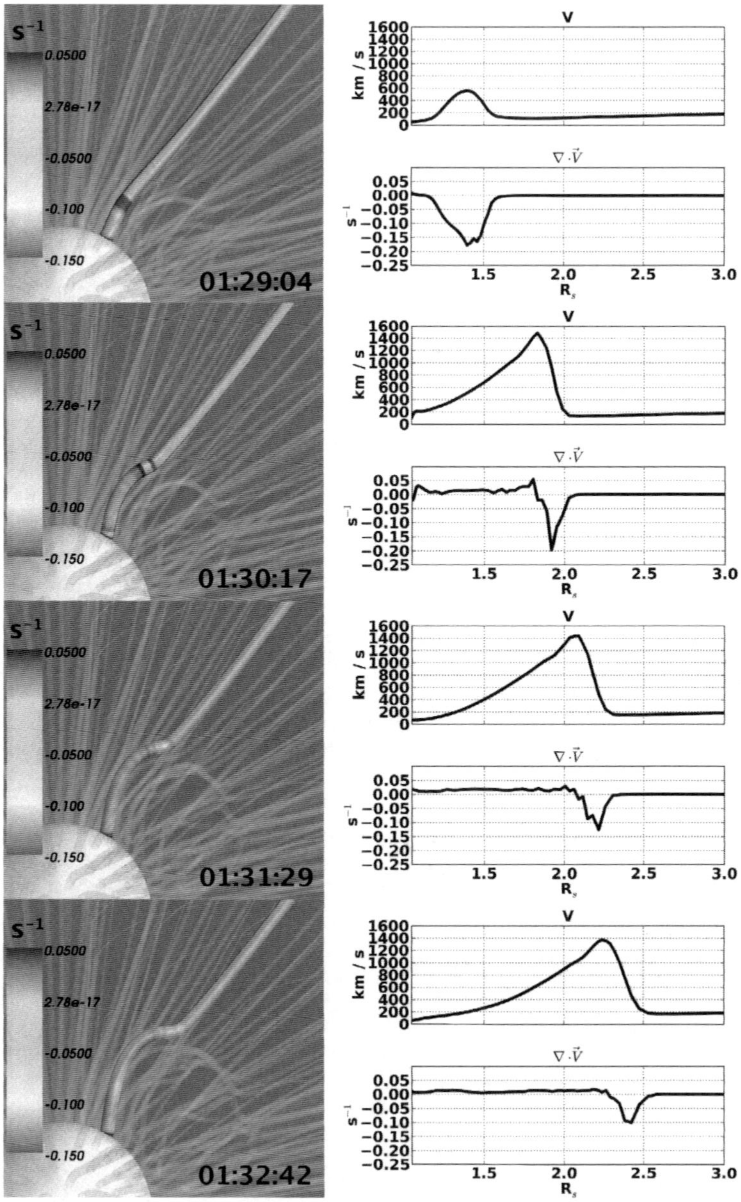

Figure 2. [From 36] The development of a compression region viewed on a bundle of field lines. Right: the curves show V and $\nabla \cdot \mathbf{V}$ along the field line bundle during CME expansion at four different instances within the simulation. Left: Configuration of the magnetic flux bundle (depicted as a tube) deformed by the CME expansion. Color-coding indicates $\nabla \cdot \mathbf{V}$. Indicated times are hh:mm:ss, where time zero corresponds to the release of the CME.

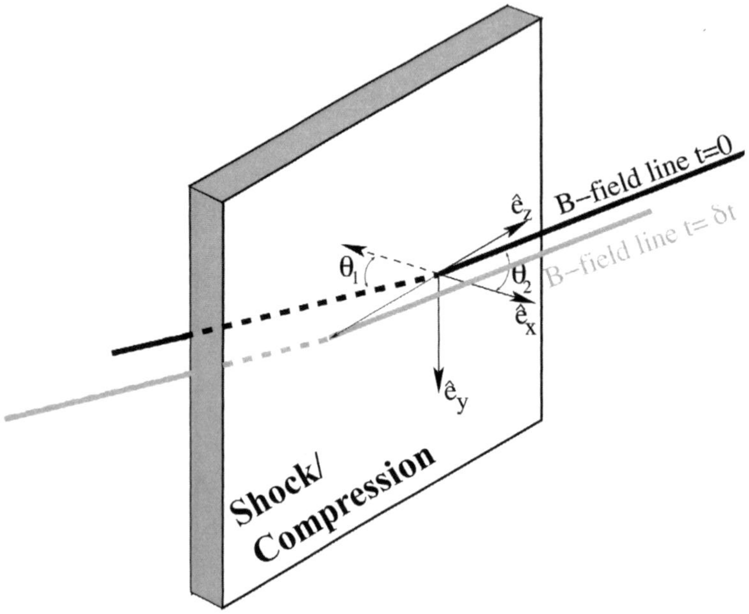

Figure 3. [From 36] Diagram of a magnetic flux bundle piercing a shock or compression region as a function of time. The initial flux bundle ($t = 0$ in black) convects through the shock or compression as the field-line footpoint (flux bundle at $t = \delta t$ in blue) moves across the structure.

the longer the flux bundle where these particles are accelerated has remained in contact with the shock or compression. A flux bundle connection point eventually moves off of the accelerator, which limits the particles' energy gain and leads to the formation of a broken power-law or exponential roll-over in the particle distribution.

We use the following expressions for the diffusion coefficient

$$\mathsf{K} = \kappa_{\|}\hat{e}_b\hat{e}_b + \kappa_{\perp}(\mathsf{I} - \hat{e}_b\hat{e}_b) \tag{4}$$

where the diffusion coefficient component projections are

$$\kappa_{xx} = \kappa_{\|}\cos^2\theta + \kappa_{\perp}\sin^2\theta \tag{5}$$
$$\kappa_{xz} = (\kappa_{\|} - \kappa_{\perp})\sin\theta\cos\theta \tag{6}$$
$$\kappa_{zz} = \kappa_{\|}\sin^2\theta + \kappa_{\perp}\cos^2\theta \tag{7}$$

Consider steady state injection of particles at $z = 0$ and $x = 0$ with momentum p_{inj}:

$$\mathbf{u}\cdot\nabla f - \nabla\cdot(\mathsf{K}\cdot\nabla f) - \frac{\nabla\cdot\mathbf{u}}{3}p\frac{\partial f}{\partial p} = Q_0\delta(x)\delta(z)\delta(p - p_{inj}) \tag{8}$$

We take (x, z) as the coplanarity plane and no dependence on y, so that cross-field drifts do not appear (see Figure 3). The scattering mean free path is $\lambda_{\|} = \lambda_{\|0}(R_g/R_{g0})^{\chi}$ where $R_g = pc/q$ is the particle rigidity and $R_{g0} = (m_p v_0 c/e)(A_0/Z_0)$ is a reference rigidity at mass $m_0 = A_0 m_p$, charge $q_0 = Z_0 e$ and speed v_0. We adopt a 1 AU mean free path of 0.14 AU and use a $1/B$ scaling into the low corona, which is typical for observationally constrained values [38]. Taking a 1 G field in the corona, we find a reference scattering mean free path of $\lambda_{\|0} = 0.015\ \mathrm{R}_s$ at $T_0 = 1$ MeV/nuc.

We now explicitly expand the differential equation,

$$
\begin{aligned}
u_x \frac{\partial f}{\partial x} &+ u_z \frac{\partial f}{\partial z} - \nabla \cdot (\kappa_\perp \mathsf{I} \cdot \nabla f) - \nabla \cdot \left[\hat{e}_b \left(\kappa_\parallel - \kappa_\perp \right) \hat{e}_b \cdot \nabla f \right] \\
&+ \frac{u_{x1} - u_{x2}}{3} \delta(x) p \frac{\partial f}{\partial p} = Q_0 \delta(x) \delta(z) \delta(p - p_{inj}). \qquad (9)
\end{aligned}
$$

and perform the Fourier transform,

$$
f(z, x, p) = \frac{1}{2\pi} \int_{-\infty}^{\infty} dk \ \exp(ikz) \ \tilde{f}(k, x, p), \qquad (10)
$$

such that

$$
\begin{aligned}
u_x \frac{\partial \tilde{f}}{\partial x} &+ iku_z \tilde{f} + k^2 \kappa_{zz} \tilde{f} - \frac{\partial}{\partial x} \left(\kappa_{xx} \frac{\partial \tilde{f}}{\partial x} \right) - \frac{\partial}{\partial x} \left[\kappa_{xz} ik \tilde{f} \right] \\
&- ik\kappa_{xz} \frac{\partial \tilde{f}}{\partial x} + \delta(x) \frac{u_{1x} - u_{2x}}{3} \frac{\partial \tilde{f}}{\partial \ln p} = Q_0 \delta(x) \delta(p - p_{inj}). \qquad (11)
\end{aligned}
$$

We take $\tilde{f} \propto \exp(\alpha x)$ for $x \neq 0$ which requires the following condition when taking into account upstream and downstream conditions:

$$
-\alpha^2 \kappa_{xx} + \alpha(u_x - 2ik\kappa_{xz}) + iku_z + k^2 \kappa_{zz} = 0 \qquad (12)
$$

The upstream solution, $j = 1$, and downstream solution, $j = 2$, are

$$
\alpha_j = \frac{u_{xj}}{2\kappa_{xxj}} - ik \frac{\kappa_{xzj}}{\kappa_{xxj}} + \frac{(-1)^{j+1}}{\kappa_{xxj}} \sqrt{u_{xj}^2/4 + \kappa_{\parallel j} \kappa_{\perp j} k^2 + ik(\kappa_{xxj} u_{zj} - \kappa_{xzj} u_{xj})}. \qquad (13)
$$

Conventionally, the signs, $(-1)^{j+1}$, are chosen for convergence so that for $k = 0$, we recover the result for a stationary shock.

Given α_j, we integrate equation (11) from the upstream to downstream side of the shock or compression. This integral extends over a narrow region centered on the shock jump. We recover the following equation:

$$
-(\kappa_{xx2}\alpha_2 - \kappa_{xx1}\alpha_1)\tilde{f} - (\kappa_{xz2} - \kappa_{xz1}) ik \tilde{f} + \frac{\Delta u}{3} \frac{\partial \tilde{f}}{\partial \ln p} = Q_0 \delta(p - p_{inj}) \qquad (14)
$$

We find the property

$$
\begin{aligned}
(\kappa_{xx2}\alpha_2 &- \kappa_{xx1}\alpha_1) + (\kappa_{xz2} - \kappa_{xz1}) ik = \\
&-\frac{\Delta u}{2} - \sum_{j=1}^{2} \sqrt{\frac{u_{xj}^2}{4} + \kappa_{\parallel j} \kappa_{\perp j} k^2 + ik(\kappa_{xxj} u_{zj} - \kappa_{xzj} u_{xj})} \qquad (15)
\end{aligned}
$$

We now express the solution for $\tilde{f}(k, p)$, and the distribution function at the shock or compression:

$$
\begin{aligned}
f_s(z, p) &= \frac{3Q_0}{2\pi \Delta u p_{inj}} \left(\frac{p}{p_{inj}} \right)^{-3/2} \int_{-\infty}^{\infty} dk \\
&\times \exp \left(ikz - \frac{3}{\Delta u} \int_{\ln p_{inj}}^{\ln p} d \ln p' \sum_{j=1}^{2} \sqrt{\frac{u_{xj}^2}{4} + \kappa_{\parallel j} \kappa_{\perp j} k^2 + ik(\kappa_{xxj} u_{zj} - \kappa_{xzj} u_{xj})} \right). \qquad (16)
\end{aligned}
$$

Here $\Delta u = u_{1x} - u_{2x}$.

Note that the square root in equation (16) equals $u_{xj}/2$ as $k \to 0$. We add and subtract $u_{xj}/2$ from the square root; the $\ln p$-integral of $u_{xj}/2$ is trivial and we obtain

$$f_s(z,p) = \frac{3Q_0}{2\pi\Delta u p_{inj}}\left(\frac{p}{p_{inj}}\right)^{-\gamma}\int_{-\infty}^{\infty} dk\ \exp(ikz) \times \exp\left(-\frac{3}{\Delta u}\int_{\ln p_{inj}}^{\ln p} d\ln p' \sum_{j=1}^{2}\left[\sqrt{\frac{u_{xj}^2}{4} + \kappa_{\|j}\kappa_{\perp j}k^2 + ik(\kappa_{xxj}u_{zj} - \kappa_{xzj}u_{xj})} - \frac{u_{xj}}{2}\right]\right) \tag{17}$$

where $\gamma = 3r_c/(r_c - 1)$ is the power law index of diffusive shock acceleration. Assuming that the protons are non-relativistic ($p \approx mv$), and noting that the square root is a function of $k(v')^{\chi+1}$, we rewrite (17) as

$$f_s(z,p) = \frac{3Q_0}{2\pi\Delta u p_{inj}}\left(\frac{v}{v_{inj}}\right)^{-\gamma}\frac{1}{v^{\chi+1}}\int_{-\infty}^{\infty} d\eta\ \exp\left[i\eta(z/v^{\chi+1}) - \frac{3}{(\chi+1)\Delta u}\right. \times \left.\int_{-\infty}^{\ln(\eta)} d\ln\xi \sum_{j=1}^{2}\left\{\sqrt{\frac{u_{xj}^2}{4} + \frac{\tau_{\|j}\tau_{\perp j}}{9}\xi^2 + i\xi\left(\frac{\kappa_{xxj}}{v^{\chi+1}}u_{zj} - \frac{\kappa_{xzj}}{v^{\chi+1}}u_{xj}\right)} - \frac{u_{xj}}{2}\right\}\right] \tag{18}$$

where $\xi = k(v')^{\chi+1}$, $\eta = kv^{\chi+1}$, $\kappa_{\|} = \tau_{\|}v^{\chi+1}/3$, $\kappa_{\perp} = \tau_{\perp}v^{\chi+1}/3$, and the lower limit of the ξ-integration is replaced by $-\infty$ since the integrand is convergent as $kv_{inj}^{\chi+1} \to 0$. **We have assumed here that $\kappa_{\perp}/\kappa_{\|}$ =constant. However, non-linear guiding center (NLGC) theory predicts that $\kappa_{\perp}$ and $\kappa_{\|}$ have different dependence on particle speed [39; 40]. Future studies will need to investigate the effects of these alternative behaviors for the differential energy spectrum, and behavior of heavy ions.**

The η-integral purely a function of $z/v^{\chi+1}$. If $v^{\chi+1}/z > \eta_0$, where η_0 is the scale for the integration decay, then the factor $\exp[i\eta(z/v^{\chi+1})] \approx 1$ and (18) yields the high-energy power-law spectrum, $f_s \propto v^{-\gamma-1-\chi}$. For small and intermediate values of $v^{\chi+1}/z$ ($v^{\chi+1}/z \ll \eta_0$ and $v^{\chi+1}/z \sim \eta_0$) we expand the integrand in powers of ξ and retain terms up to order ξ^2. The ξ-integral is evaluated,

$$f_s(z,p) = \frac{3Q_0}{2\pi\Delta u p_{inj}}\left(\frac{v}{v_{inj}}\right)^{-\gamma}\frac{1}{v^{\chi+1}} \times \int_{-\infty}^{\infty} d\eta\ \exp\left[i\eta\frac{z}{v^{\chi+1}} - \frac{3}{(\chi+1)\Delta u}(i\eta a + \eta^2 b/4)\right] \tag{19}$$

where

$$a = \sum_{j=1}^{2}\left[\frac{\kappa_{xxj}u_{zj}}{v^{chi+1}u_{xj}} - \frac{\kappa_{xzj}}{v^{\chi+1}}\right] \tag{20}$$

$$b = \sum_{j=1}^{2}\left\{2\frac{\tau_{\|j}\tau_{\perp j}}{9u_{xj}} + \frac{2}{u_{xj}}\left[\frac{\kappa_{xxj}u_{zj}}{v^{\chi+1}u_{xj}} - \frac{\kappa_{xzj}}{v^{\chi+1}}\right]^2\right\} \tag{21}$$

The remaining integration in (19) is evaluated by "completing the square". We obtain

$$f_s(z,p) \approx \frac{3Q_0}{2\pi\Delta u p_{inj}}\left(\frac{v}{v_{inj}}\right)^{-\gamma}\frac{1}{v^{\chi+1}}\sqrt{\frac{8\pi\Delta u}{3b}} \times \exp\left\{-\left[\frac{z}{v^{\chi+1}} - \frac{3a}{(\chi+1)\Delta u}\right]^2\left(\frac{(\chi+1)\Delta u}{3b}\right)\right\} \tag{22}$$

Since it takes time for ions to be accelerated to high energies, the peak of the ion distribution moves from the point of injection to progressively more distant regions on the shock or compression flank. The quantity

$$z_d = \frac{3av^{\chi+1}}{(\chi+1)\Delta u} \tag{23}$$

is the distance moved along the shock by the majority of ions in the process of being accelerated to a given energy. The average ion speed along the shock is

$$V_{sz} = \frac{dz_d}{d\tau_p} \tag{24}$$

and τ_p is the characteristic time to accelerate ions to momentum p. The quantity $d\tau_p$ is the infinitesimal time to accelerate ions from p to $p + dp$:

$$d\tau_p = \frac{3\delta x}{\Delta u}\frac{dp}{p}. \tag{25}$$

The characteristic width of the acceleration region is

$$\delta x = \frac{\kappa_{xx1}}{u_{x1}} + \frac{\kappa_{xx2}}{u_{x2}}. \tag{26}$$

Using (24) and (25), we then solve for the average particle speed along the shock

$$V_{sz} = \frac{v^{\chi+1}a}{\delta x} \tag{27}$$

The solution for f_s is written in the form,

$$f_s(z,p) \approx \frac{3Q_0}{2\pi\Delta u p_{inj}}\left(\frac{v}{v_{inj}}\right)^{-\gamma}\sqrt{\frac{\pi}{D_z}}\exp\left\{-\frac{(z-z_d)^2}{4D_z}\right\} \tag{28}$$

where the width of the Gaussian is

$$D_z = \frac{3bv^{2(\chi+1)}}{4(\chi+1)\Delta u}. \tag{29}$$

The diffusion rate is

$$K_{szz} = \frac{dD_z}{d\tau_p} = \frac{bv^{2(\chi+1)}}{2\delta x}. \tag{30}$$

Equation 28 is used as a Green's function. We integrate injection from $z' = 0$ to L. The resulting total distribution, $F_s(z,p)$, is

$$\begin{aligned} F_s(z,p) &= \frac{1}{L}\int_0^L dz_{inj} f_s'(z, z_{inj}, p) \\ &= \frac{3Q_0}{2\Delta u p_{inj} L}\left(\frac{p}{p_{inj}}\right)^{-\gamma}\left[\text{erf}\left(\frac{L+z_d-z}{2\sqrt{D_z}}\right) - \text{erf}\left(\frac{z_d-z}{2\sqrt{D_z}}\right)\right]. \end{aligned} \tag{31}$$

Figure 4 shows the spectrum in the case in which there is no escape of particles from the shock accelerator. In this case we have considered an obliquity angle of $\theta = 70°$ at 2 solar radii,

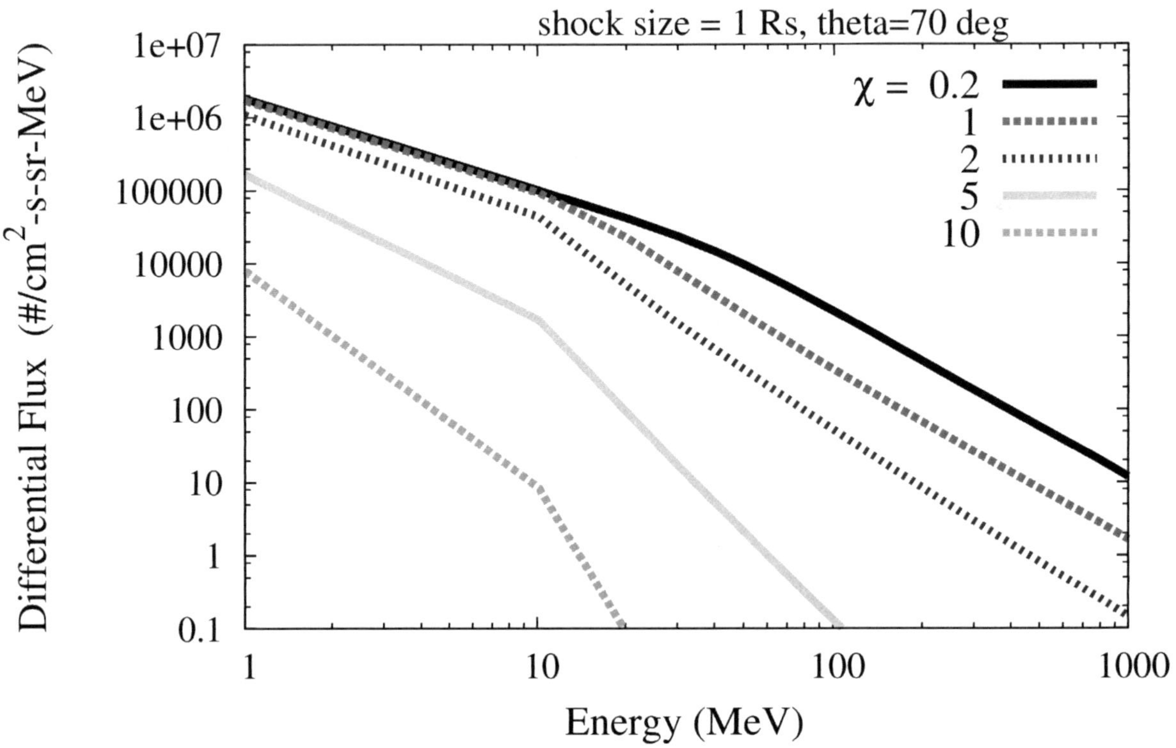

Figure 4. Differential spectra for several cases of χ based on equation 31. Note the spectral break and the relatively smooth power-law at higher energies. The spectral break drops with energy for increasing values of χ. These spectra are formed at $r \sim 2$ R_s, a shock or size of 1 R_s and an obliquity angle of $\theta = 70°$.

with a shock size of 1 R_s. We have also taken a compression ratio of 3. By comparison, Figure 5 shows a similar case with a smaller shock size of 0.25 R_s and Figure 6 shows a shock size of 1 R_s but an obliquity angle of $\theta = 30°$. In each of these cases, we have taken a ratio of perpendicular to parallel diffusion of $\kappa_\perp/\kappa_\parallel = 0.01$.

Taken together Figures 4, 5 and 6 show a number of important features. First, the variation with χ shows that stronger rigidity in the scattering mean free path leads to sharper breaks in the broken power-law distribution. It is likely that stronger rigidity dependence arises through more significant wave-particle interactions, which tend to occur when the shocks or compressions are closer to being quasi-parallel.

When we compare Figure 4 and Figure 5, it is evident that the larger shock size leads to higher energies for the spectral break. This increase in the break energy seen in Figure 4 is most pronounced in instances when the the rigidity dependence is weak. This suggests that weak rigidity dependence, likely associated with quasi-perpendicular shocks, would likely result in higher energies of the spectral break, particularly for large shocks or compressions.

When we compare Figure 4 and Figure 6, we observe that the break in the power-law drops in energy for decreased obliquity angles. As in the previous case, this decrease in the break energy is most pronounced for a weak rigidity dependence in the scattering mean free path. This shows that, regardless of the rigidity dependence, a quasi-parallel shock tends to generate lower energy spectral breaks.

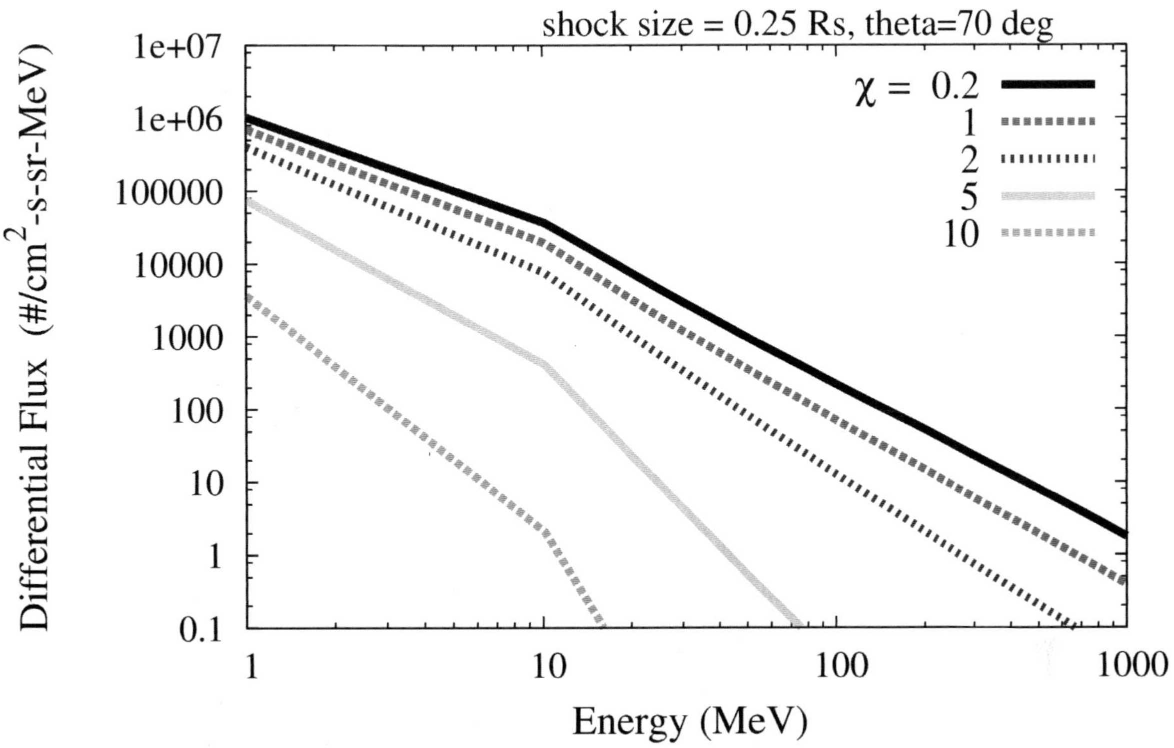

Figure 5. Differential spectra for several cases of χ based on equation 31. The conditions are the same as used in Figure 4, but the shock size is smaller, 0.25 R_s.

3. Conclusions

Recent MHD simulations of CME expansion show the formation of strong compression regions and shocks at low coronal heights, $< 2\ R_{\odot}$. These compressions and shocks form due to the rapid expansion and acceleration of the CME and then subside higher in the corona where the CME acceleration slows.

We developed a model of particle acceleration at a low coronal shock or compression with a finite size (length L) and a fixed field-flow angle. Results of the model show the pronounced effects of particle diffusion and particle escape, leading to double power-law distributions. The slope of the power-law below the spectral break is roughly consistent with steady diffusive shock acceleration. However, above the spectral break, the spectral slope is strongly controlled the rigidity dependence in the diffusion coefficients.

Thus, we find that strong compressions and shocks in the low corona may naturally give rise to broken power-law distributions that arise in large SEP events.

3.1. Acknowledgments

We thank all those who made C-SWEPA (NASA grant NNX13AI75G), EMMREM (grant NNX07AC14G), Sun-2-Ice (NSF grant AGS1135432) , DREAM (NASA grant NNX10AB17A), and DREAM2 (NASA grant NNX14AG13A) projects possible. Data for all figures can be obtained by contacting the lead author, N. Schwadron. The contributions of T.T, C.D., R.L., J.A.L., V.A.T., Z.M., and P.R. were supported by the NSF Frontiers in Earth System Dynamics program and the Center for Integrated Space Weather Modeling, and by NASA's SBIR, HTP, LWS, and SR&T programs. Computational resources were provided by the NSF supported Texas

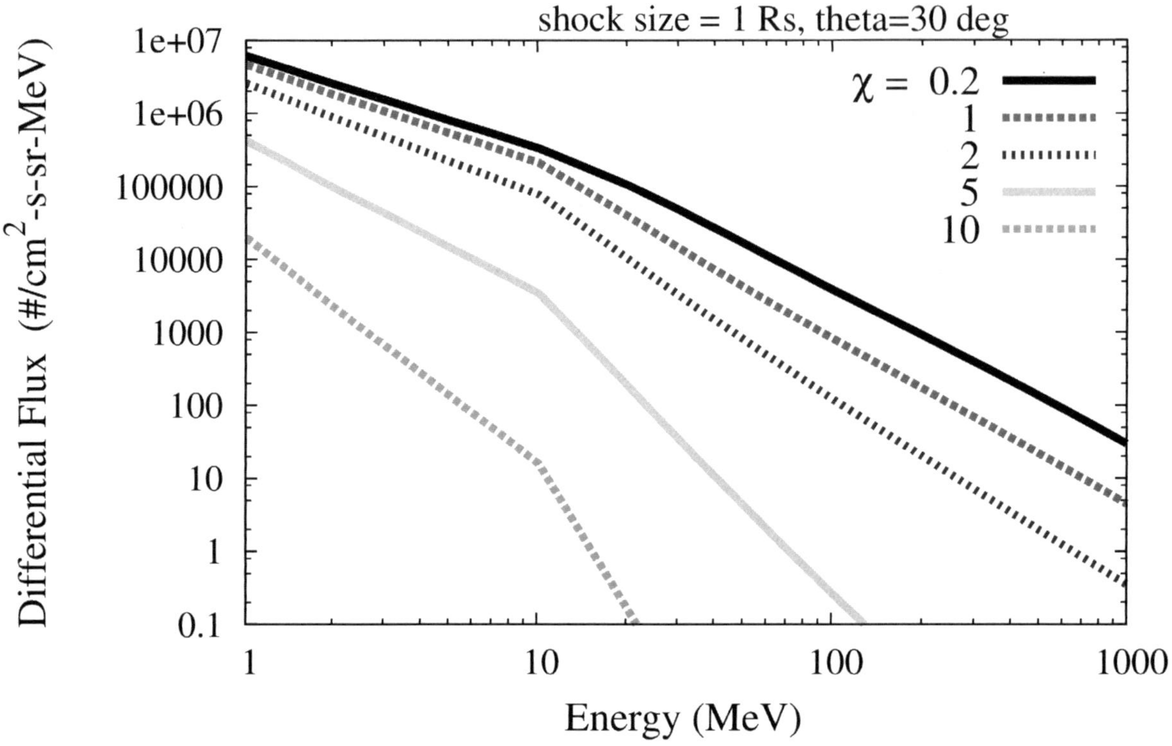

Figure 6. Differential spectra for several cases of χ based on equation 31. The conditions are the same as used in Figure 4, but the obliquity angle is $\theta = 30°$.

Advanced Computing Center (TACC) in Austin and the NASA Advanced Supercomputing Division (NAS) at Ames Research Center.

References

[1] Roussev I I, Sokolov I V, Forbes T G, Gombosi T I, Lee M A and Sakai J I 2004 *Astrophys. J. Lett.* **605** L73–L76

[2] Manchester IV W B, Gombosi T I, De Zeeuw D L, Sokolov I V, Roussev I I, Powell K G, Kóta J, Tóth G and Zurbuchen T H 2005 *Astrophys. J.* **622** 1225–1239

[3] Kocharov L, Lytova M, Vainio R, Laitinen T and Torsti J 2005 *Astrophys. J.* **620** 1052–1068

[4] Tsurutani B, Wu S T, Zhang T X and Dryer M 2003 *Astron. Astrophys.* **412** 293–304

[5] Schwadron N A, Gorby M, Török T, Downs C, Linker J, Lionello R, Mikić Z, Riley P, Giacalone J, Chandran B, Germaschewski K, Isenberg P A, Lee M A, Lugaz N, Smith S, Spence H E, Desai M, Kasper J, Kozarev K, Korreck K, Stevens M, Cooper J and MacNeice P 2014 *Space Weather* **12** 323–328

[6] Schwadron N A, Townsend L, Kozarev K, Dayeh M A, Cucinotta F, Desai M, Golightly M, Hassler D, Hatcher R, Kim M, Posner A, PourArsalan M, Spence H E and Squier R K 2010 *Space Weather* **8** 0–+

[7] Lionello R, Linker J A and Mikić Z 2009 *Astrophys. J.* **690** 902–912

[8] Riley P and Lionello R 2011 *Solar Phys.* **270** 575–592

[9] Riley P, Linker J A, Lionello R and Mikic Z 2012 *Journal of Atmospheric and Solar-Terrestrial Physics* **83** 1–10

[10] Mikić Z and Linker J A 1994 *Astrophys. J.* **430** 898–912

[11] Mikić Z, Linker J A, Schnack D D, Lionello R and Tarditi A 1999 *Physics of Plasmas* **6** 2217–2224

[12] Lionello R, Mikić Z and Linker J A 1999 *Journal of Computational Physics* **152** 346–358

[13] Gorby M J, Schwadron N A, Linker J A, Spence H E, Townsend L W and Cucinotta F A 2012 *AGU Fall Meeting Abstracts* A2216

[14] Linker J, Mikic Z, Schwadron N, Riley P, Gorby M, Lionello R, Downs C and Torok T 2014 *40th COSPAR Scientific Assembly* (*COSPAR Meeting* vol 40) p 1840

[15] Giacalone J, Jokipii J R and Kóta J 2002 *Astrophys. J.* **573** 845–850

[16] Jokipii J R, Giacalone J and Kota J 2003 *International Cosmic Ray Conference* (*International Cosmic Ray Conference* vol 6) p 3685

[17] Fermi E 1949 *Physical Review* **75** 1169–1174

[18] Drury L O 1983 *Reports on Progress in Physics* **46** 973–1027

[19] Bell A R 1978 *Mon. Not. R. Astr. Soc.* **182** 147–156

[20] Bell A R 1978 *Mon. Not. R. Astr. Soc.* **182** 443–455

[21] Jokipii J R 1982 *Astrophys. J.* **255** 716–720

[22] Jokipii J R 1986 *J. Geophys. Res.* **91** 2929–2932

[23] Jokipii J R 1987 *Astrophys. J.* **313** 842–846

[24] Lee M A, Fisk L A and Skadron G 1981 *Geophys. Res. Lett.* **8** 401–404

[25] Lee M A 1983 *J. Geophys. Res.* **88** 6109–6119

[26] Lee M A 2005 *Astrophys. J. Suppl.* **158** 38–67

[27] Mason G M, Wiedenbeck M E, Miller J A, Mazur J E, Christian E R, Cohen C M S, Cummings A C, Dwyer J R, Gold R E, Krimigis S M, Leske R A, Mewaldt R A, Slocum P L, Stone E C and von Rosenvinge T T 2002 *Astrophys. J.* **574** 1039–1058

[28] Mewaldt R A, Cohen C M S, Labrador A W, Leske R A, Mason G M, Desai M I, Looper M D, Mazur J E, Selesnick R S and Haggerty D K 2005 *Journal of Geophysical Research (Space Physics)* **110** A09S18

[29] Mewaldt R A, Cohen C M S, Mason G M, Labrador A W, Looper M L, Haggerty D E, Maclennan C G, Cummings A C, Desai M I, Leske R A, Li G, Mazur J E, Stone E C and Wiedenbeck M E 2005 *The Physics of Collisionless Shocks: 4th Annual IGPP International Astrophysics Conference* (*American Institute of Physics Conference Series* vol 781) ed Li G, Zank G P and Russell C T pp 227–232

[30] Mewaldt R A, Looper M D, Cohen C M S, Haggerty D K, Labrador A W, Leske R A, Mason G M, Mazur J E and von Rosenvinge T T 2012 *Space Sci. Rev.* **171** 97–120

[31] Desai M I, Dayeh M A, Ebert R W, McComas D J, Mason G M, Lo G, Cohen C M S, Mewaldt R A and Schwadron N A 2015 *Astrophys. J.* **To be submitted**

[32] Li G, Zank G P, Verkhoglyadova O, Mewaldt R A, Cohen C M S, Mason G M and Desai M I 2009 *Astrophys. J.* **702** 998–1004

[33] Kivelson M G and Spence H E 1988 *GRL* **15** 1541–1544

[34] Spence H E and Kivelson M G 1990 *Geophys. Res. Lett.* **17** 591–594

[35] Spence H E and Kivelson M G 1993 *J. Geophys. Res.* **98** 15487

[36] Schwadron N A, Lee M A, Gorby M, Lugaz N, Spence H E, Desai M, Kozarev K, Török T, Downs C, Linker J, Lionello R, Mikić Z, Riley P, Giacalone J, Jokipii J R and Kota J 2015 *Astrophys. J.* **submitted**

[37] Schwadron N A, Lee M A and McComas D J 2008 *Astrophys. J.* **675** 1584–1600

[38] Dröge W 2005 *Advances in Space Research* **35** 532–542

[39] Zank G P, Li G, Florinski V, Matthaeus W H, Webb G M and Le Roux J A 2004 *Journal of Geophysical Research (Space Physics)* **109** A04107

[40] Zank G P, Li G and Verkhoglyadova O 2008 *American Institute of Physics Conference Series* (*American Institute of Physics Conference Series* vol 1039) ed G Li, Q Hu, O Verkhoglyadova, G P Zank, R P Lin, & J Luhmann pp 203–213

Cosmic ray transport near the heliopause

R D Strauss[1,2], H Fichtner[3], M S Potgieter[1], J A le Roux[2] and X Luo[1,4]

[1] Center for Space Research, North-West University, South Africa
[2] The Center for Space Plasma and Aeronomic Research, University of Alabama in Huntsville, USA
[3] Institut für Theoretische Physik IV, Ruhr Universität, Bochum, Germany
[4] National Space Science Center CAS, Beijing, China

E-mail: dutoit.strauss@nwu.ac.za

Abstract. In this paper we summarize our modelling efforts for cosmic rays near the heliopause, and discuss whether galactic cosmic ray modulation beyond the heliopause is possible and present an explanation for the anisotropic nature of the observed cosmic ray intensities in the very local interstellar medium. We show that (i) modulation beyond the heliopause is possible, but highly dependent on the assumed parameters (most notable, the perpendicular diffusion coefficient). Treating the heliopause as a tangential discontinuity, significantly damps this modulation effect and leads to modelled results that are similar to Voyager 1 observations. (ii) By choosing an appropriate functional form of the perpendicular diffusion coefficient on the pitch-angle level, we are able to account for the anisotropic behaviour observed for both galactic and anomalous cosmic rays in the local interstellar medium.

1. Introduction

Anticipating the Voyager 1 (V1) crossing of the heliopause (HP), [1] questioned the use of a Dirichlet boundary condition for galactic cosmic rays (GCRs) at the HP. Previous estimates, e.g. [2], suggested that the presence of the heliosphere cannot alter the GCR distribution directly beyond the HP, so that the GCR boundary condition, the so-called local interstellar spectrum (LIS), can be prescribed directed there. The results of [1], and a follow-up paper by [3], suggested that the heliosphere can influence the GCR intensity beyond the HP significantly and that, once V1 crossed the HP, it could continue to measure a small positive radial gradient in GCR intensity.

In August 2012, V1 did indeed cross the *heliocliff* region, where anomalous cosmic ray (ACR) intensities decreased and GCR increased substantially [4, 5]. This crossing was later confirmed as the HP [6]. Since then, V1 has been sampling the very local interstellar medium (ISM) *in-situ* and is by now ~ 10 AU beyond the HP.

In this paper we summarize our GCR modelling efforts beyond the HP, after which we critically evaluate the parameters adopted in these simulations *in lieu* of the V1 GCR observations. We also discuss a plausible explanation for the observed anisotropic nature of both the ACR and GCR distributions directly beyond the HP.

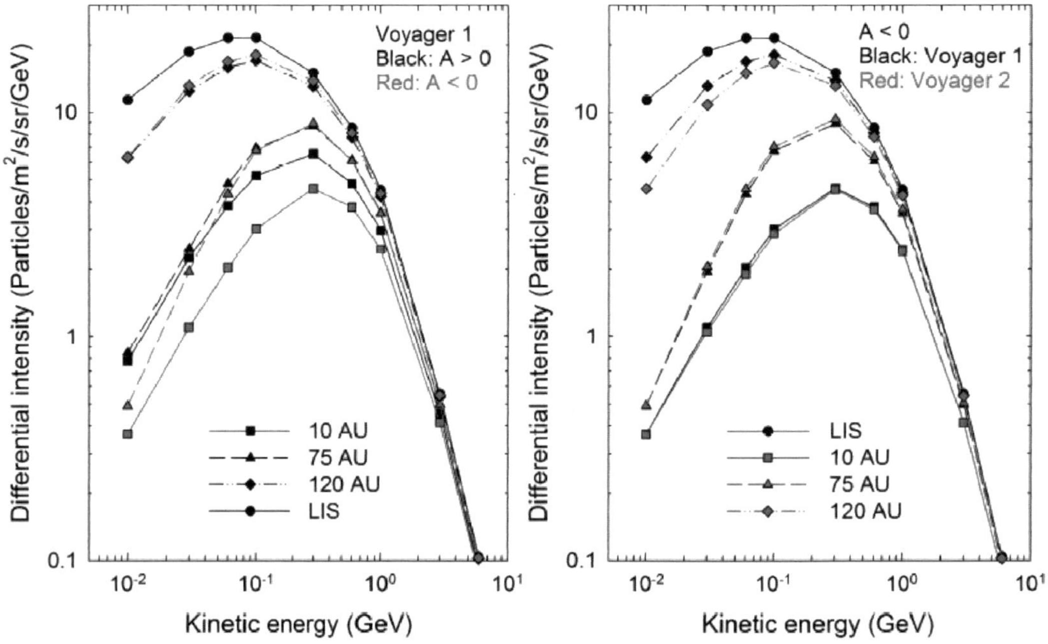

Figure 1. The left panel shows computed GCR proton energy spectra along the V1 direction for two different HMF polarity cycles, while the right panel shows energy spectra, only for the $A < 0$ drift cycle, along both V1 and V2 directions.

2. Galactic cosmic ray modulation beyond the heliopause

We make use of the *hybrid* model of [7], where the Parker transport equation [9] is solved, by means of stochastic differential equations (SDEs), on top of an MHD modelled asymmetric heliospheric environment. This allows us to capture the complex geometry of the heliosphere which cannot, very easily, be described analytically.

2.1. Generic results

Fig. 1 shows the resulting GCR energy spectra, using the model of [7], at different radial positions. The left panel is along the V1 direction and for different drift cycles, i.e. different polarities of the heliospheric magnetic field (HMF), while the right panel shows energy spectra, for only the $A < 0$ polarity cycle, along both V1 and Voyager 2 (V2) trajectories. The spectra at the HP (here corresponding to the 120 AU scenario) are significantly lower than the LIS values at low energies; at 100 MeV, the HP intensities are $\sim 25\%$ lower than the LIS values. Moreover, we find that drifts have little or no effects beyond 75 AU, while the results are also very similar along the V1 and V2 directions.

In order to examine what the main drivers of modulation in the ISM are, we performed a parameter study, varying the parallel mean free path ($\lambda_{||}$) and the ratio of the perpendicular ($\lambda_{\perp}$) to the parallel mean free path through the parameter η, defined by $\lambda_{\perp} = \eta\lambda_{||}$. The results are shown in the right panel of Fig. 2, which shows the resulting *modulation fraction*, i.e. the ratio j_{HP}/j_{LIS}, when η is kept constant and $\lambda_{||}$ is changed (the blue curve) and when $\lambda_{||}$ is kept constant and η changed (red curves). We find the modulation most sensitive to η, which is explained by the draping of the interstellar magnetic field along the HP [8], illustrated in the left panel of Fig. 2: In order for GCRs to enter the heliosphere, they must diffuse perpendicular

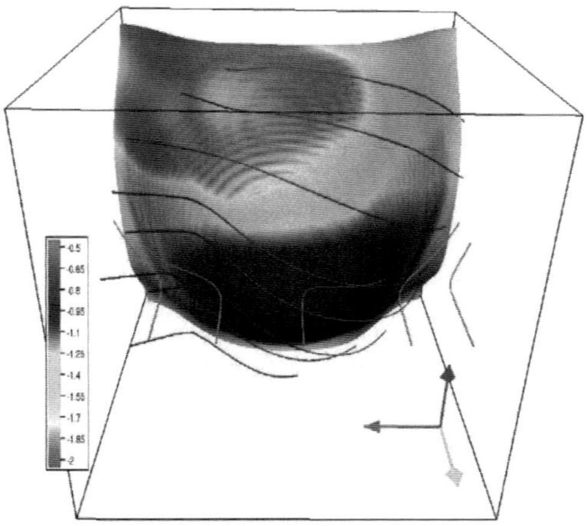

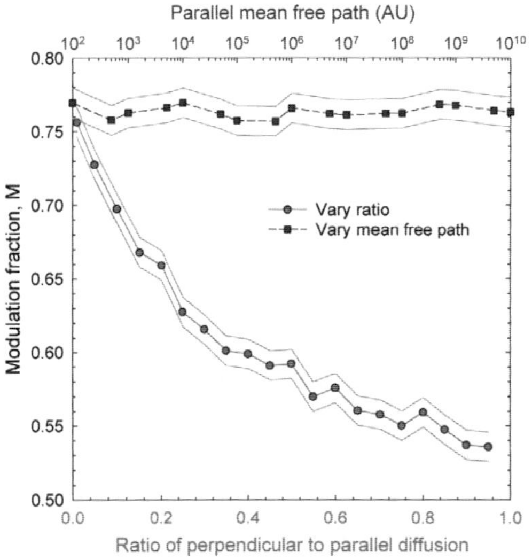

Figure 2. The left panel shows the *draping* of the interstellar magnetic field (the red lines) along the HP (the coloured surface). The right panel shows how the amount of modulation changes with varying parameters (see the text for details).

to the mean field. For larger values of η (i.e. more effective perpendicular scattering), the amount of modulation increases, and hence, the GCR gradient beyond the HP increases. We thus emphasize the importance of $\lambda_\perp$ in determining the gradient of GCRs beyond the HP.

It must be noted that these results are refuted by [10] and [11] who found that: GCR modulation is significantly less pronounced than presented here and that the modulation process in the ISM is predominantly determined by the magnitude of $\lambda_{||}$. The fact that [7] and [11] used similar models, but came to completely different conclusions, due to different results, is puzzling and is presently investigated. On the other hand, the SDE modelling of [12] is in agreement with the results of [7].

2.2. The heliopause as a tangential discontinuity

For most of the results presented in the previous section, we assumed a *generic* value of $\eta = 0.02$. Although this value is a good estimate near Earth, it is not necessarily so in the outer heliosheath and perhaps especially directly at the HP. The HP represents a tangential discontinuity, e.g. [13], separating heliospheric and interstellar plasmas, and may, as illustrated below, damp $\lambda_\perp$ significantly leading to an effective GCR propagation barrier.

To investigate the presence of this barrier, it is useful to consider other tangential discontinuities present in the heliosphere, most notable, stream interfaces associated with co-rotating interaction regions. It is well known that these stream interfaces are CR propagation barriers [14], although the exact mechanism responsible for this is still relatively unclear. Most authors postulate that the random motion of diffusing HMF lines is damped at the tangential discontinuity [15], thereby also damping the field line random walk (FLRW) process of perpendicular diffusion [16]. A schematic illustration of this is presented in Fig. 3: The left panel shows illustrative HMF lines *diffusing* under normal solar wind conditions. GCRs with gyro-radii smaller than the associated turbulence correlation length will then follow these field lines (e.g. the blue curve showing the trajectory of such a particle's guiding center), leading to motion perpendicular to the mean HMF. Near a tangential discontinuity (illustrated by the red

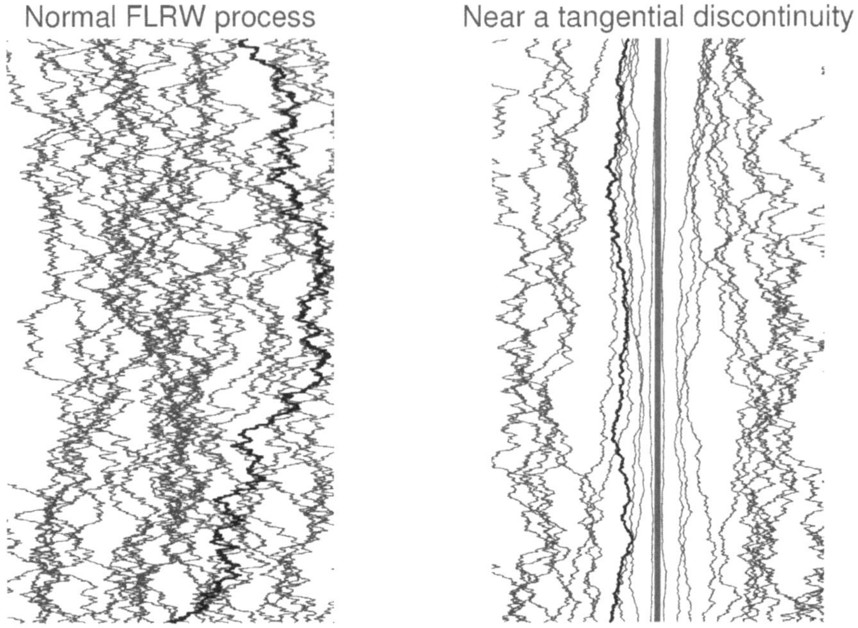

Figure 3. An illustration of the FLRW process under normal solar wind conditions (left panel) and near a tangential discontinuity, indicated by the red line in the right panel.

line in the right panel), the diffusion of the field lines is damped (inhibited) near the discontinuity and the FLRW diffusion process cannot take place effectively.

If the same process occurs at the HP, it is very conceivable that $\eta \ll 0.02$, which, as shown below, can have large consequences for GCR transport in the ISM: Referring back to Fig. 2, and extrapolating to $\eta \to 0$, it should follow that $j_{HP} \to j_{LIS}$ and no modulation takes place in the ISM. Recently, [12] studied GCR modulation beyond the HP and found similar results to [7]. They did, however, also investigate the possibility of small η values with their results summarized in Fig. 4: As η becomes smaller, the gradient beyond the HP diminishes, while for very small values $\eta \sim 10^{-10}$, a step-wise increase in GCR intensity is noticeable, directly at the HP, after which LIS values are quickly reached. This step-like increase is indeed reminiscent of the V1 data [4], which also showed no measurable gradient beyond the HP. Similar conclusions were reached by [11].

3. Cosmic ray anisotropies in the local interstellar medium

In the ISM, the observed ACR and GCR intensities show significant anisotropies [17]. Sector diagrams of these anisotropic distributions are shown in the bottom panel of Fig. 5 for ACRs (referred to as heliospheric particles) and GCRs (referred to as galactic particles), with the red arrow indicating the magnetic field direction. Note that both distributions are still isotropic inside the heliosphere. Although it is not surprising that these distributions are anisotropic in the ISM ($\lambda_{||}$ is expected to be very large there [18], which implies weak scattering, or equivalently, a small pitch-angle diffusion coefficient $D_{\mu\mu}$), it is interesting that the anisotropies are in the opposite sense for both distributions: ACRs show an enhancement at pitch-angles of 90° (or, in terms of the pitch-angle cosine, $\mu = 0$), while GCRs show a depletion near $\mu = 0$. [19] and [20] investigated the cause of the ACR anisotropies and could reproduce the observed features. Their model was, however, only applied to ACRs. We however believe that a common process

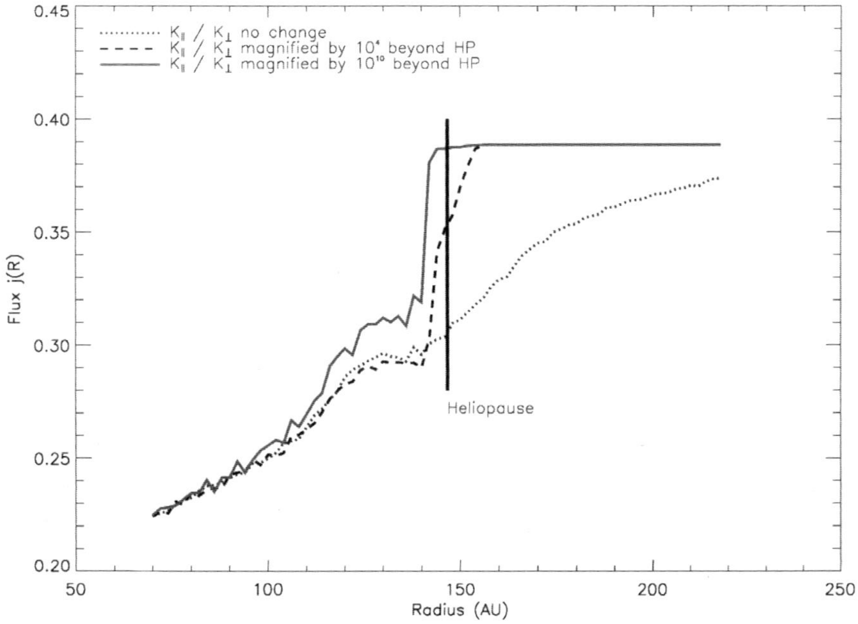

Figure 4. Results from [12], showing the modelled GCR intensity near the HP for varying values of η. For very small values of η a step-like increase in GCR intensity is found at the HP with no measurable gradient beyond it.

should be responsible for both the ACR and GCR anisotropies and such a process, as discussed by [21], is put forward in the following section.

3.1. Model results

In [21] we proposed that these anisotropies are simply due to more effective perpendicular diffusion across the HP at certain pitch-angles. By assuming that the perpendicular diffusion coefficient, on the pitch-angle level, has the functional form

$$D_\perp \sim v_\perp \sim \sqrt{1-\mu^2}, \quad (1)$$

we showed that, at least qualitatively, the anisotropic behaviour of both cosmic ray species could be explained. Results from this model are shown in the top panel of Fig. 5, in the form of modelled sector diagrams. The modelled behaviour is in agreement with the observations: The ACR distribution, in the ISM, peaks at $\mu = 0$, while the GCR distribution reaches a minimum here. The reasoning is that $\mu = 0$ particles have the highest mobility across the HP (as was the assumption when Eq. 1 was postulated): The $\mu = 0$ ACRs can thus readily *escape* across the HP, and hence, there should be an enhancement of these particles in the ISM. Similarly for GCRs, the $\mu = 0$ particles enter the heliosphere more readily and a deficiency of these particles should be observed in the ISM.

Criticism towards this explanation for the observed anisotropies is mostly related to the *ad-hoc* fashion in which the functional form of $D_\perp$ was chosen. Such a form can, however, be derived from diffusion theory as is discussed in the next section.

As will also be shown in Section 3.2, this form of $D_\perp$ results from drift effects, and as such, generally increases linearly with rigidity, $D_\perp \sim P$. The model of [21] was again used to

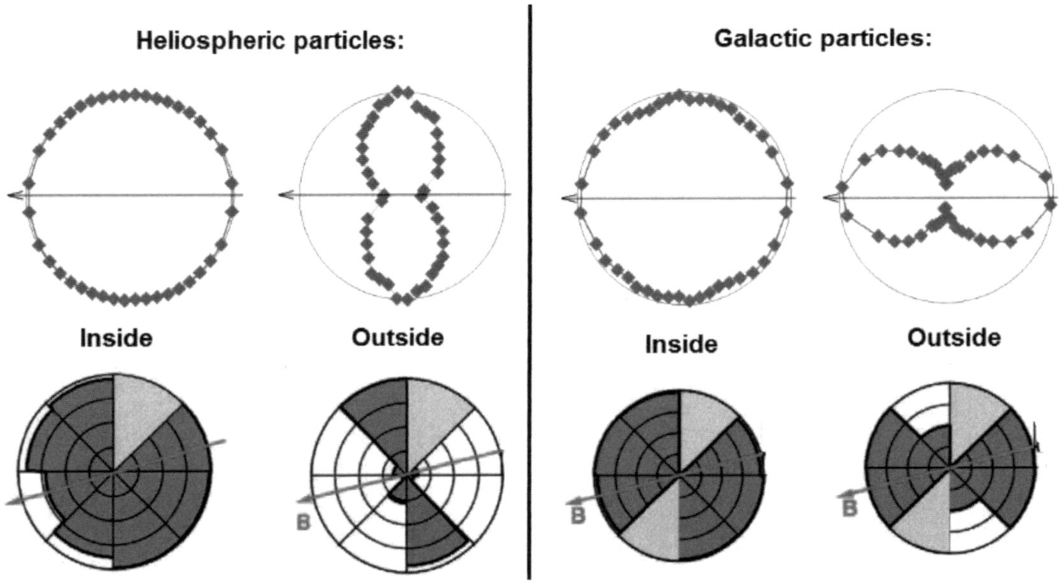

Figure 5. The top panel shows modelled sector diagrams, from [21], for ACRs and GCRs, inside and outside of the heliosphere. The bottom panels are the corresponding observations from [17].

model the transport of ACRs across the HP, but now for two different rigidities, and the results summarized in Fig. 6. The left panel shows the intensity for ACRs, of two different rigidities, across the HP, with the higher rigidity particles' intensity decreasing first in front of the HP, and remaining higher in the ISM, in accordance with the V1 observations of [20]. We believe that these results show additional, qualitative, observational support for our explanation of the observed anisotropies. Moreover, when the anisotropy level is calculated (the bottom right panel), it is clear that: The magnitude of the ACR anisotropy increases with radial distance beyond the HP, peaking at some, parameter dependent, distance beyond the HP (the same is true for the GCRs and in accordance with the observations of [17]) and that the anisotropy is generally higher for higher rigidity particles. It is unclear whether these rigidity predictions for ACRs can be tested against observations.

Interestingly, the GCR anisotropy was observed to disappear (the distribution was isotropized) and reappear from time to time in the local ISM. It is believed that times of isotropy closely coincided with the passage of transients past the spacecraft. In our modelling framework, these observations can be explained in two ways: (i) The additional scattering (enhanced turbulence levels) present in these transient structures can isotropize the distribution function. However, additional acceleration [23] will also be needed to maintain the relatively constant levels of the isotropic GCR flux. (ii) An isotropic distribution can be maintained, at the same constant level, if the effectiveness of perpendicular diffusion is diminished; without perpendicular diffusion, no GCRs will be removed from the ISM (lost to the heliosphere). Perpendicular diffusion can become less efficient if either the magnetic topology or turbulence levels change considerably with the passage of the transient (similar to a reverse Forbush decrease occurring beyond the HP – the passage of the transient can actually increase the GCR flux relative to undisturbed periods), or if the magnetic magnitude increases significantly (see Eq. 13). It is unclear which (if any or perhaps even both) of the above-mentioned isotropization processes are most viable in the interstellar medium.

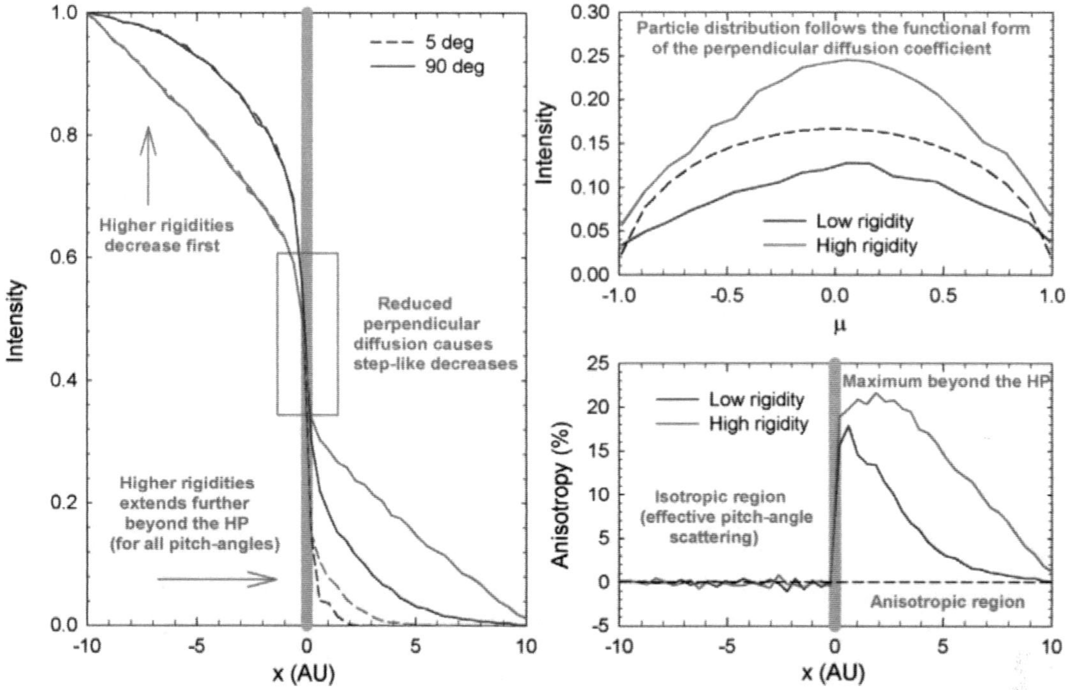

Figure 6. Applying the model of [21] to ACR transport across the HP for two different rigidities: Red curves correspond to the higher rigidity ACRs. The dashed line in the upper right panel shows the functional form of $D_\perp(\mu)$ as used in this study.

3.2. Parameter motivation

Below, we present a derivation of the applicable functional form of $D_\perp$ near the HP. The derivation follows, to some extent, the methodology presented in [22], which is valid for particles with gyro-radii much smaller than the scale of the turbulence considered; i.e. $D_\perp$ is due to non-resonant effects.

As usual, the mean and turbulent HMF is decomposed as

$$\vec{B} = B_0\hat{z} + \delta\vec{B}(x, y, z), \tag{2}$$

with B_0 constant and $\left\langle \delta\vec{B} \right\rangle = 0$.

The fluctuating 2D component, believed to be primarily responsible for perpendicular diffusion, is generalized to include a component along the mean field

$$\delta\vec{B} = \delta B_x(x, y)\hat{x} + \delta B_y(x, y)\hat{y} + \delta B_z(x, y)\hat{z}, \tag{3}$$

with weak turbulence assumed, $\delta B_{x,y,z} \ll B_0$.

In the supersonic solar wind, the fluctuations are usually observed to be transversal in nature, i.e. $\delta B_z(x, y) \approx 0$ [24]. However, observations near the HP (in the so-called stagnation region) indicate that the fluctuations are mostly longitudinal (compressional) in nature [25]. As the HP can also be considered a tangential discontinuity, it is likely that the transverse components will be strongly damped directly at the HP, thereby also damping the FLRW process. We therefore assume the 2D component to be purely longitudinal in nature, i.e. $\delta B_x, \delta B_y \ll \delta B_z$, and proceed to calculate $D_\perp$ due to non-resonant interactions, i.e. for the scenario when the particle gyro-radius is smaller than the associated turbulent correlation scales.

The mean-square displacement of particles across the mean HMF, due to random walking magnetic field lines, is defined as

$$\left\langle \Delta x_\perp^2 \right\rangle = \int_0^{\Delta t} dt' \int_0^{\Delta t} dt'' \left\langle v_\perp(t')v_\perp(t'')\right\rangle . \tag{4}$$

If homogeneous, magnetostatic fluctuations are considered, we have, at late times, a Kubo formula given by

$$D_\perp(\mu) = \lim_{\Delta t\to\infty} \frac{\left\langle \Delta x_\perp^2 \right\rangle}{2\Delta t} = \int_0^{\infty} dt' \left\langle v_\perp(0)v_\perp(t')\right\rangle . \tag{5}$$

Assuming that the correlation function decays exponentially as

$$\left\langle v_\perp(0)v_\perp(t)\right\rangle = \left\langle v_\perp^2(0)\right\rangle e^{-\frac{t}{\tau_{\mathrm{dec}}}}, \tag{6}$$

it follows that

$$D_\perp(\mu) = \left\langle v_\perp^2\right\rangle \tau_{\mathrm{dec}}. \tag{7}$$

Perpendicular motion is assumed to be due to random drift motion of the guiding center across the mean field

$$\left\langle v_\perp^2\right\rangle \to \left\langle {}^g\vec{V}_\perp^2\right\rangle . \tag{8}$$

Furthermore, the decorrelation time is taken to be the time it takes a guiding center to drift across a perpendicular correlation length, $l_\perp$,

$$\tau_{\mathrm{dec}} = \frac{l_\perp}{\left\langle {}^gV_\perp\right\rangle}, \tag{9}$$

so that $D_\perp$ becomes

$$D_\perp(\mu) = \left\langle {}^gV_\perp\right\rangle l_\perp. \tag{10}$$

The guiding center drift velocity is, in general, given by

$${}^g\vec{V} = v_{||}\hat{b} + \frac{p_\perp v_\perp}{2qB}\frac{\vec{B}\times\nabla B}{B^2} + \frac{p_{||}v_{||}}{qB}\hat{b}\times\nabla_{||}\hat{b} \tag{11}$$

where $\hat{b}$ is a unit vector pointing along $\vec{B}$ and electric fields were neglected.

For purely longitudinal fluctuations, we only get a contribution from the gradient drift term, leading to

$${}^gV_\perp = \frac{vr_L}{2l_\perp}\left(1-\mu^2\right)\frac{\delta B_z^2}{B_0^2}. \tag{12}$$

$D_\perp$ follows directly as

$$D_\perp(\mu) = \frac{vr_L}{2}\left(1-\mu^2\right)\left\langle \frac{\delta B_z^2}{B_0^2}\right\rangle , \tag{13}$$

which follows, in principle, the assumed dependence assumed *ad hoc* in the previous section. Also note the linear dependence on momentum (rigidity) contained in r_L.

4. Discussion and conclusions

Our modelling illustrates that GCR modulation beyond the HP is possible, but the amount of modulation is parameter dependent and we argue that this is mostly determined by the level of perpendicular diffusion directly at the HP. If the HP is considered as a tangential discontinuity, and the perpendicular diffusion process is effectively suppressed, it seems unlikely that any measurable amount of modulation will occur beyond the HP. The original model of [7] predicted a radial gradient of $\sim 0.25 - 0.45$ %/AU, assuming a $\sim 25 - 45$ % decrease occurring over ~ 100 AU. Currently, the presence of such a small gradient cannot be ruled out observationally. It should also be kept in mind that V1 cannot be considered yet to be in the pristine ISM, and that V2 data could show a different situation.

By choosing the appropriate functional form for the pitch-angle dependent perpendicular diffusion coefficient, we are able to reproduce a myriad of observed features in the ISM related to the anisotropy of both ACR and GCR species.

Acknowledgments

HF gratefully acknowledges financial support via the project FI 706/15-1 funded by the Deutsche Forschungsgemeinschaft (DFG). XL acknowledges the support of the post-doctoral programme of the North-West University in South Africa. This work is based on the research supported in part by the National Research Foundation (NRF) of South Africa (RDS is supported through the Thuthuka Program; Grant no. 87998). Opinions expressed and conclusions arrived at are those of the authors and are not necessarily to be attributed to the NRF. RDS acknowledges support from the Fulbright Visiting Scholar Program. We also appreciate discussions at the team meeting 'Heliosheath Processes and Structure of the Heliopause: Modeling Energetic Particles, Cosmic Rays, and Magnetic Fields' supported by the International Space Science Institute (ISSI) in Bern, Switzerland.

References

[1] Scherer K, Fichtner H, Strauss R D, Ferreira S E S, Potgieter M S and Fahr H-J 2011 *Astrophys. J.* **735** 128
[2] Jokipii, J R 2001 *COSPAR Colloquia Series* **11** 227
[3] Herbst K, Heber B, Kopp A, Sternal O and Steinhilber F 2012 *Astrophys. J.* **761** 17
[4] Stone E C, Cummings A C, McDonald F B, Heikkila B C, Lal N and Webber W R 2013 *Sci* **341** 150
[5] Webber W R and McDonald F B 2013 *Geophys. Res. Lett.* **40** 1665
[6] Gurnett D A, Kurth W S, Burlaga L F and Ness N F 2013 *Sci* **341** 1489
[7] Strauss R D, Potgieter M S, Ferreira S E S, Fichtner H and Scherer K 2013 *Astrophys. J. Lett.* **765** L18
[8] Röken C, Kleimann J and Fichtner H 2015 *Astrophys. J.* accepted
[9] Parker E N 1965 *Planet. Space Sci.* **13** 9
[10] Kóta J and Jokipii J R 2014a *Astrophys. J.* **782** 24
[11] Guo X and Florinski V 2014 *Astrophys. J.* **793** 18
[12] Luo X, Zhang M, Potgieter M S, Xueshang F and Pogorelov N V 2014 *Astrophys. J.* submitted
[13] Burlaga L F and Ness N F 1969 *Sol. Phys.* **9** 467
[14] Intrilligator D S et al. 2001 *J. Geophys. Res.* **106** 10625
[15] Crooker N U et al. 1999 *Space Sci. Rev.* **89** 179
[16] Jokipii J R 1966 *Astrophys. J.* **145** 480
[17] Krimigis, S M et al. 2013 *Sci* **341** 144
[18] Shalchi A and Büsching I 2010 *Astrophys. J.* **725** 2110
[19] Florinski V, Jokipii J R, Alouani-Bibi F and le Roux, J A 2013 *Astrophys. J. Lett* **776** L37
[20] Florinski V, Stone E C, Cummings A C and le Roux J A 2015 *Astrophys. J.* **803** 47
[21] Strauss R D and Fichtner H 2014 *Astron. Astrophys. Lett.* **572** L3
[22] Fraschetti F and Jokipii J R 2011 *Astrophys. J.* **734** 83
[23] Jokipii J R and Kóta J 2014b *Astrophys. J. Lett.* **794** L4
[24] Shalchi A 2009 *Nonlinear Cosmic Ray Diffusion Theories* Springer-Verlag Germany
[25] Burlaga L F, Ness N F, Florinski V and Heerikhuisen J 2014 *Astrophys. J.* **792** 134

Journal of Physics: Conference Series **642** (2015) 012027 doi:10.1088/1742-6596/642/1/012027

Response of Solar Wind on Extreme Solar Activity

Takeru K. Suzuki

Department of Physics, Nagoya University,Furo-cho, Chikusa-ku, Nagoya, Aichi 464-8602, Japan

E-mail: stakeru@nagoya-u.jp

Abstract. We investigate how the mass loss by the solar wind depends on the solar activity levels, particularly focusing on the solar wind during extremely high activity. We perform forward-type magnetohydrodynamical (MHD) numerical experiments for Alfvén wave-driven solar winds with a wide range of the input Poynting flux from the photosphere. Increasing the magnetic field strength and the turbulent velocity at the solar photosphere from the current solar level, the mass loss rate rapidly increases at first owing to the suppression of the reflection of the Alfvén waves. The surface materials are lifted up by the magnetic pressure associated with the Alfvén waves, and the cool dense chromosphere is extended to $\sim$ 10 % of the stellar radius. The dense atmospheres enhance the radiative losses and eventually most of the input Poynting energy from the surface escapes by the radiation. As a result, there is no more sufficient energy remained for the kinetic energy of the wind; the solar wind saturates for the extreme activity level, as observed in Wood et al. The saturation level is positively correlated with the average magnetic field strength contributed from open flux tubes. If the field strength is a few times larger than the present level, the mass loss rate could be as high as 1000 times.

1. Introduction

Young solar-type stars are generally very active in X-ray and UV radiation (1), which is probably related to the strong magnetic field with an order of kilo-Gauss (2). The mass loss rate of stars with surface convection is also larger for larger X-ray luminosity, up to $\approx$ 100 times of the mass loss rate of the Sun, but it saturates for very active ones, which is discussed from the change of magnetic topology (3; 4; 5). The relation between the stellar magnetic field and the wind has lately attracted considerable attention and has been extensively studied these days (6; 7). In this article, we briefly review our recent results on stellar winds from young active solar-type stars.

2. Simulation or the Present Solar Wind

We briefly explain our MHD simulations for the solar and stellar wind (8; 9; 10). We follow the time evolution of velocity, $\boldsymbol{v}$, density, ρ, temperature, T, and magnetic fields, $\boldsymbol{B}$, in a single open flux tube by solving MHD equations with radiation cooling (11; 12; 13) and thermal conduction by Coulomb collisions (Spitzer conductivity). The simulation region covers from the photosphere to the interplanetary space (0.3 AU), which is the biggest advantage of our setup because we can determined the mass loss rate directly from the physical properties at the photosphere where various detailed observational data are available.

We show the result for the typical high-speed solar wind introduced in (8) in Figure 1. We set up an initially static $v = 0$ and cool ($T = 10^4$ K) atmosphere and begin the simulation with

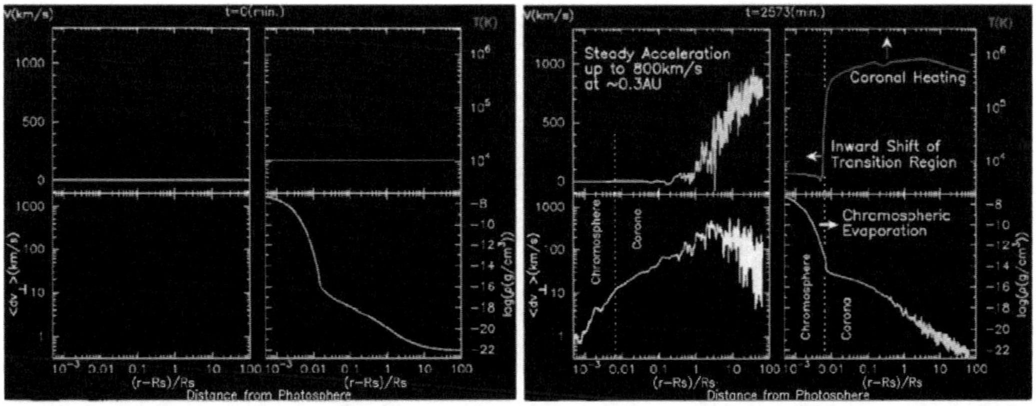

Figure 1. Time evolution of the MHD simulation for the solar wind (8). The left panel presents the initial condition and the right panel presents the snapshot at $t = 2573$ min. after the quasi-steady state condition is achieved. The four small panels in each panel shows radial velocity (v km s^{-1}; *upper left*), temperature (T K; *upper right*), root-mean-squared velocity amplitude of Alfvénic waves ($\sqrt{dv_\perp^2}$ km s^{-1}; *lower left*), and density (ρ g cm^{-3}; *lower right*).

injecting transverse fluctuations with 0.7 km s^{-1} of the surface convection from the photosphere. The figure clearly exhibits that the atmosphere is heated up to the coronal temperature with $T \approx 10^6$ K and accelerated to 700-800 km s^{-1} when the quasi-steady-state condition is obtained[1]. The transverse fluctuations drive Alfvénic waves that propagate upwardly. While a large fraction ($\sim 90\%$) of the input Poynting flux by the Alfvénic waves is reflected back downward, a sufficient energy is transmitted to the coronal region. The main dissipation channel is the nonlinear mode conversion to compressive waves; the fluctuation of the magnetic pressure of the Alfvénic waves excites longitudinal perturbation, most of which is slow MHD waves in the low-β (magnetically dominated) condition (14). These slow MHD waves eventually steepen to form shocklets and dissipate to heat up the surrounding plasma.

A characteristic feature of the excitation of compressive waves is that density perturbation could be observed. Recently, AKATSUKI, Japanese satellite originally planned as a Venus climate orbiter, observed the regions in $1.5 - 20R_\odot$ ($R_\odot$ is the solar radius) where the solar wind is accelerated (15), and detected density fluctuations there (16), whereas the simulated density amplitude exceeds the observed value (Fig. 8 of (16)) probably because of the limitation of our treatment (see below) and/or the effect of the line-of-sight integration in the observation. Because it is difficult for compressive waves to propagate from the photosphere to the coronal region owing to the dissipative character (17), these density fluctuations are supposed to be excited in situ; the nonlinear mode conversion from Alfvénic waves is one of the possible mechanisms.

Since in our simulation we consider the energy transfer in the one-dimensional flux tube ("1.5D" approximation with solving the three components of $\boldsymbol{v}$ and $\boldsymbol{B}$), we cannot treat cascading Alfvénic turbulence, which is supposed to play a role in the heating and driving the solar wind (18; 19; 20). In realistic situations, the incompressive turbulent cascade and the above mentioned generation of compressive waves both contribute to the coronal heating and solar wind acceleration in a cooperating manner. In fact, our simulation is extended to two dimensional ("2.5D") flux tubes recently (21; 22) to study both turbulent cascase and nonlinear generation compressive waves.

[1] Movie is available at www.ta.phys.nagoya-u.ac.jp/stakeru/research/swhr560_2.mpg

3. Wind from Active Stars

3.1. Rapid Increase and Saturation

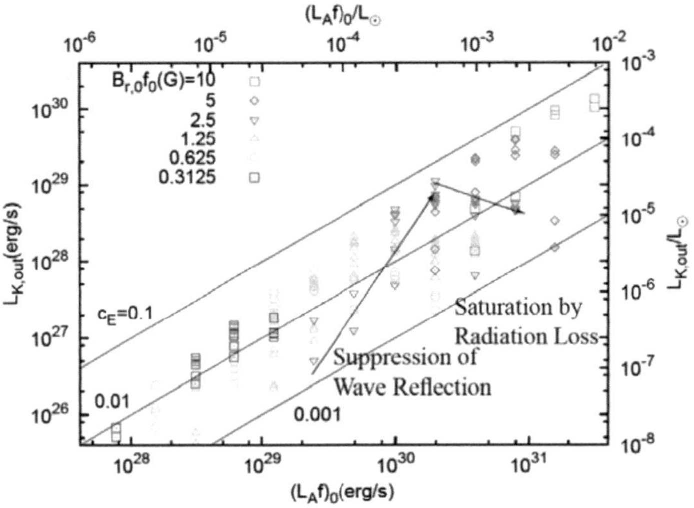

Figure 2. Final kinetic energy of the stellar wind for the input Poynting flux injected by Alfvénic waves from the photosphere. Colors indicate the average magnetic field strength contributed from open magnetic field regions, $B_{r,0}f_0$. The solid lines indicate the conversion factor, $c_E = 0.001, 0.01, 0.1$.

Our forward-type simulation can determine the mass loss rate of the wind by the properties at the photosphere. By utilizing this advantage, we investigate the response of the atmosphere on a change of the velocity fluctuation, δv, and the strength and configuration of the magnetic field. Although we have three parameters on the magnetic field, we here focus on $B_{r,0}f_0$, where $B_{r,0}$ is the radial magnetic field strength and f_0 is the areal filling factor of open magnetic field regions measured at the photosphere; $B_{r,0}f_0$ denotes the average magnetic field strength contributed from open flux regions. We performed 163 simulation runs for stellar winds from a star with the solar mass and radius by changing these parameters (see (23) for detail).

Figure 2 shows the relation between the input energy by the Alfvénic wave in units of luminosity, erg s^{-1}, from the photosphere (horizontal axis) and the kinetic energy luminosity of the stellar wind (vertical axis). Conversion factor, c_E, from the input wave energy to the output kinetic energy is also shown by solid lines. As easily expected, one can see a positive correlation between the input energy and the wind kinetic energy within a range of $c_E = 0.001 - 0.1$. However, if we focus on the data with a single $B_{r,0}f_0$, the trend is not monotonic; the trend for the data with pink ($B_{r,0}f_0 = 2.5$ G) is shown by arrows, for example; the kinetic energy rapidly increases initially with an increasing input energy but eventually saturates or even decreases.

The rapid increase of the kinetic energy in the regime of small input energy can be explained by the suppression of the reflection of Alfvénic waves. The left panel of Figure 3 compares the time-averaged density structures of a present sun case (blue) and an active young sun case with larger δv and $B_{r,0}f_0$ (red). In the active case, the decrease of the density is much slower, which indicates that the atmospheric gas is lifted up to the upper region. This is mainly because the magnetic pressure of the Alfvénic waves contributes to the hydrostatic equilibrium against the gravity in addition to the usual gas pressure gradient force. As a result, the change of the Alfvén speed (right panel) is more gradual in this case. This is in contrast to the current sun case, in which the Alfvén speed $= B_r/\sqrt{4\pi\rho}$ increases drastically from the chromosphere to the

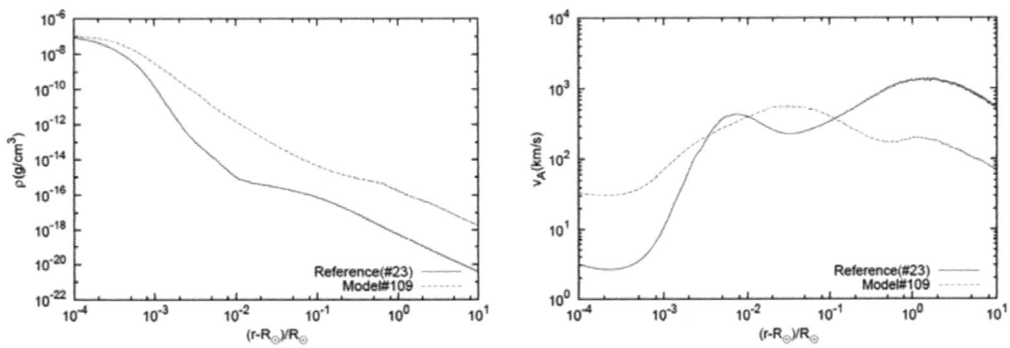

Figure 3. Comparison of the time-averaged profiles of density (left) and Alfvén speed (right) for a present sun case (blue) and an active young sun case (red).

coronal region because of the rapid decrease of the density. The profile of the Alfvén speed controls the reflection of Alfvénic waves; a rapid change of Alfvén speed generally enhances the reflection because of the deformation of the shape of propagating waves. Therefore, in the active case a larger fraction of the input energy penetrates to the corona. This argument indicates that the transmitted energy of the Alfvénic waves depends sensitively on the input energy; a small change of the injected energy leads to a large difference of the transmitted wave energy to the corona, and accordingly the final kinetic energy of the wind.

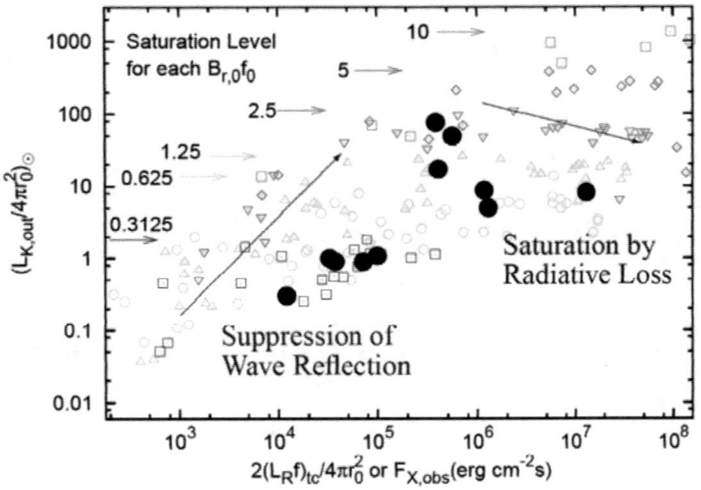

Figure 4. Comparison of the mass loss rate and the radiation flux. The color symbols are from our simulation, and the black symbols are observational data from (4).

The saturation or decrease of the kinetic energy on an increasing input energy is because of the rapid enhancement of the radiation loss. As the input energy increases, the density in the atmosphere also increases, which gives larger mass loss rate. The radiative cooling in the optically thin coronal region is proportional to the density squared. Therefore, the radiation loss rapidly increases for an increasing input energy, and eventually the sufficient energy is not remained for the kinetic energy of the wind.

In order to compare our simulation results to the observation of solar-type stars (3; 4; 5), we show the wind kinetic energy on the radiative flux in Figure 4 (see (23) for detail). The observed maximum mass loss is ≈ 100 times of the current mass loss rate from the Sun, our result shows that some cases give even larger mass losses up to ≈ 1000 times of the present solar mass loss for strong magnetic field, $B_{r,0}f_0 = 10$ G.

3.2. Extended Chromosphere

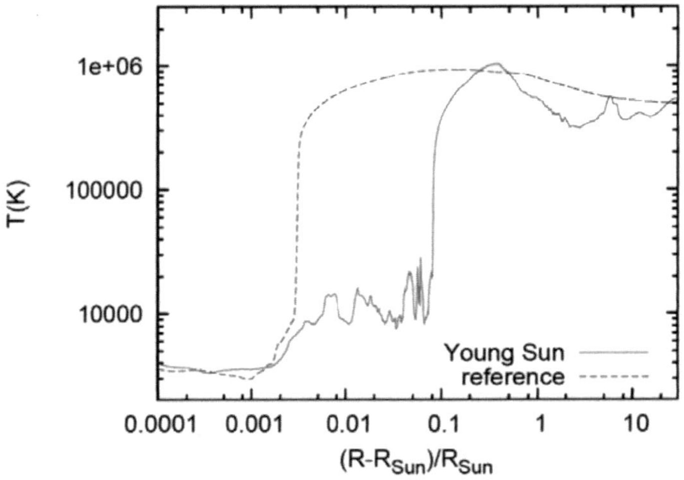

Figure 5. Comparison of the temperature structures of the current sun case (blue) and the active sun case (red).

Most of the simulation runs with a large energy input by the Alfvénic waves from the photosphere show an extended chromosphere because of the support from the magnetic pressure of the Alfvénic waves (Figure 3). Figure 5 compares the snapshot temperature structures of the current sun case and the active young sun case. In the active sun case, the chromosphere extends to ≈ 10 % of the stellar radius, which is in contrast to the very thin (< 1 %) chromosphere of the current sun. Interestingly, Czesla et al (24) reported that a young Sun-like star has a very extended chromosphere up to ≈ 15 % of the stellar photosphere based on their observation of the exoplanet system with the Rossiter-McLaughlin effect of chromospheric lines.

References

[1] Güdel M 2007 *Living Reviews in Solar Physics* **4** 3 (*Preprint* 0712.1763)

[2] Donati J F and Landstreet J D 2009 *ARA&A* **47** 333–370 (*Preprint* 0904.1938)

[3] Wood B E, Müller H R, Zank G P and Linsky J L 2002 *ApJ* **574** 412–425 (*Preprint* astro-ph/0203437)

[4] Wood B E, Müller H R, Zank G P, Linsky J L and Redfield S 2005 *ApJ* **628** L143–L146 (*Preprint* astro-ph/0506401)

[5] Wood B E, Müller H R, Redfield S and Edelman E 2014 *ApJ* **781** L33

[6] Pinto R F, Brun A S, Jouve L and Grappin R 2011 *ApJ* **737** 72 (*Preprint* 1106.0882)

[7] Matt S P, Pinzón G, Greene T P and Pudritz R E 2012 *ApJ* **745** 101 (*Preprint* 1111.6407)

[8] Suzuki T K and Inutsuka S i 2005 *ApJ* **632** L49–L52 (*Preprint* astro-ph/0506639)

[9] Suzuki T K and Inutsuka S i 2006 *Journal of Geophysical Research (Space Physics)* **111** A06101 (*Preprint* arXiv:astro-ph/0511006)

[10] Suzuki T K 2007 *ApJ* **659** 1592–1610 (*Preprint* astro-ph/0608195)

[11] Anderson L S and Athay R G 1989 *ApJ* **336** 1089–1091

[12] Landini M and Monsignori Fossi B C 1990 *A&AS* **82** 229–260

[13] Sutherland R S and Dopita M A 1993 *ApJS* **88** 253–327

[14] Kudoh T and Shibata K 1999 *ApJ* **514** 493–505

[15] Imamura T, Tokumaru M, Isobe H, Shiota D, Ando H, Miyamoto M, Toda T, Häusler B, Pätzold M, Nabatov A, Asai A, Yaji K, Yamada M and Nakamura M 2014 *ApJ* **788** 117

[16] Miyamoto M, Imamura T, Tokumaru M, Ando H, Isobe H, Asai A, Shiota D, Toda T, Häusler B, Pätzold M, Nabatov A and Nakamura M 2014 *ApJ* **797** 51

[17] Suzuki T K 2002 *ApJ* **578** 598–609 (*Preprint* astro-ph/0206239)

[18] Matthaeus W H, Zank G P, Oughton S, Mullan D J and Dmitruk P 1999 *ApJ* **523** L93–L96

[19] Verdini A and Velli M 2007 *ApJ* **662** 669–676 (*Preprint* astro-ph/0702205)

[20] Cranmer S R, van Ballegooijen A A and Edgar R J 2007 *ApJS* **171** 520–551 (*Preprint* astro-ph/0703333)

[21] Matsumoto T and Suzuki T K 2012 *ApJ* **749** 8 (*Preprint* 1109.6707)

[22] Matsumoto T and Suzuki T K 2014 *MNRAS* **440** 971–986 (*Preprint* 1402.0316)

[23] Suzuki T K, Imada S, Kataoka R, Kato Y, Matsumoto T, Miyahara H and Tsuneta S 2013 *PASJ* **65** 98 (*Preprint* 1212.6713)

[24] Czesla S, Schröter S, Wolter U, von Essen C, Huber K F, Schmitt J H M M, Reichart D E and Moore J P 2012 *A&A* **539** A150

Journal of Physics: Conference Series **642** (2015) 012028 doi:10.1088/1742-6596/642/1/012028

Solar Type III Radio Bursts: Directivity Characteristics

G. Thejappa

Department of Astronomy, University of Maryland, College Park MD 20742 USA

E-mail: thejappa.golla@nasa.gov

R. J. MacDowall

NASA, Goddard Space Flight Center, Greenbelt MD 20771 USA

E-mail: Robert.J.MacDowall@nasa.gov

Abstract. Type III radio bursts are a group of fast drifting radio emissions associated with solar flares. These radio emissions are believed to be excited at the fundamental and second harmonic of the electron plasma frequency, f_{pe} by the electron beam excited Langmuir waves through a mechanism called the plasma mechanism. This mechanism attributes the dipole and quadrupole beam patterns for the fundamental and harmonic emissions. To verify these predictions, we analyze the simultaneous observations of type III radio bursts by the STEREO A, B and Wind spacecraft located at different vantage points in the ecliptic plane, and determine their normalized peak intensities (directivity factors) at each spacecraft using their time profiles. Assuming that the sources of these bursts are located on the Parker spiral magnetic field lines emerging from the associated active regions, we estimate the angles between the magnetic field directions and the lines connecting the sources to the spacecraft (viewing angles). Based on the plots of the directivity factors versus the viewing angles, one can divide these bursts into (1) intense bursts emitted into a narrow cone centered around the tangent to the magnetic field, and (2) relatively weaker bursts emitting into a wider cone centered around the tangent to the magnetic field. We compute the distributions of ray trajectories emitted by an isotropic point source and show that the refraction focuses the fundamental and harmonic emissions into narrow and wider cones, respectively. The comparison of these distributions with observations indicates that the intense bursts visible to a narrow range of angles around the tangent to the magnetic field probably correspond to the fundamental, and the relatively weaker bursts visible to a wide range of angles probably are the harmonic emissions.

1. Introduction

Solar type III radio bursts are produced by the Langmuir waves excited by the flare accelerated electron beams. The partial conversion of Langmuir waves excited by these supra-thermal electrons into escaping radiation at the fundamental and second harmonic of the electron plasma frequency, f_{pe} (plasma mechanism) is believed to be the emission mechanism of these bursts. As shown by [18], this mechanism attributes the dipole pattern

$$\Psi_{f_{pe}} = \sin^2\theta \tag{1}$$

for the fundamental and quadrupole pattern

$$\Psi_{2f_{pe}} = \sin^2\theta\cos^2\theta \tag{2}$$

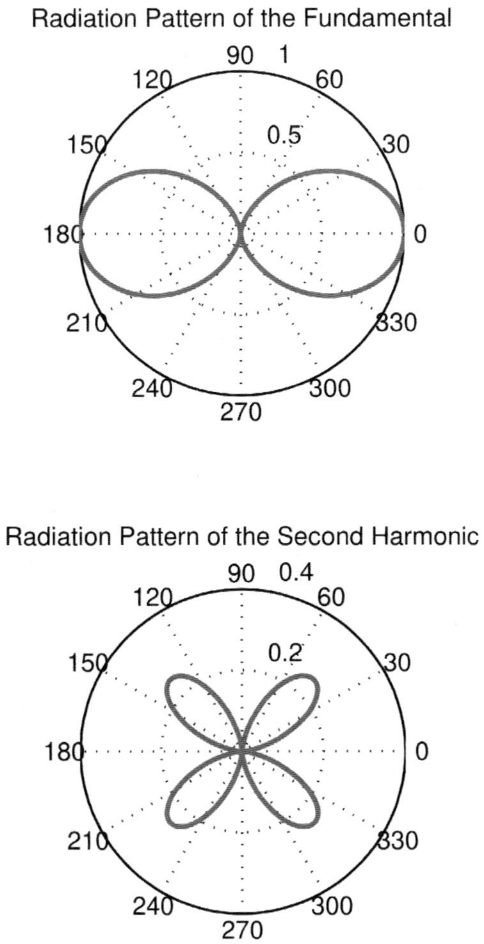

Figure 1. Directivity patterns of the fundamental and harmonic emissions

for the harmonic emission. Here θ is the angle between the wave vector $\vec{k}$ of the electromagnetic wave and the wave vector $\vec{k_L}$ of the Langmuir wave. The dipole and quadrupole nature of the fundamental and harmonic emissions is clearly depicted in Figure 1. The observational confirmation of the directivity patterns is an important element for the acceptance of the plasma mechanism as the emission mechanism as well as for understanding of many observations at f_{pe} and at $2f_{pe}$.

Some single spacecraft observations indicate a small but systematic asymmetry in the occurrence rate of type III bursts from east to west [7, 5], which has been attributed to the beam pattern of the radio emission oriented along the average magnetic field direction. Some multi-spacecraft observations appear to support these claims [6, 1]. However, there exist other observational studies, which show the number of occurrences of type III bursts peaking at the center of the disk and steadily falling at high longitudes [4]. According to [10], multi-spacecraft observations of two events show beaming patterns approximately directed along the tangent to the spiral magnetic field line at the source. [17] analyzed the stereoscopic observations obtained by the STEREO A, B and WIND spacecraft and showed that the type III emission is very intense along the tangent to the spiral magnetic field. These authors did not find east-west asymmetry. The Monte Carlo simulation techniques are also used to investigate the visibility and directivity of these bursts [12, 13, 14, 15, 16].

In this paper, we present the analysis of a relatively larger data set and extend the study conducted by [17]. We have developed a three dimensional ray tracing code for this study. We

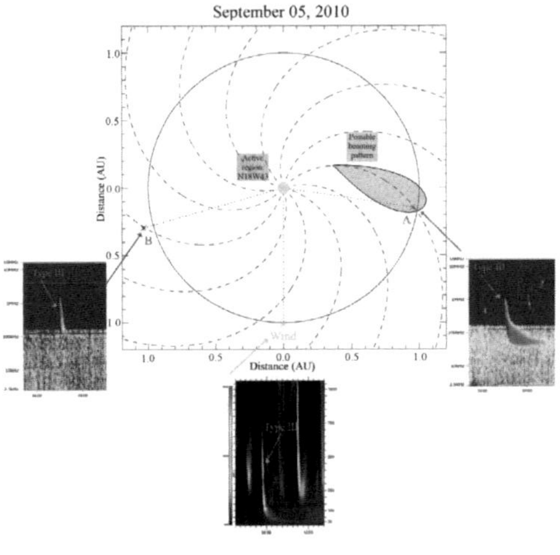

Figure 2. Simultaneous observations of a type III radio burst obtained by the STEREO A, B and WIND spacecraft. The WIND is separated from the Stereo A and B by 81.372 and -74.076 degrees, respectively. The active region associated with this event is N19W43.

compute the ray trajectory distributions emitted by an isotropic point source and compare with observations. This shows that the intense bursts visible to a narrow range of angles around the tangent to the magnetic field probably correspond to the fundamental, and the relatively weaker bursts visible to a wide range of angles probably are the harmonic emissions. In section 2, we describe the observations, and in section 3 we present the discussion and conclusions.

2. Observations

The STEREO mission consists of two identical spacecraft- one ahead of Earth in its orbit (Stereo 'A'), the other trailing behind (Stereo 'B'), each of which is equipped with identical instrumentation. The helio-longitudinal separation of these two spacecraft increases during first four years of the mission. The data obtained by the WAVES experiments of these two spacecraft and by the near-Earth Wind spacecraft [2, 3] provide an unique opportunity for stereoscopic studies of various radio emissions. We have identified several type III radio bursts detected simultaneously by all these three spacecraft for which unambiguous identification of the associated flare or active region is possible. In Figure 2, we present a typical event observed simultaneously by the STEREO A, B and Wind spacecraft. Here, the yellow circle in the center corresponds to the sun and the outer circle corresponds to the orbits of the spacecraft. The dashed lines correspond to the spiral magnetic field. The type III event at STEREO A is very intense probably because it is well connected magnetically to the source.

We define the ratio of the peak intensity at a given spacecraft to the sum of the peak intensities at all three spacecraft $R_i = \frac{I_i}{\Sigma I_i}$ as the directivity factor. We use the time profiles as given in Figure 3 to estimate these directivity factors. We identify the associated active regions using the Solar Geophysical Data (NOAA). We estimate the heliolongitudes of the sources of these bursts at different frequencies by assuming that they are located on the Parker spiral magnetic field lines emerging from the associated active regions. As far as the viewing angle is concerned, it is defined as the angle between the direction of the tangent to the magnetic field at the source and the line connecting the source to the spacecraft. These viewing angles are estimated using the heliolongitudes of the sources. A detailed procedure of estimating the viewing angles is given in [17]. We estimate the height of the sources using the empirical density model due to [8]

$$N_e = a\rho^{-2} + b\rho^{-4} + c\rho^{-6}\ cm^{-3}, \tag{3}$$

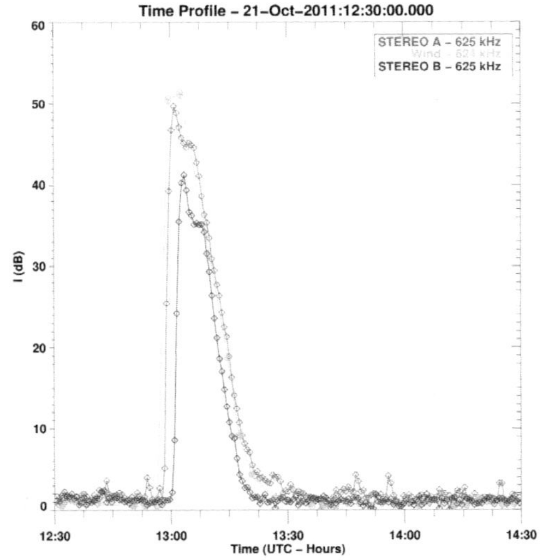

Figure 3. The time profiles of a type III radio burst observed simultaneously by the STEREO A, B and WIND spacecraft

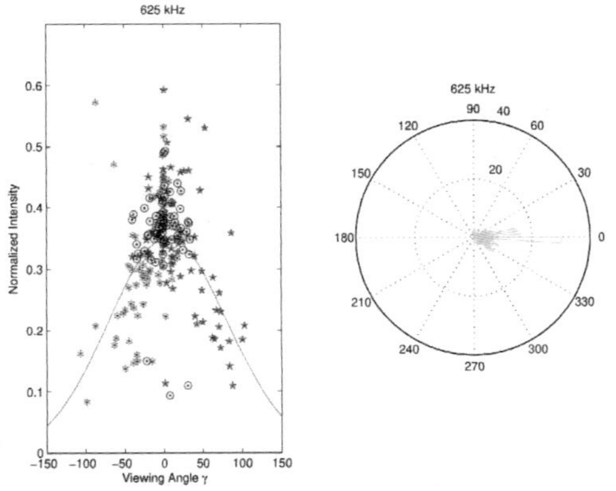

Figure 4. LHS: Normalized peak intensities of type III bursts observed by the STEREO A, B and Wind spacecraft at 625 kHz as a function of viewing angle γ. The curve fitted to the data is the Gaussian. RHS: Rose diagram of the viewing angles, γ.

where ρ is the radial distance in units of solar radius $R_\odot$, $a = 3.3 \times 10^5$, $b = 4.1 \times 10^6$, and $c = 8.0 \times 10^7$. The electron plasma frequency f_{pe}, which is related to N_e as

$$f_{pe}(kHz) = \sqrt{80.6 N_e}(cm^{-3}) \tag{4}$$

is used to estimate the heights of the sources. For example, the heights of the fundamental and harmonic sources at 625 kHz are $\sim 9R_\odot$ and $\sim 16.9R_\odot$, respectively.

In the left hand side of Figure 4, we present the plot of the directivity factor R_i as a function of the viewing angle at 625 kHz for all three spacecraft. This plot clearly shows that the directivity factor peaks around $\gamma \simeq 0$ corresponding to the tangent to the ambient magnetic field, and as γ increases to higher values the directivity factor falls. The fitted curve is the Gaussian. In the right hand side of Figure 4, we present the histogram of the viewing angles as a rose diagram at 625 kHz. In this plot, the origin coincides with the radio source and the radial direction with zero degree angle coincides with the tangent to the magnetic field at the source. From this rose diagram, it is clear that the emission of a large portion of type III events is confined to narrow range of angles around the tangent to the magnetic

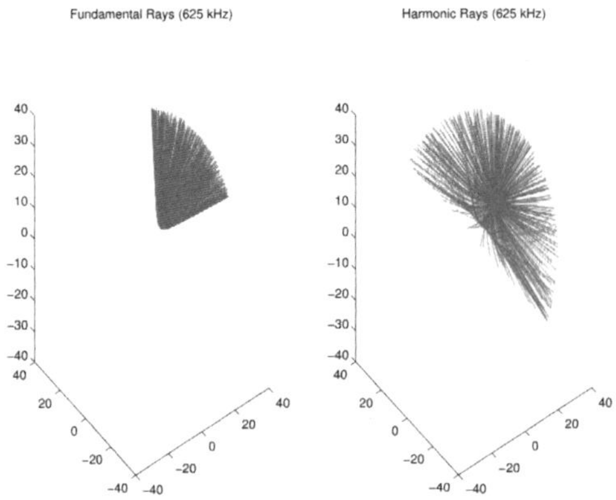

Figure 5. The distributions of ray trajectories at the fundamental (F) and harmonic (H) emissions emitted by an isotropic point source placed at the center of disk

field. This implies that the emission is mostly beamed along the tangent to the magnetic field.

3. Discussion and Conclusions

According to plasma mechanism, the Langmuir waves are the source of the solar radio type III bursts. The fundamental, which is excited at a layer where the refractive index is zero can escape only within a narrow cone normal to the surface of zero refractive index. The solid angle of this emission cone can be approximated as [18]

$$\Omega_{f_{pe}} = 3\pi \frac{v_{Te}^2}{v_b^2}, \tag{5}$$

where v_{Te} is the electron thermal speed and v_b is the speed of the electron beam. For typical values of $v_{Te} \sim 1.5 \times 10^6$ ms^{-1} and $v_b \sim 0.1c$, we estimate $\Omega_{f_{pe}} \sim 2.38 \times 10^{-2}$ steradians. In the case of second harmonic, the restrictions on the escape conditions are less severe. The solid angle of emission cone in this case is [18]

$$\Omega_{2f_{pe}} = 2\pi[1 - \cos(\sin^{-1}(\mu_{2f_{pe}}))] \simeq \pi, \tag{6}$$

where $\mu_{2f_{pe}} \sim \frac{1}{2}$ is the refractive index. Since $\Omega_{2f_{pe}} \sim 2\pi(1 - \cos\theta)$, the half widths of these emission cones can be estimated as ~ 5 and ~ 60 degrees, respectively.

We simulate the escaping conditions by tracing the rays emitted by an isotropic point source in three dimensions. Here we neglect the magnetic field since $f_{pe} >> f_{ce}$ (f_{ce} is the electron cyclotron frequency), and write the refractive index as

$$\mu^2 = 1 - \frac{f_{pe}^2}{f^2} = 1 - \frac{80.6 N_e}{f^2}, \tag{7}$$

where the frequency f is in units of kHz. To trace the rays, we use the Cartesian coordinate system with origin at the center of the Sun, and z-axis connecting the center of the Sun to one of the spacecraft. The relevant equations are the set of 6 first-order differential equations

$$\frac{d\vec{R}}{d\tau} = \vec{T} \tag{8}$$

$$\frac{d\vec{T}}{d\tau} = D(\vec{R}) = \frac{1}{2}\frac{\partial \mu^2}{\partial \vec{R}}, \tag{9}$$

where

$$T_x^2 + T_y^2 + T_z^2 = \mu \tag{10}$$

and

$$\vec{R} \equiv \begin{pmatrix} x \\ y \\ z \end{pmatrix} \& \vec{T} \equiv \begin{pmatrix} T_x \\ T_y \\ T_z \end{pmatrix}$$

are the position and direction vectors, respectively. This system of differential equations is integrated using the 4th order Runge-Kutta algorithm [11]

$$R_{n+1} = R_n + T_n\Delta\tau + \frac{1}{6}(K_1 + 2K_2)(\Delta\tau)^2 \tag{11}$$

$$T_{n+1} = T_n + \frac{1}{6}(K1 + 4K_2 + K_3)\Delta\tau \tag{12}$$

$$K_1 = D(R_n) \tag{13}$$

$$K_2 = D(R_n + \frac{1}{2}T_n\Delta\tau + \frac{1}{8}K_1(\Delta\tau)^2) \tag{14}$$

$$K_3 = D(R_n + T_n\Delta\tau + \frac{1}{2}K_2(\Delta\tau)^2). \tag{15}$$

This algorithm can be used to trace the rays in any medium with an arbitrary density distribution. Starting from a known point $(\vec{R_0}, \vec{T_0})$, one can generate successively $(\vec{R_1}, \vec{T_1})$, $(\vec{R_2}, \vec{T_2})$......$(\vec{R_n}, \vec{T_n})$. In the present case, we compute the distribution of ray trajectories emitted by an isotropic point source at 625 kHz located at a height of $9R_\odot$ and $16.9R_\odot$ for the fundamental and harmonic, respectively. We launch the rays with direction cosines

$$Tx0 = \mu_0 \sin\theta_0 \sin\phi_0 \tag{16}$$

$$T_{y0} = \mu_0 \sin\theta_0 \cos\phi_0 \tag{17}$$

$$T_{z0} = \mu_0 \cos\theta_0, \tag{18}$$

where the azimuthal and elevation angles are randomly sampled using

$$\phi_0 = 2\pi\xi_1 \tag{19}$$

$$\cos\theta_0 = 2\xi_2 - 1, \tag{20}$$

where ξ_1 and ξ_2 are the random numbers uniformly distributed between 0 and 1. This sampling scheme represents an ideal isotropic point source. As seen from Figure 5, the rays from the fundamental source escape in a narrow cone along the radial direction. In the case of second harmonic, the rays launched in the forward direction escape directly, and the rays launched in the backward direction get reflected back and escape into a broader cone, i.e., these rays are distributed over a wide range of angles. The estimated widths of these emission cones are $\sim$ 10 and $\sim$ 120 degrees as expected by the theoretical estimates. If we compare the computed ray distribution in Figure 5 with the observed directivity diagram (LHS of Figure 4), it is clear that the intense bursts observed in the range of angles around $\gamma = 0$ probably correspond to the fundamental, and relatively weaker emissions distributed over a wider range of viewing angles probably correspond to second harmonic.

Thus, the observed distribution of viewing angles as well as the variation of the directivity factors as functions of viewing angles are consistent with the computed ray trajectory distributions, i.e., the type III emissions are mainly peaked along the tangent to the magnetic field. The intense emissions emitted into a narrow cone along the tangent are probably emitted in the fundamental mode, whereas, the emissions emitted into a much wider cone are probably the harmonic emissions. Most of the observed directivity characteristics can be accounted for by the refractions. The emissions beyond the 60 degrees of the cone are probably due to scattering.

Acknowledgments
The research of T. G. is supported by the Grants NNX12AH47G from NASA. The SWAVES instruments include contributions from the Observatoire of Paris, University of Minnesota, University of California, Berkeley, and NASA/GSFC.

References
[1] Bonnin X, Hoang S and Maksimovic M 2008 *Astron. Astrophys.,* **480** 419.
[2] Bougeret J.-L. et al. 1995 *Space Sci. Rev.,* **71** 231.
[3] Bougeret et al 2008 *Space Sci. Rev.* **136** 487.
[4] Fainberg J and Stone R G 1970 *Solar Phys.* **15** 222
[5] Fitzenreiter R J, Fainberg J, Bundy R B 1976 *Solar Phys.,* **46** 465
[6] Hoang S, Poquerusse M and Bougeret J.-L. 1997 *Solar Phys.,* **172** 307.
[7] Kaiser M L 1975 *Solar Phys.,* **45** 181.
[8] Leblanc Y, Dulk G A and Bougeret J.-L 1998 *Solar Phys.,* **183** 165.
[9] Lecacheux A, Steinberg J.-L, Hoang S and Dulk G A 1989 *Astron. Astrophys.,* **217** 237.
[10] Reiner M J, Goetz K, Fainberg J, Kaiser M L, Maksimovic M, Cecconi B, Hoang S, Bale S D and Bougeret J.-L 2009 *Solar Phys.* **259** 255.
[11] Sharma A, Vizia Kumar D and Ghatak A K 1982 *Appl. Opt.* **21** 984
[12] Steinberg, J.-L, Aubier-Giraud M, Leblanc Y and Boischot A 1971 *Astron. Astrophys.,* **10** 362
[13] Steinberg J L 1972 *Astron. Astrophys.,* **18** 382
[14] Thejappa G, and MacDowall R J 1998, *Astrophys. J.* **498**, 465
[15] Thejappa G, MacDowall R J, and Kaiser M L 2007 *Astrophys. J.,* **671** 894
[16] Thejappa G, MacDowall R J, and Gopalswamy M L 2011 *Astrophys. J.,* **734** 16
[17] Thejappa G, MacDowall R J, and Bergamo M 2012 *Astrophys. J.* **745** 187
[18] Zheleznyakov V V and Zaitsev V V 1970 it Sov. Astron. **14** 250

Ulysses and IBEX Constraints on the Interstellar Neutral Helium Distribution

Brian E. Wood[1] **and Hans-Reinhard Müller**[2]

[1] Naval Research Laboratory, Space Science Division, Washington, DC 20375, USA

[2] Department of Physics and Astronomy, Dartmouth College, Hanover, NH 03755, USA

E-mail: `brian.wood@nrl.navy.mil`

Abstract. We relax the usual assumption of Maxwellian velocity distributions in the interstellar medium (ISM) in the analysis of neutral He particle data from *Ulysses* and the *Interstellar Boundary Explorer* (IBEX). For *Ulysses*, the possibility that a narrow component from heavy neutrals is contaminating the He signal is considered, which could potentially explain the lower ISM temperature measured by *Ulysses* compared to IBEX. The expected heavy element contribution is about an order of magnitude too small to resolve that discrepancy. For IBEX, we find that modest asymmetries in the ISM velocity distribution can potentially improve the quality of fit to the first two years of data, and perhaps improve agreement with the *Ulysses* measurements.

1. Introduction

The current *Interstellar Boundary EXplorer* (IBEX) and now defunct *Ulysses* spacecraft both have observed neutral helium from the interstellar medium (ISM) flowing through the inner solar system, which can be used to establish the nature of the undisturbed ISM surrounding the Sun. Analysis of these particle data generally involves a forward modeling procedure assuming a Maxwellian He velocity distribution at infinity, which is then propagated into the inner heliosphere under the influence of gravity, where the resulting distribution can then be used to confront actual observations. It is possible, however, that the particle flows encountered by *Ulysses* and IBEX may not arise from strictly Maxwellian distributions, and we here report on some initial efforts to explore whether evidence can be found for this in the data.

We consider both *Ulysses* and IBEX data here, but the nature of the non-Maxwellian behavior that we look for is very different for the two. For *Ulysses*, which unlike IBEX is not able to distinguish between different species of atomic neutrals, we investigate whether the primary He distribution might be contaminated by a weak, narrower distribution from heavy elements. Such contamination could potentially lead to an underestimate of the ISM temperature. Correcting for this would therefore increase the *Ulysses* temperature measurements, potentially improving agreement with IBEX measurements ([1]). For IBEX, with its superior signal-to-noise, we assess the detectability of asymmetries in the ISM velocity distribution.

2. Heavy Element Contamination of *Ulysses* Data

The most recent analyses of IBEX data have reported an ISM He velocity vector consistent with that measured from *Ulysses* data, but with a rather high temperature of $T = 8710^{+440}_{-680}$ K

Journal of Physics: Conference Series **642** (2015) 012029 doi:10.1088/1742-6596/642/1/012029

([1–3]), though [4] report a somewhat lower $T = 8000 \pm 1300$ K. These measurements are higher than the $T = 6300 \pm 340$ K temperature reported by [5] from *Ulysses* measurements. A more recent analysis of *Ulysses* data has revised this temperature upwards to $T = 7260 \pm 270$ K ([6]), consistent with an independent reanalysis from [7]. This is closer to the desired IBEX temperature, but still a bit low. Wood et al (2015b) [8] recently studied whether consideration of heavy element contamination to the *Ulysses* data could lead to a further revision upwards.

After He, the most abundant ISM neutrals detectable by *Ulysses* should be O and Ne. The heavier weights of these atoms will yield narrower velocity distributions than that of He. Thus, the effect of O and Ne contamination on the observed particle beam is to narrow it, thereby leading to an underestimate in the temperature. If the real ISM temperature is $T \approx 8500$ K as IBEX suggests, the question is how much O and Ne contamination there has to be in *Ulysses* data for $T = 7000 - 7500$ K to be perceived instead, when fitting the data assuming the signal is 100% He, $T = 7000 - 7500$ K being the actual temperature measured by [7] and [6].

The degree of heavy element contribution can be quantified as

$$R = \left(\frac{A_O}{A_{He}} \cdot \frac{S_O}{S_{He}} + \frac{A_{Ne}}{A_{He}} \cdot \frac{S_{Ne}}{S_{He}} \right) \frac{D_{ONe}}{D_{He}}, \tag{1}$$

where the neutral abundances of He, O, and Ne in the ISM are indicated by A_{He}, A_O, and A_{Ne}, respectively. Interstellar neutrals suffer ionization as they travel through the heliosphere, particularly photoionization by solar EUV photons, and the various S parameters in the equation are the survival probabilities for He, O, and Ne. The D_{ONe}/D_{He} factor is the ratio of detection efficiencies of O and Ne relative to He. We assume the energy dependence of detection efficiency is about the same for all these elements and that $D_{ONe}(E)$ therefore can be approximated as a simple scalar multiple of $D_{He}(E)$ ([9–10]). Assuming $D_{ONe}/D_{He} \approx 3$ for the detection efficiency ratio ([9–10]), $A_O/A_{He} = 3.2 \times 10^{-3}$ and $A_{Ne}/A_{He} = 3.9 \times 10^{-4}$ for the ISM neutral abundance ratios ([11]), and $S_O/S_{He} \approx 0.20$ and $S_{Ne}/S_{He} \approx 0.45$ for the survival probability ratios ([12]) leads to a value of $R = 0.0025$ from equation (1).

Assuming $T = 8500$ K, Figure 1 shows simulated data for a *Ulysses* observation from 2001 December 12. Maps are shown for $R = 0.003$, $R = 0.03$, and $R = 0.1$. The beam becomes noticably narrower as R increases, corresponding to more O and Ne contamination. These simulated data are fitted using the same methods as described by [6], not only for this particular *Ulysses* observation but for a total of 20 separate *Ulysses* observations from different parts of its orbit, as defined by an orbital phase ϕ, where $\phi = 0.5$ corresponds to the ecliptic plane crossing near solar perihelion (see [6]). The results are shown in Figure 2. For low R the actual $T = 8500$ K value is recovered, as there is insufficient contamination from O and Ne to affect the synthetic data, but T starts to lower significantly for $R \geq 0.003$. By $R \approx 0.03$, temperatures consistent with the measurements of the real *Ulysses* data are being observed, with some orbital phase dependence being apparent.

The question is then whether $R \approx 0.03$ is low enough to be a plausible value. This is about an order of magnitude higher than the $R = 0.0025$ value estimated above based on our best estimates of the quantities in equation (1). Thus, we conclude that heavy element contamination of *Ulysses* data is unlikely to be the primary cause of the ISM temperature discrepancy between *Ulysses* and IBEX ([8]).

3. The Effects of Velocity Distribution Asymmetries on IBEX Data

The IBEX spacecraft is designed to produce a sky map of energetic neutral atom fluxes once every 6 months. In its highly elliptical orbit around Earth, the spinning spacecraft allows its detectors to scan a roughly 14° wide swath of sky in a given orbit. The spin axis is kept pointed roughly towards the Sun, meaning data from each individual IBEX orbit provides a plot of count rate versus ecliptic latitude. After 7-8 days the orbit is changed so that the new scan plane is

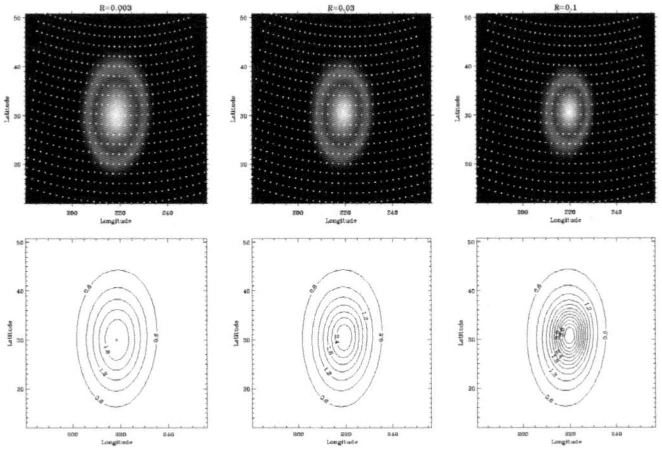

Figure 1. Simulations of an actual *Ulysses*/GAS particle beam map from 2001 December 12 (in ecliptic coordinates), shown in both image (top) and contour plot (bottom) formats ([8]). The maps are generated assuming the best-fit He flow vector from [6] and a temperature of $T = 8500$ K. The color scale in the images is rescaled in each panel so that the maximum is always bright yellow. Maps are computed assuming three different degrees of contamination from O and Ne, corresponding to $R = 0.003$ (left), $R = 0.03$ (middle), and $R = 0.1$ (right).

rotated typically $7° - 8°$ in longitude. In this way, IBEX gradually covers the entire sky every 6 months.

The interstellar neutrals are only observable at one time each year, with peak fluxes from early January to late February ([13]). Figure 3 shows the first two years of IBEX observations of the He beam. These are the same data analyzed by [13] and [14] in the first assessments of the He flow vector from IBEX. We have fitted these data with an analysis approach essentially identical to that used to analyze *Ulysses* data ([6]). Our best fit is shown in Figure 3.

The four free parameters of the fit are flow velocity (V_{ISM}), temperature (T), flow longitude (λ), and flow latitude (β). As shown in Figure 4, the four fit parameters are highly correlated, so you have long tubes in four-dimensional chi-squared space where the fits are reasonably good ([13–15]). Nevertheless, the bottom panel of Figure 4 shows that there is a well-defined χ^2 minimum, allowing clear best-fit values and error bars to be computed. Following the example of [6] in the analysis of *Ulysses* data, we use 3σ confidence intervals to define the error bounds, leading to the red error boxes in the figure, corresponding to the following values: $V_{ISM} = 22.8 \pm 1.3$ km s^{-1}, $\lambda = 77.9 \pm 1.8°$, $\beta = -5.3 \pm 0.1°$ deg, and $T = 6550 \pm 890$ K.

These values are basically in agreement with the best-fit values reported by [14], who analyzed these data with a similar forward modeling approach. They can also be compared with our *Ulysses* measurements ($V_{ISM} = 26.08 \pm 0.21$ km s^{-1}, $\lambda = 75.54 \pm 0.19°$, $\beta = -5.44 \pm 0.24°$, and $T = 7260 \pm 270$ K), which are shown as black boxes in Figure 4. The apparent inconsistency between the *Ulysses* and IBEX measurements was initially an issue of much debate, though consideration of more recent observations from IBEX seems to have yielded IBEX measurements more consistent with those of *Ulysses* ([1–3]), with the possible exception of temperature, as discussed in Section 2. The cause of the discrepancy with the first two years of IBEX data seen in Figure 4 has not been explained. However, we note here that if we used 5σ error boxes instead of 3σ, the *Ulysses* and IBEX boxes would overlap, except for V_{ISM}. In any case, our analysis combined with that of [6] suggests the *Ulysses* analysis provides more precise measurements of

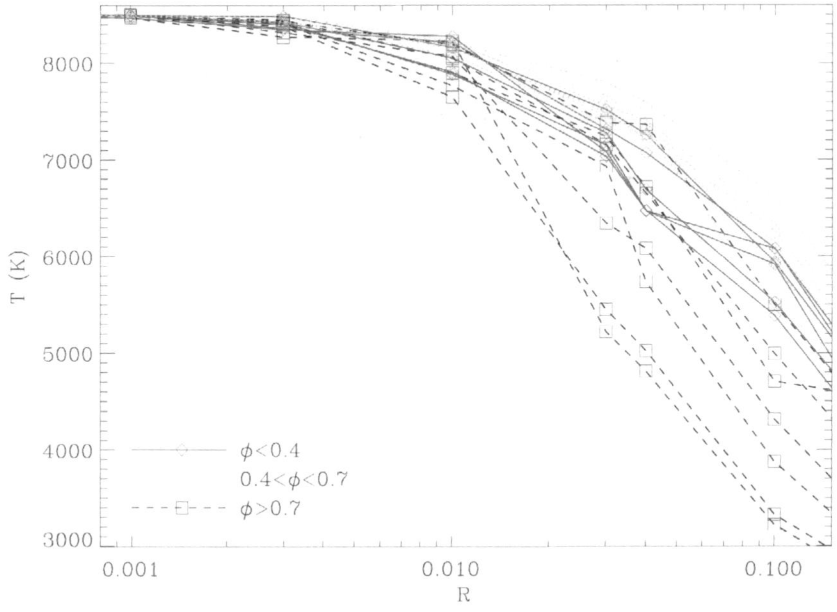

Figure 2. Single map fits have been performed on synthetic *Ulysses* particle beam maps, computed assuming various values of R (measuring contamination of O and Ne flux), with map sampling based on 20 real observations ([8]). For each of the 20 cases, the measured temperature is plotted versus R, with different symbols and line styles indicating different ranges of orbital phase, ϕ.

the flow vector than the IBEX analysis, as indicated by the smaller size of the *Ulysses* boxes relative to IBEX. The reason for this lies in the ability of *Ulysses* to observe the He beam from different locations, above and below the ecliptic. This breaks the parameter degeneracies that bedevil IBEX, which can only observe the He beam from one location ([6]).

The analysis of IBEX data just presented relies on the usual assumption that the ISM He distribution far from the Sun is a simple Maxwellian. The best fit in Figure 3 has $\chi^2 = 3.7$, well above the ideal value of $\chi^2 \approx 1$, indicating that there is significant room for improvement with regards to how well we are fitting the IBEX data. Perhaps inaccuracies in the single Maxwellian assumption may be at least partly responsible for this. A non-Maxwellian distribution could also potentially affect the best-fit He flow parameters that we have derived. Solar wind distributions are often found to have significant asymmetries ([16]), so perhaps ISM distributions might as well.

In order to explore the effects of a non-Maxwellian distribution on the IBEX analysis, we experiment with a distribution that is the sum of two closely spaced identical Maxwellians, oriented along the ISM magnetic field line. This is meant to approximate a bi-Maxwellian distribution with $T_{\parallel} > T_{\perp}$. We assume the ISM field is oriented as suggested by the IBEX ribbon, towards $(\lambda,\beta)=(221°,39°)$ ([17]). Fits are then performed to the IBEX data assuming different velocity separations for the two Maxwellians. Figure 5 shows how χ^2 varies with assumed separation. The χ^2 drops from 3.7 to 2.7 when the separation is increased from $\Delta V = 0$ to $\Delta V = 0.4$ km s^{-1}, a surprisingly large amount considering just how small an asymmetry this shift induces in the overall distribution. For comparison, the thermal width of the He distribution for an 8000 K He gas is 5.7 km s^{-1}. However, Figure 5 also shows that χ^2 starts increasing for

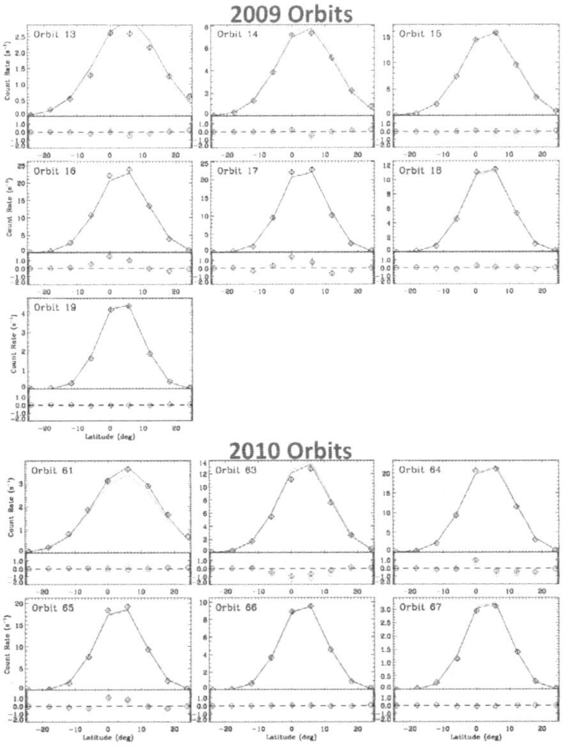

Figure 3. The first two years of IBEX observations of ISM neutral He. Each panel is a plot of count rate versus ecliptic latitude for a given orbit, with each orbit representing a different longitudinal plane. Green lines are a fit to the data assuming that the source distribution is a simple Maxwellian far from the Sun, while red lines are a fit assuming the distribution far from the Sun is the sum of two Maxwellians, with a separation between the two Maxwellians of $\Delta V = 0.4$ km s^{-1}. Residuals are shown below each panel for both fits.

even larger separations (i.e., larger asymmetries).

Figure 3 shows the fit to the data for the $\Delta V = 0.4$ km s^{-1} fit, and Figure 4 shows the new best-fit He flow parameters that result from the $\Delta V = 0.4$ km s^{-1} assumption ($V_{ISM} = 24.0$ km s^{-1}, $\lambda = 76.3°$, $\beta = -5.7°$, and $T = 7120$ K). We find a lack of robustness for fits assuming these two-Maxwellian distributions, meaning such fits do not have as clean and smooth a χ^2 curve as seen in the bottom panel of Figure 4 for the simple Maxwellian fit. The cause of this is uncertain, but as a consequence we do not try to quote error bars for the two-Maxwellian fits at this time. Nevertheless, it is interesting to note from Figure 4 that the best-fit parameters have changed significantly for the $\Delta V = 0.4$ km s^{-1} fit, and they have actually moved conveniently in the direction of the *Ulysses* values.

We conclude from this excercise that small asymmetries in the ISM velocity distribution can potentially affect both the quality of fit and the best-fit parameters in analyses of IBEX data. It is therefore worthwhile to explore further the possible effects of non-Maxwellian distributions, which may influence the consistency of parameters derived by IBEX and *Ulysses*. Future work should involve actual bi-Maxwellian distributions rather than the two-Maxwellian summation approximation.

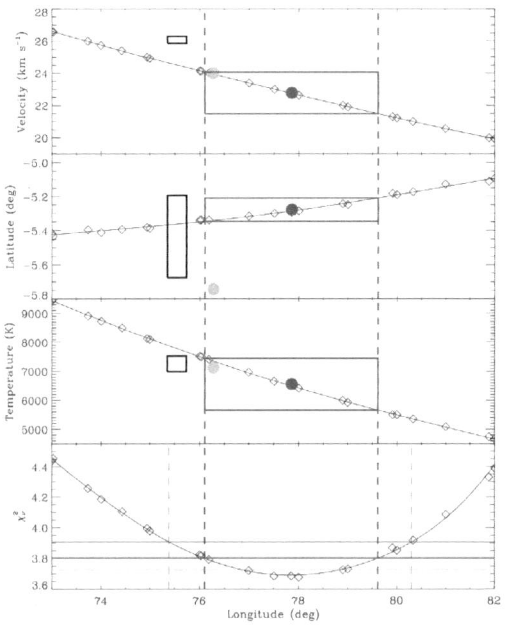

Figure 4. The diamonds show the results of various fits to the first two years of IBEX observations of the ISM He flow, assuming a simple Maxwellian distribution far from the Sun, with one of the four fit parameters held constant in each fit. The upper three panels plot ISM velocity, flow latitude, and temperature versus flow longitude, and the bottom panel plots χ^2 versus flow longitude. Horizontal lines in the bottom panel indicate 1, 3, and 5 σ confidence levels. The 3σ level is used to define the IBEX error boxes in red in the upper 3 panels, which can be compared with the *Ulysses* error boxes in black, from [6]. Red circles indicate the best single-Maxwellian IBEX fit, while the green circles indicate the best fit assuming the He flow far from the Sun is the sum of two closely-spaced Maxwellians.

Acknowledgments

This work has been supported by NASA award NNH13AV19I to the Naval Research Laboratory.

References

[1] McComas D J et al 2015 *ApJ* **801** 28
[2] Leonard T W et al 2015 *ApJ* **804** 42
[3] Möbius E et al 2015 *ApJS* submitted
[4] Schwadron N et al 2015 *ApJS* submitted
[5] Witte M 2004 *A&A* **426** 835
[6] Wood B E, Müller H R, & Witte M 2015a *ApJ* **801** 62
[7] Bzowski M, Kubiak M A, Hłond M, Sokół J M, Banaszkiewicz M, & Witte M 2014 *A&A* **569** A8
[8] Wood B E, Müller H R, Bzowski M, SokółJ M, Möbius E, Witte M, & McComas D J 2015b *ApJS* in press
[9] Banaszkiewicz M, Witte M, & Rosenbauer H 1996 *A&AS* **120** 587
[10] Yamamura Y, & Tawara H 1996 *Atomic Data and Nuclear Data Tables* **62** 149
[11] Slavin J D, & Frisch P C 2008 **A&A** **491** 53
[12] Bzowski M, SokółJ M, Kubiak M A, & Kucharek H 2013 *A&A* **557** A50
[13] Möbius E et al 2012 *ApJS* 198 11
[14] Bzowski M et al 2012 *ApJS* **198** 12
[15] McComas D J et al 2012 *Science* **336** 1291

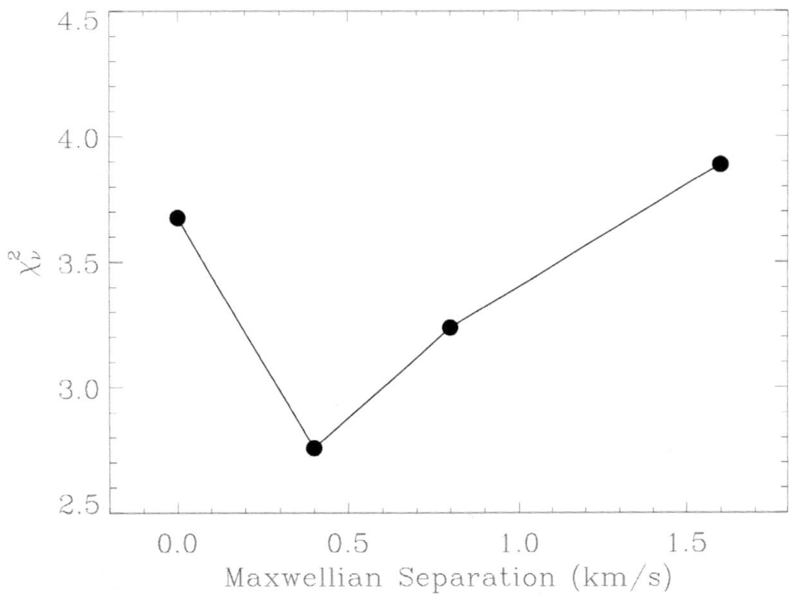

Figure 5. The χ^2 statistic is plotted versus Maxwellian velocity separation, for fits to the first two years of IBEX data assuming the He flow far from the Sun can be represented as the sum of two closely-spaced Maxwellians.

[16] Marsch E, Mühlhäuser K H, Schwenn R, Rosenbauer H, Pilipp W, & Neubauer F M 1982 *J Geophys Res* **87** 52
[17] Funsten H O et al 2009 *Science* **326** 964

Solar Wind Electron Energization by Plasma Turbulence

P. H. Yoon[*,†]

[*] *IPST, University of Maryland, College Park, MD 20742* [†] *School of Space Research, Kyung Hee University, Yongin, Gyeonggi 446-701, Korea*

E-mail: yoonp@umd.edu

Abstract. The solar wind electrons are made of the low-energy Maxwellian core, intermediate-energy halo, field-aligned *strahl*, and the highly-energetic super-halo electrons. The present paper discusses a model in which the halo electrons interact with the whistler fluctuation via cyclotron wave-particle resonance, and the super-halo electrons interact through Landau resonance with the Langmuir fluctuation, thus maintaining a local steady state.

1. INTRODUCTION

The solar wind electrons are comprised of the Maxwellian "core," suprathermal "halo," and a field-aligned component called the "*strahl*," which is typically associated with the high-speed solar wind. The "super-halo" electrons are a highly energetic component. The Maxwellian core is characterized by tens of eV energy, the halo is defined up to $\sim 10^2 - 10^3$ eV or so, and the super-halo is characterized by $10^3 - 10^5$ eV [1]. The *strahl* component overlaps the halo in energy range. Observations at 1 AU shows that the Maxwellian component makes up about $\sim 95\%$ in terms of the number density, the halo is roughly estimated to possess $\sim 4\%$ density, and the *strahl* component takes up the remaining $\sim 1\%$. The super-halo component is very low in terms of the number density (almost negligible, on the order of 10^{-9} to 10^{-6}), yet owing to their high energy, their presence can be clearly identified in velocity distribution function (VDF).

To model the solar wind electrons in a comprehensive manner, one must first address the issue of what physical processes are responsible for energizing the source electrons [2]. Then, how the source electrons escape the corona must be considered. For low corona, collisions become important [3]. Above 10 solar radii ($10R_\odot$) or so and beyond the Coulomb collision is no longer effective (for highly energetic electrons, the collision becomes insignificant even for lower altitudes), and thus it becomes necessary to include the effects of collective wave-particle acceleration and pitch-angle scattering [4]. The physics of these processes can be discussed on the basis of a formal all-embracing kinetic equation,

$$\frac{\partial f}{\partial t} + \mathbf{v}\cdot\nabla f + \left[\mathbf{g} - \frac{e}{m_e}\left(\mathbf{E} + \frac{\mathbf{v}}{c}\times\mathbf{B}\right)\right]\cdot\frac{\partial f}{\partial \mathbf{v}} = \left(\frac{\delta f}{\delta t}\right)_{\rm coll.} + \left(\frac{\delta f}{\delta t}\right)_{\rm wp}, \tag{1}$$

where f represents the total electron velocity distribution function (VDF), $\mathbf{g}$ is the solar gravitational acceleration, $\mathbf{E}$ stands for the large-scale (ambipolar) electric field, $\mathbf{B}$ designates the inhomogeneous solar magnetic field, $(\delta f/\delta t)_{\rm coll.}$ denotes the collision integral, and $(\delta f/\delta t)_{\rm wp}$

refers to the time rate of change in VDF by wave-particle interaction. The wave-particle interaction may be owing to various self-generated plasma waves, or those that are intrinsic to the solar wind.

In the present paper, we pay attention to the role played by the local wave-particle resonant interaction,

$$\frac{\partial f}{\partial t} = \left(\frac{\delta f}{\delta t}\right)_{\rm wp}. \tag{2}$$

We assume that the total electron VDF can be treated as being made of three-components, $f = f_0 + f_h + f_s$, where f_0, f_h, and f_s stand for the isotropic Maxwellian core, halo, and the super-halo VDF. We do not consider the *strahl*, since we assume isotropy at the outset. We assume that the Maxwellian core electrons are generally nonresonant with any high-frequency waves, so that they maintain the Gaussian form,

$$f_0 = \frac{n_0}{\pi^{3/2} v_{T0}^3} \exp\left(-\frac{v^2}{v_{T0}^2}\right), \tag{3}$$

where $v_{T0} = \sqrt{2T_0/m_e}$ is the thermal speed associated with the core electrons, T_0 being the core temperature, which is in the range of tens of eV. Note, however, that the core electrons may interact with low frequency turbulence such as the ion-acoustic fluctuations or kinetic Alfvén turbulence. For instance, Rudakov et al. [5] show that the interaction of low-energy electrons with kinetic Alfvén waves may lead to the flattening of the parallel electron distribution near $v \sim 0$ range. From the standpoint of the present discussion, which is concerned with the wave-particle interaction between intermediate to high energy electrons and high-frequency fluctuations, we may ignore such an effect. We assume that the halo electrons (f_h) are resonant with the whistler turbulence, while the super-halo electrons (f_s) are resonant with the Langmuir turbulence.

Pierrard et al. [6] developed a kinetic exospheric model that contains the Coulomb-collision effects, $(\delta f/\delta t)_{\rm coll.}$, as well as the wave-particle resonant interaction term, $(\delta f/\delta t)_{\rm wp}$, where the source of wave-particle resonance was assumed to arise from the pre-existing whistler fluctuations. While Ref. [6] is more general than the present limited approach of just considering the wave-particle interaction, our discussion treats the waves in a self-consistent manner, whereas Ref. [6] simply models the waves. Recently Lacombe et al. [7] identified quasi-parallel right-hand circularly polarized whistler waves for certain solar wind conditions. Even though the quasi-parallel whistler fluctuations of the type discussed by Lacombe et al. [7] may not always be easily identifiable due to the occultation by the Doppler-shifted permanent solar wind turbulence, we believe that a finite level of quasi-parallel whistler fluctuations are always present in the solar wind. Whistler waves are characterized by the frequency range $\sim 10^2$ Hz. Considering that the halo electrons have energy in the range 10^2 to 10^3 eV, one can show that the cyclotron resonance between the halo electrons and whistler waves can be satisfied.

For higher frequency range, the solar wind is permeated by the quasi-electrostatic Langmuir-wave type of fluctuations known as the quasi-thermal noise [8]. Reference [9] formulated a theory of electron kappa distribution that is in dynamical equilibrium with the quasi-thermal noise Langmuir spectrum. This theory applies to the super-halo electrons, since the Langmuir wave frequency of the range 10^5 Hz is too high for the intermediate halo electrons to effectively resonate. The asymptotic theory of super-halo electrons in Ref. [9] assumes that these energetic electrons are already present. The purpose of the theory is to provide a self-consistent explanation of how the Langmuir fluctuations scatter the electrons in order to maintain the shape of the super-halo electron distribution. How these electrons are accelerated at the source region is not understood at present.

2. ASYMPTOTIC TURBULENT STATE

We are interested in the asymptotically steady-state solution of the particle kinetic equation (2). A recent paper by Gaelzer et al. [10] derives the specific form for the right-hand side of equation (2), based upon which Kim et al. [11] discuss the steady-state solution for the halo electrons subject to whistler fluctuations,

$$f_h = C \exp\left(-\frac{\Omega_e^4}{\omega_{pe}^4} \frac{m_e c^2 \Omega_e^2}{\pi v_{T0}^2} \int \frac{dv}{v} \frac{\int_{-1}^{1} d\mu \frac{1-\mu^2}{|\mu|\mu^2}}{\int_{-1}^{1} d\mu \frac{1-\mu^2}{|\mu|} \mathcal{E}_W\left(-\frac{|\Omega_e|}{v\mu}\right)} \right), \tag{4}$$

where $\mathcal{E}_W(k)$ is the whistler fluctuation spectrum, C represents a suitable normalization constant, and $\mu = \cos\theta = v_\parallel/v$ is the cosine of the velocity pitch angle. Likewise Kim et al. [11] obtained the steady-state solution for the super-halo electrons in resonance with Langmuir wave fluctuations,

$$f_s = C \exp\left(-\frac{m_e \Omega_e^2}{4\pi v_{T0}^2} \int dv v \frac{\int_{-1}^{1} d\mu \frac{\mu^2}{|\mu|}}{\int_{-1}^{1} d\mu \mathcal{E}_L\left(\frac{\omega_{pe}}{v\mu}\right) \frac{\mu^2}{|\mu|}} \right), \tag{5}$$

where $\mathcal{E}_L(k)$ is the Langmuir wave fluctuation spectrum.

Let us first discuss the super-halo electrons. In the paper by Yoon [9] it was shown that the Langmuir wave intensity given by the form

$$\mathcal{E}_L(k) = \mathcal{E}_0 \left(1 + \frac{k_0^2}{k^2}\right), \tag{6}$$

satisfies the steady-state nonlinear wave kinetic equation for the Langmuir turbulence. Upon inserting equation (6) to equation (5), and upon identifying

$$\begin{aligned} \frac{m_e \omega_{pe}^2 \Omega_e^2}{4\pi \mathcal{E}_0 k_0^2 v_{T0}^2} &= \kappa_s + 1, \\ \frac{k_0^2}{2\omega_{pe}^2} &= \frac{1}{(\kappa_s - 3/2) v_{Ts}^2}, \end{aligned} \tag{7}$$

where $v_{Ts} = \sqrt{2T_s/m_e}$ is the thermal speed for the super-halo electrons, we obtain the desired asymptotic super-halo electron velocity distribution function, as well as the self-consistent steady-state Langmuir fluctuation spectrum,

$$\begin{aligned} f_s(v) &= \frac{n_s}{\pi^{3/2} v_{Ts}^3} \frac{\Gamma(\kappa_s + 1)}{(\kappa_s - 3/2)^{\frac{3}{2}} \Gamma(\kappa_s - 1/2)} \left(1 + \frac{1}{\kappa_s - 3/2} \frac{v^2}{v_{Ts}^2}\right)^{-\kappa_s - 1}, \\ \mathcal{E}_L(k) &= \frac{\Omega_e^2 m_e v_{Ts}^2}{8\pi v_{T0}^2} \frac{\kappa_s - 3/2}{\kappa_s + 1} \left(1 + \frac{1}{\kappa_s - 3/2} \frac{2\omega_{pe}^2}{k^2 v_{Ts}^2}\right). \end{aligned} \tag{8}$$

Here, n_s represents the super-halo electron density. In Yoon's theory [9], the value of κ_s is uniquely determined as

$$\kappa_s = \frac{9}{4} = 2.25. \tag{9}$$

This prediction, which emanates from balancing the spontaneous and induced processes in the sophisticated linear and nonlinear wave-particle interaction terms within the Langmuir wave kinetic equation, implies that for high velocity, $v \gg v_{Ts}$, the super-halo distribution should behave as an inverse power VDF

$$f_s(v) \propto v^{-2\kappa_s} \sim v^{-6.5}.$$

Wang et al. [1] made detailed statistical survey of $\sim 2 - 20$ keV super-halo electrons during quiet times in the interplanetary space over the year 2007 - 2008 time period, and based upon data from Supra Thermal Electron (STE) instrument onboard STEREO A and B spacecraft, they determined that the observed quiet-time super-halo VDF behaves as an inverse power-law,

$$f_s(v) \sim v^{-b},$$

with b ranging from 5 to 8.7, with an average of 6.69 ± 0.90, and about half the measured cases in a peak at

$$6.5 < b < 7.5.$$

Given the observational uncertainties and inherent approximate nature of the theory, the convergence between the theoretical value of $b = 6.5$ and the observed values $6.5 < b < 7.5$ is quite remarkable. We thus believe that our theory of super-halo electrons maintaining locally isotropic asymptotic power-law VDF via steady-state interaction with Langmuir-type fluctuations (i.e., quasi-thermal noise) is substantially correct. This conclusion is irrespective of how the original super-halo population is generated at the solar source. The theory of Langmuir fluctuation-mediated super-halo VDF already assumes that the seed population is available at the source and travelled to the local region.

Next we consider the asymptotic theory for the halo electrons in dynamical equilibrium with whistler turbulence. In a submitted paper, Yoon et al. [12] show that the whistler turbulence characterized by the frequency range $\Omega_p \ll \omega \ll |\Omega_e|$, where Ω_p is the proton cyclotron frequency, the electric and magnetic field fluctuation spectra of the form,

$$\mathcal{E}_W(k) = \mathcal{E}_0 \left(\frac{c^2k^2}{\omega_{pe}^2}\right)^2, \qquad \mathcal{B}_W(k) = \mathcal{E}_0 \frac{\omega_{pe}^2}{\Omega_e^2}\left(\frac{c^2k^2}{\omega_{pe}^2}\right), \tag{10}$$

where $\mathcal{B}_W(k) = \mathcal{E}_W(k)(ck/\omega_k^W)^2$, satisfies the steady-state nonlinear wave kinetic equation for whistler waves propagating parallel to the ambient magnetic field. The solution, however, does not take into account the pre-existing level of whistler turbulence, which might be the product of quasi-parallel cascade of intrinsic solar wind turbulence. To account for the influence of the intrinsic level of whistler fluctuation, we generalize (10) by

$$\mathcal{E}_W(k) = \mathcal{E}_0 \left(\frac{c^2k^2}{\omega_{pe}^2}\right)^{2-\gamma}, \qquad \mathcal{B}_W(k) = \mathcal{E}_0 \frac{\omega_{pe}^2}{\Omega_e^2}\left(\frac{c^2k^2}{\omega_{pe}^2}\right)^{1-\gamma}, \tag{11}$$

where γ is a free parameter, but it represents the typical power-law index associated with the intrinsic solar wind turbulence in the kinetic whistler frequency range. Making use of this solution, and substituting to equation (4), we may construct the desired halo electron VDF,

$$\begin{aligned} f_h(v) &= \frac{n_h(2-\gamma)}{2\pi v_{Th}^3 \Gamma\left(\frac{3}{4-2\gamma}\right)} \exp\left[-\left(\frac{v^2}{v_{Th}^2}\right)^{2-\gamma}\right], \\ \mathcal{E}_W(k) &= \frac{m_e c^2 \Omega_e^2}{4\pi v_{T0}^2}\left(\frac{\Omega_e}{\omega_{pe}}\right)^4 \left(\frac{k v_{Th}}{\Omega_e}\right)^{4-2\gamma}, \\ \mathcal{B}_W^{(1)}(k) &= \frac{T_h}{\pi}\frac{\Omega_e^2}{2v_{T0}^2}\left(\frac{k v_{Th}}{\Omega_e}\right)^{2-2\gamma}, \end{aligned} \tag{12}$$

where n_h is the halo density. Here, we have defined the halo electron "temperature," $T_h \equiv m_e v_{Th}^2/2$, and the associated with the "halo thermal speed" v_{Th},

$$\frac{1}{v_{Th}^{4-2\gamma}} \equiv \frac{1}{4-2\gamma}\frac{m_e c^2 \Omega_e^2}{\pi v_{T0}^2 \mathcal{E}_0}\frac{\Omega_e^{2\gamma}}{\omega_{pe}^{2\gamma}}, \tag{13}$$

and identified

$$\mathcal{E}_0 \equiv \frac{\Omega_e^2 m_e c^4 v_{Th}^{4-2\gamma}}{2\pi(2-\gamma)v_{T0}^2}\left(\frac{\Omega_e^2}{\omega_{pe}^2}\right)^{\gamma}. \tag{14}$$

Note that the quantity v_{Th}, which we conveniently termed the "halo thermal speed" and the related quantity T_h, which we called the "halo temperature," do not strictly follow from the correct thermodynamic definition for temperature. The rigorous definition for "effective" halo electron temperature is given by

$$\frac{3}{2}n_h T_h^{\rm eff} \equiv \int d\mathbf{v}\frac{m_e v^2}{2} f_h(v). \tag{15}$$

Inserting $f_h(v)$ from equation (12) to the above we obtain

$$T_h^{\rm eff} = \frac{3\Gamma\left(\frac{3}{4-2\gamma}\right)}{\Gamma\left(\frac{5}{4-2\gamma}\right)} T_h. \tag{16}$$

We now bring together all the results we have discussed thus far and construct the total solar wind electron VDF near 1 AU. We compare our theoretical result with the left-hand panel of Figure 5 of Ref. [1], where the statistical survey of ~2 – 20 keV super-halo electron observations during quiet times in the interplanetary medium in 2007 - 2008, from the Supra Thermal Electron (STE) instrument onboard the STEREO A & B spacecraft are shown.

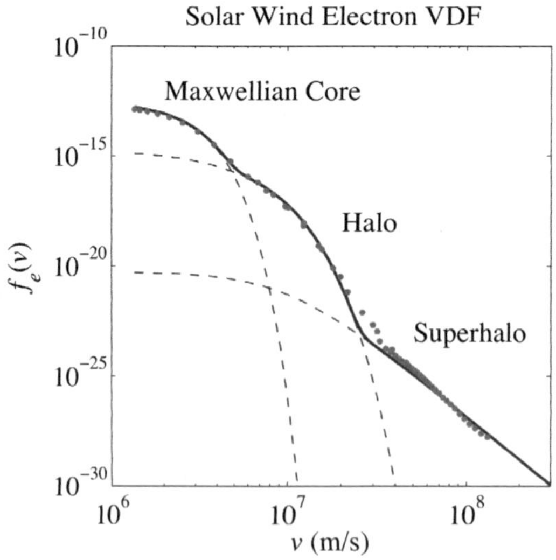

Figure 1. Theoretical construction of the solar wind electron VDF $f_e = f_0 + f_h + f_s$ and comparison against observation [1]. The red dots are the observed VDF values.

Shown in Figure 1 is the theoretical solar wind electron VDF superposed with the observed VDF with red dots. In order to facilitate direct comparison with observed VDF we have multiplied an arbitrary constant 10^7 to the theoretical VDF. We assume $n_0/n_e = 0.96$, $v_{T0}/c = 0.006$, $n_h/n_e = 0.039999$, $v_{T0}/v_{Th} = 0.7$, $n_s/n_e = 1 - n_0/n_e - n_h/n_e = 10^{-6}$, and take $v_{T0}/v_{Ts} = 0.16$. Here, the total electron number density is $n_e = n_0 + n_h + n_s$. In the numerical construction of the halo distribution, we chose $\gamma = 1.35$, which is close to the observed solar wind turbulence magnetic field spectrum in the nominal whistler frequency range. Note how the theoretical model closely fits the observed electron VDF.

3. CONCLUSIONS

In the present paper we presented a theory for asymptotically steady-state solar wind electron VDF. By treating the local solar wind electron VDF as a superposition of the low-energy Maxwellian core electrons with tens of eV energy range, which is considered to be non-resonant with any high-frequency waves, the intermediate $\sim 10^2$ to 10^3 eV energy-range halo electrons that are treated as being in steady-state wave-particle resonance with the whistler turbulence, and the high $\sim 10^3$ to 10^5 eV-range super-halo electrons, which is assumed to maintain dynamical steady-state equilibrium with the Langmuir fluctuations, we have provided a theoretical model for the quiet-time solar wind electrons.

Our model of the steady-state solar wind electrons relies on the presence of spontaneously generated whistler/Langmuir fluctuations, which might coexist with the intrinsic turbulence. The spontaneously emitted Langmuir turbulence is also known as the quasi-thermal noise [8]. However, the presence of spontaneously generated whistler noise has not been thoroughly investigated thus far. The present paper invokes the recent observation by Lacombe et al. [7] as perhaps the evidence for such an example, but obviously more systematic observational research program is called for.

The present model of the solar wind electrons leaves out the important *strahl* component, since it is our belief that the *strahl* is the consequence of inhomogeneity and magnetic mirror effect balancing the wave-induced pitch-angle scattering. The present theory lacks the inhomogeneity. We have also assumed isotropy at the outset. Including inhomogeneity and other effects will make the theory more complete but such a task is beyond the scope of the present paper.

Acknowledgments
This work was supported in part by the BK21 plus program through the National Research Foundation (NRF) funded by the Ministry of Education of Korea. P.H.Y. acknowledges NSF grants AGS1138720 and AGS1242331 to the University of Maryland. Part of this work was carried out while P.H.Y. was visiting Ruhr University Bochum, Germany, which was made possible by support from the Ruhr University Research School PLUS, funded by Germany's Excellence Initiative (DFG GSC 98/3).

[1] L. Wang et al., ApJ **753**, L23 (2012).
[2] C. Vocks and G. Mann, ApJ **593**, 1134 (2003); C. Vocks, G. Mann, and G. Rausche, A&A **480**, 527 (2008); H. Che et al., PRL **112**, 061101 (2014); H. Che and M. L. Goldstein, ApJL **795**, L38 (2014).
[3] J. D. Scudder and S. Olbert, JGR **84**, 2755 (1979); JGR **84**, 6603 (1979); J. D. Scudder, ApJ **398**, 299 (1992); ApJ **427**, 446 (1994); M. V. Canullo et al., ApJ **462**, 1005 (1996); Ø. Lie-Svendsen et al., JGR **102**, 4701 (1997); Ø. Lie-Svendsen and E. Leer, JGR **105**, 35 (2000); V. Pierrard et al., JGR **104**, 17021 (1999); JGR **106**, 29305 (2001); S. Landi et al., ApJ **760**, 143 (2012).
[4] C.-Y. Ma and D. Summers, GRL **25**, 4099 (1998); C. Vocks et al., ApJ **627**, 540 (2005); S. P. Gary and S. Saito, GRL **34**, L14111 (2007); S. Saito and S. P. Gary, GRL **34**, L01102 (2007); JGR **112**, A06116 (2007); J. Pavan, ApJ **769**, L30 (2013).
[5] L. Rudakov, et al., POP **18**, 012307 (2011).
[6] V. Pierrard et al., Solar Phys. **269**, 421 (2011).
[7] C. Lacombe et al., ApJ **796**, doi: 10.1088/0004-637X/796/1/5 (2014).
[8] N. Meyer-Vernet, JGR **84**, 5373 (1979); N. Meyer-Vernet and C. Perche, JGR **94**, 2405 (1989); I. Zouganelis, JGR **113**, A08111 (2008); G. Le Chat, POP **16**, 102903 (2009).
[9] P. H. Yoon, JGR **119**, 7074 (2014).
[10] R. Gaelzer et al., POP **22**, 032310 (2015).
[11] S. Kim et al., ApJ, **806**, 32 (2015).
[12] P. H. Yoon, et al., Asymptotic state of high-frequency whistler turbulence, POP, submitted (2015).

Particle acceleration by combined diffusive shock acceleration and downstream multiple magnetic island acceleration

G P Zank[1,2], P Hunana[1], P Mostafavi[1,2], J A le Roux[1,2], Gang Li[1,2], G M Webb[1] and O Khabarova[3]

[1] Center for Space Plasma and Aeronomic Research (CSPAR), University of Alabama in Huntsville, Huntsville, AL 35805, USA

[2] Department of Space Science, University of Alabama in Huntsville, Huntsville, AL 35805, USA

[3] Heliophysical Laboratory, Institute of Terrestrial Magnetism, Ionosphere and Radiowave Propagation RAS (IZMIRAN), Troitsk, Moscow 142190, Russia

E-mail: garyp.zank@gmail.com

Abstract. As a consequence of the evolutionary conditions [28; 29], shock waves can generate high levels of downstream vortical turbulence. Simulations [32–34] and observations [30; 31] support the idea that downstream magnetic islands (also called plasmoids or flux ropes) result from the interaction of shocks with upstream turbulence. Zank et al. [18] speculated that a combination of diffusive shock acceleration (DSA) and downstream reconnection-related effects associated with the dynamical evolution of a "sea of magnetic islands" would result in the energization of charged particles. Here, we utilize the transport theory [18; 19] for charged particles propagating diffusively in a turbulent region filled with contracting and reconnecting plasmoids and small-scale current sheets to investigate a combined DSA and downstream multiple magnetic island charged particle acceleration mechanism. We consider separately the effects of the anti-reconnection electric field that is a consequence of magnetic island merging [17], and magnetic island contraction [14]. For the merging plasmoid reconnection-induced electric field only, we find i) that the particle spectrum is a power law in particle speed, flatter than that derived from conventional DSA theory, and ii) that the solution is constant downstream of the shock. For downstream plasmoid contraction only, we find that i) the accelerated particle spectrum is a power law in particle speed, flatter than that derived from conventional DSA theory; ii) for a given energy, the particle intensity peaks downstream of the shock, and the peak location occurs further downstream of the shock with increasing particle energy, and iii) the particle intensity amplification for a particular particle energy, $f(x, c/c_0)/f(0, c/c_0)$, is not 1, as predicted by DSA theory, but increases with increasing particle energy. These predictions can be tested against observations of electrons and ions accelerated at interplanetary shocks and the heliospheric termination shock.

1. Introduction

Energetic particles are thought to be accelerated at shock waves by a process known as diffusive shock acceleration (DSA). DSA has been invoked to explain energetic particle acceleration throughout the universe, from gradual solar energetic particle events [1–9] to galactic cosmic rays (e.g., [10]). For DSA, the particle intensity at a given energy increases exponentially ahead of the shock, with a scale length $L^{-1} \simeq U_1/\kappa$, where U_1 is the upstream flow speed and κ is a spatial diffusion coefficient describing the scattering of charged particles in magnetic turbulence. The accelerated particle distribution function at and downstream of the shock is constant spatially and a power law in momentum p^{-w}, where $w = 3r/(r-1)$ depends only on the shock compression ratio r.

The DSA of electrons is difficult because electrons of moderate energy cannot resonantly scatter of background Alfvénic fluctuations associated with the proton streaming instability at a shock [11]. It is possible that some other mechanism may therefore be responsible for either pre-energizing or fully accelerating electrons in the vicinity of a shock. [12] surveyed proton and electron intensity-time profiles of 168 forward interplanetary shocks observed by the ACE spacecraft between September 1997 and December 2001, finding that there were almost as many events corresponding to classical DSA events as events that had either step-like or irregular post-shock increases in the proton intensity. However, step-like post-shock increases are much more common for electrons (energy range 38 - 53 keV) than protons. [12] conclude that there is a clear trend towards the peak intensities occurring in the downstream region of the shock, confirmed by [13] with a more extended study. Complex magnetic field geometrical effects do not explain the post-shock peaking of the observed electron intensity-time profiles. The frequent observed peaking of proton and electron intensity-time profiles some distance/time behind the shock is quite different from the predictions of DSA.

The acceleration of electrons and ions by magnetic island reconnection-related processes has been proposed by [14–17], and extended by [18; 19]. Reconnection associated with merging and contracting plasmoids or flux ropes can lead to the first-order Fermi energization of electrons and ions of initially moderate energies trapped in the islands. Simulations often reveal the generation of a rather hard power law spectrum in a "sea of magnetic islands" (e.g., [17; 20]). The transport theory developed by [18] and [19] incorporates energy changes to the charged particle distribution function (ions or electrons) by magnetic island contraction and merging. For a fixed steady source of particles injected into a super-Alfvénic flow filled with dynamically interacting magnetic islands, the general (Greens function) solution is a power law-like spectrum with an index that depends on the Alfvén Mach number and the ratio τ_{diff}/τ_c of the charged particle diffusion time scale τ_{diff} and the magnetic island/plasmoid/flux rope contraction time scale τ_c [18] .

In situ evidence that electrons and possibly ions are energized in solar wind regions filled with magnetic islands or plasmoids has emerged in the past few years. The distribution of solar wind small-scale discontinuities has been associated with small-scale magnetic islands or flux ropes [21; 22], and increases in the electron and ion heating and heat flux [23; 24]. Shock waves observed in the vicinity of the heliospheric current sheet appear to exhibit electron heating and energization [25; 26] related to the presence of magnetic islands. Further possible evidence of electron acceleration by magnetic islands may have been found by [27], with superhalo (~ 2 - 20 keV) electrons observed during quiet time periods. Quiet time superhalo electrons were observed i) almost exclusively in the slow solar wind, ii) to be isotropically distributed, and iii) to possess power law spectra in particle speed $c^{-\gamma}$, with $\gamma \in [5, 8.7]$. Nearly half the spectra lie in the interval $\gamma \in [6.5, 7.5]$. We have suggested that charged particles can be accelerated during quiet times in the vicinity of the heliospheric current sheet [18; 26], where a large population of magnetic islands may be expected. Power law distributions for accelerated electrons with spectral indices ranging from ~ 6 - 7 between ~ 0.8 - 2 AU for the simplest case of magnetic island merging can be derived from the theory of [18] (see Figure 6 of [18]). It is possible that the origin of the quiet time electron superhalo spectra observed by [27] may well be due to stochastic particle acceleration by distributed magnetic islands in the neighborhood of the heliospheric current sheet.

The shock evolutionary conditions [28; 29] show that downstream vortical modes are generated by the transmission of upstream fluctuations through the shock. Observed [30; 31] and simulated [32–34] shock waves generate substantial levels of MHD turbulence, vortices, and structures downstream of shocks. Downstream magnetic islands will contract and merge dynamically as they advect away from the shock. [18] speculated that because shocks typically generate significant levels of vortical turbulence downstream, reconnection associated with plasmoid contraction and merging may be partly responsible for the acceleration of charged particles at collisionless shock waves. This may be particularly important for electrons.

Based primarily on theory and simulations that indicate that shocks generate vortical turbulence, we present a preliminary extension of the basic test particle theory of DSA [35–39] to include particle energization by downstream magnetic field islands. We consider fast mode shocks only and solve two special cases.

2. Coupled DSA and magnetic island acceleration of charged particles

Three processes can increase the energy of test particles interacting with a dynamical "sea" of magnetic islands. (1) As an elongated island contracts, trapped particles experience repeated reflections at either end of the contracting plasmoid. Particles are energized by a first-order Fermi process [14] due to curvature drift in the direction of the induced electric electric field generated at the strongly curved magnetic field at the endpoints of a contracting flux rope, or in the outflow regions of reconnection sites between merging plasmoids [14; 18; 19; 40; 41]. (2) Depending on whether the magnetic island contraction is compressible or incompressible, the magnetic field strength will either increase or decrease. In the former case, betatron acceleration of the particle will contribute to the energy gain, whereas in the latter case, betatron deceleration will result, both a consequence of charged particle conservation of magnetic moment [18; 19]. The compressible plasmoid contraction case yields a first-order Fermi energization mechanism [18; 19], whereas the incompressible mechanism corresponds to a second-order Fermi energization mechanism [16; 18; 19]. (3) Particles trapped inside an island merger will repeatedly interact with the "anti-reconnection" electric field induced by the merging process[17; 42–45], leading to electron heating and energization. This is a form of direct acceleration by the reconnection-induced electric field component parallel to the magnetic field. Test particle simulations by [17] and [45] concluded that particle acceleration by the induced electric field associated with small-scale magnetic island merging may be the dominant energization process for particles in reconnection layers. However, test particle simulations of electron acceleration in a region of cascading magnetically reconnecting magnetic islands [41] suggest that electron energization is due primarily to magnetic field curvature and gradients, i.e., by either compressible or incompressible magnetic island contraction. These two sets of somewhat contradictory simulations do not make it clear which process is likely to dominate the energization of charged particles in flux rope-related reconnection processes. Accordingly, we retain both mechanisms in our transport theory [18; 19] and explore the effects of both.

2.1. *Transport formalism*

[18] derived a transport equation for a gyrotropic distribution of particles experiencing pitch-angle scattering and energization via the three reconnection-related processes described above for a super-Alfvénic plasma flow. The gyrophase-averaged or focused transport equation can be simplified further by assuming a nearly isotropic particle distribution [18], yielding an advection-diffusion transport equation that resembles the familiar Parker-Gleeson-Axford cosmic ray transport equation (e.g., [46]). [18] and [19] make the important implicit assumption that the time scale over which the particle distribution is averaged is much longer than the trapping time τ_{trap} for particles trapped in individual plasmoids.

The first-order correct transport equation can be expressed as

$$\frac{\partial f}{\partial t} + (U_i + 3|V_E|b_i)\frac{\partial f}{\partial x_i} - \frac{c}{3}\frac{\partial U_i}{\partial x_i}\frac{\partial f}{\partial c} + \frac{1}{c^2}\frac{\partial}{\partial c}\left(\frac{c^3}{3}2\eta_c f\right) = \frac{\partial}{\partial x_i}\left(K_{ij}\frac{\partial f}{\partial x_j}\right) - c\frac{\partial}{\partial x_i}\left(|V_E|_i\frac{\partial f}{\partial c}\right) - |V_E|_i c\frac{\partial^2 f}{\partial x_i \partial c}, \quad (1)$$

where $f = f(\mathbf{x}, t, c)$ is the charged particle distribution function in terms of particle position $\mathbf{x}$, time t, and speed c. $\mathbf{U}$ is the background large-scale plasma flow velocity, $\mathbf{b} \equiv \mathbf{B}/|\mathbf{B}|$ is the unit vector in the direction of the magnetic field, $|V_E| \equiv 1/3q/m\delta E_3\tau_s$ relates the anti-reconnection electric field along the $\mathbf{b}$ direction δE_3 to particle scattering via τ_s, the characteristic pitch-angle scattering time, and q and m are the particle charge and mass respectively ([17; 18; 45] - see also [47]). The term η_c describes the characteristic contraction rate of magnetic islands [16; 18; 19; 40; 41]. The second-order Fermi energization term associated with merging magnetic island line shortening or plasmoid incompressibility [16; 18; 19] enters equation (1) only at the second-order and is therefore neglected here. The term $\mathbf{K}$ describes the spatial diffusion of the nearly isotropic distribution of particles due to pitch-angle scattering.

Three energization terms are present in (1), these being (i) the divergence of the large-scale background flow velocity; (ii) a similar term for magnetic island contraction, and (iii) two mixed derivative terms with coefficient $|V_E|$ that incorporate the anti-reconnection electric field.

The phase space conservation form of the transport equation (1) is given by

$$\frac{\partial f}{\partial t} + \frac{\partial S_i}{\partial x_i} + \frac{1}{c^2}\frac{\partial}{\partial c}\left(c^2 J_p\right) = 0, \quad (2)$$

where

$$\mathbf{S} \equiv -\mathbf{K}\cdot\nabla f - \frac{c}{3}\frac{\partial f}{\partial c}\left(\mathbf{U} - 3|\mathbf{V}_E|\right),$$

is the energetic particle streaming in space, and

$$J_p \equiv \frac{c}{3}\left(\mathbf{U} + |\mathbf{V}_E|\right)\cdot\nabla f + \frac{c}{3}2\eta_c f,$$

is the streaming in momentum space. This formulation is necessary to determine the energetic particle boundary conditions at a shock wave.

3. DSA at fast-mode shocks with downstream magnetic islands

Since equations (1) and (2) are valid for super-Alfvénic flows, they can be applied to fast-mode shocks. Any upstream perturbation incident on a shock generates vortical fluctuations [28; 48]. [32–34; 49; 50] considered the effect of preexisting, large-scale, broadband turbulent fluctuations on propagating hydromagnetic shock waves, using numerical simulations that solve the 2D MHD equations. Upstream density fluctuations induce rippling of the shock surface that generates vorticity and swirling in the downstream flow, stretching and folding the entrained magnetic field. Their simulations show evidence of magnetic islands in the super-Alfvénic downstream flow. We shall assume that collisionless shock waves generate vortical turbulence that evolves dynamically as it is advected away from the shock. For simplicity, we shall further assume that vortical turbulence can be neglected upstream of the shock.

Assuming a planar shock allows us to use a spatially 1D form of the transport equation (1),

$$\frac{\partial f}{\partial t} + (U + 3|V_E|)\frac{\partial f}{\partial x} - \frac{c}{3}\frac{\partial U}{\partial x}\frac{\partial f}{\partial c} + \frac{1}{c^2}\frac{\partial}{\partial c}\left(\frac{c^3}{3}2\eta_c f\right) = \frac{\partial}{\partial x}\left(K\frac{\partial f}{\partial x}\right) - 2|V_E|c\frac{\partial^2 f}{\partial x \partial c}, \quad (3)$$

for constant upstream and downstream flow fields, U_1 and U_2, where 1 denotes upstream and 2 downstream quantities. Upstream of the shock, in the absence of magnetic islands, the steady-state form of equation (3) reduces to

$$U_1 \frac{df}{dx} - \frac{d}{dx}\left(K\frac{df}{dx}\right) = 0,$$

which yields as usual [35]

$$f(x,c) = f_1(c)e^{(U_1/K)x}, \qquad x < 0 \tag{4}$$

on assuming, for convenience, that $f(-\infty, c) = 0$. This upstream solution will be used for both cases considered below.

To determine the downstream solution that must be matched to (4), we consider two idealized problems. For the first, we set $\eta_c = 0$ and consider the role of the anti-reconnection electric field term $|V_E|$ exclusively, i.e., magnetic island contraction is neglected. This approach assumes that the energization is due primarily to anti-reconnection electric fields [17; 45]. For the second, we include magnetic island contraction only [41] and neglect the island reconnection-induced electric field contribution, i.e., $|V_E| = 0$. In so doing, we isolate the specific physical effects that each energization mechanism engenders. In both examples, we assume for simplicity that the diffusion coefficient K is constant and equal upstream and downstream of the shock. Changing K from one constant to another upstream and downstream is simple but having a non-constant K renders the problem analytically challenging.

3.1. Role of magnetic island anti-reconnection electric field

On assuming that magnetic island contraction can be neglected ($\eta_c = 0$), the downstream steady-state transport equation is then given by

$$\frac{\partial^2 f}{\partial x^2} - 2\frac{|V_E|}{K}\frac{\partial^2 f}{\partial x \partial \xi} - \frac{U_2 + 3|V_E|}{K}\frac{\partial f}{\partial x} = 0, \tag{5}$$

where we have introduced the variable $\xi \equiv \ln c/c_0$, and c_0 denotes the lower speed (energy) that is described appropriately by the transport formalism (1). Since $-\infty < \xi < \infty$, $x > 0$, we introduce a Fourier transform in ξ,

$$\begin{aligned} F(\lambda) &= \frac{1}{\sqrt{2\pi}}\int_{-\infty}^{\infty} e^{i\lambda\xi} f(\xi) d\xi; \\ f(\xi) &= \frac{1}{\sqrt{2\pi}}\int_{-\infty}^{\infty} e^{-i\lambda\xi} F(\lambda) d\lambda, \end{aligned}$$

which transforms (5) to

$$\frac{d^2 F}{dx^2} - \frac{U_2 + 3|V_E| - i2|V_E|}{K}\frac{dF}{dx} = 0. \tag{6}$$

On imposing boundedness as $x \to \infty$, the solution to (6) is simply

$$F(x, \lambda) = G(\lambda), \tag{7}$$

where $G(\lambda)$ is constant in x and to be determined by the boundary conditions imposed by the conservation form (2). The first boundary condition is simply continuity of the isotropic distribution function $f(x, \xi)$ across the shock, i.e., $[f] = f(0^+) - f(0^-) = 0$, where $0^\pm$ denotes the value of $f(x, \xi)$ on either side of the shock discontinuity located at $x = 0$. Use of (4) and (7) yields

$$f_1(\xi) = g(\xi), \tag{8}$$

Journal of Physics: Conference Series **642** (2015) 012031 doi:10.1088/1742-6596/642/1/012031

where $g(\xi) = \mathcal{F}^{-1}[G(\lambda)]$ denotes the inverse Fourier transform of $G(\lambda)$. The Fourier transform of the second boundary condition,

$$\left[K\frac{\partial f}{\partial x} + (U - 3|V_E|)\frac{c}{3}\frac{\partial f}{\partial c}\right] = Q\delta(c - c_0), \tag{9}$$

where $Q \equiv \dot{n}/4\pi c_0^2$ ($\dot{n}$ denotes the injection rate of particles into the acceleration process and c_0 is the injected particle speed). Use of (4) and (8) yields

$$G(\lambda) = -\frac{i}{\sqrt{2\pi}}\frac{3\bar{Q}}{U_1 - U_2 + 3|V_E|}\frac{1}{\lambda + i3r/(r - 1 + 3|V_E|/U_2)}, \tag{10}$$

where $r \equiv U_1/U_2$ is the shock compression ratio, $\bar{Q} \equiv \dot{n}/(\sqrt{2\pi}4\pi c_0^3)$. Equation (10) is easily inverted using the residue theorem, yielding the energetic charged particle spectrum at and downstream of the shock as

$$g(\xi) = \frac{3\bar{Q}}{U_1 - U_2 + 3|V_E|}\left(\frac{c}{c_0}\right)^{-3r/(r-1+3|V_E|/U_2)}. \tag{11}$$

To illustrate the role of $|V_E|$, a crude estimate of $|V_E|$ can be derived [18] if we assume that δE_3 can be approximated as $V_A B$, B the local magnetic field, and suppose that the scattering time τ_s is inversely proportional to the gyrofrequency $\Omega \equiv |q|B/m$. Then, $|V_E| = |q||\delta E_3|\tau_s/m \sim (|q|/m)V_A B/\Omega = V_A$, allowing us to express $|V_E|/U_2 \sim M_{A2}^{-1}$, where $M_{A2} \equiv U_2/V_{A2}$ is the downstream Alfvén Mach number. If we adopt the relation $|V_E| \sim V_{A2}$, then

$$\tilde{q} \equiv \frac{3r}{r - 1 + 3|V_E|/U_2} \simeq \frac{3r}{r - 1 + 3M_{A2}^{-1}}. \tag{12}$$

Although offering insight into the meaning of $|V_E|$ and the ratio $U_2/|V_E|$, the relationship $U_2/|V_E| = M_{A2}$ is likely too simplistic, as it relates the parameters $|V_E|$ and the compression ratio r through the downstream Alfvén Mach number, which is not obviously apparent. Unless otherwise indicated, we will continue to regard the parameters $U_2/|V_E|$ and r as independent.

Equations (11) and (12) modify slightly the standard result from DSA for which the spectral index is given by $w = 3r/(r - 1)$. The spectral index is harder as a result of additional particle acceleration downstream of the shock via the merging induced magnetic island electric field, although like conventional DSA, the downstream particle intensity for a fixed speed c is spatially constant.

The three panels of Figure 1 show two sets of solutions (11), one assuming that $|V_E|/U_2$ and the compression ratio r are independent, and the second that $|V_E|/U_2 \sim V_{A2}/U_2 = M_{A2}^{-1}$. In the latter case, the efficiency of particle acceleration via downstream reconnection induced electric fields is directly related to the shock strength as measured by M_{A2}. The left panel assumes $U_2/|V_E| = 3$ and r varies between 1.5 and 4, and the middle panel assumes $r = 3.5$ and varies $|V_E|/U_2$. The right panel assumes $|V_E| \sim V_{A2}$. Since $3U_2/|V_E|$ is typically the dominant term in (12), varying r makes little difference to the spectral slope (Figure 1, left). For the same reason, holding r fixed and varying $U_2/|V_E|$ leads to significant changes in the spectral slope of the accelerated particles. Also plotted are power law spectra from standard DSA at shocks with the same compression ratio r. As $U_2/|V_E| \to \infty$, the solution converges to the standard DSA spectrum. The additional acceleration associated with the anti-reconnection electric field hardens the accelerated spectrum, and weak shocks have harder spectra (because of the reduced escape efficiency due to the smaller downstream Alfvén Mach number) than strong shocks. As the shock strength increases, the spectral index $\tilde{q}$ tends to -4 from above, unlike the standard DSA spectral index that tends to -4 from below.

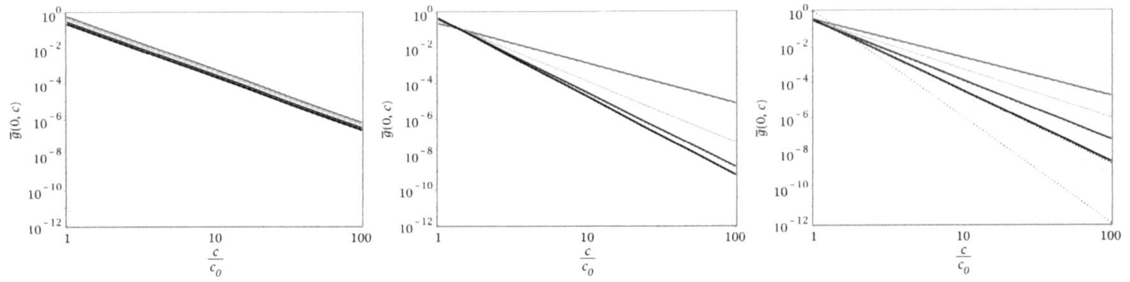

Figure 1. Plot of the charged particle spectrum equation (11), resulting from diffusive shock acceleration in the presence of downstream reconnecting magnetic islands in which the anti-reconnection electric field only is included as an energization term in the transport equation (i.e., the island contraction term $\eta_c = 0$). **Left**: $(|V_E|/U_2)^{-1} \sim 3$ and the shock compression ratio r is varied as $r = 1.5$ (red), $r = 2.0$ (green), $r = 3.0$ (blue), $r = 4.0$ (black). **Middle**: $r = 3.5$ and $(|V_E|/U_2)^{-1} = 1.5$ (red), $(|V_E|/U_2)^{-1} = 5.0$ (green), $(|V_E|/U_2)^{-1} = 20.0$ (blue), $(|V_E|/U_2)^{-1} = 100.0$ (black). **Right**: Here, we assume that $|V_E| \sim V_{A2}$. Three representative solutions are shown for shock compression ratios $r = 2.01$ or $\tilde{q} = 2.15$ (solid red line), $r = 3.03$ or $\tilde{q} = 2.74$ (solid green), $r = 3.80$ or $\tilde{q} = 3.35$ (solid blue). In addition, one solution corresponds to the limit $M_{A2}^2 \to \infty$, which yields $r = 4.0$ or $\tilde{q} = 4.0$ (solid black line). The dotted lines identify the DSA power law solutions $\propto (c/c_0)^{-w}$, where $w = 3r/(r-1)$. For these cases we have $r = 2.01$ and $w = 5.97$ (dotted red line), $r = 3.03$ and $w = 4.48$ (dotted green), $r = 3.8$ and $w = 4.07$ (dotted blue). The reconnection mediated spectrum becomes steeper with increasing compression ratio r, whereas the DSA spectrum becomes flatter with increasing r.

3.2. Particle acceleration by DSA and magnetic island contraction only

On assuming magnetic island contraction only downstream of the shock, the steady form of the transport equation (3) reduces to

$$\frac{\partial^2 f}{\partial x^2} - \frac{U_2}{K}\frac{\partial f}{\partial x} - \frac{2}{3}\frac{\eta_c}{K}\frac{\partial f}{\partial \xi} - \frac{2\eta_c}{K} f = 0, \tag{13}$$

and we assume again that f is bounded as $x \to \pm\infty$. The Laplace transformed equation (13) is

$$\frac{d^2 \bar{f}}{dx^2} - \frac{U_2}{K}\frac{d\bar{f}}{dx} - \frac{2\eta_c}{3K}(s+3)\bar{f} = 0, \tag{14}$$

and we assume injection of particles at the shock ($x = 0$) only, which implies that $f(x, 0^+) = 0$. Following a procedure similar to that outlined above but using now a Laplace transform approach, the boundary conditions yields the full solution for $x \in \mathcal{R}$ as

$$f(x,\xi) = \begin{cases} g(\xi)e^{(U_1/K)x}, & x < 0; \\ \frac{1}{2\sqrt{\pi}}\sqrt{\frac{2\eta_c}{3K}}\, x e^{(U_2/2K)x} \int_0^\xi g(\tau) \frac{e^{-s_0(\xi-\tau)}\exp\left[-\frac{2\eta_c}{3K}x^2/4(\xi-\tau)\right]}{(\xi-\tau)^{3/2}} d\tau. & x > 0 \end{cases}, \tag{15}$$

The spectrum $g(\xi)$ is given by

$$g(\xi) = \frac{3\bar{Q}}{U_1 - U_2}\frac{1}{\alpha - 3}\left[\left((\bar{q}-3) - \beta\sqrt{s_0 - 3}\,\mathrm{erf}\left(\sqrt{s_0 - 3}\xi^{1/2}\right)\right)\left(\frac{c}{c_0}\right)^{-3}\right.$$
$$\left. + \left((\alpha - \bar{q}) + \beta\sqrt{s_0 - \alpha}\,\mathrm{erf}\left(\sqrt{s_0 - \alpha}\xi^{1/2}\right)\right)\left(\frac{c}{c_0}\right)^{-\alpha}\right], \tag{16}$$

where

$$\begin{aligned}
\bar{q} &= \frac{3}{2}\frac{2U_1 - U_2}{U_1 - U_2} = \frac{3}{2}\frac{2r-1}{r-1}, \quad r = \frac{U_1}{U_2}; \\
\beta &\equiv \frac{\sqrt{6\eta_c K}}{U_1 - U_2} = \frac{\sqrt{6\tau_{diff}/\tau_c}}{r-1}; \\
\alpha &= \frac{3}{r-1}\left(r - 2\frac{\tau_{diff}/\tau_c}{r-1}\right); \\
s_0 &\equiv 3\left(1 + \frac{U_2^2/K}{8\eta_c}\right) = 3\left(1 + \frac{\tau_c}{8\tau_{diff}}\right).
\end{aligned}$$

Two characteristic time scales enter (15) and (16), these being the diffusion time scale $\tau_{diff} \equiv K/U_2^2$ and the plasmoid contraction time scale $\tau_c \equiv \eta_c^{-1}$. $\bar{Q}$ is defined as before.

Unlike the standard DSA solution, the downstream contracting magnetic island solution (15) is not constant and instead will continue to increase for a given energy, eventually peaking and then decaying when the e^{-x^2} term dominates. The particle intensity will peak further from the shock with increasing particle energy since there will be more time for particles to experience acceleration by the contracting plasmoids. This effect will eventually be limited by the decay of the post-shock turbulence, for which we have not accounted in our simple model.

Normalized spectra $\bar{g}(c/c_0)$ for the plasmoid contraction-only case for compression ratios $r = 3$ are shown in the top left panel of Figure 2, and the ratio τ_{diff}/τ_c is varied between 0.1 and 5. The spectra are slightly concave at low energies but harden quite significantly at high energies. The solutions are essentially power laws in particle speed, with the hardest spectra corresponding to $\tau_{diff}/\tau_c > 1$. Overplotted is the standard DSA spectrum predicted for a shock with compression ratio 3. The contracting magnetic island spectra are harder and, as $\tau_{diff}/\tau_c \to 0$, approach the DSA solution from above. The combination of classical DSA and particle acceleration by plasmoid contraction (only) produces power law spectra with indices harder than predicted by DSA alone. Shown in the top right panel of Figure 2 are plots of the DSA-plasmoid contraction-only spectra (normalized) at different distances downstream of the shock. The ratio $\tau_{diff}/\tau_c = 0.5$. For small values of τ_{diff}/τ_c, the spectrum does not change much as the distance from the shock increases, nor does the spectrum change much when close to the shock (within a few diffusion scale lengths downstream of the shock) regardless of the size of τ_{diff}/τ_c. However, as the contraction rate η_c increases relative to the particle diffusion time, the accelerated particle spectrum changes quite markedly with increasing distance from the shock. This is due to the greater gain in energy that a particle experiences as magnetic islands are convected away from the shock. In Figure 2, bottom left, we hold the particle energy fixed and plot the particle intensity profile as a function of distance for $\tau_{diff}/\tau_c = 0.1 - 5$ and compression ratio $r = 3$. All intensity profiles exhibit the exponential increase in particle intensity ahead of the shock (located at $x = 0$), which connects to a downstream solution that continues to increase with increasing distance from the shock. The intensity profile eventually peaks, after which it decays smoothly to zero. The color refers to various choices of τ_{diff}/τ_c. The peak of the intensity profile is located at an increasing distance from the shock as the particle energy increases. The cyan curve shows that the solution converges to the DSA solution when τ_{diff}/τ_c becomes very small i.e., as $\eta_c \to 0$, the downstream solution is approximately constant with increasing heliocentric distance.

Two important predictions for coupled DSA-reconnection particle acceleration emerge that distinguish this process from the conventional DSA model. These are that the particle intensity peaks downstream of the shock and that the peak occurs further downstream of the shock with increasing charged particle energy. To illustrate this prediction, we plot the spatial profile in the bottom right panel of Figure 2 for magnetic island contraction only, using fixed values of

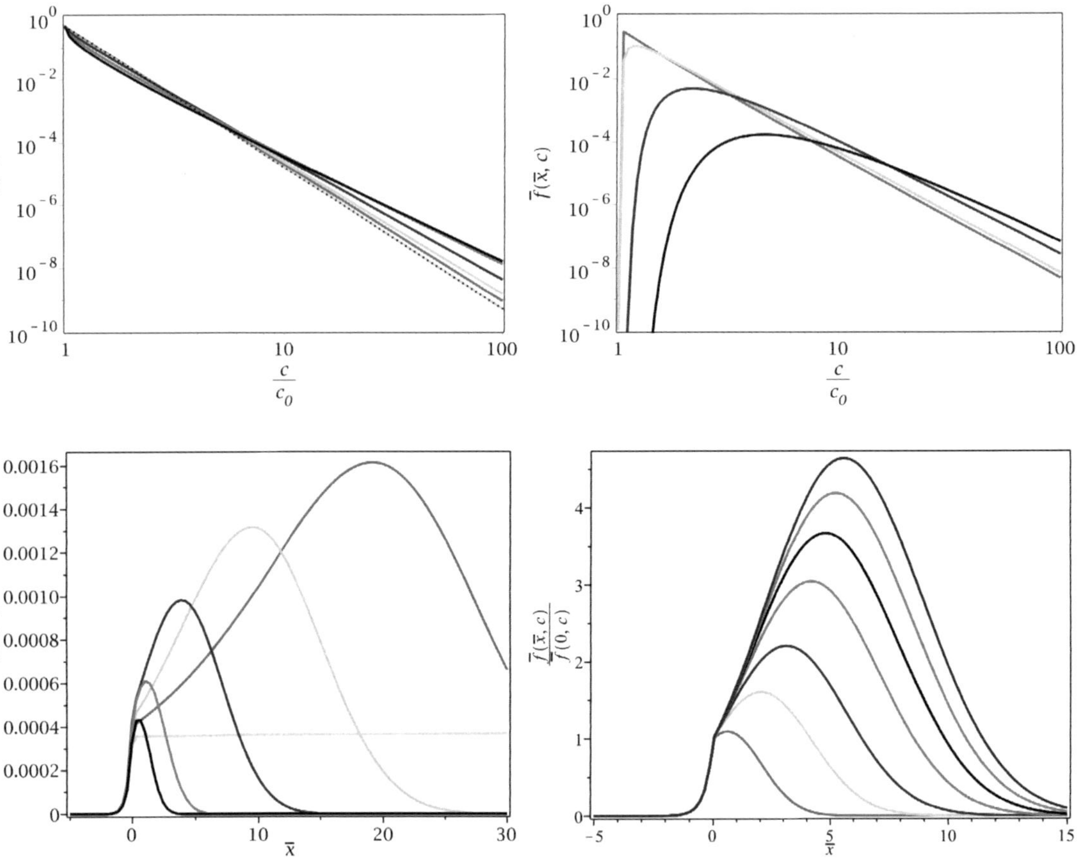

Figure 2. **Top Left**: Normalized energetic particle spectrum $\bar{g}(c/c_0)$, equation (16), for a compression ratio $r = 3$ in the limiting case of magnetic island contraction only. The colored curves correspond to assumed values of $\tau_{diff}/\tau_c = 0.001$ (cyan line), $\tau_{diff}/\tau_c = 0.1$ (red line), $\tau_{diff}/\tau_c = 0.2$ (green line), $\tau_{diff}/\tau_c = 0.5$ (blue line), $\tau_{diff}/\tau_c = 2.0$ (magenta line), $\tau_{diff}/\tau_c = 5.0$ (black line). The dotted black lines correspond to the DSA power laws $\propto (c/c_0)^{-3r/(r-1)}$ $(r = 3)$. As $\tau_{diff}/\tau_c \to 0$ the spectra converge to the DSA spectrum and with increasing values of τ_{diff}/τ_c, the spectra flatten. **Top Right**: Plots of the normalized accelerated particle spectra downstream of the shock i.e., the solution (15) $\bar{f}(\bar{x} = \text{const.}, c)$ at different normalized distances $\bar{x} = 0.1$ (red line), $\bar{x} = 1.0$ (green line), $\bar{x} = 5.0$ (blue line), $\bar{x} = 10.0$ (black line). A compression ratio $r = 3$ is assumed and $\tau_{diff}/\tau_c = 0.5$. **Bottom Left**: Normalized particle intensity plot for a fixed energy $c/c_0 = 5$ as function of position. Here $\tau_{diff}/\tau_c = 0.001$ (cyan line), $\tau_{diff}/\tau_c = 0.1$ (red line), $\tau_{diff}/\tau_c = 0.2$ (green line), $\tau_{diff}/\tau_c = 0.5$ (blue line), $\tau_{diff}/\tau_c = 2.0$ (magenta line), $\tau_{diff}/\tau_c = 5.0$ (black line). The cyan line corresponds to a solution that is very close to the DSA limit and so remains almost constant downstream of the shock with increasing distance. A compression ratio $r = 3$ is assumed. **Bottom Right**: Particle intensity as a function of position with particle intensities normalized to the value at the shock $\bar{f}(0, c/c_0)$, assuming a ratio $\tau_{diff}/\tau_c = 1.01$. Each colored curve corresponds to a particular normalized (square root of the) energy c/c_0: $c/c_0 = 2$ (red line), $c/c_0 = 5$ (green line), $c/c_0 = 10$ (blue line), $c/c_0 = 20$ (magenta line), $c/c_0 = 30$ (black line), $c/c_0 = 40$ (gold line), $c/c_0 = 50$ (violet line). $r = 3$.

τ_{diff}/τ_c at different energies. We normalize the plots to the value of the intensity $f(0, c/c_0)$ as measured at the shock, so the figures show an amplification factor relative to the value at the shock. Conventional DSA predicts that the particle intensity downstream of the shock is constant as a function of distance and, when normalized to the intensity $f(0, c/c_0)$, the intensity is 1 for all energies. Contrast this with the curves illustrated in the bottom right panel of Figure 2. Each of the colored curves in the panels corresponds to a particular normalized (square root of the) energy c/c_0, from $c/c_0 = 2$ (red) to 50 (black). The lowest energies peak close to the shock (located at $x = 0$) and have the lowest amplification factor and higher energies peak further from the shock and have a larger amplification factor. The distance of the particle intensity peak from the shock front is ordered by energy, with the higher energies peaking systematically further from the shock. Similarly, the amplification factor is ordered by energy, with the higher energies having the larger amplification factor. Depending on the assumed value of τ_{diff}/τ_c, the amplification of the particle intensity downstream of the shock can be substantial. As τ_{diff}/τ_c decreases, although the amplification factor increases, the peak location moves further from the shock so that the solution gradually converges towards the DSA limit of no amplification behind the shock.

4. Conclusions

Zank et al. [18] speculated that because shock waves typically generate vortical turbulence, particle acceleration at shocks may be due to a combination of DSA and downstream reconnection processes associated with the dynamical interaction of magnetic islands. We present a simple test particle model, based on the charged particle transport theory developed by [18] and [19], that describes the coupled and simultaneous acceleration of particles at both a shock and downstream interacting merging and contracting magnetic islands.

The transport equation and appropriate boundary conditions for charged particles interacting with a collisionless fast-mode shock and downstream magnetic island turbulence are solved for two cases. Considering the merging plasmoid reconnection-induced electric field only, we find i) that the particle spectrum is a power law in particle speed with an index that depends on the shock compression ratio r and the parameter $U_2/|V_E|$; and ii) that the solution is constant downstream of the shock. The accelerated particle spectrum is flatter than that derived from conventional DSA theory. In the case that only plasmoid contraction is considered, we find that i) the accelerated particle spectrum is a power law in particle speed with an index that depends on the shock compression ratio r and the ratio of the diffusion and the contraction time scales τ_{diff}/τ_c; ii) the plasmoid contraction-only spectrum is harder than the corresponding DSA spectrum to which it converges as $\tau_{diff}/\tau_c \to 0$; iii) for a given energy, the particle intensity peaks downstream of the shock, and the peak location occurs further downstream of the shock with increasing particle energy, and iv) the particle intensity amplification for a particular particle energy, $f(x, c/c_0)/f(0, c/c_0)$, is not 1, as predicted by DSA theory, but increases with increasing particle energy.

These two sets of predictions can be tested against observations of electrons and ions accelerated at interplanetary shocks and the heliospheric termination shock.

Acknowledgments

We acknowledge the partial support of NASA grants NNX08AJ33G, Subaward 37102-2, NNX09AG70G, NNX09AG63G, NNX09AJ79G, NNG05EC85C, Subcontract A991132BT, NNX09AP74A, NNX10AE46G, NNX09AW45G, NNX14AF43G, NNN12AA012, NNX15AI65G, and NSF grant ATM-0904007. OK was supported by RFBR grant no. 14-02-00769.

References

[1] Zank G P, Rice W K M and Wu C C 2000 *J. Geophys. Res.* **105** 25079–25096

[2] Zank G P, Li G and Verkhoglyadova O 2007 *Space Sci. Rev.* **130** 255–272

[3] Li G, Zank G P and Rice W K M 2003 *J. Geophys. Res.* **108** 1082–+

[4] Rice W K M, Zank G P and Li G 2003 *J. Geophys. Res.* **108** 1369–+

[5] Verkhoglyadova O P, Li G, Zank G P, Hu Q, Cohen C M S, Mewaldt R A, Mason G M, Haggerty D K, von Rosenvinge T T and Looper M D 2010 *Journal of Geophysical Research (Space Physics)* **115** A12103

[6] Verkhoglyadova O P, Li G, Ao X and Zank G P 2012 *Ap. J.* **757** 75

[7] Verkhoglyadova O P, Zank G P and Li G 2015 *Phys. Rep.* **557** 1–23

[8] Reames D V 1999 *Space Sci. Rev.* **90** 413–491

[9] Reames D V 2013 *Space Sci. Rev.* **175** 53–92 (*Preprint* 1306.3608)

[10] Axford W I 1981 *International Cosmic Ray Conference* **12** 155–203

[11] Lee M A 1983 *J. Geophys. Res.* **88** 6109–6119

[12] Lario D, Ho G C, Decker R B, Roelof E C, Desai M I and Smith C W 2003 *Solar Wind Ten* (*American Institute of Physics Conference Series* vol 679) ed Velli M, Bruno R, Malara F and Bucci B pp 640–643

[13] Ho G C, Lario D, Decker R B, Smith C W and Hu Q 2008 *American Institute of Physics Conference Series* (*American Institute of Physics Conference Series* vol 1039) ed Li G, Hu Q, Verkhoglyadova O, Zank G P, Lin R P and Luhmann J pp 184–189

[14] Drake J F, Swisdak M, Che H and Shay M A 2006 *Nature* **443** 553–556

[15] Drake J F, Swisdak M, Schoeffler K M, Rogers B N and Kobayashi S 2006 *Geophys. Res. Lett.* **33** L13105

[16] Drake J F, Swisdak M and Fermo R 2013 *Ap. J. Lett.* **763** L5 (*Preprint* 1210.4830)

[17] Oka M, Phan T D, Krucker S, Fujimoto M and Shinohara I 2010 *Ap. J.* **714** 915–926 (*Preprint* 1004.1154)

[18] Zank G P, le Roux J A, Webb G M, Dosch A and Khabarova O 2014 *Ap. J.* **797** 28

[19] le Roux J A, Zank G P, Webb G M and Khabarova O 2015 *Ap. J.* **801** 112

[20] Guo F, Liu Y H, Daughton W and Li H 2015 *ArXiv e-prints* (*Preprint* 1504.02193)

[21] Greco A, Matthaeus W H, Servidio S, Chuychai P and Dmitruk P 2009 *Ap. J. Lett.* **691** L111–L114

[22] Greco A, Matthaeus W H, Servidio S and Dmitruk P 2009 *Phys. Rev. E* **80** 046401

[23] Osman K T, Matthaeus W H, Greco A and Servidio S 2011 *Ap. J. Lett.* **727** L11

[24] Chasapis A, Retinò A, Sahraoui F, Vaivads A, Khotyaintsev Y V, Sundkvist D, Greco A, Sorriso-Valvo L and Canu P 2015 *Ap. J. Lett.* **804** L1

[25] Tessein J A, Matthaeus W H, Wan M, Osman K T, Ruffolo D and Giacalone J 2013 *Ap. J. Lett.* **776** L8

[26] Khabarova O, Zank G P, Li G, le Roux J A, Webb G M, Dosch A and Malandraki O E 2015 *Ap. J.* **808** 24 (*Preprint* 1504.06616)

[27] Wang L, Lin R P, Salem C, Pulupa M, Larson D E, Yoon P H and Luhmann J G 2012 *Ap. J. Lett.* **753** L23

[28] McKenzie J F and Westphal K O 1968 *Physics of Fluids* **11** 2350–2362

[29] Westphal K O and McKenzie J F 1969 *Physics of Fluids* **12** 1228–1236

[30] Hu Q, Ao X, Peltzer R and Zank G P 2012 *American Institute of Physics Conference Series* (*American Institute of Physics Conference Series* vol 1500) ed Hu Q, Li G, Zank G P, Ao X, Verkhoglyadova O and Adams J H pp 192–197

[31] Hu Q, Zank G P, Li G and Ao X 2013 *American Institute of Physics Conference Series* (*American Institute of Physics Conference Series* vol 1539) ed Zank G P, Borovsky J, Bruno R, Cirtain J, Cranmer S, Elliott H, Giacalone J, Gonzalez W, Li G, Marsch E, Moebius E, Pogorelov N, Spann J and Verkhoglyadova O pp 175–178

[32] Giacalone J and Jokipii J R 2007 *Ap. J. Lett.* **663** L41–L44

[33] Lu Q, Hu Q and Zank G P 2009 *Ap. J.* **706** 687–692

[34] Guo F, Li S, Li H, Giacalone J, Jokipii J R and Li D 2012 *Ap. J.* **747** 98 (*Preprint* 1112.6373)

[35] Axford W I, Leer E and Skadron G 1977 *International Cosmic Ray Conference* **11** 132–137

[36] Krymskii G F 1977 *Akademiia Nauk SSSR Doklady* **234** 1306–1308

[37] Bell A R 1978 *MNRAS* **182** 147–156

[38] Bell A R 1978 *MNRAS* **182** 443–455

[39] Blandford R D and Ostriker J P 1978 *Ap. J. Lett.* **221** L29–L32

[40] Bian N H and Kontar E P 2013 *Physical Review Letters* **110** 151101 (*Preprint* 1302.6090)

[41] Zhou X, Büchner J, Barta M, Gan W and Liu S 2015 *ArXiv e-prints* (*Preprint* 1504.06486)

[42] Egedal J, Fox W, Katz N, Porkolab M, ØIeroset M, Lin R P, Daughton W and Drake J F 2008 *Journal of Geophysical Research (Space Physics)* **113** A12207

[43] Pritchett P L 2008 *Physics of Plasmas* **15** 102105

[44] Tanaka K G, Yumura T, Fujimoto M, Shinohara I, Badman S V and Grocott A 2010 *Physics of Plasmas* **17** 102902

[45] Le A, Karimabadi H, Egedal J, Roytershteyn V and Daughton W 2012 *Physics of Plasmas* **19** 072120

[46] Zank G P 2014 *Transport Processes in Space Physics and Astrophysics, Lecture Notes in Physics, Berlin Springer Verlag* vol 877 (Springer New York Heidelberg Dordrecht London)

[47] le Roux J A, Zank G P and Matthaeus W H 2002 *Journal of Geophysical Research (Space Physics)* **107** 1138

[48] McKenzie J F and Westphal K O 1969 *Planet. Space Sci.* **17** 1029–1037

[49] Mizuno Y, Pohl M, Niemiec J, Zhang B, Nishikawa K I and Hardee P E 2011 *Ap. J.* **726** 62 (*Preprint* 1011.2171)

[50] Fraschetti F 2013 *Ap. J.* **770** 84 (*Preprint* 1304.4956)

Journal of Physics: Conference Series **642** (2015) 012032 doi:10.1088/1742-6596/642/1/012032

Particle acceleration in 3D single current sheets formed in the solar corona and heliosphere: PIC approach

Valentina V. Zharkova [1]& **Taras Siversky**[2]

[1] Department of Mathematics and Information Sciences, Northumbria University, Newcastle upon Tyne, NE2 1XE, UK

[2] Department of Radiophysics, T.Shevchenko National University of Kyiv, 8 Glushkov prosp., Kyiv, 03022, Ukraine

E-mail: `valentina.zharkova@northumria.ac.uk`

Abstract. Acceleration of protons and electrons in a reconnecting current sheet (RCS) is investigated with the test particle and particle-in-cell (PIC) approaches in a 3D magnetic topology. PIC simulations confirm a spatial separation of electrons and protons with respect to the midplane depending on the guiding field. Simulation reveals that the separation occurs in magnetic topologies with strong guiding fields and lasts as long as the particles are kept dragged into a current sheet. This separation produces a polarisation electric field induced by the plasma feedback to a presence of accelerated particles, which shape can change from symmetric towards the midplane (for weak guiding field) to fully asymmetric (for strong guiding field). Particles are found accelerated at a midplane of any current sheets present in the heliosphere to the energies up to hundred keV for electrons and hundred MeV for protons. The maximum energy gained by particles during their motion inside the current sheet is defined by its magnetic field topology (the ratio of magnetic field components), the side and location from the X-nullpoint, where the particles enter a current sheet. In strong magnetic fields of the solar corona with weaker guiding fields, electrons are found circulating about the midplane to large distances where proton are getting accelerated, creating about the current sheet midplane clouds of high energy electrons, which can be the source of hard X-ray emission in the coronal sources of flares. These electrons are ejected into the same footpoint as protons after the latter reach the energy sufficicent to break from a current sheet. In a weaker magnetic field of the heliosphere the bounced electrons with lower energies cannot reach the midplane turning instead at some distance D before the current sheet midplane by 180 degrees from their initial motion. Also the beams of accelerated transit and bounced particles are found to generate turbulent electric fields in a form of Langmuir waves (electrons) or ion-acoustic waves (protons).

1. Introduction

In the present paper we investigate particle acceleration by a drifted (reconnection) electric field generated by a reconnecting current sheet (RCS). The first analytical study of this acceleration mechanism was proposed by Speiser[1] who found that the particles can be accelerated to rather high energies after they enter the configuration with reconnecting magnetic field lines forming an RCS and a perpendicular (drifted) electric field. This theory was extended by Litvinenko and Somov [2, 3] who found that the guiding magnetic field can significantly increase the energy gains by accelerated particles.

Simulation of particle trajectories by using test particle approach in a 3D RCS was carried for a constant reconnection electric field [4] or for a variable electric field enhanced near the X-nullpoint due to anomalous resistivity [6]. Zharkova and Gordovskyy [4] showed that the trajectories of particles with the opposite charges (electrons or protons) can be either fully symmetric or strongly asymmetric towards the midplane of the RCS depending on the ratio between the magnetic field components.

As a result, some fraction of the released magnetic energy is transformed into kinetic energy of accelerated particles. The energy spectra of these particles also depend on a magnetic field topology, electric field strength and the dependence of transverse magnetic field on a distance from the X-nullpoint. Particles are accelerated at a current sheet midplane (X=0)[4, 5], the longer they gyrate about the Y and Z components the higher energy they gain from a reconnection electric field E_y (see Fig.1). The particles can break from a current sheet only after they gain the critical energy defined by a given magnetic field topology[7].

Accelerated particles gain energies up to 100 keV for the electrons and up to 100 MeV for the protons [6, 8, 9]. Spectral indices of energy spectra gained at acceleration vary from 2 for electrons and 1.5 for protons if the transverse magnetic field increases with the distance linearly. The spectral indices for both electrons and protons become much higher if the transverse magnetic field $B_x(z)$ increases exponentially with a distance z from the X-nullpoint towards its full length a (e.g $B_x(z) = B_0 \cdot (\frac{z}{a})^n$ having index n>1 near the ratio z/a) [5, 9].

More realistic approach considering plasma feedback to accelerated particles in current sheet in the solar corona is achieved with PIC simulation for reduced proton-to-mass ratios allowing to reproduce electro-magnetic fields generated by accelerated particles in solar flares [10, 11, 12]. These studies reproduced well magnetic reconnection rates [10] and some key features of accelerated particles; power law energy spectrum and energy gains up to 100 keV.

Recent study of particle acceleration during a magnetic reconnection with PIC approach in a 3D single current sheet by Siversky and Zharkova [12] for conditions of the solar corona and by Zharkova and Khabarova [13] for conditions of the heliospheric current sheet have shown that magnetic field topology defines the energy gains, pitch angle and density distributions of electrons and protons after their passage through the current sheet. Of course, the findings are derived only for a single current sheet while Drake et al [14] have shown that during a magnetic reconnection a current sheet can be broken by tearing instability into a few magnetic islands where particle acceleration scenario can change.

However, for the current study we continue to investigate particle passage through and gaining energy in a single current sheet and compare particle characteristics in the current sheets with stronger magnetic field and denser plasma (close to the solar corona) and in the current sheets with low density and low magnetic field magnitudes relevant to the heliosphere.

2. Description of the model

In this paper we apply the PIC approach for the simulation of particle acceleration in a 3D reconnecting current sheet with a simple 3D magnetic field topology. We assume that the current sheet plasma has typical density of either the solar corona or the heliosphere, which sustain the equilibrium with the background magnetic field. However, in order to avoid the problem with the small Debye length in PIC approach, only a small fraction of the plasma particles (with density of $10^{10} m^3$) is included in the PIC simulation for the solar corona and $1 - 10^6$ m^3 in the heliosphere. Electric and magnetic fields generated by the accelerated particles in the PIC approach are used in the TP approach to investigate particle trajectories and densities.

2.1. Magnetic field topology

Since the acceleration time electrons or protons is much shorter than the time of a reconnecting magnetic field variation as shown by Priest and Forbes [15], one can assume the background

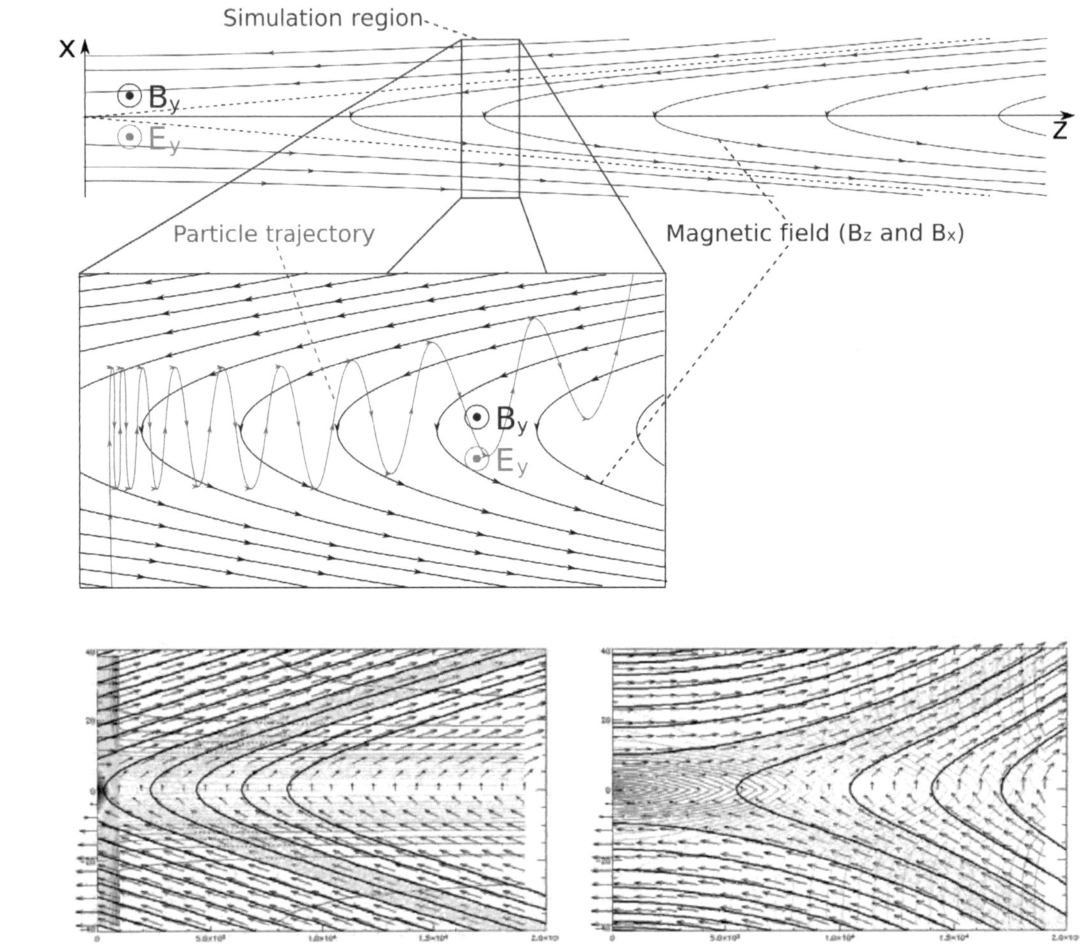

Figure 1. The model of 3D current sheet used in the PIC simulations for a sliding simulation region shown by a rectangular (the top plot). The current sheet topologies for different values of the parameter α describing the variations of transverse magnetic field $B_x(z) = B_{x0}(\frac{z}{L_z})^{\alpha}$: $\alpha = 0.5$ (the bottom left plot), 1.5 (the bottom right plot).

magnetic field to be stationary. Also, from the TP simulations one found that travel distances of accelerating particles along the RCS are of the order of 10 km at most (for the protons) [9] and this length is much shorter than the length scale of magnetic field variations along the current sheet. In addition, we assume that the magnetic field variations across the current sheet have much shorter length scale than along the current sheet, i.e. $L_x \ll L_z$, L_y.

The simulation domain is a small part of the RCS (see Fig. 1 from [12]), but still large enough to contain the full trajectories of accelerated particles if the periodic boundary conditions are used. Previously, it was shown[9] that electrons normally travel the distances of about a few hundred meters for the coronal conditions (with proton gyroradius is about 1 m) while protons travel a few thousand meters before they gain energy sufficient to get ejected from a current sheet.

The main component B_z depends on x as follows: $B_z(x) = -B_{z0}\tanh\left(\frac{x}{L_x}\right)$. Similar to Zharkova and Gordovskyy [5], it is assumed that the B_x component varies inside the simulation domain with z as follows, i.e. $B_x(z) = -B_{x0}(\frac{z}{L_z})^{\alpha}$ where α >1. In Fig. 1 (bottom two plots) the views of magnetic field topologies in reconnecting current sheet are presented for two magnitudes of the spectral parameter α. The guiding (out-of-plane) magnetic field B_y is maximal in the midplane and vanishes outside the RCS: $B_y(x) = B_{y0}\text{sech}\left(\frac{x}{L_x}\right)$. We show below that this parameter α defines spectral indices of energy spectra for both kinds of particles: electrons and protons, depending in the ratio between magnetic field components.

Contrary to the TP simulations, in the PIC simulations the accelerated particles are considered to generate their own electric and magnetic fields following the plasma feedback according to the Maxwell's equations, which is self-consistently taken into account as described by Siversky and Zharkova [12].

2.2. *Reconnection electric field*

In order to provide the inflow of plasma in the simulation domain we set up a background electric field, as the one drifted in by a magnetic diffusion process with velocity V_{in} as shown by [15]: $E_{y0} = V_{in} \times B_{z0} + \frac{1}{\sigma\mu}\frac{\partial B_z}{\partial x}$, where V_{in} is the inflow velocity, σ is the ambient plasma conductivity, μ is magnetic permeability. On the boundaries of an RCS we ignore a gradient of the magnetic component B_z over x-coordinate by putting the second term to zero.

Simulations are carried out for the following current sheet parameters: the main component of the magnetic field $B_{z0} = 10^{-3}$ T , the current sheet half-thickness $L_x = 1$ m, the drifted electric field $E_{y0} = 250$ V/m and the guiding, B_{y0}, and transverse, B_{x0}, components of the magnetic field are selected to range from $(0.1 - 10) \times 10^{-4}$ T, in order to cover acceleration at the various parts of RCS in the solar corona.

3. Results

3.1. *Polarization electric field induced by accelerated particles*

The PIC simulations have shown that the magnetic field $\tilde{\mathbf{B}}$ induced by the plasma feedback to a sepration of accelerated particles in the ambient plasma is much smaller than the background $\mathbf{B_z}$. On the other hand, the induced electric field $\tilde{\mathbf{E}}_\mathbf{x}$, called a polarization electric field, induced by the separation of electrons from protons is much larger than the reconnection (drifted) field E_y [9, 12].

In this subsection we consider this polarization electric field $\tilde{E}_x$, which is directed across the current sheet along X-axis and plotted in Fig. 2 as a function of x averaged over the z coordinate for various magnetic field parameters of B_{x0} and B_{y0}. The appearance of polarization electric field leads to a local non-neutrality of the plasma, which becomes stronger if the transverse magnetic field B_{x0} decreases or the guiding magnetic field B_{y0} increases.

The polarisation electric field defines the motion of charged particles in the vicinity of a current sheet, which can be captured from the ion velocity profiles by in-situ observations with the spacecrafts passing through the heliospheric current sheet (or crossing the sector boundary)[13].

3.2. *Proton trajectories*

In order to reconstruct the particle trajectories, we use a TP approach, where the induced polarisation electric field $\tilde{E}_x$ obtained from PIC is added to the background electro-magnetic configuration described above. The trajectories of the two protons in the x-V_z phase plane entering current sheet from the opposite directions are shown in Fig. 3 for magnetic field topology forming a polarisation field from Fig. 2 (the left plot).

First, we note that the "bounced" protons have wider orbits of their gyration about the midplane than the "transit" ones. Second, the presence of polarisation electric field makes very peculiar velocity profiles of protons and ions which follow the profile of this polarisation field.

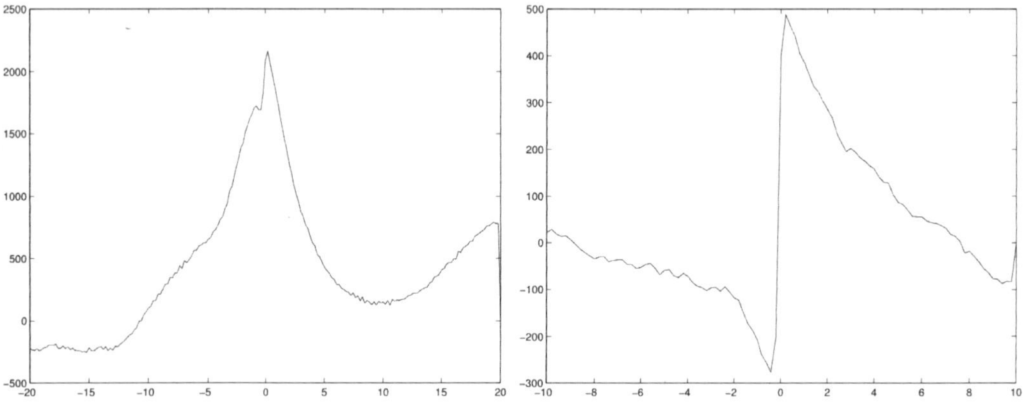

Figure 2. Polarisation electric field $\tilde{E}_x$ (Y-axis, V/m) induced by accelerated particles as derived from the PIC simulations in the corona for $B_{z0} = 10^{-3}$ T and the following values of other components: $B_{x0} = 10^{-4}$ T and $B_{y0} = 10^{-6}$ T (left plot) and $B_{x0} = 10^{-5}$ T and $B_{y0} = 2 \times 10^{-3}$ T (right plot). The X-axis represents distances from a current sheet midplane (X=0). Note that the reconnection electric field is equal to 100 V/m and a current sheet width is 2 m (1 m to each side from the midplane X=0).

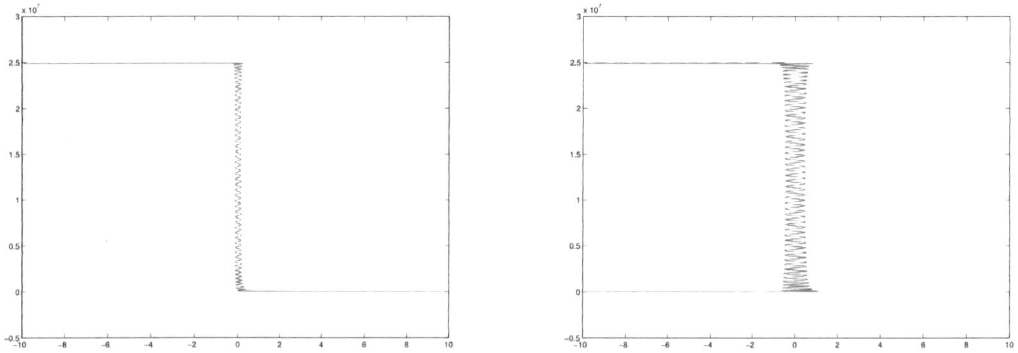

Figure 3. Trajectories and velocities (Y-axis, m/s) of the transit (left plot) and bounced (right plot) protons entering a current sheet from the 3D PIC simulation for the coronal magnetic field topologies: $B_{z0} = 10^{-3}$ T, $B_{x0} = 10^{-4}$ T and $B_{y0} = 10^{-4}$ T. The X axis presents a distance (in gyroradius of protons, 1 m in the coronal field) from the current sheet midplane (X=0).

These trajectories are rather similar to those obtained in the TP simulations without $\tilde{E}_x$ with the "bounced" proton during the acceleration phase has a wider orbit while the "transit" proton has a narrower one producing energy distributions with two peaks discussed below.

3.3. Electron trajectories and formation of electron clouds

Electrons entering from the $x < 0$ semispace have a dynamics similar to those of the "transit" protons, e.g. the electrons drift towards the midplane, become accelerated and ejected to the $x > 0$ semispace. However, the polarization field $\tilde{E}_x(x)$, which extends beyond the current sheet and has a component parallel to the magnetic field B_z, decelerates the ejected electrons. For chosen magnitudes of B_x and B_y, the majority of electrons are unable to escape away from the RCS, instead they are dragged back towards the current sheet and become undistinguishable

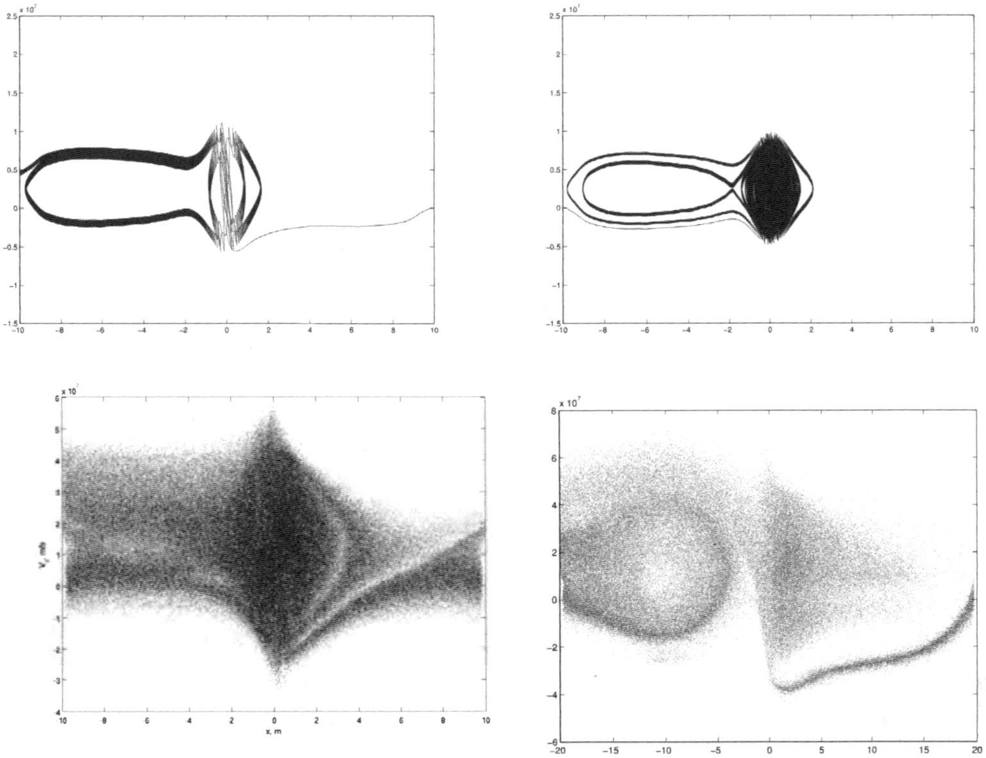

Figure 4. Electron velocity distribution (Y-axis, m/s) showing formation of the clouds from transit electrons about the midplane and from bounced electrons far away from the midplane as deduced from the 3D PIC simulation. Top plots: test particle trajectories of transit electron (left plot) and bounced electron (right plot) during their circulation about the midplane for the coronal magnetic field: $B_{z0} = 10^{-3}$ T, $B_{x0} = 10^{-4}$ T and $B_{y0} = 10^{-4}$ T. The bottom plots: the PIC simulations of electron motion and formation of clouds from the transit (the left plot) and bounced (theright plot) electrons calculated for the magnetic field topology: $B_{z0} = 10^{-3}$ T, $B_{x0} = 10^{-4}$ T and $B_{y0} = 10^{-5}$ T.

from the other electrons entered from the $x > 0$ semispace.

The electrons entering from the positive x side which are "bounced" from RCS in the absence of $\tilde{E}_x$, are able to reach its midplane if polarization field is present. In the vicinity of the midplane, the electrons become unmagnetized and oscillate with the gyrofrequency determined by B_y until they gain sufficient energy to break from the magnetic field. Although, if the electron initial velocity is small, it can be quasi-trapped inside the RCS (Fig. 4, top plots).

Such electrons are accelerated at the midplane, ejected from it, then decelerated outside the RCS and returned back to it because of the polarization electric field appearing due to the protons still being accelerated in the RCS midplane. This cycle is repeated for many times, forming the electron clouds around the current sheet (Fig. 4, top plots for TP, bottom plots for PIC) which can be observed as a coronal source in solar flares. These electron clouds exist until the protons gyrate about the midplane and gain sufficient energy to leave the current sheet. Since a magnitude of the polarisation field $\tilde{E}_x(x)$ is smaller in the current model at $x < 0$ than at $x > 0$ (see Fig. 2, the bottom left plot), for electrons it is easier to escape to the $x < 0$ semispace, where the protons are ejected.

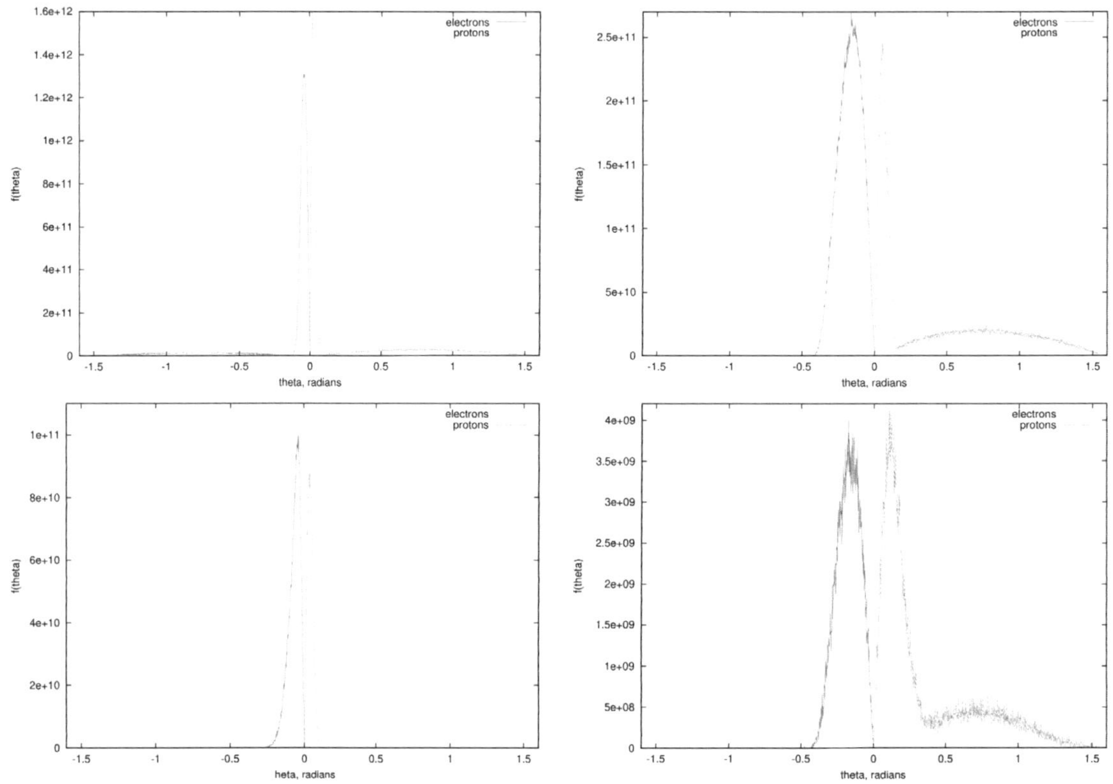

Figure 5. Pitch angles (in radians) of electrons (blue plots) and protons (red plots) calculated for weak guiding field ($B_y/B_z < 10^{-4}$) with the symmetric polarisation electric field (left column) and strong guiding field ($B_y/B_z = 0.1$ (top plot) and $B_y/B_z = 2$ (bottom plot)) with the strongly asymmetric polarisation electric field (right column).

3.4. Energy spectra of electrons and protons

The spectral indices of energy spectra for electrons and protons accelerated by the drifted electric field are found to be strongly dependent on the magnetic field topology [9, 5]. For the accepted variations of B_x being proportional to $(\frac{z}{L_z}))^{\alpha}$ and for the assumed linear increase of the density with z as $(\frac{z}{L_z}))^{\lambda}$ in a vicinity of the X-nullpoint the spectral indices of particles are defined as follows[9] :

For electrons in RCS with weak guiding field, $B_x >> B_y$:

$$\gamma = 1 + \frac{2+\lambda}{2\alpha} \tag{1}$$

For electrons in RCS with strong guiding field, B_y:

$$\gamma = \frac{1}{2} + \frac{1+\lambda}{2\alpha} \tag{2}$$

For protons in RCS with any magnetic field:

$$\gamma = 1 + \frac{1+\lambda}{2\alpha} \tag{3}$$

It can be spotted that for strong guiding fields the spectral indices of electrons and protons become rather close, while for a weaker guiding field the spectral indices of electrons are higher by unit or so than those of protons.

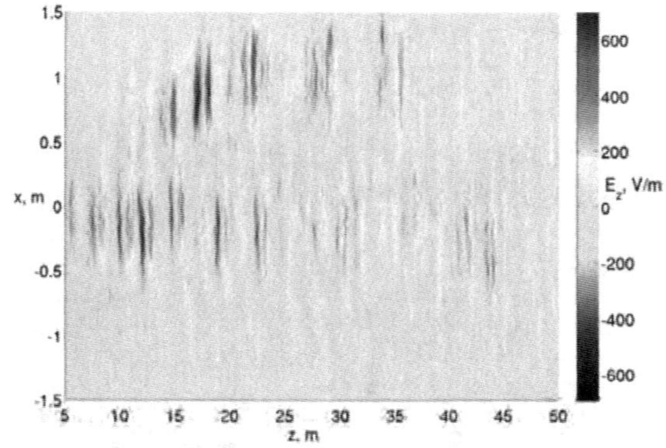

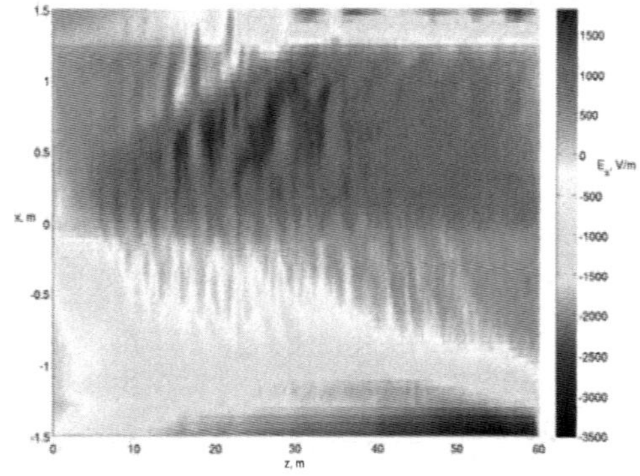

Figure 6. The turbulent electric field induced by the transit and bounced electron beams along Z axis $\tilde{E}_z$ (the left plot) and X axis $\tilde{E}_x$ (the right plot) induced by particles in PIC simulation ($B_{z0} = 10^{-3}$ T, $B_{y0} = 10^{-4}$ T, $B_{x0} = 4 \cdot 10^{-5}$ T, $E_{y0} = 250$ V/m, $m_p/m_e = 10$, $n = 10^{12}$ m^{-3}.

3.5. Pitch-angle distributions of particles at ejection from a current sheet

There are interesting pitch angle distributions of accelerated particles with respect to B_z after their ejection from the simulation region shown in Fig. 5. Pitch angle dispersion of accelerated particles increases the case of a strong guiding field (Fig. 5, the right column of plots) while being rather narrow (focused along the component B_z) (Fig. 5, the left column of plots).

At the same time, the decrease of the transverse magnetic field B_x (Fig. 5, the bottom row plots) leads to wider pitch angle distributions compared to a higher B_x (located far away from the X-nullpoint) (Fig. 5, the top row plots).

3.6. Turbulence generated by bounced and transit beams

As it can be seen from the PIC simulation shown in Fig. 6, we detect formation of turbulence with the induced electric field components, $\tilde{E}_z$ and $\tilde{E}_x$ being structured with a characteristic length scale of about $\lambda_{wave} \approx 2m$ for the conditions of the solar corona or $2 \times 10^5 m$ for the heliosphere at 1 AU. This structure propagates in time in the positive direction of the z axis.

The speed of this propagation is about $V_{wave} \approx 1.3 \cdot 10^7 m/s$, which makes the temporal period of oscillations about $T_{wave} \approx 1.5 \cdot 10^{-7}$ s. Since the plasma frequency in the corona for $n = 10^{12}$ m^{-3} is $9.1 \cdot 10^6$ Hz, the generated wave is the Langmuir wave (see Fig. 6). The same applies to the conditions of the heliosphere. Note also that the oscillating component of the excited wave $\tilde{E}_z$ is parallel to the direction of propagation, which indicates a linear polarisation of Langmuir wave generated.

The plots a V_z distribution function of electrons (see Fig. 7) in the region where the instability occurs (some rectangular region in Fig. 6) indicates that the electrons have the typical unstable distribution with "bump-in-tail". The range of velocities V_z for which the derivative df/dV_z is positive is from $1.3 \cdot 10^7$ to $2 \cdot 10^7$ m/s, these values correspond to the phase velocity of the Langmuir wave V_{wave} deduced earlier.

The generation of Langmuir waves is not the only possible instability to occur in an RCS but can also lead to excitation of whistler or ion-acoustic waves as was shown by Shay et al [16, 17]. This statement is supported by the distributions of accelerated protons (see Fig. 7, red lines) showing a presence of 'bump-in-tail in the proton distributions.

However, in our current PIC simulations we can only observe the waves with the length scales less than the size of the simulation domain. Also, since the current simulation system is 2D in space (and 3D in velocities), some ion waves, which are essentially 3D cannot be generated in

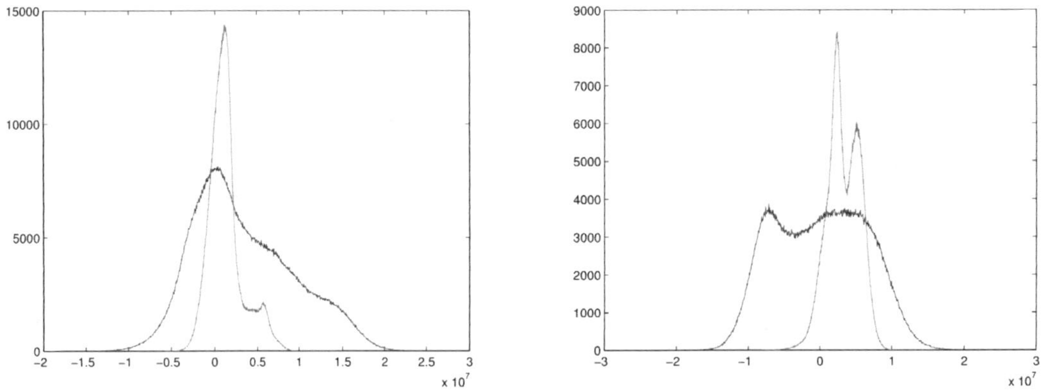

Figure 7. Density distributions (Y axis, arbitrary units) of electrons (blue) and protons (red) versus velocities (X axis, m/s) showing the double maxima defined by the difference in acceleration scenario for transit and bounced by electrons/protons. These distributions lead to the two beam instability and formation of turbulence, examples of which (formed by the transit and bounced electron beams) are shown in Fig. 6.

the current model. The investigation of other instabilities, apart from Langmuir waves, can be a scope of the forthcoming study.

4. Conclusions

The PIC simulation has shown that accelerated particles produce a strong polarisation field, $\tilde{E}_x$, caused by the separation of accelerated protons from electrons across the current sheet midplane. This polarisation field is shown to have much higher magnitudes that the reconnection electric fiel that leads to essential modification of the trajectories of electrons, and, to some extent, of protons.

The polarisation electric field can be essentially skewed towards one or another semiplane, with respect to the midplane, where the magnetic field components are weaker, or become strongly asymmetric, if the guiding field is strong and comparible with the main component B_z.

Electric polarisation field affects in different ways the "transit" particles (those which enter from and are ejected to the opposite semispaces towards the midplane) and "bounced" particles (those which enter from and are ejected to the same semispace towards the midplane). First, we note that the "bounced" protons have wider orbits of their gyration about the midplane than the "transit" ones. Second, the presence of polarisation electric field makes very peculiar velocity profiles of protons and ions which follow the profile of this polarisation field.

Moreover, the electrons, which are "bounced" in the TP approach, in the PIC simulations are dragged by the polarisation field $\tilde{E}_x$ back towards the midplane. Depending on the magnetic field magnitudes, ratio of the component and the ambient plasma density, the bounced electrons can either reach the midplane (in the case of the coronal parameters)[12] or are turned back from a current sheet (in the case of the HCS parameters) [13].

At the same time, the "transit" electrons, which are ejected to the opposite semispace from protons in the TP approach travel from the current sheet to a distance up to 10 (or larger) sizes of RCS thickness and then are dragged back to the RCS by the polarisation field $\tilde{E}_x$. This process results in formation of electron clouds around current sheet midplane, which exist as long as protons are being accelerated on that midplane. As result, in the PIC model with a weaker polarization field the protons are often found ejected along with electrons to the same side from the midplane, contrary to the full separation detected in pure TP simulations.

Electrons and protons gain energy from the current sheets forming power law energy spectra with either the close spectral indices (for magnetic field topologies with strong guiding field) or with electrons spectral indices being by unity plus higher than those of protons.

Both types of particles, electrons and protons, produce turbulence cause by two beam instabilities (transit and bounced beams) which moves with the beam outside from the X-nullpoint.

These characteristics of energetic particles gained during their passage of a reconnecting current sheets can explain many puzzling incosistencies in the in-situ observations of the solar wind in the vicinity of the heliospheric current sheet [13] and during passage of interplanetary Coronal Mass Ejections [18].

The current model is concentrated on the parameters of accelerated particles in a single current sheet while observations often show that current sheets can be broken into a number of magnetic islands [14, 18]. The investigation of more complex models with magnetic islands will be done in the forthcoming paper.

[1] Speiser T W 1965 *Journal of Geophysics Research* **70** 4219–+
[2] Litvinenko Y E and Somov B V 1993 *Solar Physics* **146** 127–133
[3] Litvinenko Y E 1996 *Astrophysical Journal* **462** 997–+
[4] Zharkova V V and Gordovskyy M 2004 *Astrophysical Journal* **604** 884–891
[5] Zharkova V V and Gordovskyy M 2005 *Space Science Reviews* **121** 165–188
[6] Wood P and Neukirch T 2005 *Solar Physics* **226** 73–95
[7] Oreshina A V and Somov B V 2006 *Astronomical and Astrophysical Transactions* **25** 261–273
[8] Zharkova V V and Gordovskyy M 2005 *Monthly Notices of the Royal Astronomical Society* **356** 1107–1116
[9] Zharkova V V and Agapitov O V 2009 *Journal of Plasma Physics* **75** 159–+
[10] Birn J, Drake J F, Shay M A, Rogers B N, Denton R E, Hesse M, Kuznetsova M, Ma Z W, Bhattacharjee A, Otto A and Pritchett P L 2001 *Journal of Geophysics Research* **106** 3715–3720
[11] Tsiklauri D and Haruki T 2007 *Physics of Plasmas* **14** 2905–+ (*Preprint* `arXiv:0708.1699`)
[12] Siversky T V and Zharkova V V 2009 *Journal of Plasma Physics* **75** 619–+
[13] Zharkova V V and Khabarova O V 2012 *Astrophysical Journal* **752** 35
[14] Drake J F, Swisdak M, Che H and Shay M A 2006 *Nature* **443** 553–556
[15] Priest E and Forbes T 2000 *Magnetic Reconnection* (Magnetic Reconnection, by Eric Priest and Terry Forbes, pp. 612. ISBN 0521481791. Cambridge, UK: Cambridge University Press, June 2000.)
[16] Shay M A, Drake J F, Denton R E and Biskamp D 1998 *Journal of Geophysics Research* **103** 9165–9176
[17] Shay M A and Drake J F 1998 *Geophysics Research Letters* **25** 3759–3762
[18] Khabarova O, Zank G, Li G, le Roux J A, Webb G M, Dosch A, Malandraki O E and Zharkova V V 2015 *AAS/AGU Triennial Earth-Sun Summit* (*AAS/AGU Triennial Earth-Sun Summit* vol 1) p 20501

Journal of Physics: Conference Series **642** (2015) 012033 doi:10.1088/1742-6596/642/1/012033

Dynamical small-scale magnetic islands as a source of local acceleration of particles in the solar wind

Olga V. Khabarova[1], Gary P. Zank[2,3], Gang Li[2], Jakobus A. le Roux[2,3], Gary M. Webb[2], Olga E. Malandraki[4], Valentina V. Zharkova[5]

1. Heliophysical Laboratory, Pushkov Institute of Terrestrial Magnetism, Ionosphere and Radiowave Propagation RAS (IZMIRAN), Troitsk, Moscow 142190, Russia
2. Center for Space Plasma and Aeronomic Research (CSPAR), University of Alabama in Huntsville, Huntsville, AL 35805, USA
3. Department of Space Science, University of Alabama in Huntsville, Huntsville, AL 35899, USA
4. IAASARS, National Observatory of Athens, GR-15236 Penteli, Greece
5. Department of Mathematics and Information Sciences, Northumbria University, Newcastle upon Tyne, NE2 1XE, UK

E-mail: habarova@izmiran.ru

Abstract. We present observations of energetic particle flux increases up to 1 MeV at 1 AU, which cannot be associated with ordinary mechanisms of particle acceleration, such as acceleration at shocks or at the Sun. Such unusual energetic particle events very likely have a local origin. Multi-spacecraft observations show that numerous cases of energetic particle flux enhancements and spikes correspond to passages of spacecraft through areas filled with magnetic islands with a typical width ~0.01–0.001AU that experience dynamical merging or/and contraction. The presence of magnetic islands inside magnetically confined cavities in the solar wind may lead to local particle energization, especially in the case when the particles have already been pre-accelerated to keV energies, for example, at shocks or due to magnetic reconnection at the heliospheric current sheet. We consider different magnetic configurations that provide favourable conditions for both the appearance of small-scale magnetic islands and their confinement.

1. Introduction

Small-scale magnetic islands with a typical size of l ~0.01 AU or less are commonly observed in the solar wind plasma at 1 AU. The crossing of a certain magnetic island lasts from minutes to hours, but, in contrast to large-scale magnetic islands (or magnetic clouds), small-scale magnetic islands exhibit some grouping. Areas filled with small-scale magnetic islands, sometimes called flux ropes, can be observed for days [1]. The interplanetary magnetic field (IMF) strength increases and the plasma density decreases inside magnetic islands. The main features of these structures are a rotating magnetic field and the presence of small-scale current sheets at their borders.

Magnetic islands are often related to magnetic reconnection. The probability of their observation increases as the heliospheric current sheet (HCS) is approached [1], which is not surprising. Indeed, there are some signatures of magnetic reconnection that re-currently occur in many places on the HCS and determine specific properties of the plasma and magnetic field in the vicinity of the HCS [2].

Recently, we showed that magnetic islands near the HCS can experience merging and contraction [3], which may serve as a source of particle acceleration (as discussed in theoretical studies [4, 5, 6]). Magnetic island merging dominates near the HCS if it is undisturbed by high-speed flows or corotating interaction regions (CIRs). A specific rippled profile of the HCS confines magnetic islands and forms a tokamak-like structure that provides the possibility of additional energization of pre-accelerated particles [3]. Such an approach solves the problem of hypothetical low effectiveness of particle acceleration directly at 1 AU. Particle acceleration by magnetic island dynamics seems to play a significant role, because some crossings of the HCS are associated with energetic particle flux enhancements up to 1 MeV, which can be observed for many hours around the sector boundary crossing. We supposed that the energy gained by particles depends both on the reconnection rate and the local profile of the HCS. The reconnection rate increase, which produces more magnetic islands, may occur due to the impact of interplanetary coronal mass ejections (ICMEs) or corotating interaction regions (CIRs). The small-scale wavy, rippled and distorted HCS confines magnetic islands and accelerate particles more effectively than the plain or large scale wavy-shaped HCS. The combination of these features can produce strong particle acceleration.

Contraction of magnetic islands can take place everywhere, because the solar wind is a dynamic medium, but this process is most effective during HCS-ICME interactions. In this case there is (i) strong compression: (ii) plasma confinement between the ICME front and the HCS, and (iii) intensification of magnetic reconnection at the HCS due to its interaction with the ICME. As a result, one can observe strong energetic particle flux enhancements, sometimes comparable with energies typical for strong solar energetic particle SEP events [3].

The way to identify small-scale magnetic islands was shown in [1, 7]. The Grad-Shafranov reconstruction of flux ropes can help with understanding the local structure of magnetic fields (see ftp://ftp.iwf.oeaw.ac.at/pub/moestl/publicgscode/GS_handbook_june_2014.pdf and [8]), but it has some limitations, is not easy to perform, and sometimes is not exact. At the same time, a hodogram method [3] can help by providing a quick visual inspection of the analyzed events, simply using regression plots to reveal the IMF rotation inside magnetic islands. This supposes that if the rotation is obvious, one can see a half-circle in one of the planes B_r-B_t, B_t-B_n or B_r-B_n during the crossing of one island. Here and below, indices r, t and n correspond to radial, tangential and normal directions in the RTN coordinate system. At 1 AU, $r = -x$ in the GSE system, which is related to the ecliptic plane. The crossing of two islands or a chain of islands usually gives a full circle or many circles. The anti-correlation of a pair of the IMF components in combination with changing correlation between the density and the IMF during several hours may indicate a passage through the chain of magnetic islands as well.

We provide here support for theory and simulations [4, 5, 6], showing signatures of particle energization related to the occurrence of magnetic islands inside specifically confined regions of expanding solar wind. An explanation of observed time-intensity profiles of energetic particle fluxes through particle energization in merging/contracting magnetic islands for the case of the HCS-ICME interaction as well as for the crossing of the isolated HCS can be found in the recent publication [3]. In this work we will show cases of HCS-CIR and unusual HCS-ICME interactions. The shock fronts of CIRs which have fully developed beyond ~ 2 AU have been typically considered as the main source of energetic particle enhancements observed at 1 AU, with the particles, associated with the Reverse CIR shock beyond the spacecraft to be streaming towards the Sun [9, 10].

Magnetic islands are often observed in the turbulent area behind interplanetary shocks, which may represent a source of additional acceleration of particles in this area. Theoretical aspects of this phenomenon are considered in [4, 6]. Observations show that interplanetary shocks are usually accompanied by current sheets, and the shock itself always coincides with a sharp change in the IMF clock (azimuthal) angle. We will discuss a case when an interplanetary shock was observed close to the HCS and smaller current sheets.

As a whole, the work is aimed at showing the importance of taking into account the local structure of the IMF and the presence of dynamical small-scale magnetic islands, which are poorly-investigated natural accelerators of particles at the Earth's orbit.

2. Observations of energetic particle flux enhancements associated with the occurrence of small-scale magnetic islands

In our previous study [3] we showed that energetic particle flux increases are observed simultaneously with the occurrence of magnetic islands near the HCS. Here we will extend our multi-spacecraft study to cases when magnetic islands occur in between magnetic walls (represented by current sheets) and plasma boundaries (CIRs, ICMEs, shocks).

The local configuration of the IMF at 1 AU determines acceleration processes to a higher degree than is usually expected [3]. Therefore, it is important to analyse the whole picture of the ICME (or CIR) interaction with the HCS. The SMEI (Solar Mass Ejection Imager) can provide us with the necessary nearly whole sky data for the period from 2003 to 2011. A detailed description of the restored velocity/IMF plots obtained from STELab interplanetary scintillation data as well as the solar wind density plots from SMEI will be given below. The role of current sheets in the confinement of magnetic islands and the impact of high density (or the total pressure) on observed time-intensity profiles of energetic particle fluxes will be discussed as well.

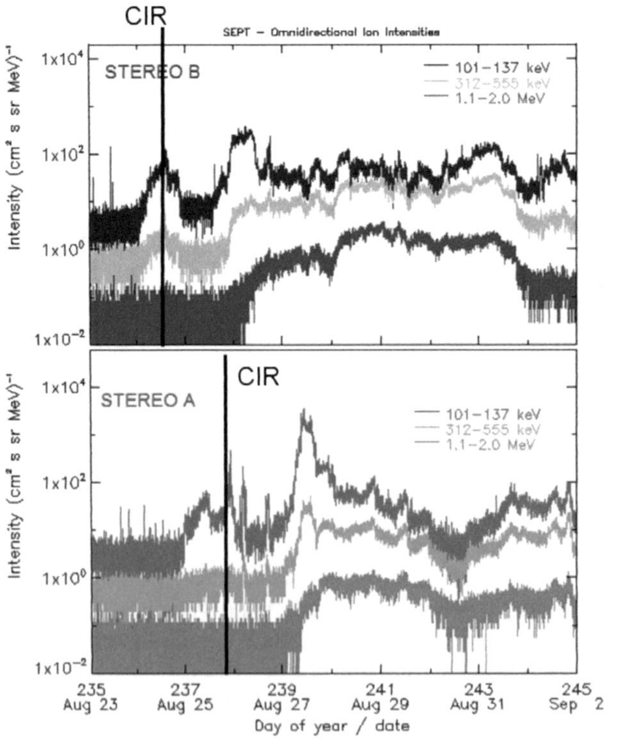

Figure 1. An increase in ion flux with energies up to 500 keV before the CIR as measured by STEREO A and STEREO B.

2.1. *The HCS – CIR interaction*

Let us explore the role of local CIR-associated particle acceleration on the formation of time-intensity profiles of suprathermal particle fluxes. We investigate whether the bounding of HCS- associated magnetic islands by the HCS from one side and the CIR front from the other (which also contains local current sheets) may produce significant particle acceleration.

The energetic particle flux increases in August 2007 observed by the STEREO pair separated by 27 degrees (figure 1) were analyzed in [11]. These were not classical SEP events due to flares or acceleration at the ICME-related shock, but flux enhancements related to CIRs. The interesting point is that energetic particle flux enhancements were observed both before the first CIR's approach and between the two CIRs. The CIRs originated from (1) a long-lived low-latitude coronal hole, and (2) weak CMEs that did not hit the Earth directly (details can be seen in **1.avi** and **2.avi**). A classic explanation of the event, at which particles are supposed to be accelerated at reverse/forward shocks behind the Earth, was given in [11].

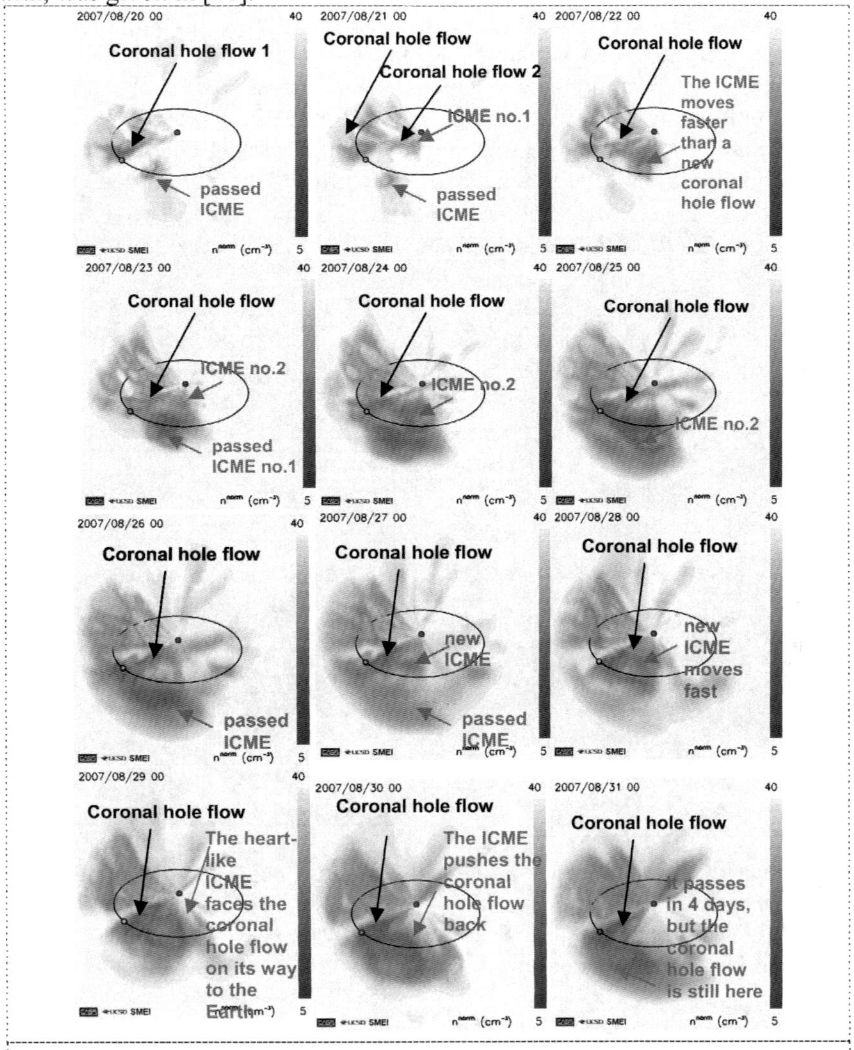

Figure 2. Daily snapshots of the solar wind density from the Sun to the Earth as observed by SMEI for the time period corresponding to Figure 1. The energetic particle flux profiles shown in Figure 1 are determined by interaction of the coronal hole flow and ICMEs that produce CIRs detected by the STEREO pair. The Earth's orbit is shown with the black ellipse, the Earth is the blue dot, and the Sun is the red thick dot in the centre. See 2.avi movie.

We are mainly interested here in the nature of the ion flux enhancements that occurred before the first CIR, which was not examined in [11]. The leading edge of the CIR, indicated by vertical lines in Figure 1, was observed in sequence by STEREO B and STEREO A. The separation angle between the STEREO pair was not very large, but, in contrast to typical SEP events, CIR-related energetic particle increases were detected by STEREO B and STEREO A with a time delay of many hours that corresponded to the rotation of the structure from the first spacecraft to the second one. Therefore, the observed changes in time-intensity energetic particle flux profiles were apparently determined by local configurations of the IMF.

A spatial distribution of the IMF from the Sun to the Earth's orbit may be reconstructed for the investigated period from the analysis of ground-based interplanetary scintillation data from the Solar Terrestrial Environment Laboratory (STEL), because the velocity profile approximately corresponds to the picture of the IMF spatial variations (see **1.avi**). One can find the corresponding 3D tomographic reconstructions performed with a 102-minute cadence and resolutions of 0.05 AU in height and 1°x1^{o} in latitude and longitude on the page of The Solar Mass Ejection Imager (SMEI): http://smei.ucsd.edu/new_smei/data&images/data&images.html. SMEI itself viewed nearly the whole sky in visible light and provided the solar wind density profiles (for example, see **2.avi**). Generally, the solar wind density and velocity/IMF plots show different patterns: the plasma flows nearly radially outward from the Sun (**2.avi**), and the velocity/IMF pictures exhibit both rotation and expansion (**1.avi**).

During the period shown in figure 1, long-lived low latitude coronal holes were found side by side with active regions that produced weak coronal mass ejections (see the corresponding images at http://spaceweather.com/ and CME movies at http://sidc.oma.be/cactus/catalog.php). As a result, the flows interfaced in the solar wind and formed a sequence of CIRs and confined regions having delta (or nabla) shapes.

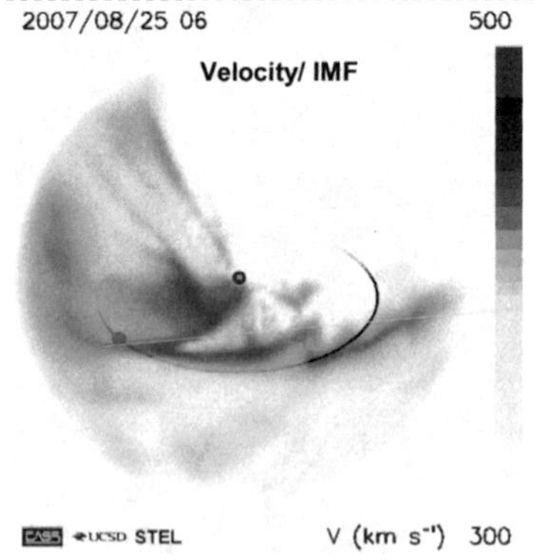

Figure 3. STEL measurements of the solar wind speed. ∇-shape cavities in the solar wind. See the 1.avi movie.

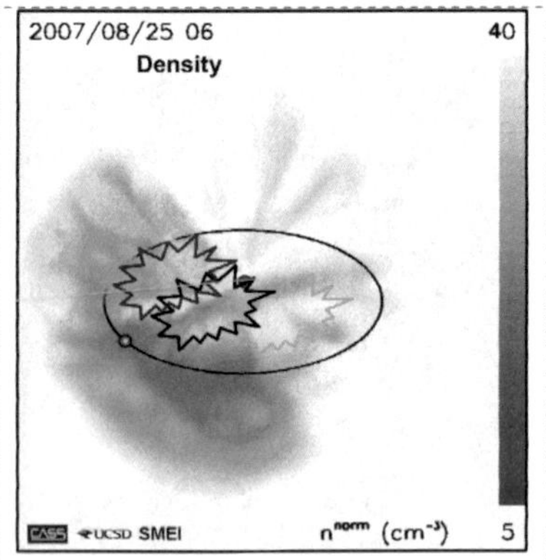

Figure 4. SMEI measurements of the solar wind density profiles that correspond to the speed/IMF cavities in figure 3. See the 2.avi movie.

The ∇-shape IMF cavities can be seen in the **1.avi** movie. Most streams are flowing from low-latitude coronal holes and sometimes face weak ICMEs. The streams of both kinds have different shapes, different speeds and different ways of propagation from the Sun to 1 AU. ICMEs propagate nearly radially and look like expanding balls (see figure 2), and streams from coronal holes resemble expanding tubes. As a result, at least three unusual ∇-shape IMF structures can be seen between the Sun and the 1 AU sphere at the same time (figure 3). Note that the upper side of ∇ is closer to the Sun than the narrow part, in contrast with a common view of the flows as V-shape structures that expand from the Sun [11].

The borders of each cavity shown in figure 3 are formed by the HCS and current sheets that belong to two different CIRs - passed and arriving ones (see the **1.avi** movie and figure 2). As we show below, such a magnetic cavity can confine dynamically changing magnetic islands, which results in local particle acceleration.

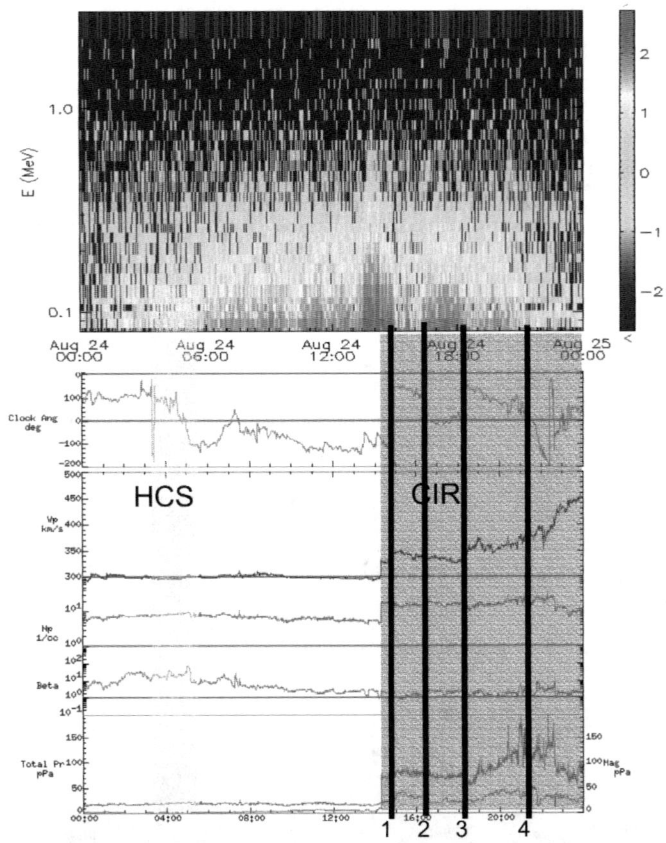

Figure 5. STEREO B observations showing the HCS crossing before the CIR. From top to bottom: energetic ion flux spectrogram, the IMF clock (azimuth) angle, the solar wind speed, the density, the plasma beta, the total (green) and magnetic (red) pressure. Energetic particle flux enhancements are seen in between the HCS (yellow stripe) and the CIR (pink stripe) as well as between current sheets inside the CIR (vertical lines).

Figure 2 shows SMEI measurements of the solar wind density in the visible light. SMEI was designed to observe ICMEs. It can show dense borders of ICMEs or flows from coronal holes (which are stream interaction regions – SIRs or CIRs). It cannot look inside ICMEs until they crossed the Earth orbit. CIRs from coronal holes are stable and conic-like, but ICMEs (even from weak CMEs), move faster, and produce ball-like "CIR-shells". Sometimes an ICME can move slower than plasma inside a coronal hole flow, but since the latter has a permanent source at the Sun, the entire picture looks like the ICME moves against the background of a stable slowly rotating stream. The changing density is partially responsible for what we see in blue velocity/IMF pictures (compare figure 3 and figure 4). On the other hand, velocity determines the IMF shape, so the whole picture of a particular event can be achieved by analysing both the combination of density and velocity plots.

Let us examine in detail the pre-CIR energetic flux enhancement shown in figure 1. Subsequent crossings of the HCS and the CIR are shown in figure 5. The IMF changed its direction approximately 10 hours before the CIR arrival (see the corresponding changes of the IMF clock (azimuth) angle). This was accompanied by a very high plasma beta and an increased plasma density in a background of low velocity. The yellow stripe shows the plasma sheet. The pink stripe indicates the CIR body. STEREO B detected a weak shock first, followed by significant density and velocity increases. The upper panel shows the energy of omni-directional ions (spectrogram).

Remarkably, the energetic ion flux enhancements occurred between the HCS and the CIR. Spikes in the spectrogram correspond to local magnetic islands. Even more remarkable is that the maximum increases in the flux are observed not exactly at the CIR-related shock, but in between local current sheets. Most strong current sheets inside the CIR are identified by numbers.

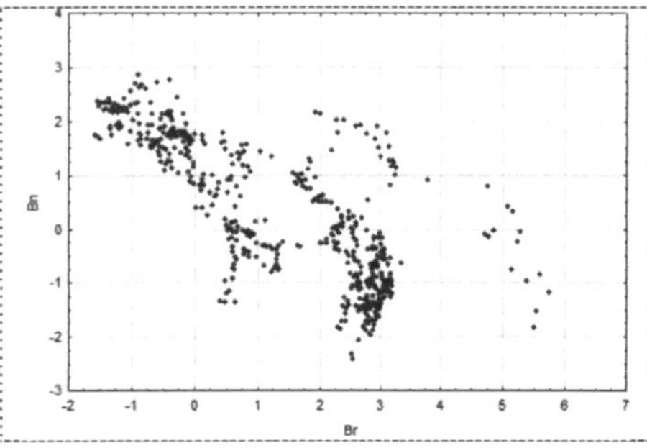

Figure 6. Hodogram of the IMF vector rotation in the B_n-B_r plane during the period of observations from the HCS crossing to the CIR crossing. 1-min STEREO-B IMF data, nT.

The rotation of the IMF vector inside magnetic islands located in the magnetic cavity between the HCS and the first strong CIR-related current sheet is shown in figure 6. In figure 5, one can see that the second prominent increase of the energetic ion flux occurs between the second and the third current sheet inside the CIR.

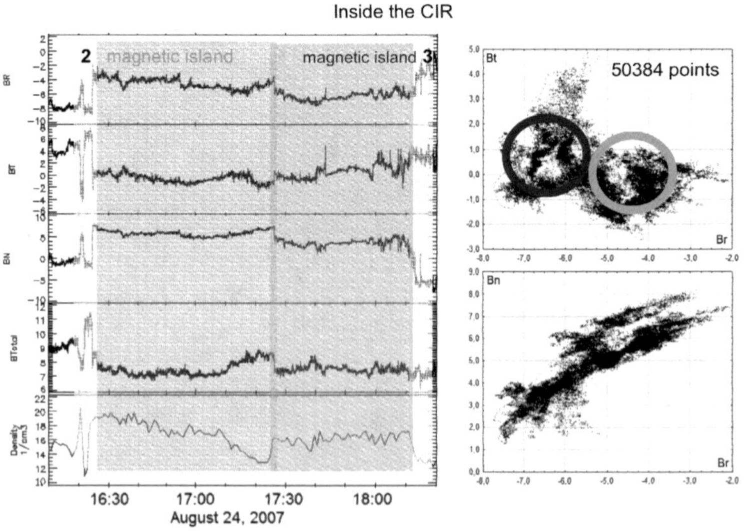

Figure 7. The IMF and density properties during the ion flux increase observed between current sheets 2 and 3 (see figure 5). Left figure, from top to bottom: three components of the IMF, the total B and the solar wind density. The right panel: similar to figure 6, but for R-T and R-N planes. High resolution data from STEREO B.

The behaviour of three IMF components, the total B and the solar wind density inside the magnetic islands as well as in the current sheets is shown in the left panel of figure 7. In the right panel of

Figure 7 we plot the hodogram of the IMF that shows the magnetic field vector rotation in two planes (B_r-B_t and B_n-B_r) inside two relatively large magnetic islands identified by the green and blue shading. The current sheets numbered 2 and 3 in figure 5 are shown in detail and marked with yellow stripes in figure 7. The magnetic islands are separated by a small-scale current sheet as well (the dark blue stripe).

The ACE spacecraft was located at the Lagrangian point in front of the Earth, and the separation angle with STEREO B was of 12°. ACE detected the passage of the whole structure several hours later. The HCS crossing, which is shown by the yellow stripe in figure 8, lasted longer in comparison with STEREO B, but the energetic ion flux behaviour was very similar. There was an area filled with magnetic islands in between the HCS and the first current sheet that belonged to CIR (line 1 in figure 8). An ion flux increase is observed in this area. Energetic ion fluxes measured by LEMS 120 on ACE show an increase many hours earlier than the CIR arrival at the Earth. The maximum is related rather to current sheet 1 than to the CIR shock.

STEREO A detected the CIR later than STEREO B, but the ion flux increase observed before the CIR leading front is even clearer and lasts longer (see figure 1). The bottom panel of figure 1 shows the omni-directional ion fluxes. However, if looked at more closely, the ion flux exhibits strong anisotropy. The upper panel of figure 9 shows fluxes of ions directed from the Sun. The bottom panel shows variations of the sunward ion flux. The HCS/plasma sheet crossing is indicated by the yellow stripe and the CIR shock is marked with the vertical line.

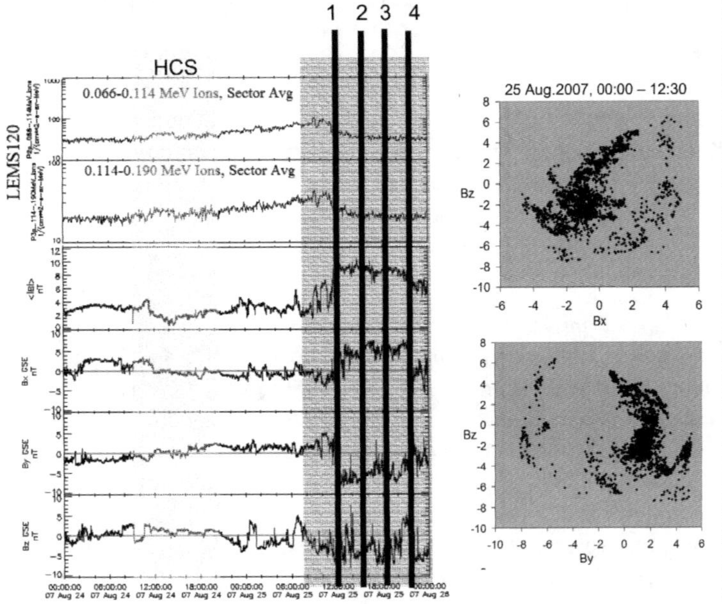

Figure 8. ACE observations during the event shown in figure 5. Left panel: energetic particle fluxes and the IMF components. Right panel: a hodogram of magnetic field rotation in the area between the HCS and the CIR (similar to figure 6 and figure 7, right panel).

The time-intensity ion flux profile is similar to that discussed above: an energetic ion flux increase is observed between the HCS and the leading edge of the CIR. It is clear that the first increase in energetic ion flux is due to 101-137 keV anti-sunward ions. This would not be possible if the particles originated from the CIR forward shock, as the dominant diffusive shock acceleration paradigm asserts, given that the shock is located at roughly 2-3 AU. Sunward accelerated ions, which could be from the forward shock, were detected many hours later. The interesting point is that the maximum sunward ion flux occurs after the shock crossing, which may be well related to results discussed in [6].

In summary, for the case that there are magnetic islands in the region in between the HCS and the CIR, the HCS-CIR interaction may produce energetic particle flux enhancements comparable with those observed during solar energetic particle (SEP) events. In the case discussed here, the magnetic islands were confined by the HCS and CIR-associated current sheets and compressed from at least one side by the approaching CIR.

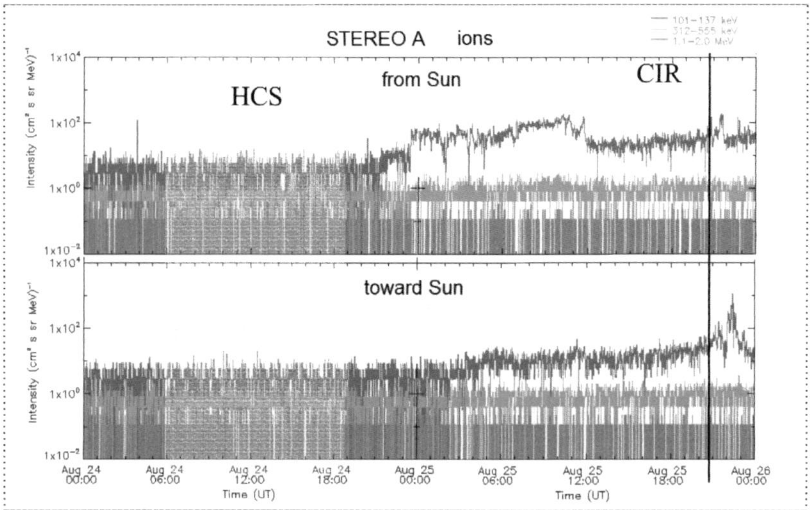

Figure 9. STEREO A observations of energetic particle flux enhancement before the CIR arrival. There was an approximately one day delay between the arrival of accelerated anti-sunward and sunward ions at the Earth.

2.2. The HCS-ICME interaction.

The HCS-CIR case examined above is rather similar to the HCS-ICME case discussed in [3]. The observational evidence points to the same underlying paradigm for particle acceleration that is favored because particles and magnetic islands are spatially confined by the HCS the approaching CIR or ICME. Two cases of the HCS-ICME interaction were considered in [3], terms of possible scenarios [12] of (i) the pre-existing HCS and (ii) the post-ICME crossing of the HCS, which was quickly restored after the passage of the ICME.

Sometimes we can observe at 1 AU very complicated cases with multiple CMEs that impact the HCS so strongly that it can no longer possesses a simple anymore, being rippled, and disrupted to a high degree or reconnected along the leading front of ICME. In STEL 3D velocity/IMF plots this appears as a temporary absence of the HCS in some places or the presence of multiple current sheets that resemble rose-leaves (see the **3.avi** movie). Let us consider cases when the HCS structure was very complicated due to a sequence of strong ICMEs that separated the HCS into many single current sheets.

The list of ICME events observed by STEREO A and B can be found at http://www-ssc.igpp.ucla.edu/forms/stereo/stereo_level_3.html . Event number 6 is an ICME detected by STEREO A on November 19, 2007. The magnetic obstacle (~ flux rope) starting time is indicated as 22:00 UT (the left vertical line in figure 10), and the end time is on November 21, 2007, 3:17 UT (the right line in figure 10). This period approximately corresponds to the growth and the main increase in the total pressure (figure 10, bottom panel).

The classification of the event as a clear ICME is arguable, because the velocity increase occurs after the "official" end of the magnetic cloud identified by Lan Jian, which is not typical for ICMEs, though the IMF rotation in the magnetic cloud is very clear. However, one can see that neither the very high solar wind density nor the total pressure enhancement is associated with the energetic particle flux increase. On the contrary, two strong energetic ion flux enhancements are observed both before and after the passage of the large ICME-associated magnetic cloud characterised by the increased IMF strength and the solar wind density.

The first enhancement is most interesting, because there is absolutely nothing remarkable in the behaviour of parameters, which are usually considered to be responsible for particle acceleration: the solar wind speed decreases after the passage of the previous ICME, and there is no strong density increase during the energetic particle flux enhancements in different energy ranges as figure 10 shows. At the same time, the clock angle is changing frequently (the third panel from the top), which indicates the presence of multiple current sheets.

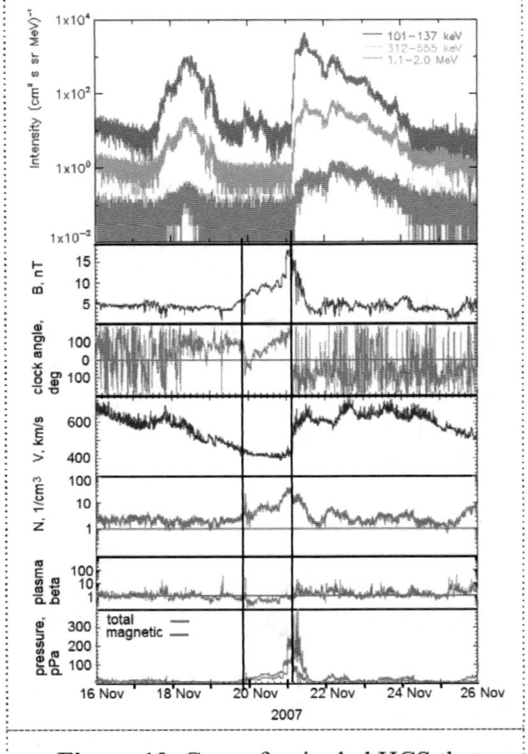

Figure 10. Case of a rippled HCS that surrounded an ICME. The energetic particle flux enhancements are observed not during the ICME, but in the areas filled with magnetic islands.

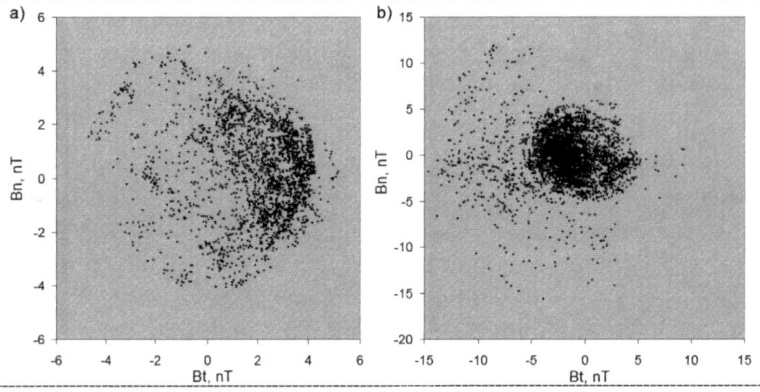

Figure 11. Rotation of the IMF that coincides with two strong energetic particle flux enhancements shown in figure 10 . STEREO A 1-min data.

We find that the observed frequent change of the IMF direction is due to the occurrence of multiple ripples on the HCS (see the **3.avi** movie). According to the 3.mpeg movie, the ICME was surrounded by the strongly disrupted and rippled HCS. The occurrence of small-scale magnetic islands inside the ripples is obvious from hodograms shown in figure 11.

Figure 11a corresponds to the first energetic ion flux increase seen in figure 10, and figure 11b shows the IMF vector rotation during the second energetic ion flux increase, which occurred after the passage of the ICME-related magnetic cloud. These enhancements were obviously local, because STEREO B detected them with some time shift, and the time-intensity profiles of the energetic ion fluxes were slightly different (not shown).

Do summarize, we emphasize again the importance of local IMF structure for confining and accelerating particles in the solar wind. In this case, particles accelerated at the ICME-related shock most likely could not propagate freely in the solar wind, being trapped by magnetic islands and confined by a very complicated HCS configuration. As a result, at 1 AU the picture became "inverted": areas where energetic ion flux enhancements are usually observed had no SEPs, but the surrounding HCS-related magnetic islands were full of accelerated particles.

2.3. The HCS-shock interaction

Interplanetary shocks (ISs) at 1 AU are usually associated with SIRs/CIRs or ICMEs, as seen from any shock list. An example can be found in paragraph 2.1. Thus ISs are followed by structures that may impact the HCS and change its configuration. This increases the reconnection rate at the HCS, produces more magnetic islands in some areas of the solar wind and, potentially, leads to local particle acceleration. The HCS is sensitive to the passage of an isolated interplanetary shock as well. Theoretically, the HCS is transparent for interplanetary shocks. At the same time, shocks may disrupt the HCS [13].

Observationally, it appears that the HCS tries to keep its form, but, being compressed by an interplanetary shock, it bends, moves forward and is then restored the shock passes. This process is clearly seen in STEL movies. If a spacecraft is close to the HCS, it detects the HCS crossing first, and then the IS crossing. The HCS restoration is seen as a second crossing of the HCS some time after the CS passage. This process has been poorly investigated observationally, although some simulations have been performed [13, 14]. Many questions regarding the state of the solar wind and energetic particles after IS passage throughout the HCS remain unanswered.

We know that ISs are followed by a turbulent wake that contains numerous magnetic islands [6, 14, 15]. If this is the case, the highly disturbed HCS may (i) retain the magnetic islands that are formed downstream of the IS, and (ii) produce more magnetic islands because of the increased reconnection rate. In any event, the traversal of the HCS by an IS may result in a prolonged highly turbulent state of the solar wind behind the shock (much more prolonged than in the case of an IS passage that does not interact with the HCS). An example of an IS interacting with the HCS is presented in figure 12.

STEREO A observed the crossing of an isolated interplanetary shock at 9:18 UT on May 19, 2008 (see the STEREO shock list at http://www-ssc.igpp.ucla.edu/forms/stereo/stereo_level_3.html). It is marked in figure 12 by the black vertical line. The solar wind density and speed indicate any presence of a structure that could contaminate the case significantly, except for the presence of the HCS. The main crossings of the HCS are indicated by the vertical red lines, and the areas of fluctuating IMF direction, which, similar to previous events, may be treated as the plasma sheet, are identified by yellow stripes.

The HCS crossing occurred several hours before the IS arrival at 1 AU. One can see the clock angle changes sharp and the plasma beta is very high (more than 100) during the HCS crossing. The IS pushed the HCS back and the spacecraft found itself again in the same sector as before the HCS crossing. The restoration of the HCS was detected several hours later.

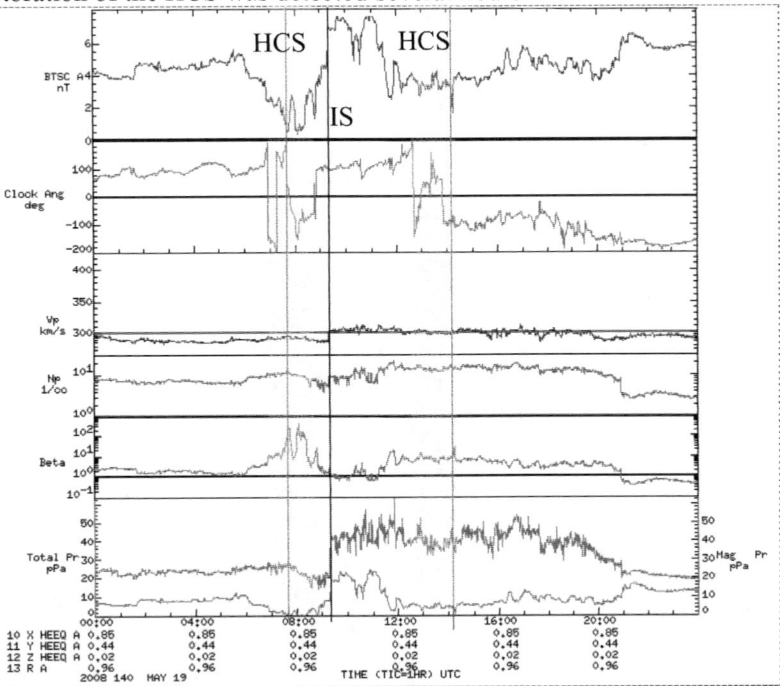

Figure 12. Plasma and magnetic field parameters for the case of the HCS-shock interaction (similar to figure 10). The shock is identified by the black line, and the main crossings of the HCS are indicated by red lines. STEREO A data.

The behaviour of suprathermal electrons (pitch angle distribution function PAD), the energetic ion dynamic spectrum, and the IMF component variations during the time period shown in figure 12 are illustrated in figure 13. The upper panel of figure 13 shows the running correlation coefficient between the tangential B_t and the vertical B_n components of the IMF (the time-window where the coefficient is calculated is 180 minutes, and the consequent shift is 1 minute).

The PAD function shows a wide area of fluctuations in the suprathermal electron propagation direction and a prolonged flux dropout that coincides with the HCS/plasma sheet crossing. It was discussed before [3] that such strong pitch angle scattering in the vicinity of the HCS is a signature of the occurrence of magnetic islands. The most prominent rotation of the IMF vector is observed in the T-N plane, and the negative correlation coefficient between B_t and B_n shows an approximate position of the islands along the time axis. It should be noted that periods of strong negative correlation coefficient coincide with the dropouts and periods of multidirectional propagation of suprathermal electrons.

Figure 13 clearly shows that the energetic ion fluxes increase most significantly behind the shock. Moreover, the maximum occurs exactly during the period of long-lasting electron flux dropout and pitch-angle direction isotropy observed after the second HCS crossing. This period is characterised by the presence of HCS-associated magnetic islands and rotating magnetic field as seen in figure 14.

This example shows that many unusual energetic particle events may be related to the interaction of the IS and the HCS due to the ability of the disrupted HCS to produce and confine magnetic islands in its vicinity. All the effects discussed in the theoretical models [4, 5, 6, 16] may be applied to this case, but the details of the HCS-shock interaction and peculiarities of particle acceleration that occur during this rather long process should be investigated further.

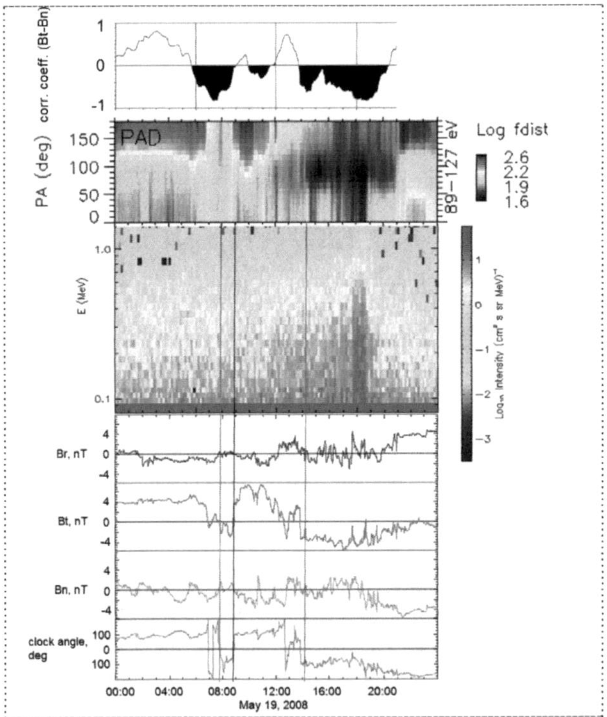

Figure 13. From top to bottom: The running correlation coefficient between B_t-B_n components, the suprathermal electron pitch-angle distribution function (spectrogram), the energy of ions (spectrogram). The main energetic ion flux enhancement occurs behind the shock (black line), which coincides with the electron heat flux dropout and corresponds to the largest magnetic island near the HCS. Measurements from STEREO A.

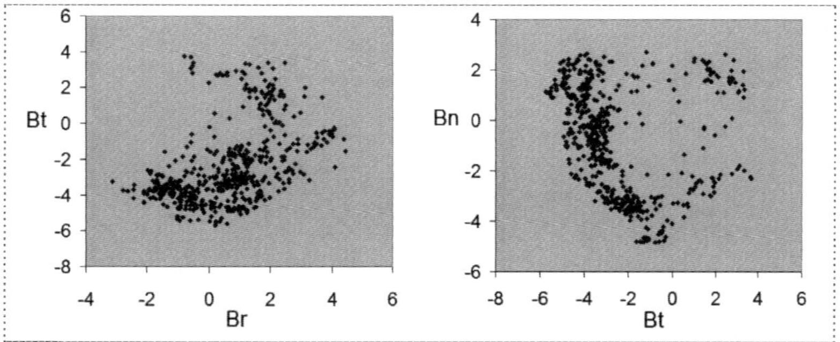

Figure 14. Hodogram of the magnetic field rotation behind the shock shown in figures 12 and 13.

3. Discussion and summary

In this case-study paper, we have shown examples of energetic particle enhancements observed near the Earth orbit that cannot be easily explained in terms of the dominant paradigm for the acceleration of particles to keV-MeV energies in the solar wind. The commonly accepted view for accelerating particles to these energies includes acceleration either at shocks or at the Sun. In both cases this occurs far from 1 AU, so an observer will see merely the result of various distant processes. However, numerous discrepancies between observations and the theoretical perspective stimulated us to seek other mechanisms that could explain energetic particle flux enhancements, as a result of local processes that occur directly at 1 AU and that can be traced by different spacecraft.

The application of a theory for particle energization by magnetic islands through their dynamical compression and merging [3, 4, 5, 6, 16] to observations introduces the possibility for solving many problems that traditionally arise in addressing the particle acceleration in the solar wind. Recently, we have shown that the heliospheric current sheet is a place near which magnetic islands permanently exist and evolve, experiencing both contraction and merging [3]. Both processes may occur simultaneously, but the dominant evolution depends on the local IMF configuration and such circumstances as approaching ICMEs, interplanetary shocks or CIRs. All shown examples of the HCS interaction with various structures suppose strong compression of magnetic islands either locked between two current sheets (the HCS and the strongest current sheet inside the ICME or the CIR) or produced by the IS impact.

In summary, the presence of the HCS near ICMEs, CIRs and ISs increases the probability of observing energetic particles accelerated up to 1 MeV.

(i) We have found evidence for significant particle energization in the ∇-shape magnetic cavity formed due to a stream interface. Single-point measurements show energetic ion flux enhancements that occur in an area filled with magnetic islands in between the HCS and the strongest current sheet inside the CIR.

(ii) Acceleration of particles to keV-MeV energies appears to occur inside CIRs as well if there are magnetic islands in between CIR-associated current sheets.

(iii) An increased solar wind density/total pressure is not a factor that determines strong particle acceleration (as seen from the HCS-ICME interaction case).

(iv) We confirmed that the occurrence of magnetic islands near the heliospheric current sheet results in electron pitch-angle scattering and prolonged heat flux dropouts.

(v) The most intensive energetic ion flux enhancements are observed not exactly at interplanetary shocks, but somewhere behind them. The maximum energetic ion flux increase is associated with current sheets surrounded by magnetic islands in the CIR case and with the post-shock restoration of the HCS and following magnetic islands in the case of an isolated shock.

(vi) The presence of the HCS makes the turbulent area behind the interplanetary shock wider because it keeps and additionally produces magnetic islands in its vicinity.

3.1. Acknowledgements

SMEI and STELab data were taken from the official website http://smei.ucsd.edu/new_smei/data&images/data&images.html. The corresponding movies (1.avi, 2.avi and 3.avi) can also be downloaded from http://files.olgakhabarova.webnode.com/200000049-c00d8c108d/1.avi, http://files.olgakhabarova.webnode.com/200000050-49c684ac0c/2.avi, http://files.olgakhabarova.webnode.com/200000051-3245f343a0/3.avi. The Solar Electron and Proton Telescope STEREO data were obtained from the http://www2.physik.uni-kiel.de/STEREO/index.php (University of Kiel, Germany), and other STEREO data were provided by the STEREO science center: http://stereo-ssc.nascom.nasa.gov/data.shtml. High-resolution ACE data were taken from the official Goddard Space Flight Center OMNIweb plus Web site: http://omniweb.gsfc.nasa.gov.

We acknowledge the partial support of NASA grants NNX08 AJ33G, Subaward 37102-2, NNX09AG70G, NNX09AG63G, NNX09AJ79G, NNG05EC85C, Subcontract A991132BT, NNX09AP74A, NNX10AE46G, NNX09AW45G, and NNX14 AF43G, and NSF grant ATM-0904007. O.V.K. was supported by RFBR grant 14-02-00769 and partially by RFBR grant 14-02-00308. JAlR acknowledges support from NASA grants NNX14AF43G and NNX15AI65G. O.E.M. received funding from the European Union's Horizon 2020 research and innovation program under grant agreement No 637324.

References

[1] Cartwright M L and Moldwin M B 2010 *J. Geophys. Res.* **115** A08102

[2] Zharkova V V and Khabarova O V 2012 *Astrophys. J.* **752** 35

[3] Khabarova O, Zank G P, Li G, le Roux J A, Webb G M, Dosch A, Zharkova V V and Malandraki O E 2015 *Astrophys. J.* in press

[4] Zank G P, le Roux J A, Webb G M, Dosch A, and Khabarova O 2014 *Astrophys. J.* **797** 28

[5] le Roux J A, Zank G P, Webb G M and Khabarova O 2015 *Astrophys. J.* **801** 112

[6] Zank G P, Hunana P, Mostafavi P, le Roux J A, Li G, Khabarova O, Cummings A, and Stone E *J. Phys.:* Conf Series this volume

[7] Moldwin M B, Ford S, Lepping R, Slavin J and Szabo A 2000 *Geophys. Res. Lett.* **27** 57

[8] Hu Q and Sonnerup B U Ö 2001 *Geophys. Res. Lett.* **28** 467

[9] Malandraki O E, Marsden R G, Tranquille C, Forsyth R J, Elliott H A, Lanzerotti L J, and Geranios A 2007 *J. Geophys. Res.* **112**, A06111

[10] Gómez-Herrero R, Malandraki O, Dresing N, Kilpua E, Heber B, Klassen A, Müller-Mellin R, and Wimmer-Schweingruber R F 2011 *JASTP* **73** 551

[11] Wu Z, Chen Y, Li G, , Zhao L L, Ebert R W, , Desai M I, Mason G M, Lavraud B, Zhao L, Y , Liu C-M, Guo F, Tang C L, Landi E and Sauvaud J **2014** *Astrophys. J.* **781** 17

[12] Manchester W B IV, van der Holst B and Lavraud B 2014, *Plasma Phys. Control. Fusion* **56** 064006

[13] Hu Y Q and Jia X Z *J.* 2001 *Geophys. Res.* **A12** 29299

[14] Odstrcil D, Dryer M and Smith Z 1996 *J. Geophys. Res.* **101** 19973

[15] Tessein J A, Matthaeus W H, Wan M, Osman K T, Ruffolo D and Giacalone J 2013 *Astrophys. J. Lett.* **776** L8

[16] le Roux J A , Webb G M, Gary P Zank and Khabarova O *J. Phys.*: Conf. Series this volume